D0202413

Soil Mechanics

Principles and Practice

G. E. Barnes

First published 1995 by
MACMILLAN PRESS LTD
Houndmills, Basingstoke, Hampshire RG21 6XS
and London
Companies and representatives throughout the world

ISBN 0–333–59654–4

A catalogue record for this book is available from the British Library.

10 9 8 7 6 5 4 3 2 1
04 03 02 01 00 99 98 97 96 95

Printed and bound in Malaysia

Acknowledgements

Extracts from British Standards are reproduced with the permission of BSI. Complete copies can be obtained by post from BSI Sales, Linford Wood, Milton Keynes, MK14 6LE. Tables 13.1 and 13.7 are reproduced with the permission of the Controller of Her Majesty's Stationery Office. Material from ASTM publications is reproduced with their permission. Full versions can be obtained from American Society for Testing and Materials, 1916 Race Street, Philadelphia, Pa. 19103-1187, USA. Figure 12.10 is reproduced with permission from the Boston Society of Civil Engineers Section, American Society of Civil Engineers. The Journal of the Boston Society of Civil Engineers is now known as 'Civil Engineering Practice'. Figures 5.6, 9.7, 9.18, 9.19, 10.9 and 10.10 have been reproduced with permission from the American Society of Civil Engineers. Figures 5.2, 8.19, 8.20, 10.9, 10.10, 10.11, 10.20, 12.4, 12.17, 13.11 and Table 9.7 have been reproduced with permission from John Wiley and Sons, Inc., New York. Figure 3.15 is reproduced with the permission of McGraw-Hill, Inc. Figure 5.8 is reproduced with the permission of Engineering Publications Office, University of Illinois. Figure 7.13 is reproduced with the permission of The Royal Society and Professor P.W. Rowe. Figure 9.1 is reproduced with the permission of Transportation Research Board, National Research Council, Washington, D.C. Figure 9.5 is reproduced with the permission of Mr. F.G. Butler. Figures 10.15, 10.16, 10.18 and 10.19 are reproduced with the permission of Dr T. Whitaker.

The author wishes to record his thanks to all of the other organisations who have granted permission to use material from their sources.

Contents

4 Effective Stress and Pore Pressure

5 Contact Pressure and Stress Distribution

Contact pressure

Stress distribution

6 Compressibility and Consolidation

Compressibility

Consolidation

7 Shear Strength

General

Shear strength of sand

Shear strength of clay

10 Pile Foundations

11 Lateral Earth Pressure and Design of Retaining Structures

12 Slope Stability

13 Earthworks and Soil Compaction

14 Site Investigation

Preface

The main aims of this book are to provide an understanding of the nature of soil, an appreciation of soil behaviour and a concise and clear presentation of the basic principles of soil mechanics.

The subject of soil mechanics attempts to provide a framework for understanding the behaviour of the ground by considering the principles which apply to soils. The geotechnical engineer must then use judgement to determine how to apply these principles in real situations.

It is often said that soil mechanics is a 'black art' because these principles may not apply universally, and there is a considerable amount of empirical knowledge which has been built up over the years but which still serves the engineer well. This is also probably a result of trying to apply a purely scientific approach to a material which has not been controlled during its manufacture by human interference. Instead the ground is a natural material, variable, unique, not fully understood and sometimes surprising in its behaviour.

The book is intended as a main text for undergraduate civil and ground engineering students to provide the basic principles and to illustrate how, why and with what limitations these principles can be applied in practice. It is also intended to be retained by these students when they become practitioners and for professionals already in practice as a reference source providing guidance and information for the solution of real geotechnical problems.

It is assumed that the reader will have a basic understanding of mathematics and science and a good understanding of applied mechanics. In civil engineering undergraduate degree courses there is often insufficient emphasis on the need to provide a sound knowledge of geology and, in particular, the superficial geology, in other words, the soils! This material too often gets in the way for many geologists who are mainly interested in the rocks.

A good geotechnical engineer will have a knowledge of mathematics, science and be proficient with soil mechanics but a knowledge of geology, soil profiles and groundwater conditions is fundamental to the application of soil mechanics. For this reason the book aims to consider soil mechanics with more emphasis on its application in the ground and less emphasis on the behaviour of soils in the unnatural environment of the laboratory.

The book contains a range of worked examples to assist the learning of the subject and illustrate the applications of the various analytical approaches. To consolidate this understanding, problem exercises have been included for students to attempt themselves.

I am most grateful to all those researchers, writers and practising engineers who have investigated the subject and collected information over the last seventy years or so, without whom no standard text-book could be written. In particular, I wish to record my thanks to those publishers, organisations and individuals who have granted permission to use material from their publications.

I wish to express my gratitude to Dr Bob Saxton from Plymouth University for reviewing the draft manuscript and making valuable comments. I wish to thank Professor Clive Melbourne, Head of School of Civil Engineering at Bolton Institute for his support and encouragement. Thanks are are also due to Miss Joanne Carney for typing the draft manuscript. Finally, my sincere thanks go to my wife, Linda, for her support and understanding during the preparation of the book.

Graham Barnes

List of symbols

A	Activity
A	Area
A'	Effective area
A	Pore pressure parameter
A_c	Ash content
A_b	Pile base area
A_f	Pore pressure parameter at failure
A_r	Area ratio
A_s	Pile shaft area
A_v	Air voids content
a	Slope stability coefficient
B	% of particles passing maximum size
B	Width of foundation
B	Pore pressure parameter
$\bar{B}$	Pore pressure parameter
B'	Effective width
b	Slope stability coefficient
C_N	Correction for overburden pressure
C_W	Correction for water table
C_c	Compression index
C_c	Coefficient of curvature
C_α	Coefficient of secondary compression
C_s	Soil skeleton compressibility
C_s	Swelling index
C_w	Compressibility of pore water
CD	Consolidated drained
CU	Consolidated undrained
CI	Consistency index
CSL	Critical state line
c_a	Adhesion
c_b	Adhesion at underside of foundation
c_r	Remoulded undrained cohesion
c_u	Undrained cohesion
c_v	Coefficient of consolidation, vertical direction
c_H	Coefficient of consolidation, horizontal direction
c_w	Adhesion between soil and wall
c'	Cohesion in effective stress terms
D	Depth of foundation
D	Depth factor of slip circle
d	Diameter, depth of penetration, particle size
d	Length of drainage path
d_0	Initial depth of embedment
D_r	Relative density
E'	Young's modulus in terms of effective stress (drained condition)
E_p	Pressuremeter modulus
E_u	Young's modulus in terms of total stress (undrained condition)
E	Lateral force on side of slice
ESP	Effective stress path
e	Eccentricity
e	Void ratio
e_0	Initial void ratio
e_f	Final void ratio
e_{max}	Void ratio at loosest state
e_{min}	Void ratio at densest state
F	Factor of safety, length factor
F	Force
F_d	Enlargement factor
F_B	Correction for roughness
F_D	Correction for depth of embedment
f	Shape factor or intake factor
f_0	Slope stability correction factor
f_s	Skin friction, sleeve friction
f_s	Shape factor
f_t	Correction for time
f_t	Permissible tensile strength of reinforcement
f_y	Yield factor
f_l	Thickness factor
G	Shear modulus
G_s	Specific gravity of particles
g	Gravitational acceleration (9.81 m/s^2)
g	Soil constant for the Hvorslev surface
H	Height, thickness, horizontal force
H_c	Constant head above the water table
H_0	Initial head above the water table
H_t	Head at time t
h	Head difference
h_c	Capillary rise
h_m	Mean head
h_p	Pressure head
h_s	Fully saturated capillary zone
h_w	Depth to water table
h_z	Elevation or position head
I	Influence value or factor

ICL Isotropic normal consolidation line
I_c Compressibility index
I_p Plasticity index (or PI)
I_z Strain influence factor
i Hydraulic gradient
i_c Critical hydraulic gradient
i_e Exit hydraulic gradient
i_m Mean hydraulic gradient
J Seepage force
K Absolute or specific permeability
K_0 Coefficient of earth pressure at rest
K_0CL K_0 normal consolidation line
K_a Coefficient of active earth pressure
K_{ac} Earth pressure coefficient
K_p Coefficient of passive pressure
K_{pc} Earth pressure coefficient
K_s Coefficient of horizontal pressure
k Coefficient of permeability
k Coefficient for modulus increasing with depth
L, l Length, lever arm
L' Effective length
LL Liquid limit
LI Liquidity index
M Moment
M Gradient of the critical state line on p'- q' plot
MCV Moisture condition value
m Mass
m Slope stability coefficient
m_v Coefficient of volume compressibility
N Normal total force
N Stability number
N Specific volume at $p' = 1.0$ kN/m² on ICL
N Standard penetration test result, No. of blows
N' Corrected SPT value
N' Normal effective force
N_c Bearing capacity factor
N_q Bearing capacity factor
N_γ Bearing capacity factor
NC Normally consolidated
N_0 Specific volume at $p' = 1.0$ kN/m² on K_0CL
N_s Stability number
n Porosity, number of piles
n Slope stability coefficient
n Ratio R/r_d
n_d Number of equipotential drops
n_f Number of flow paths
O_c Organic content
OC Overconsolidated
OCR Overconsolidation ratio
PL Plastic limit
PI Plasticity index (or I_p)
P Force
P_a Resultant active thrust or force
P_p Resultant passive thrust or force
P_{an} Normal component of active thrust
P_{pn} Normal component of passive thrust
P_w Horizontal water thrust
p Pressure, contact pressure
p Stress path parameter (Total stress)
p' Stress path parameter (Effective stress)
p_c' Preconsolidation pressure
p_c' Initial isotropic stress
p_0' Present overburden pressure (Effective stress)
p_0 Total overburden pressure
Q Steady state quantity of flow
Q_{ult} Ultimate load
Q_s Ultimate shaft load
Q_b Ultimate base load
Q Line load surcharge
q Flow rate
q Uniform surcharge
q Stress path parameter (Total stress)
q' Stress path parameter (Effective stress)
q_a Allowable bearing pressure
q_{app} Applied pressure (or q)
q_b End bearing resistance
q_c Cone penetration resistance
q_{max} Maximum bearing pressure
q_s Safe bearing capacity
q_{ult} Ultimate bearing capacity
R Resultant force, distance
R Dial gauge reading
R Radius of influence of drain
R_f Friction ratio
R_s Pile group settlement ratio
R_3, R_t Time correction factors
R_T Correction for temperature
r Radial distance, or radius
r_d Radius of well or drain
r_u Pore pressure ratio
S_r Degree of saturation
SL Shrinkage limit
s Spacing of drains, spacing of piles, anchors
s Stress path parameter (Total stress)
s' Stress path parameter (Effective stress)
T Shear force, surface tension force, torque

T Tensile force in reinforcement
TSP Total stress path
T_v Time factor for one-dimensional consolidation
T_R Time factor for radial consolidation
t Time
t Stress path parameter (Total stress)
t' Stress path parameter (Effective stress)
U Water force
U, U_c Uniformity coefficent
U_c Combined or overall degree of consolidation
U_R Degree of radial consolidation
U_v Degree of one-dimensional consolidation
$\overline{U}_v$ Average degree of consolidation
UU Unconsolidated undrained
u Horizontal displacement
u_a Pore air pressure
u_w Pore water pressure
V Volume
V_a Volume of air
V_0 Initial volume
V_s Volume of solids
V_T Total vertical laod
V_v Volume of voids
V_w Volume of water
v Velocity
v Specific volume
v_κ Specific volume on isotropic swelling line at $p' = 1.0$ kN/m^2
$v_{\kappa 0}$ Specific volume on anisotropic swelling line at $p' = 1.0$ kN/m^2
v_s Seepage velocity
w Water content or moisture content
w_c Saturation moisture content of particles
w_e Equivalent moisture content
W Weight
W_p Weight of pile
W_t Total weight
W_w Weight of water
X Shear force on side of slice
x, y, z Coordinate axes
Z Dimensionless depth
Z_I Depth of influence
z Depth
z_a Height above the water table
z_c Critical depth
z_c Depth of tension crack
z_0 Depth of negative active earth pressure
α Angle, angular strain
α Shaft adhesion factor
α_F Settlement interaction factor
α_p Peak adhesion factor
β Angle, relative rotation
β Skin friction factor
χ Proportion of cross-section occupied by water
δ Angle of wall friction, base sliding, piles
Δ delta, change in, increment of
Δ Relative deflection
$\delta\rho$ Differential settlement
$\delta\rho_h$ Differential heave
ε_α Coefficient of secondary compression
Φ Potential function
ϕ Friction angle
ϕ_1 ϕ before pile installation
ϕ_u Angle of failure envelope, undrained condition
ϕ_{cv} ϕ at constant volume
ϕ_μ Particle-particle friction angle
ϕ_m Mobilised friction angle
ϕ_r ϕ at residual strength
Γ Specific volume at $p' = 1.0$ kN/m^2 on *CSL*
γ Unit weight
γ_b Bulk unit weight
γ_d Dry unit weight
γ_{min} Dry unit weight in loosest state
γ_{max} Dry unit weight in densest state
γ_{sat} Saturated unit weight
γ_{sub} Submerged unit weight
γ_w Unit weight of water
η Efficiency, viscosity of fluid
κ Slope of overconsolidation line
λ Slope of normal consolidation line
λ Pile adhesion coefficient
θ Rotation, inclination of a plane
μ Interparticle friction
μ Vane correction factor
μ Correction for consolidation settlement
μ_I Influence factor
μ_0 Correction for depth
μ_r Correction for rigidity
ν Poisson' ratio
ρ Mass density
ρ_b Bulk density
ρ_d Dry density
ρ_f Fluid density
ρ_s Particle density
ρ_w Density of water

ρ_{all} Allowable settlement
ρ_i Immediate settlement
ρ_c Consolidation settlement
ρ_h Heave
ρ_s Secondary settlement
ρ_t Consolidation settlement at time t
ρ_T Total settlement
ρ_y Immediate settlement including yield
σ Total stress
σ_N Normal total stress
σ' Effective stress
σ_N' Normal effective stress
σ_m Mean stress
$\sigma_1, \sigma_2, \sigma_3$ Major, intermediate and minor principal total stresses
$\sigma_1', \sigma_2', \sigma_3'$ Major, intermediate and minor principal effective stresses
σ_H, σ_H' Total and effective horizontal stresses
σ_v, σ_v' Total and effective vertical stresses
τ Shear stress
τ_y Yield stress
ω Tilt, correction for strength of fissured clays
Ψ Flow function

Notes on units

SI Units

The International System of units (SI) has been used throughout in this book. A complete guide to the system appears in ASTM E-380 published by the American Society for Testing and Materials. The following is a brief summary of the main units.

The base units used in soil mechanics are

Quantity	Unit	Symbol
length	metre	m
mass	kilogram	kg
time	second	s

Other commonly used units are:

for length:

micron (μm)
millimetre (mm)

for mass:

gram (g)
megagram (Mg)
1 Mg = 1000 kg = 1 tonne or 1 metric ton

for time:

minutes (min)
hours, days, weeks, years

Mass, force and weight

Mass represents the quantity of matter in a body and this is independent of the gravitational force. Weight represents the gravitational force acting on a mass.

Unit force (1N) imparts unit acceleration (1 m/s^2) to unit mass (1 kg). Newton's Law gives

$$\text{Weight} = \text{mass} \times \text{gravitational constant}$$

The acceleration due to gravity on the earth's surface (g) is usually taken as 9.81 m/s^2 so on the earth's surface1 kg mass gives a force of 9.81 N.

The unit of force is the newton (N) with multiples of

kilonewton (kN) = 1000 N
meganewton (MN) = 10^6 N

Measuring scales or balances in a laboratory respond to force but give a measurement in grams or kg, in other words, in mass terms.

Stress and pressure

These have units of force per unit area (N/m^2). The SI unit is the pascal (Pa).

1 N/m^2 = 1Pa
1 kN/m^2 = 1 kPa
(kilopascal or kilonewton per square metre)
1 MN/m^2 = 1 MPa

Density and unit weight

Density is the amount of mass in a given volume and is best described as mass density (ρ). The SI unit is kilogram per cubic metre (kg/m^3). Other units are megagram per cubic metre (Mg/m^3).

Density is commonly used in soil mechanics because laboratory balances give a measure of mass.

Unit weight (γ) is the force within a unit volume where

$$\gamma = \rho g$$

The common unit for unit weight is kilonewton per cubic metre (kN/m^3) or sometimes MN/m^3.

Unit weight is a useful term in soil mechanics since it gives vertical stress directly when multiplied by the depth.

Other titles of interest to Civil Engineers

Understanding Hydraulics
Les Hamill

Prestressed Concrete Design by Computer
R. Hulse and W. H. Mosley

Reinforced Concrete Design by Computer
R. Hulse and W. H. Mosley

Reinforced Concrete Design, Fourth Edition
W. H. Mosley and J. H. Bungey

Civil Engineering Contract Administration and Control, Second Edition
I. H. Seeley

Civil Engineering Quantities, Fifth Edition
I. H. Seeley

Understanding Structures
Derek Seward

Fundamental Structural Analysis
W. J. Spencer

Surveying for Engineers, Third Edition
J. Uren and W. F. Price

Engineering Hydrology, Fourth Edition
E. M. Wilson

Civil Engineering Materials, Fifth Edition
Edited by N. Jackson and R. K. Dhir

Timber – Structure, Properties, Conversion and Use, Seventh Edition
H. E. Desch and J. M. Dinwoodie

Highway Traffic Analysis and Design
R. J. Salter and N. B. Hounsell

Plastic Methods for Steel and Concrete Structures, Second Edition
S. S. J. Moy

1 Soil Formation and Nature

This chapter has been divided into three sections:

1 Soil formation
2 Soil particles
3 Soil structure.

Soil formation

Introduction

Soils in the engineering sense are either naturally occurring or man-made. They are distinguished from rocks because the individual particles are not sufficiently bonded to be considered rocks.

Man-made soils

These are described as made ground or fill (BRE Digest 274:1991). The main types of made ground are:

- waste materials
- selected materials.

Waste materials

These include surplus and residues from construction processes such as excavation spoil and demolition rubble, from industrial processes such as ashes, slag, PFA, mining spoil, quarry waste and other industrial by-products and from domestic waste in landfill sites. They can be detrimental to new works through being soluble, chemically reactive, contaminated, hazardous, toxic, polluting, combustible, gas generating, swelling, contaminated, compressible, collapsible or degradable.

All made ground should be treated as suspect because of the likelihood of extreme variability and compressibility (BS 8004:1986). These deposits have usually been randomly dumped and any structures placed on them will suffer differential settlements. There is also increasing concern about the health and environmental hazards posed by these materials.

Selected materials

These are materials selected because they have none or very few of the detrimental properties above.

They are used to form a range of highway structures such as highway embankments and earth dams, backfilling around foundations and behind retaining walls. They are spread in thin layers and are well-compacted. This gives high shear strength and low compressibility, to give adequate stability and ensure that subsequent volume changes (settlements) are small.

Contaminated and polluted soils

Due to past industrial activities many sites comprising naturally-occurring soils have been contaminated (there are potential hazards) or polluted (there are recognisable hazards) by careless or intentional introduction of chemical substances. These contaminants could comprise metals (arsenic, cadmium, chromium, copper, lead, mercury, nickel, zinc), organics (oils, tars, phenols, PCB, cyanide) or dusts, gases, acids, alkalis, sulphates, chlorides and many more compounds.

Several of these may cause harm to the health of people, animals or plants occupying the site and some may cause degradation of building materials such as concrete, metals, plastics or timber buried in the ground.

Naturally-occurring soils

The two groups of naturally occurring soils are those formed *in situ* and those transported to their present location. There are two different types of soils formed *in situ*: weathered rocks and peat. Transported soils are moved by the principal agents of water, wind and ice although they can also be formed by volcanic activity and gravity.

In situ soils – weathered rocks

Weathering produces the decomposition and disintegration of rocks. Disintegration is brought about largely by mechanical weathering, which is most intense in cold climates and results in fragmentation or fracture of the rock and its mineral grains. Chemical alteration results in decomposition of the hard rock minerals to softer clay minerals and is most intense in a hot, wet climate such as in the tropics.

A scale of weathering grades for rock masses is given in BS 5930:1981 and summarised in Table 1.1. It can be seen that several of the grades could behave as a soil rather than a rock but the soil/rock interface could be placed at different levels, as shown in Table 1.2, according to the engineering application. Thus, a 'moderately weathered' rock may behave as a soil!

Table 1.1 *Scale of weathering grades (From Anon, 1977)*

Grade	Symbol	Term	Comment
VI	RS	residual soil	100% soil, mass structure destroyed
V	CW	completely weathered	100% soil, mass structure intact
IV	HW	highly weathered	> 50% soil, rock present as discontinuous framework or as corestones
III	MW	moderately weathered	< 50% soil, rock present as continuous framework or as corestones
II	SW	slightly weathered	rock, discoloured only, weaker than fresh
I	F	fresh	rock, no visible sign of weathering

Table 1.2 *Soil/rock interface*

Grade	Symbol	Tropical soils*	General engineering applications	Stricter engineering applications
VI	RS	Solum		e.g. settlements, bearing capacity, swelling/shrinkage, erosion
V	CW	Saprolite		
				←
IV	HW		e.g. excavatability, slope stability	
		←		
III	MW	Weathered bedrock		
			←	
II	SW			
I	F	Bedrock		

← soil/rock interface

*After Anon (1990)

In situ soils – peat
Almost entirely organic matter, peats are referred to as cumulose soils and may occur as high or moor peats comprising mostly mosses, raised bogs consisting of sphagnum peat and low or fen peat composed of reed and sedge peat. Moor and bog peats tend to be brown or dark brown in colour, fibrous and lightly decomposed while fen peats are darker, less fibrous and more highly decomposed.

Landva and Pheeney (1980) suggested a suitable classification for peats based on genera, degree of humification, water content and the content of fine fibres, coarse fibres and wood and shrub remnants. This is summarised in Table 2.9 in Chapter 2.

Water-borne soils *(Figure 1.1)*
For soils to be deposited they have to be first removed from their original locations or eroded and then transported. During these processes the particles are also broken down or abraded into smaller particles. The most erosive locations are in the highland or mountainous regions and upper reaches of rivers, especially during flood conditions, and along the coastline, particularly at high tides and during storms. Cliff erosion can produce a wide variety of particles which are sorted into beach materials (sands, gravels etc.) and finer materials which are carried out to sea.

Soils that are deposited by water tend to be named according to the deposition environment, as shown in Figure 1.1, e.g. marine clays. Whether a particle can be lifted into suspension depends on the size of the particle and the water velocity, so that further downstream in rivers, where the velocity decreases, certain particles tend to be deposited out of suspension progressively. However, various geological processes such as meandering, land emergence, sea level changes and flooding tend to produce a complex mixture of different soil types.

Glacial deposits
During the Pleistocene era, which ended about 10 000 years ago, the polar ice caps extended over a much greater area than at present, with ice sheets up to several hundred metres thick and glaciers moving slowly over the earth's surface eroding the rocks, transporting rock debris and depositing soils of wide variety over Northern Europe, United States, Canada and Asia.

The deposits are generally referred to as glacial drift but can be separated into:

- soils deposited directly by ice
- soils deposited by melt-waters.

Soils deposited by ice
These are referred to as till. Lodgement till was formed at the base of the glaciers and is often described as boulder clay. Unless the underlying rock was an argillaceous shale or mudstone, the fine fraction consists of mostly rock flour or finely ground-up debris with the proportion of clay minerals being low. Gravel, cobble and boulder-size lumps of rock are embedded in this finer matrix. These deposits have been compressed or consolidated beneath the thickness of ice to a much greater stress than at present and are overconsolidated,

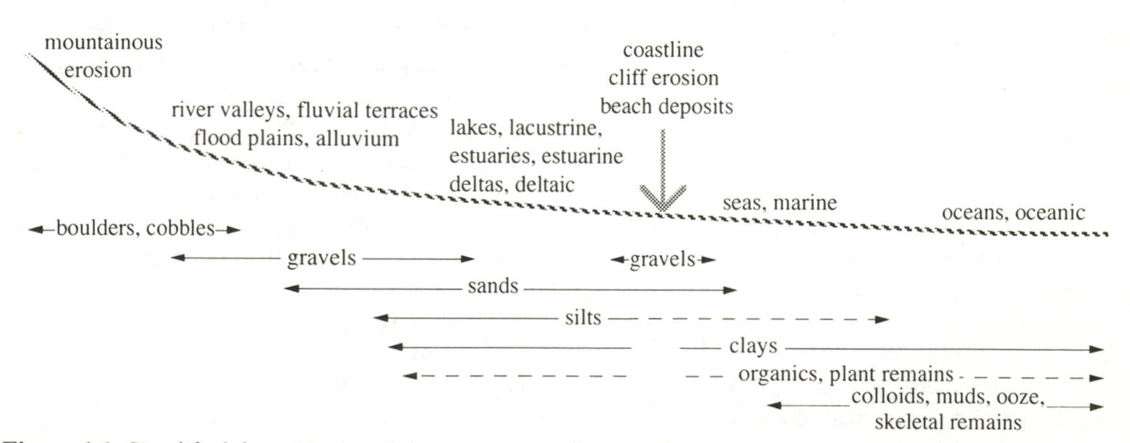

Figure 1.1 *Simplified deposition environment – water-borne soils*

which makes them stiff and relatively incompressible. The deposits have been left as various landforms; oval-shaped mounds of boulder clay called drumlins are a common variety.

Ablation till was formed as debris on the ice surface and then lowered as the ice melted. It typically consists of sands, gravels, cobbles and boulders with little fines present but, because of its mode of formation, it is less dense and more compressible. Melt-out till was formed in the same way but from debris within the ice.

Soils deposited by melt-waters

These may be referred to as outwash deposits, stratified deposits or fluvio-glacial drift. Close to the glaciers the coarser particles (boulders, cobbles, gravels) will have been deposited as ice contact deposits but streams will have provided for transport of particles, sorting and rounding producing various stratified or layered deposits of sands and gravels (outwash and fluvio-glacial deposits).

As the glaciers melted and retreated leaving many large lakes, the finer particles of clays and silts were deposited (glacio-lacustrine deposits) producing laminated clays and varved clays. In deeper waters or seas where more saline conditions existed glacio-marine deposits are found.

From the soil mechanics point of view, the term 'till' is of limited use since it can describe soils of any permeability (very low to very high), any plasticity (non-plastic to highly plastic) with cohesive or granular behaviour. Although glacial soils are often considered to be varied and mixed it may still be possible at least within a small site to identify a series of layers or beds of different soil types and it is worth attempting this during the site investigation.

Post-depositional changes

These have altered glacial and many other soils in the following ways:

- freezing/thawing – this tends to destroy the structure of the soil so that the upper few metres of laminated or varved clays have been made more homogeneous. Associated with desiccation, vertical prismatic jointing (columnar) has been produced in many boulder clays.
- fissures – these may be produced in boulder clays due to stress relief on removal of ice. Where they have opened sufficiently they may be filled with other clay minerals making them much weaker, or with silt and sand particles making them more permeable. Due to chemical changes they may be gleyed (light grey or blue and softer).
- shear surfaces – due to moving ice shear stresses may have produced slip surfaces in the clay soil which can be grooved, especially if gravel particles are present, slicken-sided or polished.
- weathering – oxidation will change the colours in the upper few metres, especially of clays, and leaching of carbonates is likely.
- leaching – where this has been extensive in post-glacial marine clays such as in Norway, Sweden and Canada, the removal of some of the dissolved salts in the original pore water by the movement of fresh water has resulted in a rearrangement of the particle structure which is potentially unstable. This structure can support a high void content or high moisture content (usually greater than the liquid limit) and can be fairly strong (soft, firm or stiff) when undisturbed but when it is disturbed the soil structure collapses and with the excess of water present the soil liquefies. The reduction in strength is called sensitivity and when the reduction or sensitivity is high the soils are referred to as quick clays.

Wind-blown soils

Wind action is most severe in dry areas where there is little moisture to hold the particles together and where there is little vegetation and consequently no roots to bind the soil together. Wind-blown or aeolian soils are mostly sands and occur near or originate from desert areas, coastlines and periglacial regions at the margins of previously glaciated areas.

There are basically two forms of wind-blown soils.

- *Dunes* These are mounds of sand having different shapes and sizes. They have been classified with a variety of terms such as ripples, barchan dunes, seif dunes and draas (which are found in desert areas) and sand-hills (found in temperate coastal regions). They are not stationary mounds but will move according to wind speed and direction with some desert dunes moving at over 10 m/year and coastal sand-hills moving at a slower rate.

  Sand sizes are typically in the medium sand range (0.2 – 0.6 mm) with coarser particles forming the smaller mounds (ripples). In coastal regions, vegetation (marram grass) binds the sand together and stabilises the dunes for coastal protection.

- *Loess* Silt mixed with some sand and clay particles is stirred up by the wind to form dust-clouds which can be large and travel several thousand kilometres. For example, the loess found in Russia and Central Europe is believed to have originated from the deserts of North Africa. Loess deposits cover large areas of the earth's surface especially United States, Asia and China.

  During deposition a loose structure is formed but loess has reasonable shear strength and stability (standing vertically in cuts) due to clay particles binding the silt particles and, to a lesser extent, secondary carbonate cementation. Loess is typically buff or light brown in colour but inclusion of organic matter gives it a dark colour as in the 'black earth' deposits of the Russian Steppes. Fossil root-holes provide greater permeability, especially in the vertical direction, than would be expected of a silt, so making the soil drain more easily. However, when loess is wetted, the clay binder may weaken causing collapse of the metastable structure and deterioration of the soil into a slurry. Therefore, loess is very prone to erosion on shallow slopes.

Soil particles

Nature of particles

The nature of each individual particle in a soil is derived from the minerals it contains, its size and its shape. These are affected by the original rock from which the particle was eroded, the degree of abrasion and comminution during erosion and transportation, and decomposition and disintegration due to chemical and mechanical weathering. A discussion on particle size, shape and density and the tests required to identify these parameters is given in Chapter 2.

The mineralogy of a soil particle is determined by the original rock mineralogy and the degree of alteration or weathering. Particles could be classed as:

- hard granular – grains of hard rock minerals especially silicates, from silt to boulder sizes
- soft granular – coral, shell, skeletal fragments, volcanic ash, crushed soft rocks, mining spoil, quarry waste, also from silt to boulder sizes
- clay minerals – see below
- plant residues – peat, vegetation, organic content.

These are discussed in Soil Formation, above.

Most granular particles are easy to identify with the naked eye or with the aid of a low magnification microscope after washing off any clay particles present. The hard granular particles consist mostly of quartz and feldspars and are roughly equidimensional. The quartz particles, in particular, have stable chemical structures and are very resistant to weathering and abrasion so these minerals comprise the bulk of silt and sand deposits. Gravel, cobble and boulder particles are usually worn-down fragments of the original rock.

Soft granular particles will produce a more compressible soil since the particles can be easily crushed and they are more likely to be loosely packed.

Clay minerals *(Figure 1.2)*

The term 'clay' can have several meanings.

1 Clay soil

The soil behaves as a 'clay' because of its cohesiveness and plasticity even though the clay mineral content may be small.

2 Clay size

Most classification systems describe particles less than 2 μm as 'clay' which is a reasonably convenient size. However, some clay minerals may be greater than 2 μm and some soils less than 2 μm, such as rock flour, may not contain many clay minerals at all.

3 Clay minerals

These are small, crystalline substances with a quite distinctive sheet-like structure, producing plate-shaped particles.

Clay minerals are complex mineral structures but they can be visualised and classified by considering the basic 'building blocks' which they comprise, as shown in Figure 1.2. The octahedral sheet and the tetrahedral sheets combine to form layered units, either two-layer (1:1) or three layer (2:1) units. The octahedral sheets are not electrically neutral and therefore do not exist alone in nature. However, the minerals gibbsite and brucite are stable.

The oxygen and hydroxyl ions dominate the mineral structure because of their numbers and their size; they are about 2.3 times larger than an aluminium ion and about 3.4 times larger than a silicate ion. Even if their negative charges are satisfied because the O^{2-} and OH^- ions exist on the surface of the sheets they will impart a slightly negative character.

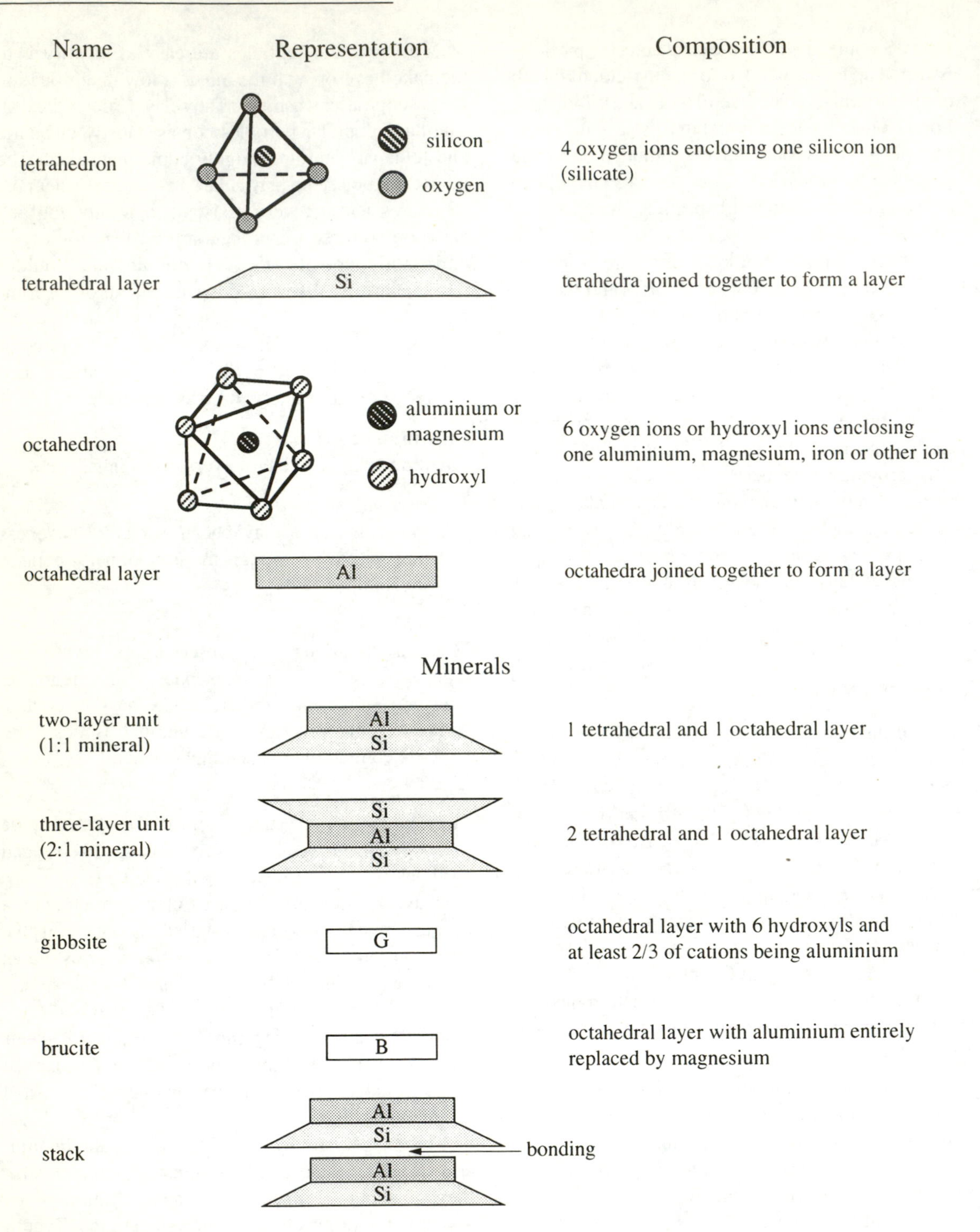

Figure 1.2 *Clay minerals*

If substitution of the cations has occurred, for example, Al^{3+} for Si^{4+} or Mg^{2+} for Al^{3+} because these ions were more available at the time of formation, then there will be a greater net negative charge transmitted to the particle surface. Isomorphous substitution refers to the situation when the ions substituted are approximately of the same size. Base-exchange or cation-exchange capacity is the ability of a clay mineral to exchange the cations within its structure for other cations and is measured in milli-equivalents per 100 grams of dry soil.

The resulting negative charges are neutralised by adsorption on the mineral surfaces of positive ions (cations) and polar water molecules (H_2O) so that with various combinations of substituted cations, exchangeable cations, interlayer water and structural layers or stacking, a wide variety of clay mineral structures is possible.

The structure of the more common types is illustrated in Table 1.3 and the nature of the clay particles is described in Table 1.4. Most clays formed by sedimentation are mixtures of kaolinite and illite with a variable amount of montmorillonite, whereas clays formed by chemical weathering of rocks may also contain chlorites and halloysites.

Table 1.3 *Structure of clay minerals*

Mineral	Layer structure	Stack structure	Bonding between layers	Base exchange capacity me/100g
kaolinite	1:1	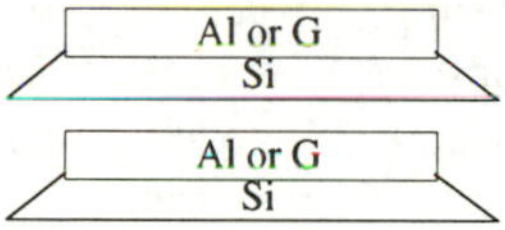 Al or G / Si / Al or G / Si	hydrogen bonds (strong)	3 – 15
halloysite	1:1	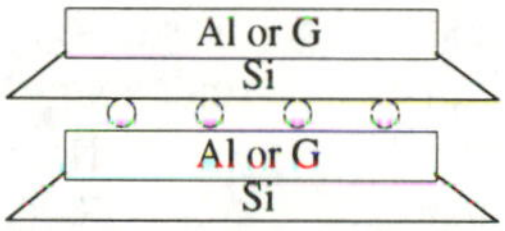 Al or G / Si / (water molecules) / Al or G / Si	hydrated with water molecules	6 – 12
illite	2:1	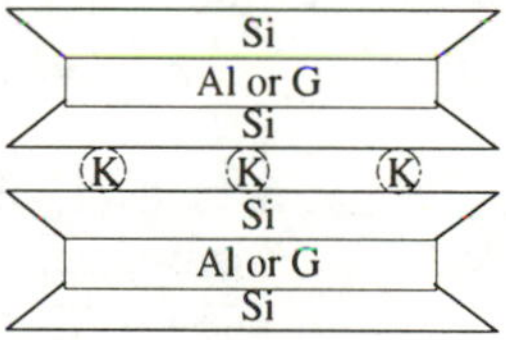 Si / Al or G / Si / K K K / Si / Al or G / Si	potassium ion (strong)	10 – 15
montmorillonite	2:1	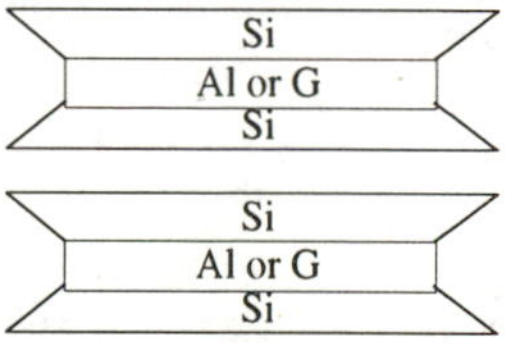 Si / Al or G / Si / Si / Al or G / Si	van der Waal's forces (weak) exchangeable ions water molecules	80 – 140
chlorite	2:1:1	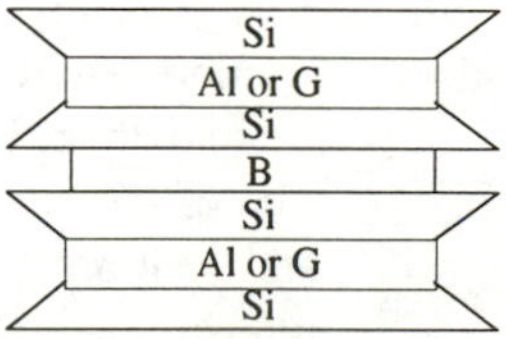 Si / Al or G / Si / B / Si / Al or G / Si	brucite sheet	20

Table 1.4 *Nature of clay mineral particles*

Mineral	Diameter: thickness ratio	Surface area m^2/gram	Nature
kaolinite	10 – 20	10 – 70	Hydrogen bond prevents hydration and produces stacks of many layers (up to 100 per particle). Particle size up to 3 μm diameter, low shrinkage/swelling
halloysite	–	40	Two water layers between stacks when fully hydrated ($4H_2O$) distort structure to a tubular shape. Low unit weight. Water (in crystal) irreversibly driven off at 60 – 75°C affecting moisture content, classification and compaction test results
illite	20 – 50	80 – 100	Common mineral but varies in chemical composition. Particles flaky, small, diameter similar to montmorillonite but thicker. Moderate susceptibility to shrinkage/swelling
montmorillonite (smectite)	200 – 400	800	High surface area due to small (<1 μm) and thin (<0.01 μm) particles produced by water molecules and exchangeable ions entering between layered units and separating them. A good lubricant. Water readily attracted to mineral causing very high susceptibility to expansion, swelling and shrinkage

Cation exchange is important in that the nature and behaviour of the clay minerals is altered. This can occur as a result of depositional environment, weathering after deposition, leaching due to sustained groundwater flow and following chemical stabilisation for engineering purposes, e.g. addition of lime (calcium hydroxide) to strengthen a soil for road-building.

Soil structure

Introduction

The way in which individual particles arrange themselves in a soil is referred to as soil structure. This structure is sometimes referred to as a soil skeleton.

Granular soils *(Figure 1.3)*

For a granular soil (coarse silts, sands, gravels) the soil structure will depend on:

- the size, shape and surface roughness of the individual particles
- the range of particle sizes (well-graded or uniformly graded)
- the mode of deposition (sedimented, glacial)
- the stresses to which the soil has been subjected (increasing effective stresses with depth, whether the soil is normally consolidated or overconsolidated)
- the degree of cementation, presence of fines, organic matter, state of weathering.

The state of packing of a granular soil, whether loose or dense, can be visualised in a number of ways, see Table 1.5. The limits of packing can be illustrated by considering the void spaces within an ideal soil consisting of equal-sized spherical particles as illustrated on Figure 1.3. In practice, typical values of porosity lie within the limits of about 35 – 50% with the lowest porosity and greatest density produced for:

- larger particle sizes
- a greater range of particle sizes
- more equidimensional particles
- smoother particles.

The difference in porosity between the densest state and the loosest state of a granular soil is typically about nine or ten per cent.

Table 1.5 *Packing of granular particles*

Effect of packing	Loose	Dense
distance between particles	furthest apart	closest together
void space	maximum (maximum *e* and *n*)	minimum (minimum *e* and *n*)
density	minimum	maximum
particle contacts	least	most
freedom of movement of particles	most	least

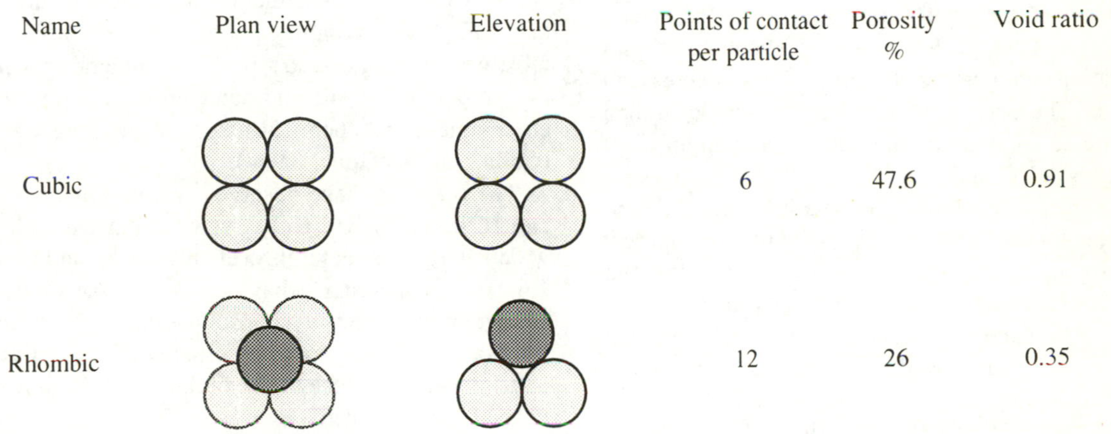

Name	Plan view	Elevation	Points of contact per particle	Porosity %	Void ratio
Cubic			6	47.6	0.91
Rhombic			12	26	0.35

Figure 1.3 *Particle packing*

Relative density

Because of the variables discussed above, porosity or void ratio are not good indicators of the state of packing so relative density, D_r is often used:

$$D_r = \frac{e_{max} - e_{in\ situ}}{e_{max} - e_{min}} \quad (1.1)$$

where:

e_{max} = maximum void ratio
e_{min} = minimum void ratio
$e_{in\ situ}$ = *in situ* void ratio

or:

$$D_r = \frac{\gamma_{in\ situ} - \gamma_{min}\,\gamma_{max}}{\gamma_{max} - \gamma_{min}\,\gamma_{insitu}} \quad (1.2)$$

where:

γ_{max} = maximum dry unit weight
γ_{min} = minimum dry unit weight
$\gamma_{in\ situ}$ = *in situ* dry unit weight

Values of densities can be used instead of unit weights. Descriptive terms for the state of packing are given in Table 1.6.

Table 1.6 *Terms for state of packing*

Term	Relative density %	SPT *'N'*
very loose	0 – 15	0 – 4
loose	15 – 35	4 – 10
medium dense	35 – 65	10 – 30
dense	65 – 85	30 – 50
very dense	85 – 100	> 50

The minimum and maximum densities (BS 1377: Part 4:1990) are determined for oven-dried soils so the dry density of the soil *in situ* must be determined.

The minimum density test involves measuring the maximum volume a 1 kg sample of clean, dry sand can occupy. The sand is placed in a measuring cylinder and sealed with a rubber bung. The cylinder is then shaken and inverted to loosen the sand, and returned to its upright position. The volume of sand in the cylinder is measured. The maximum volume is assessed by repeating this procedure at least ten times.

The maximum density is determined by placing a water-soaked sample in three layers in a 1 litre mould with water in the mould above the top of the layers. Each layer is compacted for at least two minutes with a vibrating hammer until there is no further decrease in volume. The dry mass of soil in the mould is then determined by oven-drying and the maximum (dry) density is calculated.

The *in situ* density can be obtained from the sand replacement test (BS 1377: Part 4:1990). A cylindrical hole, approximately 100 mm in diameter and 150 mm deep, is excavated on a levelled ground surface, without disturbing the soil around the sides of the hole, and the mass of soil removed is determined. A clean, dry uniform sand of known or calibrated bulk density is poured into the hole from a pouring cylinder and the volume of the hole is determined from the difference in mass of the pouring cylinder before and after pouring.

The bulk density is obtained from the mass of soil removed divided by the volume of the hole and from the moisture content or dry mass of soil the dry density is obtained.

Relative density is not a widely used parameter apart from laboratory research and perhaps compaction control since:

1. The tests should not be carried out on slightly cohesive sands or with more than about 10% silt fines, or on particles which are crushable.
2. The range of density values between minimum and maximum can be quite small and the accuracy of the minimum, maximum and in-situ density tests can be poor so the errors in the calculation for relative density are compounded.
3. The sand replacement test can only be carried out at or near to ground surface. Sampling of sand at depth from pits or boreholes is prone to disturbance affecting any *in situ* density determination. Relative density at depth is normally assessed by relation to the Standard Penetration Test, Table 1.6.

Cohesive soils *(Figure 1.4)*
Clay mineral particles are too small to be seen by the naked eye so their arrangements are referred to as microstructure or microfabric and our knowledge of particle structure comes largely from electron microscope studies.

Clay mineral particles (see above) have electrically charged surfaces (faces and edges) which will dominate any particle arrangement. The microstructure of clay soils is very complex but appears to be affected mostly by the amount and type of clay mineral present, the proportion of silt and sand present, deposition environment and chemical nature of the pore water (Collins and McGown, 1974).

These authors have observed within a number of natural normally or lightly overconsolidated soils a wide variety of structural forms illustrated on Figure 1.4. The engineering behaviour of clay soils (shear strength, compressibility, consolidation, permeability, shrinkage, swelling, collapse, sensitivity etc.) will be better understood if the nature of the microstructure of the soil is appreciated.

The macro-structure of a clay soil comprises the structure which can be seen with the naked eye and generally consists of features produced during deposition such as inclusions, partings, laminations, varves and features produced after deposition such as fissures, joints, shrinkage cracks, root holes.

Moisture content is a commonly used parameter to represent the structural nature of a clay soil since it is related to the open-ness of the micro-structure, provided the soil is fully saturated. If the soil is partially saturated then void ratio, porosity or specific volume are also used.

The state of packing of a clay soil cannot be represented by a relative density approach since maximum and minimum densities cannot be sensibly defined. The liquidity index and consistency index are parameters sometimes used to represent the structural state of a clay soil since they compare the natural moisture content with two limits, the plastic limit and the liquid limit.

This is described further in Chapter 2.

A) Elementary Particle Arrangements

Face to face groups of particles arranged as:

dispersed – mostly face-face arrangement of groups

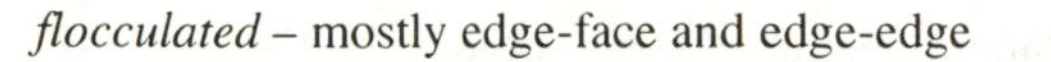

flocculated – mostly edge-face and edge-edge

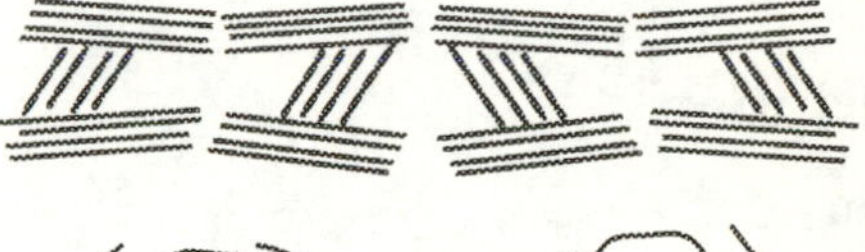

partly discernible – no strong structural tendency, difficult to distinguish

B) Particle Assemblages

clay coating
on silt and sand particles

connectors
'bridges' of mostly clay particles between silt and sand particles

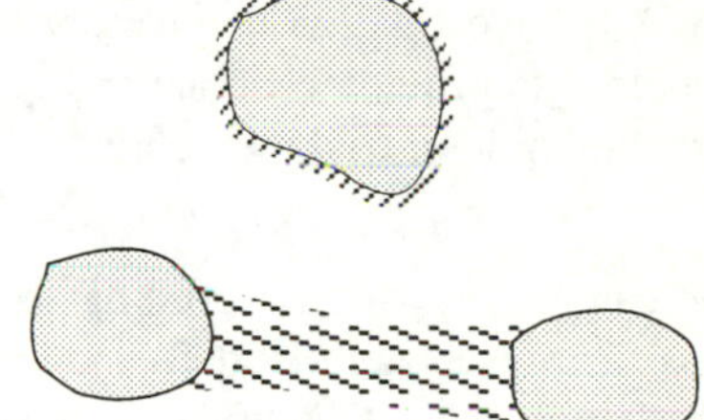

aggregations
silt to fine sand size mixtures of elementary particle arrangements

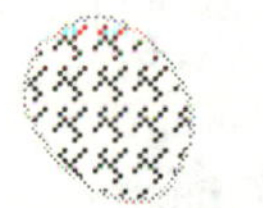

interweaving bunches
strips of clay particles interwoven around each other and around silt particles

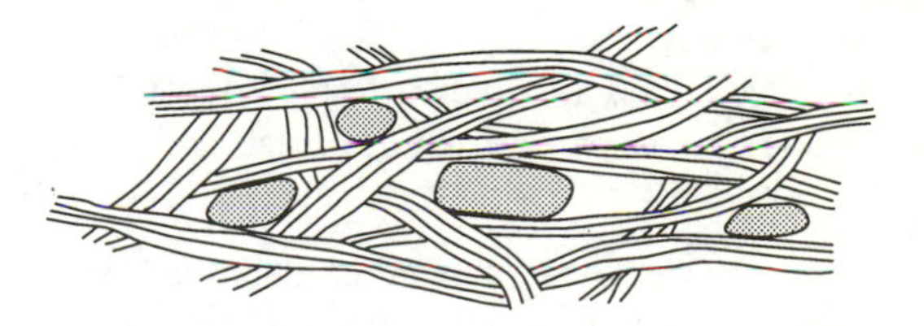

particle matrix
present where clay content is high binding other assemblages together

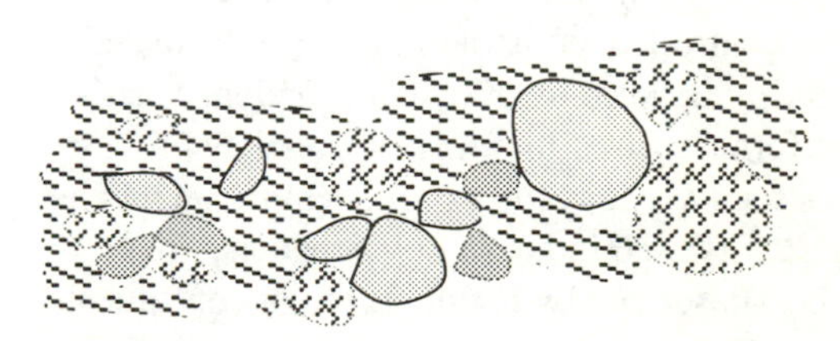

Figure 1.4 *Clay particle arrangements (From Collins and McGown, 1974)*

2 Soil Description and Classification

This chapter has been divided into two sections:

1 Soil description
2 Soil classification

Soil description

Introduction

Designers of geotechnical works and the contractors constructing the works will often have seen little, if any, of the soils they intend to build on or work with. Soil description must convey sufficient information to enable the designers and constructors to appreciate the nature and properties of the soils and to anticipate the likely behaviour and potential problems. It is then essential that the site investigation company's soils engineers and geologists provide accurate and sufficient detail on the borehole or trial pit records.

Classification

In the UK soils can be classified according to the British Soil Classification System (BSCS) and in the United States the Unified Soil Classification System (USCS) is adopted. In the sections below the British system for soil description has been followed primarily, using the scheme given in BS 5930:1981 and as suggested by Norbury *et al* (1986). Various simple tests can be carried out to help in classifying soils and these are described in the second part of this Chapter, Soil Classification.

The major soil types are given in Table 2.1, classified according to their particle nature or size.

Made ground

This is a man-made layer of material deposited or dumped over the natural ground. It can either be a carefully controlled construction (suitable material spread and compacted in layers) or a random, variable material formed from dumping of a variety of waste materials such as excavation spoil, demolition rubble, domestic refuse and industrial by-products.

It is best described by listing its major constituents with some estimate of relative proportions, followed by its minor constituents. The major constituents such as soil, ashes, rubble, refuse and degradable materials, together with the age of deposition, will give some indication of its compressibility, both under its own weight and superimposed loads, and its suitability for improvement.

The minor constituents can be of equal if not greater importance, since these could be combustible, hazardous and toxic and could produce harmful by-products such as methane gas, which could explode, and leachates which could pollute watercourses or aquifers and permeate into natural soils, producing contaminated land.

Topsoil

Humus is formed from the microbial breakdown of plant and animal tissues in the soil. Topsoil comprises an accumulation of humus-rich soil covering the natural inorganic soils or rocks and provides support to plant life. It is usually no more than 100 – 300 mm thick although, in the Tropics, where intense vegetation occurs and erosion is limited, topsoil thicknesses over one metre can exist.

Its major use in civil engineering works is for landscaping and supporting erosion protection.

Table 2.1 *BSCS Group Symbols (From B.S. 5930:1981)*

Soil Type	BSCS Group Symbol
MADE GROUND OR FILL	None
TOPSOIL	None
CLAY	C
SILT	M
SAND	S
GRAVEL	G
COBBLES	Cb
BOULDERS	B
ORGANIC SOILS	O
PEAT	Pt

Fines
These are particles smaller than 60 μm and comprise silt and clay particles.

Clay
According to the BSCS this is a a soil comprising 35 – 100% fines where the clay particles predominate to produce cohesion, plasticity and low permeability.

The description of a clay soil is commonly given in the following order:

strength / mass structure / colour / soil NAME / of – plasticity / with other structure

followed by the geological formation or type of deposit, e.g. London Clay, Estuarine Clay.

Strength terms according to laboratory tests and hand identification are given in Table 2.2.

Mass structure can consist of:

- bedding – terms for spacing are given in Table 2.3.
- interstratified deposits – such as partings where bedding surfaces separate easily, e.g. silt dusting in laminated clay, or interlaminated or interbedded silt and clay, e.g. varved clay.
- discontinuities – types are joints and fissures; bedding is treated separately, above. Shear planes and faults are best described individually. Joints and fissures can be described in a number of ways:
  1 intensity – highly fissured, fissured, poorly developed
  2 spacing – use a term (Table 2.3) or give range of values
  3 block size – relate to particle size e.g. cobble size blocks.

Table 2.2 *Strength terms*

Term	Undrained shear strength c_u kN/m^2	Field identification
very soft	< 20	exudes between fingers when squeezed in hand
soft	20 – 40	moulded easily by finger pressure
soft to firm	40 – 50	
firm	50 – 75	can be moulded by strong finger pressure
firm to stiff	75 – 100	
stiff	100 – 150	cannot be moulded by fingers but can be indented with thumb
very stiff	150 – 300	can be indented by thumb nail
hard	> 300	broken with difficulty

Table 2.3 *Discontinuity terms*
(From B. S. 5930:1981)

Bedding Term	Mean spacing mm	Other discontinuities Term
very thickly bedded	over 2000	very widely spaced
thickly bedded	600 to 2000	widely spaced
medium bedded	200 to 600	medium spaced
thinly bedded	60 to 200	closely spaced
very thinly bedded	20 to 60	very closely spaced
thickly laminated	6 to 20	extremely closely spaced
thinly laminated	less than 6	

Other important factors should be reported such as the following:

- tightness or aperture – use a term (slightly, moderately, very open) or give a range of sizes
- orientation – give direction and angle of dip and whether planar, curved, undulating, wavy
- surface texture – terms such as rough, stepped, ridged, grooved, striated, smooth, slicken-sided, polished, or condition if weaker than the mass
- infilling – peat, silt, sand-filled fissures.

• weathering – this is less important for soils than for rocks but it can be appropriate for tropical and residual soils. Weathering grade schemes have been devised for Keuper Marl (Chandler and Davis, 1973), Lias Clay (Chandler, 1972), London Clay (Chandler and Apted, 1988), Glacial Clay (Eyles and Sladen, 1981). These are generally based on the proportion of matrix (homogeneous clay) to remnant corestones or lithorelics (original rock), fissure intensity, degree of oxidation and Atterberg limits.

Colour terms can be obtained from Table 2.4.

Table 2.4 *Colour terms*
(From B.S. 5930:1981)

Tone	Shade	Colour
light	pinkish	pink
dark	reddish	red
mottled	orangeish	orange
variegated	yellowish	yellow
	greenish	green
	brownish	brown
	olive	olive
	blueish	blue
	purplish	purple
		white
	greyish	grey
		black

The soil NAME is given in capitals. Examples are

CLAY
Silty CLAY
Sandy CLAY (sand 35 – 65%)
Very Silty CLAY
Very Sandy CLAY (sand 65%+)

It is, however, possible for a soil to contain less than 35% clay particles but to have sufficient cohesion, plasticity or a low permeability so that it still behaves as a clay soil. This soil should be described as a clay since it is the engineering behaviour which is paramount, not the particle size distribution.

Plasticity terms are given in Table 2.5. Other structure could consist of:

- discrete pockets, lenses or layers of peat, clay, silt, sand, gravel
- inclusions of gravel or cobbles in a matrix of the clay. Table 2.6 gives terms which could be used to describe the proportions present.
- root, plant, organic, peat inclusions.

Table 2.5 *Plasticity terms*
(From B.S. 5930:1981)

Liquid limit	Plasticity term	Group symbol	
		Clay	Silt
< 20 (or PI < 6)	non-plastic	--	M
20 – 35	low	CL	ML
35 – 50	intermediate	CI	MI
50 – 70	high	CH	MH
70 – 90	very high	CV	MV
> 90	extremely high	CE	ME

Silt

This is a soil comprising 35–100% fines where the silt particles predominate to produce marked dilatancy and fairly low permeability but little cohesion or plasticity. Two types of silt could be distinguished – non-plastic and plastic.

Non-plastic Silt could be described in the following order

density / mass structure / colour / grain size / non-plastic / soil NAME / with other structure

Density terms based on the Standard Penetration Test are given in Table 1.6, in Chapter 1.

The mass structure and any other structure may not be as distinct as for clays but any present may be better observed by allowing a sample of the silt to partially dry. Grain size can be reported as fine, medium or coarse based on the results of a sedimentation test.

The soil NAME is given in capitals as:

SILT (sand 0 – 35%)
sandy SILT (sand 35 – 65%)

with any gravel or cobbles described according to Table 2.6.

Table 2.6 *Gravel and cobble inclusions (From B.S. 5930:1981)*

Term	Approx. % of inclusions
with a little gravel or *occasional* cobbles	5
with gravel or *with* cobbles	5 - 20
with much gravel or *with many* cobbles	20 - 40
and gravel or *and* cobbles	50 - 65

Gravel should be described according to size and shape e.g. with fine to medium subangular gravel. Cobbles could be small or large with shape described.

Plastic Silt will display some cohesion and plasticity with inhibited dilatancy and can be described in the same way as a clay. The soil name should reflect the degree of cohesion or plasticity with terms such as

clayey SILT – moderate cohesion, plasticity, inhibited dilatancy
SILT – (sand 0 – 35%)
sandy SILT – (sand 35 – 65%) little cohesion, greater dilatancy.

Silt soils will tend to lie below the A-line on the Casagrande plasticity chart (Figure 2.7) and so can be distinguished from clay soils in this way. However, other soil types such as organic, micaceous, diatomaceous, volcanic soils and halloysite rich soils may also lie below the A-line.

Sand and gravel

According to BS 5930 : 1981 this is a soil containing up to 35% fines. However, if these fines are clayey then they will probably dictate the behaviour of the soil so that it may still be better described as a clay. Care and judgement are required in this situation. The order of description for a sand and/or gravel could be:

density / mass structure / colour / minor constituent / grain size / grain shape / soil NAME / with other structure

Colour terms are given in Table 2.4 and density terms in Table 1.6.

The mass structure and any other structure may not be as distinct as for clays and will only be preserved in field exposures, but terms given for the clay descriptions may be appropriate.

Grain size can be assessed by visual inspection supplemented by sieving tests and described as fine, medium and coarse sand or gravel. Grain shape can also be assessed visually. Grain size and shape should refer to the major constituent so it is preferable to split the soil name e.g. sandy fine to medium subangular GRAVEL.

Soil names as given in BS 5930 : 1981 are summarised in Table 2.7. The percentages are based on the whole sample less any cobbles and boulders. The terms used are satisfactory but strict adherence to the proportions can lead to conflict, especially for very clayey SAND or GRAVEL of which the mass behaviour could be more like that of a clay, when this soil would be best described as a clay.

Table 2.7 *Soil NAME - Sand and Gravel*
(From B.S. 5930:1981)

Term	Fines content %	Term	Sand content %
slightly silty or clayey	up to 5	slightly sandy GRAVEL	up to 5
silty or clayey	5 – 15	sandy GRAVEL	5 – 20
very silty or clayey	15 – 35	very sandy GRAVEL	20 – 40
		SAND AND GRAVEL	40 – 60
		very gravelly SAND	60 – 80
		gravelly SAND	80 – 95
		slightly gravelly SAND	> 95

If gravel is present, conflict can be avoided by treating the soil as a clay matrix, since this largely dictates its properties, with inclusions of gravel, see Table 2.6. The gravel will only tend to dictate the mass behaviour when it is in sufficient proportion so that the gravel particles interfere with each other. The name claybound gravel has been used to describe this sort of material.

Cobbles and boulders

These are particles greater than 200 mm and 600 mm, respectively and are classed as very coarse soils. They are not usually included in a particle size distribution test by sieving. A minimum mass of about 50 kg is required for a representative sample of small cobbles. This is impractical for most site investigation purposes. Instead they are removed from a sample by hand and their proportion is estimated.

Soils containing very coarse particles and finer material can be described using the terms given in Table 2.8 with the finer material described as a separate soil, e.g. gravelly SAND with occasional cobbles; small to medium COBBLES with some finer material (silty very sandy gravel).

Table 2.8 *Cobbles and Boulders*
(From B.S. 5930:1981)

Term	Cobble or boulder content %
COBBLES or BOULDERS with a *little* finer material	> 95
COBBLES or BOULDERS with *some* finer material	80 – 95
COBBLES or BOULDERS with *much* finer material	50 – 80
FINER MATERIAL with *many* cobbles or boulders	20 – 50
FINER MATERIAL with *some* cobbles or boulders	5 – 20
FINER MATERIAL with *occasional* cobbles or boulders	up to 5

Peat and organic soils

Peats are easily distinguished by their dark brown to black colour, high organic content, high moisture content and lightweight nature, especially when dried.

When significant inorganic particles (extraneous sediments) are present the soils are referred to as organic soils. A useful classification of these soils is given by Landva *et al* (1983) and this is summarised in Table 2.9. This is based largely on the organic content ($O_c\%$) which is obtained from an ash content ($A_c\%$) test where:

$$O_c\% = 100 - A_c\% \qquad (2.1)$$

Table 2.9 *Classification of peats and organic soils (after Landva et al, 1983)*

Soil type	Peats	Peaty Organic Soils	Organic Soils	Soils with Organic Content
Group Symbol	Pt	PtO	O	MO or CO
Ash content %	< 20	20 – 40	40 – 95	95 – 99
Organic content %	> 80	60 – 80	5 – 60	1 – 5
Particle density	< 1.7	1.6 – 1.9	> 1.7	> 2.4
Moisture content %	200 – 3000	150 – 800	100 – 500	< 100
Liquid Limit %	difficult test to perform		> 50	< 50
fibre content %	> 50	< 50	insignificant	—
Degree of decomposition (von Post)	H1 – H8	H8 – H10	H10	

von Post Classification based on degree of humification of peat

H1	no decomposition	entirely unconverted mud-free peat
H5	moderate decomposition	fairly converted but plant structure still evident
H8	very strong decomposition	well converted, plant structure very indistinct
H10	complete decomposition	completely converted

These authors recommend that the moisture content be determined by oven-drying at a temperature no greater than 90° C to expel water, but also to avoid charring or oxidation of the organic matter. The ash content can then be determined by igniting the oven-dried soil at a temperature of 440° C for about five hours to remove the combustible organic matter.

Organic soils can be distinguished from inorganic soils by their grey, dark grey or black colour and their distinctive odour which can be enhanced by gentle heating. Another approach is to determine whether there is a significant difference in liquid limit values of specimens tested from the natural state and on oven-dried specimens.

Peats and Peaty Organic soils (see Table 2.9) could be described with the same order as for clays, whereas Soils with organic content should be described according to the type of inorganic soil present. The intermediate Organic soils group would probably be best decribed according to their prominent engineering behaviour.

Types of description

There are basically two types of description. One is of a piece or discrete unit of the soil, such as a hand specimen, and the other is of a thick stratum, layer or deposit. The former should depict the classification and basic intrinsic material properties of the soil such as permeability, compressibility and shear strength, while the latter should demonstrate the variability of these properties within a particular stratum or with depth (in a borehole) or in plan or section (in a trial pit or exposure).

Some examples of typical soil descriptions used on borehole or trial pit records are given in Table 2.10.

Soil classification

Introduction

Various laboratory tests are available to classify soils. The main tests relate to the nature of the particles (particle density, shape, size distribution, packing or structure) and the relationship with water (moisture content, shrinkage, plastic and liquid limit).

Table 2.10 *Examples of soil descriptions*

Soil type	Soil description (on borehole or trial pit record)
MADE GROUND	Brick rubble, concrete pieces and lumps of clay with some slate, timber, plaster, plastic (Demolition and excavation waste)
PEAT	Soft medium brown fibrous PEAT High fibre content, some wood and other plant remains, moderate decomposition (von Post H5) (Moor Peat)
CLAY	Very soft grey organic silty CLAY of intermediate plasticity with fine root inclusions Weathered in upper horizon with oxidised root holes (Estuarine Clay) Firm reddish brown sandy CLAY of low plasticity with much fine to medium subangular to subrounded gravel and occasional cobbles Thin lenses (< 100 mm thick) of fine to medium sand between 5.50 and 8.50 m (Glacial Clay) Stiff fissured greyish brown silty CLAY of high plasticity with occasional bands (< 100 mm) of silt and fine sand near base of stratum Fissures planar, smooth, generally tight, mostly 50 – 100 mm apart (London Clay)
SILT	Medium dense very thinly bedded light brown medium to coarse non-plastic sandy SILT (Glacial Silt) Firm pinkish brown clayey SILT of low plasticity with many thin (<2 mm) bands of silty clay of high plastisity (Laminated Silt/Clay)
SAND AND GRAVEL	Medium dense brown slightly silty very sandy fine to coarse GRAVEL Sand mostly fine to medium (Flood Plain Gravel) Loose yellowish brown very silty fine to medium gravelly SAND Gravel fine to coarse subrounded Structure not distinguished (Glacial Sand)

Particle density *(Figure 2.1)*
Soil particles consist of a range of mineral grains, of differing molecular weights, with the minerals present in any soil determined by geological processes. Particle density is the mass per unit volume of the mineral grains, so for soils containing a wide range of minerals only an average value will be obtained.

Particle density can be useful in identifying the presence of certain minerals but since, for most soils, it has a small range of values (2.6 – 2.8) it is of limited use. Its main use is in the determination of other soil properties such as void ratio and degree of saturation, and for tests such as sedimentation tests and the air voids lines on a compaction test result.

Particle density should be quoted in units of Mg/m^3 when it will be numerically equal to specific gravity. Specific gravity is the ratio of the unit weight of soil particles to the unit weight of distilled water and, as it is a ratio, it has no units.

The test requires the determination of the volume of a mass of dry soil particles. This is obtained by placing the soil particles in a glass bottle filled completely with de-aired distilled water. To remove all of the air trapped between the soil particles the bottle and its contents are either shaken vigorously, for coarser-grained soils, or placed under vacuum, for finer-grained soils. This procedure is the most critical part of the test.

The volume of the soil particles is determined from differences in mass (see Figure 2.1), assuming the specific gravity of water to be unity so masses in grams or Mg are equivalent to volumes in cubic centimetres or cubic metres, respectively. See the worked example 2.1.

Some typical values of particle density are:

Peat	< 1 to 1.2
Feldspars	2.55 to 2.70
Quartz	2.65
Calcite	2.75 to 2.9
Micas	2.7 to 3.1
Haematite	5.0 to 5.2

For light-coloured sand a particle density of 2.65 is usually adopted, with slightly higher values for darker-coloured sands. For most clays a value of 2.70 to 2.72 is usually adopted, in the absence of test results.

Particle shape
The shape of clay minerals is typically platy with the thickness of the grains being an order of magnitude less than its other dimensions, see Clay minerals, above.

For coarser-grained soils, silts to gravels, particle shape can affect the properties of the soil. The more the particle shape deviates from that of a perfect sphere, the greater the angle of shearing resistance, ϕ', and the greater the range between the maximum and minimum densities.

Various methods are available to quantify particle shape using the laws of granulometry but, for most practical purposes, it is sufficient to use general terms based on:

- origin – such as crushed rock (angular, rough surfaces) or natural sands and gravels (more rounded and smoother)
- closest geometrical form – spherical, cubic, elliptic, prismatic, disc, cylindrical, plate, blade, needle-shaped

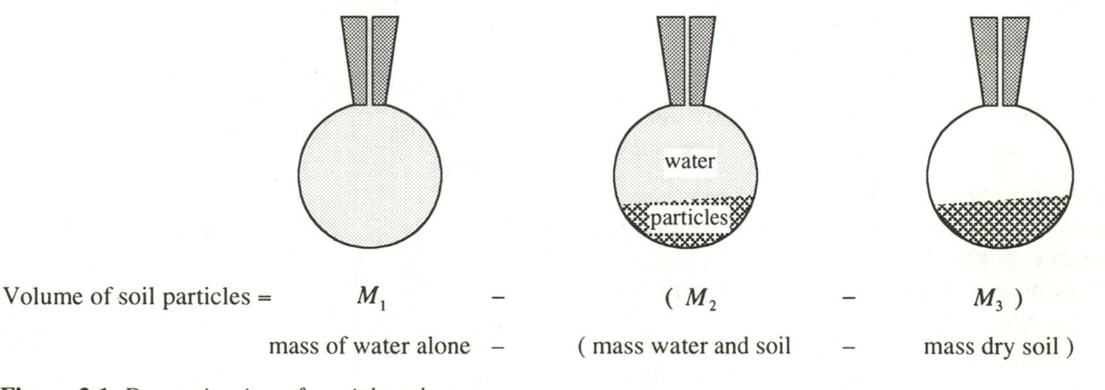

Figure 2.1 *Determination of particle volume*

- surface asperity or particle edges – angular, sub-angular, sub-rounded, rounded, irregular.

Particle size distribution

Particle sizes vary considerably, from those measured in microns (clays) to those measured in metres (boulders) – a factor of one million! Most natural soils are composite soils, mixtures of particles of different sizes. The distribution of these sizes gives very useful information about the engineering behaviour of the soil.

The distribution is determined by separating the particles using two processes – sieving and sedimentation.

Sieving *(Figure 2.2)*

Sieves separate particles in the range between 75 mm and 63 μm (gravel and sand) and sedimentation separates particles less than 63 μm (silt and clay). Particles greater than 75 mm are not usually included in the tests, they are removed before testing and an estimate is made of their proportion. The test can be a lengthy and involved procedure as shown in Figure 2.2, for three reasons:

- to ensure separation of particles. Wet sieving is preferred, where the soil particles are soaked and washed through the sieves. If the soil fines are cohesive a dispersant should be added to the wash water.
- to commence sieving and sedimentation with a known mass of dry soil. Oven-drying after wet sieving at several stages will then be necessary.
- to prevent overloading of the sieves. Each sieve can support only a limited weight so smaller but representative sub-samples should be obtained by riffling at several stages.

Fortunately, not all soils contain the full range of particle sizes so the test can be simplified. Soils that are non-cohesive may only require dry sieving. It is usually considered that the sedimentation procedure is not necessary if the soil contains less than 10% fines.

Sedimentation *(Figure 2.3)*

This is based on Stokes' Law which states that a smooth spherical particle suspended in a fluid (water and dispersant solution) will settle under gravity at a velocity given by:

$$v = \frac{d^2}{18\eta}\left(\rho_s - \rho_f\right)g \qquad (2.2)$$

where:
d = diameter of particle
ρ_s = particle density
g = gravitational constant, 9.81 m/s^2
ρ_f = density of fluid (water and dispersant)
η = viscosity of fluid.

It is essential that the soil particles are separated from each other before sedimentation so a pre-treatment procedure is carried out, where any small amounts of organic matter are removed with hydrogen peroxide, followed by dispersion comprising mechanical shaking for several hours in a dispersant such as a sodium hexametaphosphate solution.

Two test methods are available, the pipette and hydrometer methods. The pipette method involves taking a small sample of the soil suspension (about 10 ml) from 100 mm below the fluid surface at a particular time, as illustrated on Figure 2.3. From the above expression, using velocity = 100/time t, it can be seen that time $t \propto 1/d^2$ so that if the samples are taken at fixed times these will relate to the standard particle sizes. The times chosen will also depend on the particle density but for typical silts and clays, ρ_s lies between 2.65 and 2.75.

If a sample is taken by the pipette at a time of about 4 minutes then the coarse silt particles (> 20 μm) will not be present and the proportion of this size range can

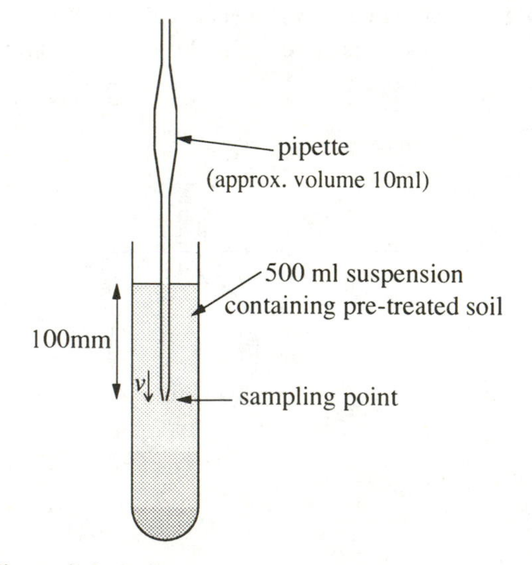

Figure 2.3 *Sedimentation test*

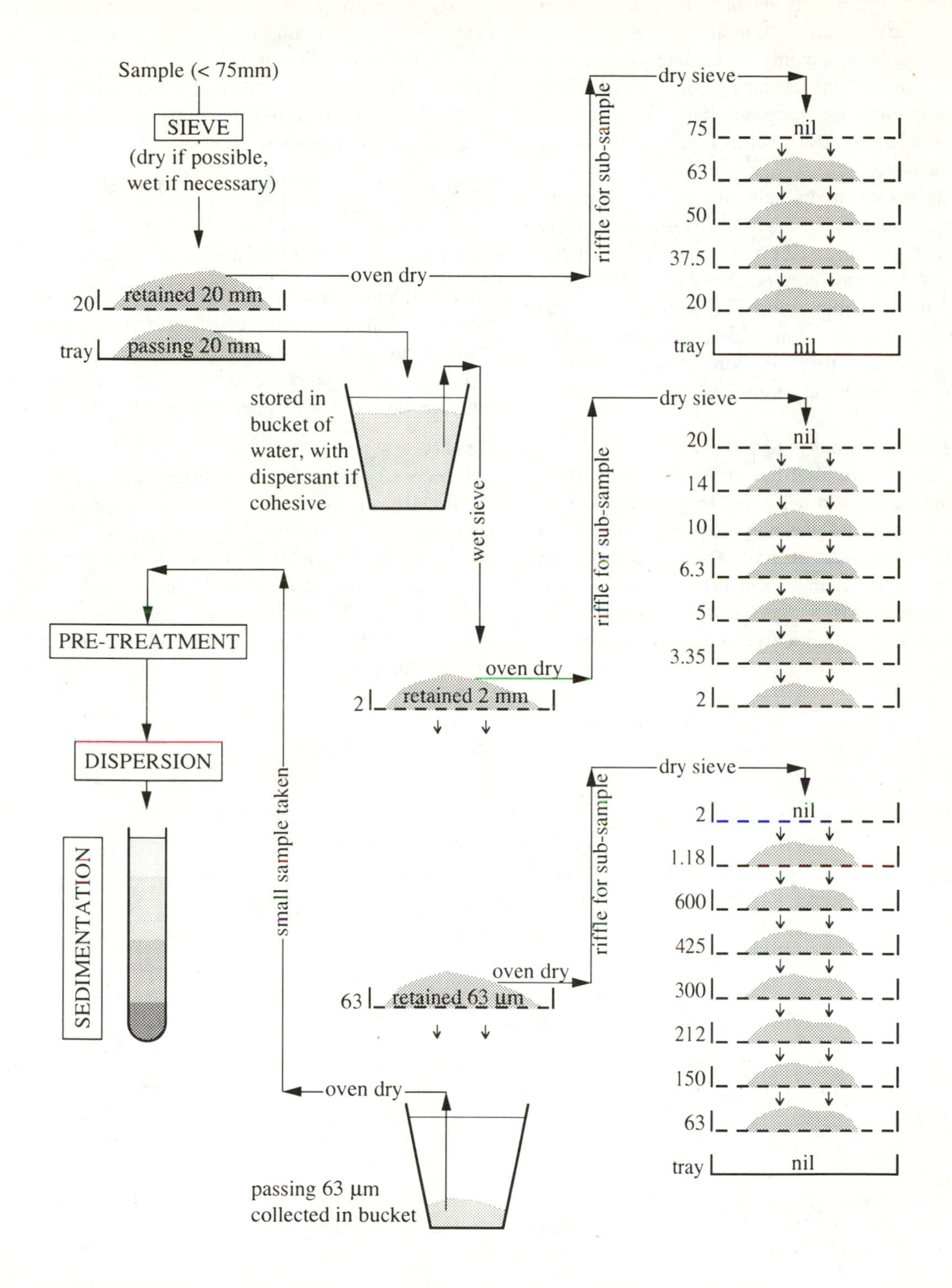

Figure 2.2 *Particle size distribution test*

be obtained by difference. The other times are chosen to represent the medium silt, fine silt and clay sizes, but remember that these are only the equivalent spherical diameters. For silts this test may produce an adequate guide for engineering purposes, but where a high proportion of flaky clay minerals is present the result will not be as accurate.

The alternative hydrometer method is based on measuring the density of the fluid and soil suspension as it reduces with time, as the particles settle around the hydrometer. More readings can be taken with this test than the pipette method but it can be more prone to errors, since difficulties can be experienced in reading the meniscus around the hydrometer and calibration of the hydrometer scales is required.

Grading characteristics *(Figures 2.4 and 2.5)*
Clay particles are distinct in their small size, flaky shape and mineralogy and they give a soil its important property of plasticity. However, there is no internationally agreed size for these particles (see Figure 2.4), so clay contents will differ according to the scheme adopted. Of perhaps less significance, but equally confusing, the grades of sands and gravels do not agree.

It is generally recognised that the properties of a composite soil containing a wide range of particle sizes are dictated by the finer particles, the coarser particles often simply acting as a filler in a finer matrix. The British Soil Classification System (BSCS) gives more recognition to this phenomenon:

Soil group	USCS	BSCS
fine soil	> 50% fines	> 35% fines
coarse soil	< 50% fines	< 35% fines

Some 'strategic' particle sizes are defined in Figure 2.5 from which the uniformity coefficient and the coefficient of curvature are determined. These parameters represent the shape of the grading curve and denote whether the soil is well-graded or poorly graded.

Density
It is necessary to distinguish between density and unit weight. Density (or mass density) is the mass of a material in a unit volume. It is denoted by ρ and has units of kg/m^3, Mg/m^3 or gram/cm^3. The latter two are numerically equal. Bulk density ρ_b is the total mass of soil (solid particles, water, air) in a given volume. Dry density ρ_d is the mass of just the solid particles in a given volume.

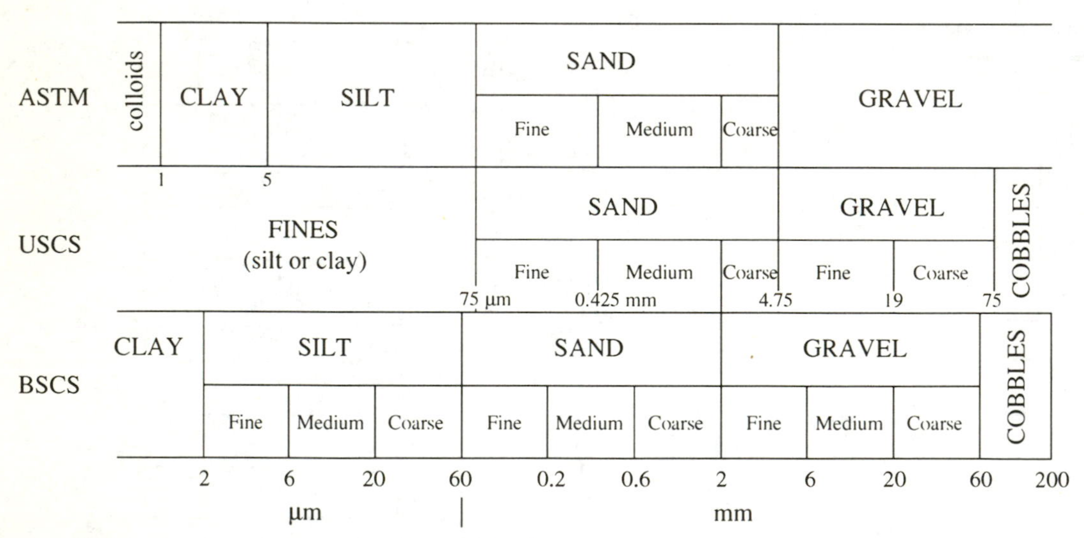

ASTM – American Society for Testing and Materials
USCS – Unified Soil Classification System (US Bureau of Reclamation, Corps. of Engineers)
BSCS – British Soil Classification System (BS 5930 :1981)

Figure 2.4 *Soil classification systems*

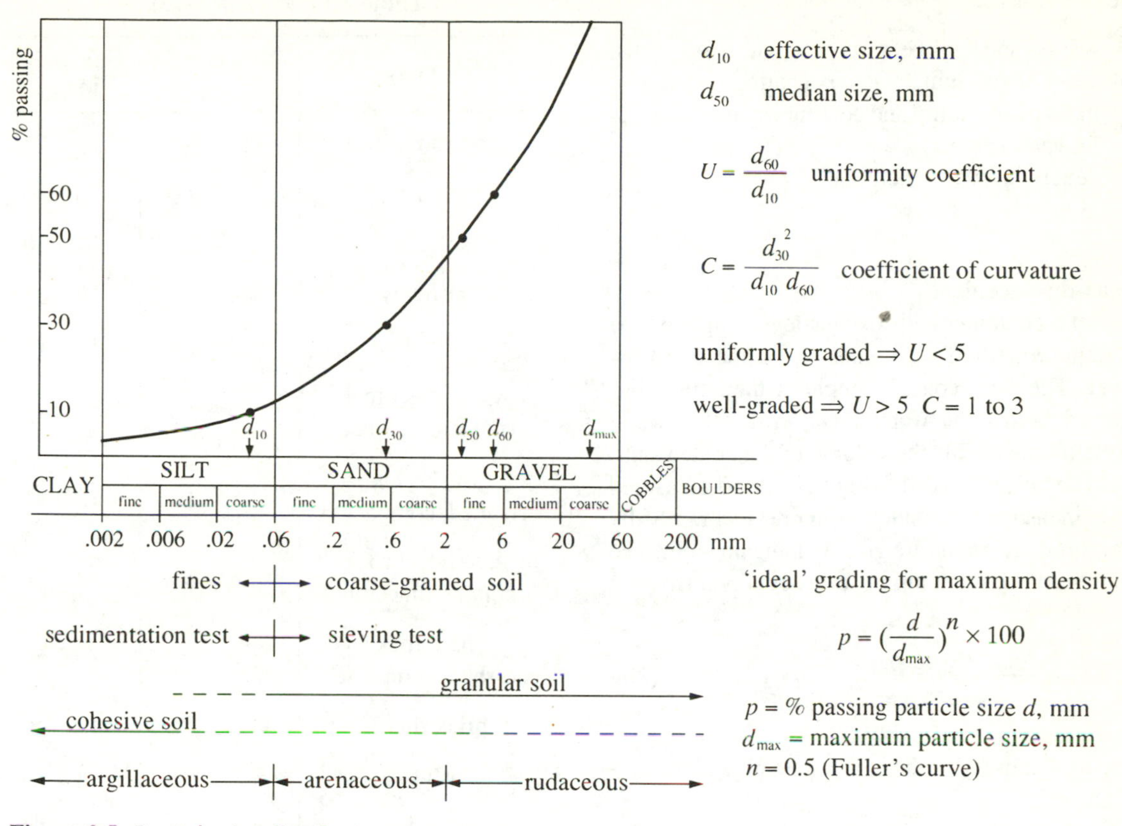

Figure 2.5 *Particle size distribution parameters*

Unit weight (or weight density) is the weight of soil in a unit volume. It is denoted by γ and has units kN/m^3. Weight is the force exerted by a mass due to gravity:

weight = mass × g

where g is the acceleration due to gravity (m/s^2). Unit weight is related to mass density, therefore, by

$$\gamma = \rho \times g$$

(kN/m^3) = (Mg/m^3) × (9.81)

Unit weight is a useful parameter since it gives the vertical stress in the ground directly from

$$\sigma_v = \gamma_b \times z$$

(total) vertical stress = bulk unit weight × depth

To determine the density of soil it is necessary to measure the volume of a sample (as well as its mass). The simplest procedure is to use volumes of regular shapes, such as the right cylinder used for a triaxial specimen. Its volume can then be obtained from linear measurement. This approach is only suitable for soils of a cohesive nature and which are little affected by sample preparation. For soils which cannot be easily formed without breaking or structural disturbance, then the volume of a lump of soil can be found by first coating it with a layer of molten paraffin wax and then using either:

- immersion in water
  The coated lump is placed on a cradle, is weighed in air and then weighed fully immersed in water. The volume of the lump of soil is obtained from:

$$V = m_w - m_g - \frac{m_w - m_s}{\rho_p} \qquad (2.3)$$

volume of soil = mass (= volume) of water displaced − volume of wax where:

m_s = mass of soil lump before coating with wax (grams)
m_w = mass of soil lump and wax coating (grams)
m_g = mass of soil lump and coating when suspended in water (grams)
ρ_p = density of wax (gram/cm^3)

or

- water displacement
  A metal container with an overflow or siphon tube is required, filled with water up to the overflow level. The wax coated sample is then fully immersed below the water level so that the water overflowing equals the volume of the coated sample. The volume overflowing V_o could be measured in a measuring cylinder or more accurately by weighing in grams to give volume in cm^3. The volume, V, of the lump of soil is obtained from:

$$V = V_o - \frac{m_w - m_s}{\rho_p} \tag{2.4}$$

The bulk density of the soil is then given by:

$$\rho_b = \frac{m_s}{V} \tag{2.5}$$

The bulk density of a soil will depend on the particle density, the distances between the particles or state of packing and the degree of saturation or the amount of water and air in the void spaces. Some typical values for saturated soils are given in Table 2.11.

Table 2.11 *Typical densities*

Soil type	Bulk density Mg/m^3	Unit weight kN/m^3
sand and gravel	1.6 – 2.2	16 – 22
silt	1.6 – 2.0	16 – 20
soft clay	1.7 – 2.0	17 – 20
stiff clay	1.9 – 2.3	19 – 23
peat	1.0 – 1.4	10 – 14
weak intact rock (mudstone, shale)	2.0 – 2.3	20 – 23
weak rock crushed, compacted	1.8 – 2.1	18 – 21
hard intact rock (granite, limestone)	2.4 – 2.7	24 – 27
hard rock crushed, compacted	1.9 – 2.2	19 – 22
brick rubble	1.6 – 2.0	16 – 20
ash fill	1.3 – 1.6	13 – 16
PFA	1.6 – 1.8	16 – 18

Moisture content or water content
Apart from soils in dry desert areas, the voids within all natural soils contain water. Some soils may be fully saturated with the voids full of water, some only partially saturated with a proportion of the voids containing air as well as water. Moisture content (or water content) is simply the ratio of the mass of water to the mass of solid particles and is an invaluable indicator of the state of the soil and its behaviour.

If a sample of clay is taken from the ground the water in the voids is held in the sample by suction stresses so a true 'natural' moisture content can be determined.

If a sample of sand (or silt, gravel, some fibrous peats) is taken from the ground below the water table, then the voids will contain some 'free' water which will drain (or flow) under gravity out of the sample. The remaining capillary-held water or films of water on the particle surfaces will then be driven off in a moisture content test but the result will represent a partially drained sample, not the fully saturated soil. It is therefore important to exercise caution with 'free-draining' soils sampled below the water table.

The test is carried out by oven drying a sample at a temperature between 105°C and 110°C until all of the 'free' moisture is expelled. This usually means leaving the sample overnight, so a result would not be available on the same day as sampling. Microwave ovens may be faster but are not recommended, due to the lack of temperature control and the possibility of alteration of the solid particles, especially clays and peats.

Some soils contain minerals which, on heating, lose their water of crystallisation, producing a moisture content higher than the 'free' moisture content. Examples are soils containing gypsum and certain tropical soils, for which maximum temperatures of 80°C and 50°C, respectively, are suggested.

The 'natural' moisture content result is often compared with the liquid and plastic limit test results (the Atterberg limits are both moisture contents) and the optimum moisture content from a compaction test. For soils containing coarse particles (sandy clays, gravelly clays), a correction to the 'natural' moisture content should be made since the latter tests are carried out only on a selected or 'scalped' portion of the total sample thus

Atterberg limits – on particles less than 425 µm
Compaction tests – on particles less than 20 mm

If a clay soil contains sand (> 425 µm) or gravel (> 20 mm) particles, these are likely to have little, if any, moisture associated with them.

The natural moisture content w_n% of the whole soil sample should be corrected to give the equivalent moisture content w_e% of that portion of the sample without the oversize particles:

$$w_e = \frac{w_n}{P} \times 100 \quad (2.6)$$

where:

P = % passing 425 mm (for comparison with the Atterberg limits)
P = % passing 20 mm (for comparison with the compaction test result)

If the sand or gravel particles contain moisture due to their porous nature then the saturation moisture content w_c% of these particles should be determined and the equivalent moisture content can then be obtained from:

$$w_e = \frac{100}{P}\left[w_n - w_c\left(\frac{100 - P}{100}\right)\right] \quad (2.7)$$

Moisture contents of natural soils will depend on their degree of saturation but can vary considerably from less than 5% for a 'dry' sand to several hundred per cent for a montmorillonitic clay or a peat. As a guide, some typical values of natural moisture content are given in Table 2.12.

Table 2.12 *Typical moisture contents*

Soil Type	moisture content %
moist sand	5 – 15
'wet' sand	15 – 25
moist silt	10 – 20
'wet' silt	20 – 30
NC clay *low plasticity*	20 – 40
NC clay *high plasticity*	50 – 90
OC clay *low plasticity*	10 – 20
OC clay *high plasticity*	20 – 40
organic clay	50 – 100
extremely high plasticity clay	100 – 200
peats	100 – > 1000

NC – normally consolidated
OC – overconsolidated

Consistency and Atterberg limits *(Figure 2.6)*
Soils containing fines (silt and clay) display the properties of plasticity and cohesiveness, where a lump of soil can have its shape changed or remoulded without the soil changing in volume or breaking up. This property depends on the amount and mineralogy of the fines and the amount of water present, or moisture content.

As the moisture content increases, a clayey or silty soil will become softer and stickier until it cannot retain its shape, when it is described as being in a liquid state. If the moisture content is increased further, then there is less and less interaction between the soil particles and a slurry, then a suspension is formed.

If the moisture content is decreased the soil becomes stiffer as shown in Figure 2.6 until there is insufficient moisture to provide cohesiveness, when

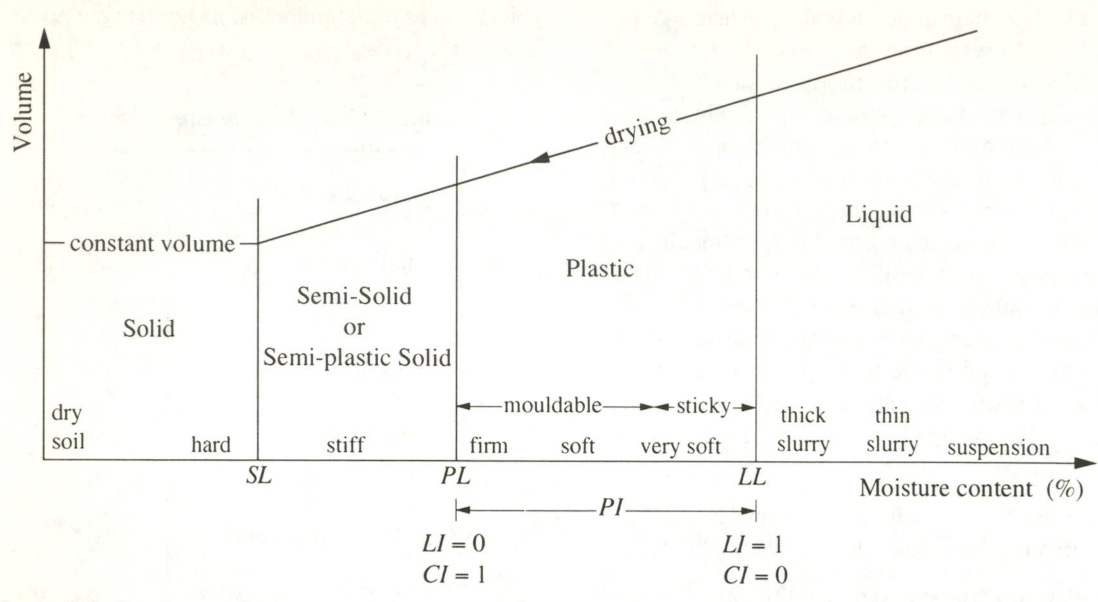

Figure 2.6 *Consistency limits*

the soil becomes friable and cracks or breaks up easily if remoulded. This state is described as semi-plastic solid or semi-solid. If the moisture content is decreased further, there is a stage when the physico-chemical forces between the soil particles will not permit them to move any closer together and the soil is then described as a solid.

The transitions between these states is gradual rather than abrupt but it is very useful to define moisture contents at these changes as:

liquid limit - *LL*
plastic limit - *PL*
shrinkage limit - *SL*

even though these moisture contents may be somewhat arbitrary.

Tests for the liquid and plastic limits were devised by a Swedish soil scientist named Atterberg and they are now known together as the Atterberg limits. Both tests are carried out on the portion of a soil finer than 425 μm.

Liquidity index (LI)
This is defined as:

$$LI = \frac{w - PL}{LL - PL} \qquad (2.8)$$

and is a good indicator of where the natural moisture content, *w*, lies in relation to the Atterberg limits.

Consistency index (CI)
This is defined as:

$$CI = \frac{LL - w}{LL - PL} \qquad (2.9)$$

but is used less often than the liquidity index.

Liquid limit
The definitive method for the determination of the liquid limit is the cone penetrometer method (BS 1377:1990). This test consists of a 30° cylindrical cone, with a smooth, polished surface and a total mass of 80 grams, allowed to fall freely into a cup of very moist soil which is near to its liquid limit.

The penetration of the cone and the moisture content of the soil are measured. As the moisture content increases the penetration increases, so the test is repeated with the same soil but with further additions of distilled water. Then a graph of cone penetration versus moisture content is obtained. The liquid limit of the soil is taken as the moisture content at a penetration of 20 mm.

An alternative method developed many years ago

by A. Casagrande can also be used, where a pat of soil is placed in a circular brass cup and a groove is made in the soil to separate it into two halves. The cup is then lifted and allowed to drop onto a hard rubber base. The number of such blows needed to cause the two soil halves to come together over a distance of 13 mm is related to the moisture content of the soil, and the liquid limit is defined as the moisture content when this condition is achieved after 25 blows.

However, this method can have poor reproducibility since it is more sensitive to operator error, requires judgement concerning the closing of the gap and it has been found that the results are affected by the hardness of the rubber base on which the cup is dropped. Comparative tests have shown that for liquid limits less than about 100 the two methods give similar results.

It has been found that the liquid limit can be affected by the amount of drying a sample of soil undergoes before the test is carried out, so the current British Standard recommends that the test is carried out on soil wetted up from its natural state. Table 2.13 gives some indication of the effects of drying on index properties. These effects are most marked for tropical soils where the clay minerals aggregate to form larger particles and less plastic soils.

Plastic limit

For reasons given above, it is preferable to carry out this test on material prepared from the natural state. The soil is dried to near its plastic limit by air drying, moulding it into a ball and rolling it between the palms of the hands. When the soil is near its plastic limit a thread about 6 mm in diameter and about 50 mm long is rolled over the surface of a smooth, glass plate beneath the fingers of one hand with a backward and forward movement, and just enough rolling pressure is applied to reduce the thread to a diameter of 3 mm. If the soil does not shear at this stage then it is dried further and the test is repeated until the thread crumbles or shears both longitudinally and transversely. The soil is now considered to be at its plastic limit and its moisture content is determined.

It will be appreciated that the test can be prone to variability since it is dependent on the individual operator's fingers, hand pressure and judgement concerning the achievement of the crumbling condition at the required diameter.

Soils also behave differently during the test. Higher plasticity (heavy) clays may not crumble easily, becoming quite tough at this low moisture content whereas for lower plasticity soils it may be difficult to produce a 3 mm thread without premature crumbling.

Table 2.13 *Effect of drying on index properties, Tropical soils (From Anon, 1990)*

Soil type	Test	natural state	air-dried	oven-dried
Red clay Kenya	LL	101	77	65
	PL	70	61	47
	CC	79		47
Weathered shale Malaysia	LL	56	48	47
	PL	24	24	23
	CC	25	36	34
Weathered granite Malaysia	LL	77	71	68
	PL	42	42	37
	CC	20	17	18
Weathered basalt Malaysia	LL	115	91	69
	PL	50	49	49
	CC	80	82	63
Volcanic ash Vanuata	LL	261	192	N-P
	PL	184	121	
	CC	92	57	6

CC = clay content, % < 2µm

Plasticity index (PI)

This is defined as:

$$PI = LL - PL \qquad (2.10)$$

Plasticity chart *(Figure 2.7)*

The Atterberg limits are useful in identifying the type of clay mineral present using the plasticity index and the liquid limit. The British Soil Classification System gives a plasticity chart (Figure 2.7) distinguishing fine-

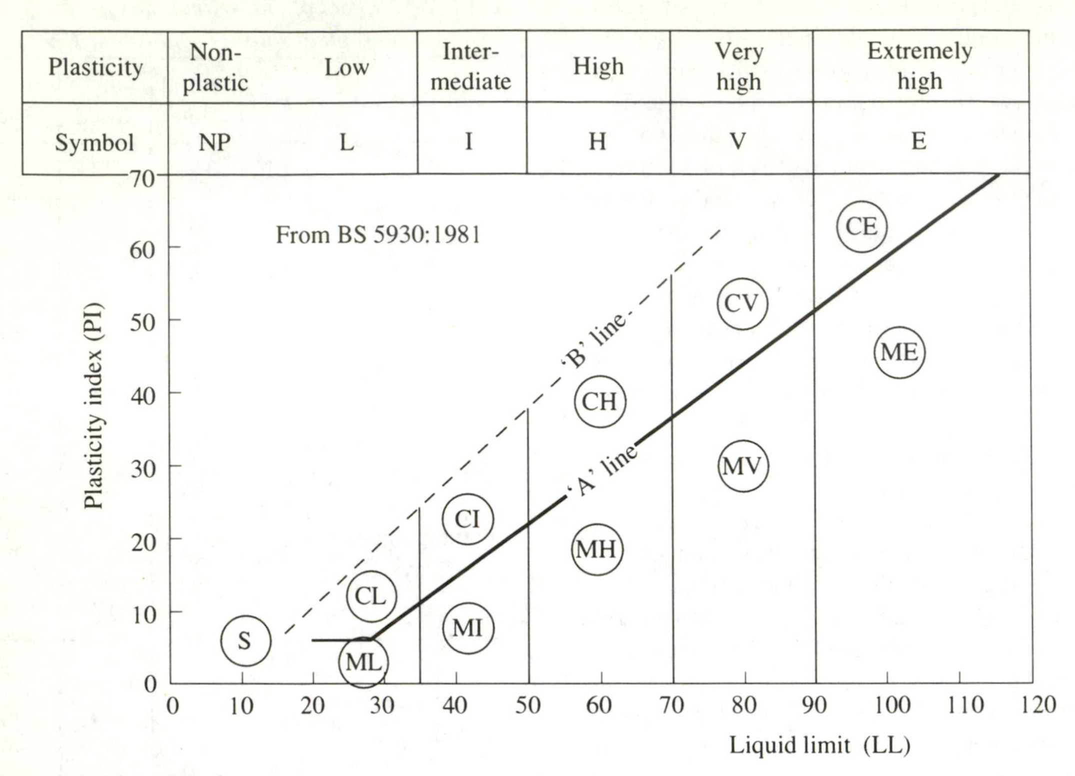

Figure 2.7 *Plasticity chart*

grained soils on the basis of predominantly clays (C) or silts (M) lying above or below the A-line and varying degrees of plasticity from low (LL < 35%) to extremely high (LL > 90%) with symbols for each type of soil.

Organic soils usually lie below the A-line and are given the symbol O or Pt for peat. Most soils are found below the B-line.

Activity

The moisture content of a clay soil is affected not only by its particle size and mineral composition but also by the amount of clay present. Silt and sand particles will be present in a clay soil when carrying out the plasticity tests and will affect the moisture content value (see moisture content, above) but will have little – if any – effect on the plasticity properties of the soil since the clay particles dominate.

Activity was defined by Skempton (1953) as:

$$\text{Activity} = \frac{PI}{C} \tag{2.11}$$

where:

PI = plasticity index %
C = clay content %

and from Equation 2.6 it can be seen that activity represents the plasticity index of the clay minerals alone. Four groups of activity are defined (inactive to highly active) and some typical values for different soils are given in Table 2.14.

Shrinkage limit

Shrinkage refers to the reduction of volume as moisture content decreases. Silts and sands are not particularly susceptible to shrinkage. A clay soil will reduce significantly in volume if its moisture content is reduced.

The amount of shrinkage will depend on the clay content and its mineralogy but it will also be significantly affected by the structural arrangement of the particles, and the forces between them, as produced in the soil's natural state. For this reason, this test is carried out on an undisturbed specimen.

Table 2.14 *Activity of clays*

Groups		Typical values	
Description	Activity	Soil/mineral	Activity
inactive	< 0.75	kaolinite	0.4
		Lias Clay	0.4 – 0.6
		Glacial Clays	0.5 – 0.7
		illite	0.9
normal	0.75 – 1.25	Weald Clay	0.6 – 0.8
		Oxford, London Clay	0.8 – 1.0
		Gault clay	0.8 – 1.25
active	1.25 – 2.0	calcium montmorillonite	1.5
		organic alluvial clay	1.2 – 1.7
highly active	> 2	sodium montmorillonite (bentonite)	7

The test consists of measuring the volume of a cylindrical specimen of soil as its moisture content decreases. The shrinkage limit is determined as the moisture content below which the volume ceases to decrease.

The cylindrical specimen, which may be 38 – 50 mm in diameter and 1 – 2 diameters high, is placed in a cage and lowered into a tank containing mercury until it is fully submerged. Volume measurements are carried out by observing the rise in level of the mercury with a micrometer, which is adjusted until its tip is in contact with the surface, when an electrical circuit illuminates a lamp. The mass of the specimen is determined for each volume reading so that moisture content for each reading can be obtained.

The shrinkage limit of clay soils from temperate climates typically lies within the small range of 10 – 15% so it is difficult to distinguish soils on the basis of this test. If the natural moisture content lies above the shrinkage limit then the soil will shrink on drying and the gradient of the volume-moisture content graph will indicate the amount of shrinkage.

Soil model *(Figure 2.8)*

The soil model, illustrated in Figure 2.8 considers the masses and volumes of the three constituents of a soil – solids, water and air – as separate entities. This is a useful way to determine relationships between the basic soil properties.

Table 2.15 lists the main properties used in soil mechanics to denote, for example:

- how much mass is present in a given volume – density
- the weight or force applied by a given volume – unit weight
- how much void space is present – void ratio, porosity, specific volume
- how much water is present – moisture content
- how much air is present – degree of saturation, air voids content.

Another purpose of the classification tests is to derive values for these properties, using the expressions given in Table 2.15. For example, in order to determine the degree of saturation of a sample of soil three classification tests are required:

- moisture content
- bulk density
- particle density.

It can be seen that many of the properties are ratios. These are dimensionless and, therefore, useful in comparing results and plotting data.

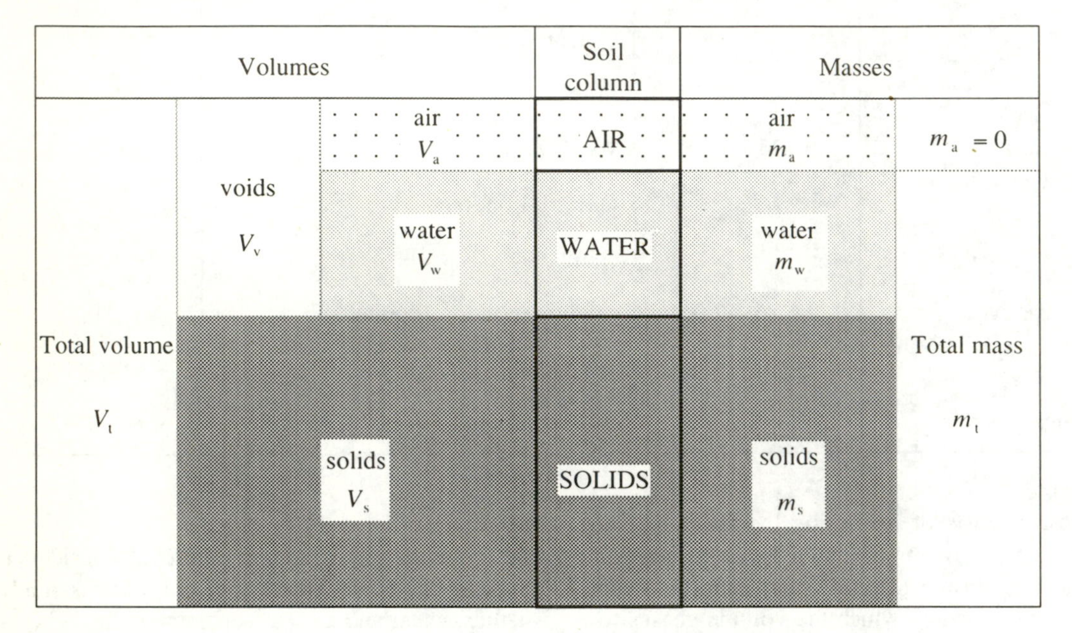

Figure 2.8 *Soil model*

Table 2.15 *Soil model expressions*

Term	Symbol	Units	Expression		Formulae
moisture content	w	%	$\frac{\text{mass water}}{\text{mass solids}}$	$\frac{m_w}{m_s}$	w is a fraction in formulae below
void ratio (partially saturated)	e	ratio	$\frac{\text{volume voids}}{\text{volume solids}}$	$\frac{V_a+V_w}{V_s}$	$e = \frac{n}{1-n} = \frac{wG_s}{S_r}$
void ratio (fully saturated)	e	ratio	$\frac{\text{volume water}}{\text{volume solids}}$	$\frac{V_w}{V_s}$	$e = wG_s$
porosity	n	ratio	$\frac{\text{volume voids}}{\text{total volume}}$	$\frac{V_a + V_w}{V_t}$	$n = \frac{e}{1+e} = \frac{wG_s}{S_r + wG_s}$
specific volume	v	ratio	$\frac{\text{total volume}}{\text{volume solids}}$	$\frac{V_a + V_w + V_s}{V_s}$	$v = 1 + e$
degree of saturation	S_r	%	$\frac{\text{volume water}}{\text{volume voids}}$	$\frac{V_w}{V_a+V_w}$	$S_r = \frac{\rho_b wG_s}{\rho_w G_s(1+w) - \rho_b} \times 100$
air voids content	A_v	%	$\frac{\text{volume air}}{\text{total volume}}$	$\frac{V_a}{V_t}$	$A_v = n(1 - S_r)$
particle density	ρ_s	Mg/m^3	$\frac{\text{mass solids}}{\text{volume solids}}$	$\frac{m_s}{V_s}$	$G_s \rho_w$
specific gravity	G_s	ratio	$\frac{\text{solids density}}{\text{water density}}$	$\frac{m_s}{V_s}\frac{1}{\rho_w}$	$\frac{\rho_s}{\rho_w}$
water density	ρ_w	Mg/m^3	$\frac{\text{mass water}}{\text{volume water}}$	$\frac{m_w}{V_w}$	$\rho_w = 1.0$ Mg/m^3
bulk density (partially saturated)	ρ_b	Mg/m^3	$\frac{\text{total mass}}{\text{total volume}}$	$\frac{m_s + m_w}{V_a + V_w + V_s}$	$\rho_b = \frac{G_s(1+w)\rho_w}{1+e}$
bulk density (fully saturated)	ρ_{sat}	Mg/m^3	$\frac{\text{total mass}}{\text{total volume}}$	$\frac{m_s + m_w}{V_s + V_w}$	$\rho_{sat} = \frac{(G_s + e)\rho_w}{1+e}$
dry density	ρ_d	Mg/m^3	$\frac{\text{mass solids}}{\text{total volume}}$	$\frac{m_s}{V_t}$	$\rho_d = \frac{\rho_b}{1+w}$
bulk unit weight (partially saturated)	γ_b	kN/m^3	$\frac{\text{total weight}}{\text{total volume}}$	$\frac{m_t g}{V_t}$	$\gamma_b = \frac{G_s(1+w)\gamma_w}{1+e}$
bulk unit weight (fully saturated)	γ_{sat}	kN/m^3	$\frac{\text{total weight}}{\text{total volume}}$	$\frac{m_t g}{V_s + V_w}$	$\gamma_{sat} = \frac{(G_s + e)\gamma_w}{1+e}$
dry unit weight	γ_d	kN/m^3	$\frac{\text{weight solids}}{\text{total volume}}$	$\frac{m_s g}{V_t}$	$\gamma_d = \frac{\gamma_b}{1+w}$
unit weight water	γ_w	kN/m^3	$\frac{\text{weight water}}{\text{volume water}}$	$\frac{m_w g}{V_w}$	$\gamma_w = \rho_w g = 9.81$ kN/m^3

Worked Example 2.1 *Particle density of sand*
From the following masses determine the particle density of a sample of sand.

mass of bottle (with stopper) $m_1 = 27.464$ grams
mass of bottle and sand $m_2 = 33.660$ grams
mass of bottle, sand and water $m_3 = 84.000$ grams
mass of bottle and water $m_4 = 80.135$ grams

$$\text{Particle density} = \frac{\text{mass of sand particles}}{\text{volume of sand particles}}$$

mass of sand particles = $m_2 - m_1 = 33.660 - 27.464 = 6.196$ g
volume of sand particles = $M_1 - (M_2 - M_3)$ (From Figure 2.1)

$$M_1 = m_4 - m_1$$
$$M_2 = m_3 - m_1$$
$$M_3 = m_2 - m_1$$

volume of sand $= (m_4 - m_1) - (m_3 - m_2)$
$= (80.135 - 27.464) - (84.000 - 33.660) = 2.331$ g (of water, $\rho_w = 1$ g/cm^3)
$= 2.331$ cm^3

Particle density $\rho_s = \frac{6.196}{2.331} = 2.658$, say 2.66 g/cm^3 or Mg/m^3
Specific gravity $G_s = 2.66$

Worked Example 2.2 *Particle size distribution*
The total mass of a sample of dry soil is 187.2 grams before being shaken through a series of sieves. From the masses retained on each of the following sieves determine the percentages passing and give a description of the soil tested.

Sieve size (mm or µm)	6.3	2	1.18	600	425	300	212	150	63	tray
	0	6.8	4.7	7.2	27.8	31.3	33.1	28.7	26.5	21.1

Sieve size	Mass retained	Mass passing	% passing
6.3	0	187.2	100
2	6.8	180.4	96.4
1.18	4.7	175.7	93.9
600	7.2	168.5	90.0
425	27.8	140.7	75.2
300	31.3	109.4	58.4
212	33.1	76.3	40.8
150	28.7	47.6	25.4
63	26.5	21.1	11.3
tray	21.1	–	

The soil could be described as a silty (or clayey) fine and medium SAND with a little fine gravel. Alternatively, it could be described as a silty (or clayey) slightly gravelly fine and medium SAND.

Worked Example 2.3 *Density, unit weight and moisture content*
A triaxial specimen of moist clay has a diameter of 38 mm, length of 76 mm and mass of 183.4 grams. After oven drying the mass is reduced to 157.7 grams. Determine for this soil:
i) bulk density and bulk unit weight
ii) dry density and dry unit weight
iii) moisture content

$$\text{Volume of specimen} = \frac{\pi \times 38^2}{4} \times \frac{76}{10^9} = 8.62 \times 10^{-5} \text{ m}^3$$

$$\text{i) Bulk density} = \frac{183.4 \times 10^5}{8.62 \times 10^3} = 2127.6 \text{ kg/m}^3 \text{ or } 2.13 \text{ Mg/m}^3$$

$$\text{Bulk unit weight} = 2127.6 \times \frac{9.81}{1000} = 20.87 \text{ kN/m}^3$$

$$\text{ii) Dry density} = \frac{157.7 \times 10^5}{8.62 \times 10^3} = 1829.5 \text{ kg/m}^3 \text{ or } 1.83 \text{ Mg/m}^3$$

$$\text{Dry unit weight} = 1829 \times \frac{9.81}{1000} = 17.95 \text{ kN/m}^3$$

$$\text{Moisture content} = \frac{183.4 - 157.7}{157.7} = 16.3\%$$

Alternatively, dry density can be obtained from the soil model, see Table 2.15.

$$\rho_d = \frac{2127.6}{1 + 0.163} = 1829.4 \text{ kg/m}^3$$

The difference in values is a result of the degree of accuracy adopted in the calculations.

Worked Example 2.4 *Equivalent moisture content*
In an earthworks contract it is stated that the natural moisture content of a gravelly clay soil must not exceed the optimum moisture content (OMC) obtained from a standard compaction test plus 2% (OMC + 2%). Otherwise, it is unacceptable and cannot be used. Determine whether the following soil should be classified as acceptable or not.

Natural moisture content = 18.5% (including particles > 20 mm)
Optimum moisture content = 19% (for particles < 20 mm)
% passing 20 mm = 83%
From equation 2.6

$$\text{Equivalent moisture content} = \frac{18.5 \times 100}{83} = 22.3\%$$

At first glance, it would appear that the clay soil is acceptable, its natural moisture content being numerically less than the optimum moisture content. However, the equivalent moisture content of the soil, excluding the coarse gravel (> 20 mm) is much greater than the OMC so the soil should be treated as unacceptable.

Worked Example 2.5 *Classification tests*
The following test results were obtained for a fine-grained soil.
LL = 48% *PL = 26%*
Clay content = 25%
Silt content = 36%
Sand content = 39%
Natural moisture content = 29%
Determine appropriate parameters to classify the soil.

Plasticity index = *LL* – *PL* = 48 – 26 = 22%

According to the plasticity chart, Figure 2.7, the soil would classify as CI, clay of intermediate plasticity. The large proportion of silt and sand (75% in total) will have little effect on the engineering behaviour, particularly the shear strength.

$$\text{Activity} = \frac{22}{25} = 0.88$$

$$\text{Liquidity index} = \frac{29-26}{22} = 0.14$$

$$\text{Consistency index} = \frac{48-29}{22} = 0.86$$

The clay is of normal activity and should be of a firm to stiff consistency.

Worked Example 2.6 *Void ratio, degree of saturation, air voids content*
For the triaxial specimen in Worked example 2.3 determine:
i) void ratio and porosity
ii) degree of saturation
iii) air voids content
given the particle density of 2.72 Mg/m³.

From Table 2.15 use

$$\rho_b = \rho_s\left(\frac{1+w}{1+e}\right)$$

i) Void ratio $e = \frac{\rho_s}{\rho_b}(1+w) - 1 = \frac{2.72}{2.128}(1+0.163) - 1 = 0.487$

Porosity $n = \frac{0.487}{1.487} = 0.328$

ii) Degree of saturation $S_r = \frac{wG_s}{e} = \frac{0.163 \times 2.72}{0.487} \times 100 = 91.0\%$

Air voids content $A_v = n\,(1 - S_r) = 0.328\,(1 - 0.91) \times 100 = 3\%$

Exercises

2.1 The results of a particle density test on a sample of gravel were as follows:
mass of empty gas jar = 845.2 g
mass of gas jar full of water = 1870.6 g
mass of gas jar with soil sample = 1608.7 g
mass of gas jar with soil sample and full of water = 2346.0 g.
Determine the particle density of the gravel. All of the above masses include the mass of the cover lid. Assume the density of the water to be 1.0 Mg/m³ or 1.0 g/cm³.

2.2 The results of a dry sieving test are given below. The total mass of the sample of dry soil was 2105.4 g. Determine the percentages passing, plot a particle size distribution curve and give a description for the soil.

Sieve sizes (mm or µm)	20	14	10	6.3	3.35	2	1.18	600
Mass retained (g)	0	18.9	67.4	44.2	75.8	122.1	193.7	240.0

425	300	212	150	63	tray
282.2	242.1	233.7	265.3	240.0	80.0

2.3 An irregular lump of moist clay with a mass of 537.5 g was coated with paraffin wax. The total mass of the coated lump (in air) was 544.4 g. The volume of the coated lump was found to be 250 ml by displacement in water. After carefully removing the paraffin wax, the lump of clay was oven-dried to a dry mass of 479.2 g. The specific gravity of the hardened wax is 0.90. Determine the water or moisture content w, the bulk density ρ_b, the bulk unit weight γ_b, dry density ρ_d and the dry unit weight γ_d of the clay.

2.4 For the sample of clay in Exercise 2.3 determine the void ratio, porosity, degree of saturation and air voids content. The specific gravity of the (dry) clay particles was 2.72.

2.5 The results of a liquid limit test and a plastic limit test on a sample of soil are given below.
Liquid limit

Cone penetration (mm)	14.8	16.9	19.1	21.2	23.2	24.7
Water content %	50.8	52.9	54.2	56.0	57.3	58.7

Plastic limit
Water content of threads % 26.6 and 27.3%
Determine the liquid limit, plastic limit, plasticity index and soil classification.

2.6 The natural water content of the clay in Exercise 2.5 is 35%. Determine the liquidity index and the consistency index.

3 Permeability and Seepage

Introduction *(Figure 3.1)*
As a result of the hydrological cycle (rainfall, infiltration) it is inevitable that the voids between soil particles will fill with water until they become fully saturated. There then exists a zone of saturation below ground level the upper surface of which is called the water table.

The water table generally follows the shape of the ground surface topography but in a subdued manner. A sloping water table surface is an indication of the flow of groundwater or seepage in the direction of the fall. Water tables change with varying rates of infiltration so that in winter they can be expected at high levels and at lower levels in summer.

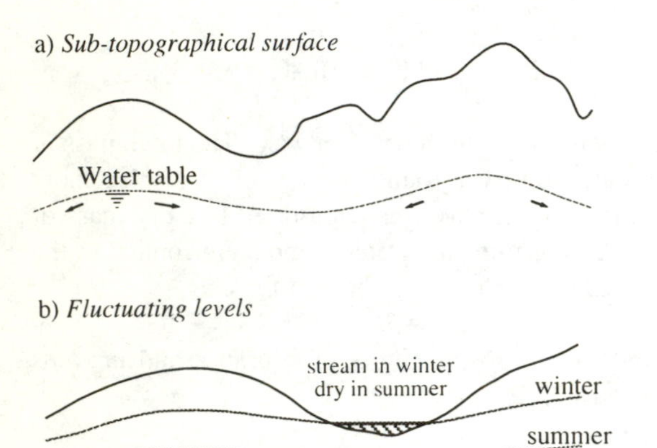

Figure 3.1 *Fluctuating water tables*

Groundwater *(Figure 3.2)*
The voids of permeable deposits such as sands will fill up easily and also allow this water to flow out easily, so they are called aquifers (bearing water). The void spaces in a clay will also contain water but these void spaces are so small that flow of water is significantly impeded, making a clay impermeable. Clay deposits will then act as aquicludes (confining water). The location and state of groundwater in soil deposits is often determined by the stratification of sand-clay or permeable-impermeable sequences. Some commonly used terms are described in Figure 3.2.

Flow problems
In nature groundwater may be flowing through the ground but this flow will not normally be large enough to cause instability, so the ground is stable or in equilibrium. Ground engineering works, particularly excavations, will disturb this equilibrium and alter the pattern of flow. It is the responsibility of the geotechnical engineer to identify where a problem may be encountered and how to ensure that stability of the works is maintained.

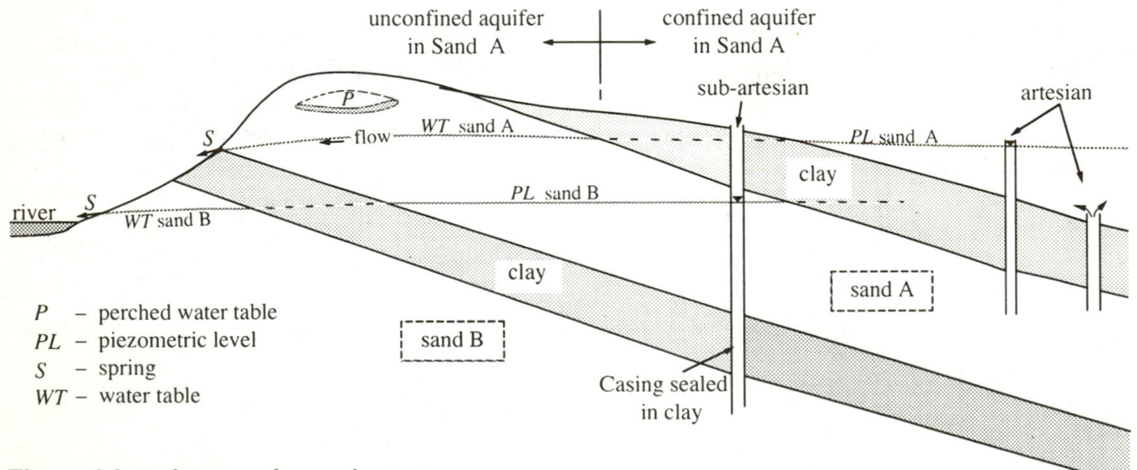

Figure 3.2 *Definitions of groundwater terms*

Flow into excavations *(Figure 3.3)*
An estimate of the quantity of flow must be obtained so that adequate numbers and capacities of pumps can be provided to remove inflows. This technique is referred to as sump pumping and relies on the groundwater emanating from the soil, leaving the soil behind. This is only feasible for open, clean gravel deposits under small heads of water, otherwise stability problems arise, see below.

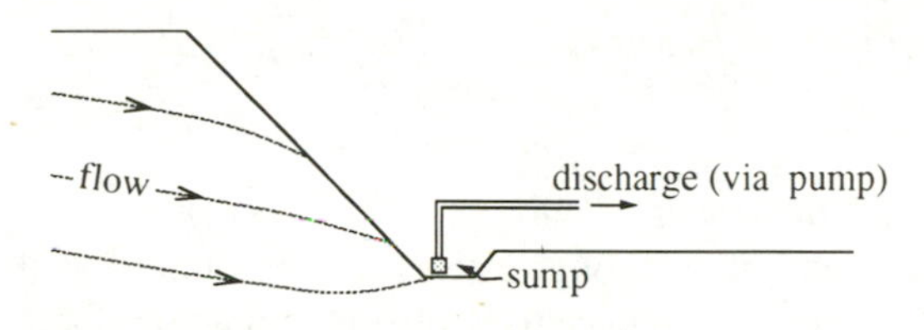

Figure 3.3 *Flow into excavations*

Flow around cofferdams *(Figure 3.4)*
The required pumping capacity must be assessed but also the quantity of flow can be altered by changing the length of sheet piling and increasing the length of seepage path.

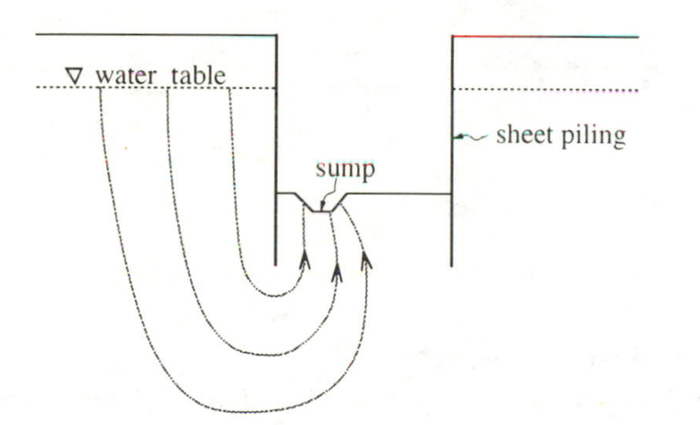

Figure 3.4 *Flow around cofferdams*

Dewatering *(Figure 3.5)*
Where the risk of sump pumping cannot be accepted (which is the case for most permeable soils) then it is far more suitable to lower the water table temporarily before excavating, so that excavation can be carried out in the 'dry'. This is achieved by using dewatering techniques such as pumping from well-points inserted below the water table. A knowledge of the rate of flow of water through the soil is then required to determine the numbers, depths and spacing of these well-points.

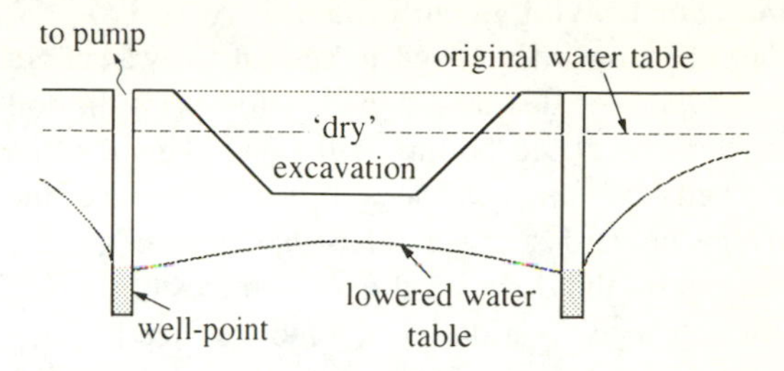

Figure 3.5 *Dewatering excavations*

Flow through earth structures *(Figure 3.6)*
An earth structure impounding water will allow water to flow through it or beneath it, if it is permeable. Measures to minimise these flows through the dam (central clay core) and beneath it (grout curtain) may have to be incorporated. Alternatively the earth structure may only be required to contain water for a short period, such as for flood banks alongside rivers where these measures would be uneconomical. Then flows may be permitted but are controlled using drainage layers. These drains must allow water to enter and pass through unimpeded, so they require protection by filters which prevent soil particles being carried into the drainage layers and blocking them.

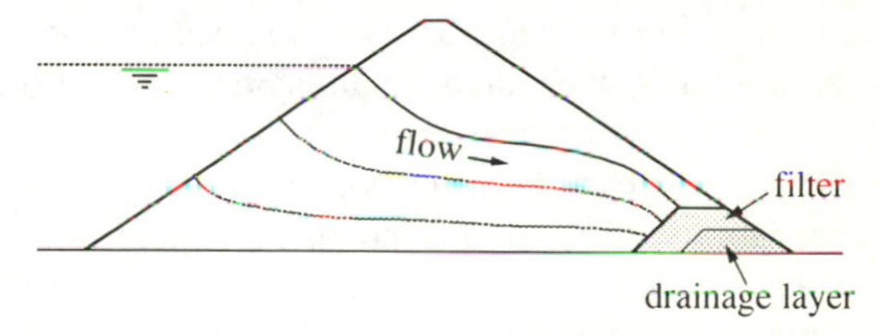

Figure 3.6 *Flow through earth structures*

Stability problems – 'running sand' *(Figure 3.7)*
It is incorrect to state that 'running sand' exists naturally on a site. Sand and groundwater exist naturally but the 'runs' are man-made by allowing water to flow out of the sides of an excavation, carrying with it soil particles. This results in the excavation sides slumping, soil above being undermined and collapse of the excavation.

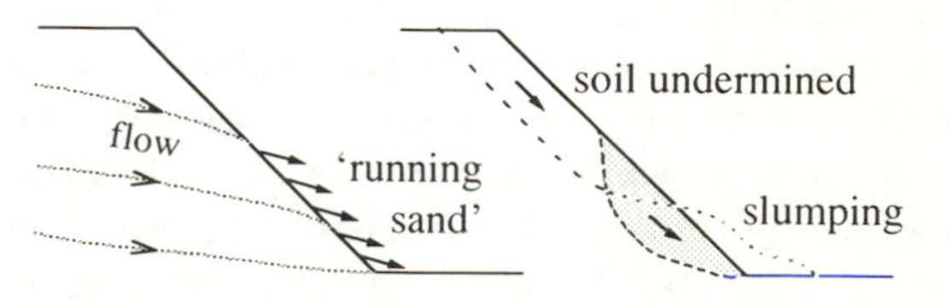

Figure 3.7 *The 'running sand' condition*

Boiling or heaving in cofferdams *(Figure 3.8)*
When the upward seepage forces (in the water) are greater than the downward gravity forces (in the soil particles), localised boiling will occur. This is often observed as small 'volcanoes' of soil. Whenever the average upward seepage forces are not sufficiently balanced by the downward gravity forces of the soil mass below excavation level, a more severe heaving condition can occur. This results in separation of the particles, increased permeability and increased seepage resulting in progressive and rapid loss of passive resistance in front of the sheet piling, followed by complete collapse of the cofferdam.

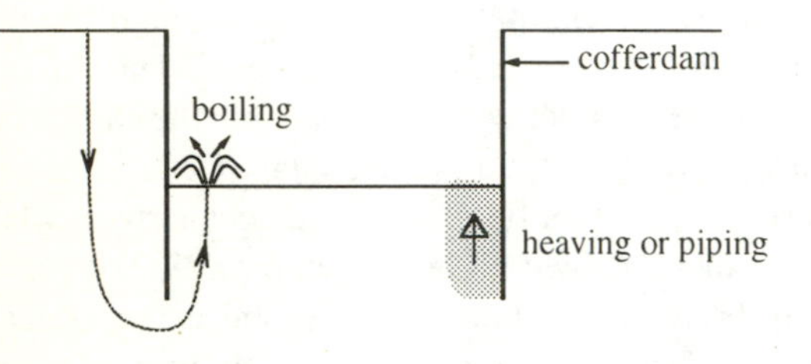

Figure 3.8 *'Boiling' or heaving in cofferdams*

Piping *(Figure 3.9)*
The erosive force produced by water passing out of an earth structure can be large, resulting in the formation of 'pipes' which increase in size and flow capacity, leading to progressive erosion, undermining and eventual instability. Certain soils (non-cohesive, fine soils) have a low resistance to erosion and should not be used in earth dams, flood banks or canal linings unless they are protected by other resistant materials.

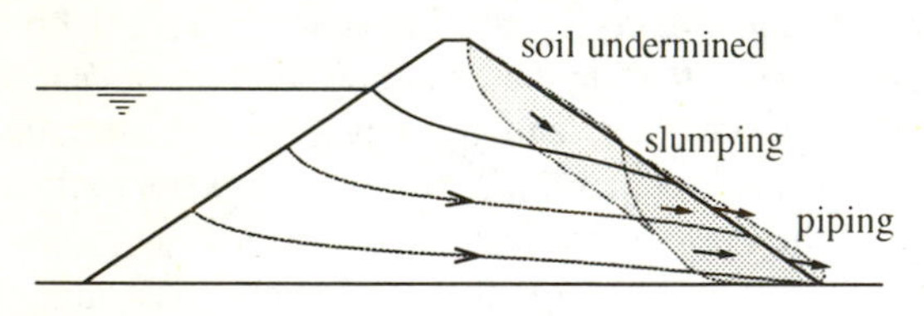

Figure 3.9 *Piping through an earth dam*

Heaving beneath a clay layer *(Figure 3.10)*
The groundwater and the pressure within it can be confined beneath a relatively impermeable stratum (clay or silt) producing an uplift pressure. If the downward pressure from the clay at the base of an excavation is insufficient to balance the uplift pressure then the clay will be lifted, causing it to crack and allowing water to flow through. This will carry soil particles upwards, leading to erosion of the sand, undermining of the clay and a severely disturbed excavation.

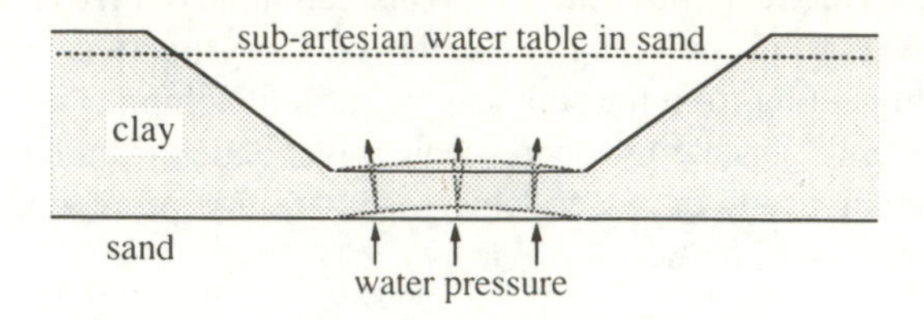

Figure 3.10 *Heaving of clay layer at the base of an excavation*

Uplift pressures *(Figure 3.11)*
Permitting flow through a permeable stratum beneath a structure will reduce the hydrostatic pressure in the water due to energy losses. However, the pore water pressures remaining will produce uplift beneath any impermeable structure. There must be sufficient deadweight (or anchors, if necessary) to balance these remaining uplift pressures.

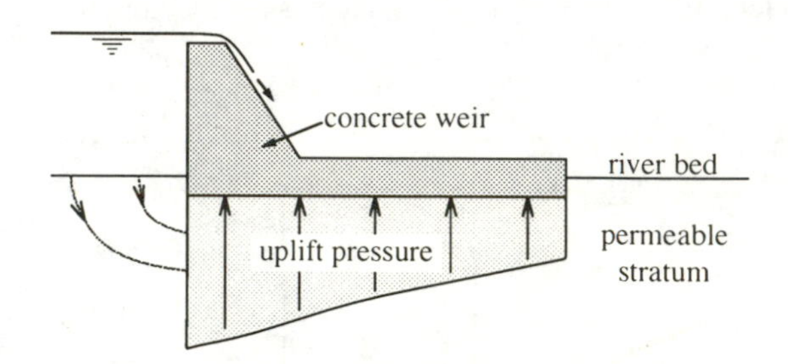

Figure 3.11 *Uplift pressure beneath structures*

Soil voids *(Figure 3.12)*
All soils contain voids or pores and can be described as porous. However, to allow water to flow, at least some of the voids must be continuous and then the soil can be described as permeable. Permeability is not dependent

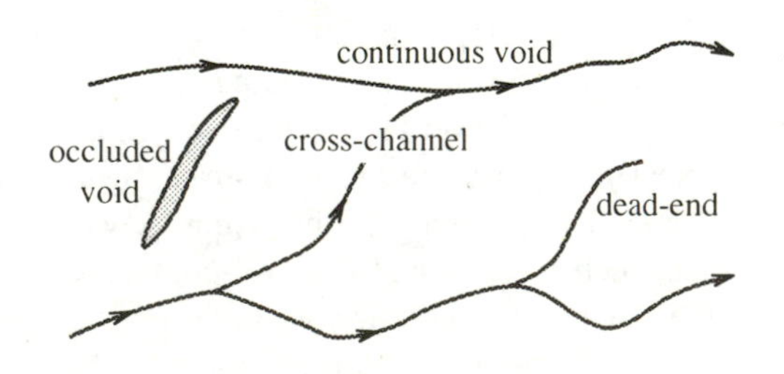

Figure 3.12 *Soil voids*

on the amount of voids present. The porosity of a clay is usually greater than that of a sand but the clay could be a million times less permeable than the sand. Some typical values are given below.

Soil	**Porosity**	**Typical *k* m/s**
clay	0.3 – 0.5	10^{-9}
sand	0.2 – 0.4	10^{-3}

This difference is due to the nature and size of the voids. In sands, particularly clean sands without fines, most of the voids are continuous, more direct and relatively large, while in clays the voids are smaller, more tortuous and with several stagnant voids. Three types of void are described in Table 3.1 (macropores, capillaries and micropores) showing the influence on flow of their size (as in flow through pipes) but also the retarding effects of the particles' surfaces (mineralogy, adsorption, orientation) as water flows over and around them.

The types of voids present in different clay soils will determine their permeability, as shown in Table 3.2.

Table 3.1 *Influence of voids on flow*

Effect on flow of:		Void name	Typical soil
particle surface	void size		
negligible	negligible	macropores	gravel
negligible	moderate	macropores and capillaries	sand
small	significant	capillaries	silt
significant	very significant	micropores	clay (intact)

Table 3.2 *Influence of voids in clays*

Clay type	Voids present
fissured clays desiccated clays compacted clays (dry of optimum)	macropores and micropores
intact clays compacted clays (wet of optimum)	micropores

Pressure and head *(Figure 3.13)*

Pressure is the force per unit area (kN/m^2) acting at a point in the water. Head is a measure of pressure but in terms of a height (metres of water). The total head at a point such as point B in Figure 3.13 is given by Bernoulli's equation:

$$H = h_Z + h_B + v^2/2g \qquad (3.1)$$

Total head (at B) = position head (at B) + piezometric head (at B) + velocity head (negligible)

The velocity head can be ignored since the velocity of flow of water through soil is quite small. If the water levels in the piezometric tubes *a* and *b* are the same then no flow will occur. Flow will only occur if there is a total head difference between *a* and *b*. This head difference is known as head loss or hydraulic head.

Darcy's Law *(Figure 3.13)*

This states that the discharge velocity, v of water is proportional to the hydraulic gradient, i

$$\frac{q}{A} = v = k\,i \qquad (3.2)$$

where:

k = Darcy coefficient of permeability, m/s

The hydraulic gradient i is the ratio of the head loss h over a distance l.

The discharge velocity v is defined as the quantity of water, q percolating through a cross-sectional area A in unit time. This is not the same as the velocity of the water percolating through the voids of the soil which is known as the seepage velocity.

Since porosity $n = \dfrac{\text{volume of voids}}{\text{total volume}}$

area of voids $A_s = nA$

$$\text{seepage velocity} = v_s = \frac{q}{A_s} = \frac{v}{n} \qquad (3.3)$$

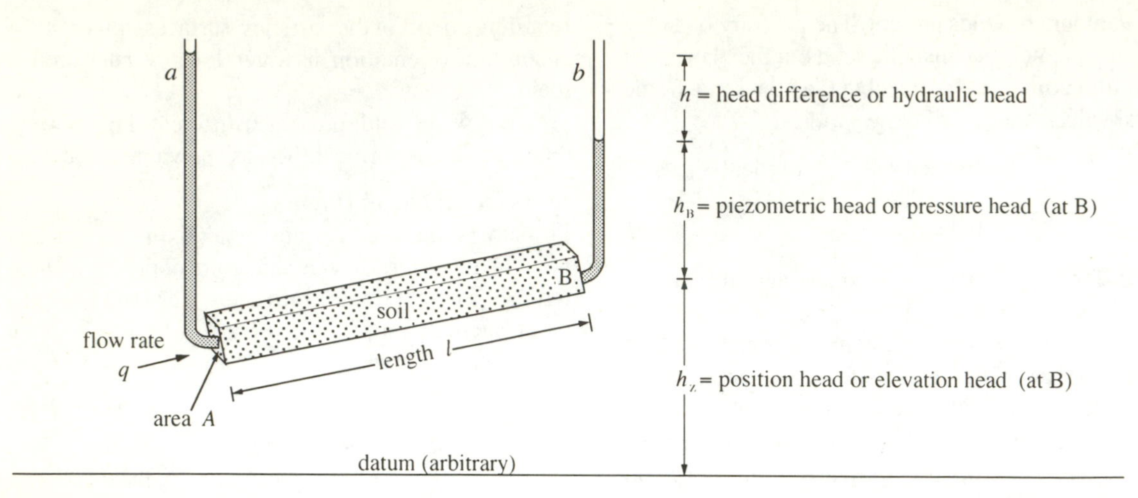

Figure 3.13 *Darcy's Law*

The flow of water through most soils is laminar and Darcy's Law applies to most soils. However, with open, large-void gravels, flow can become turbulent when Darcy's Law may not be valid.

The coefficient of permeability, k, is dependent on the nature of the voids (see above) and the properties of the fluid, particularly its viscosity. It is given by:

$$k = K\frac{\gamma_w}{\eta} \tag{3.4}$$

where:

γ_w = unit weight of fluid
η = viscosity of fluid
K = absolute or specific permeability (m^2).

Effect of temperature *(Figure 3.14)*
Unit weight, γ_w is affected very little by changes in temperature but viscosity is affected. It is conventional to report k at a temperature of 20°C so a correction factor, R_T, should be applied if the test is carried out at a temperature of T °C.

$$k_{20} = R_T k_T \qquad R_T = \frac{\eta_T}{\eta_{20}} \tag{3.5}$$

where:

k_{20} = permeability at 20°C
k_T = permeability at T °C
R_T = Correction factor, see Figure 3.14

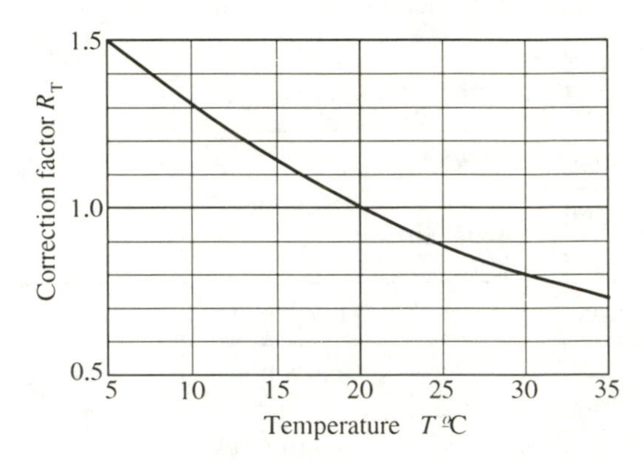

Figure 3.14 *Temperature correction*

Empirical correlations for *k* *(Figure 3.15)*
Typical values of k for various soil types are given in Table 3.3 illustrating the vast range of values obtainable.

The component of the coefficient of permeability, K or absolute permeability has been found to be related to the macro-pore sizes between the soil particles (for sands and gravels) which are in turn related to:

- particle size – d_{10}
  (larger particles produce larger voids)
- grading of particles – U_c
  (smaller particles clog larger voids)
- density or void ratio – e
  (closeness and orientation of particles).

Empirical correlations relating all three of these factors have been suggested in the form:

$$k = Cd_{10}{}^{a}U_c{}^{b}e^{c} \tag{3.6}$$

Table 3.3 *Typical values of k*

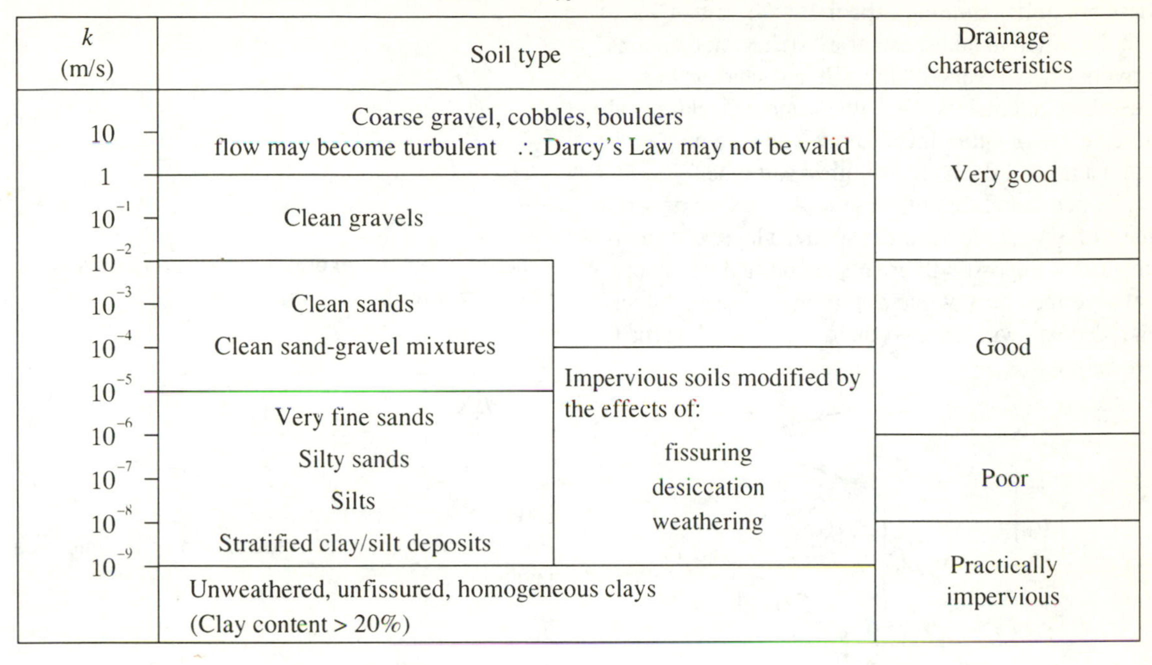

k (m/s)	Soil type		Drainage characteristics
10	Coarse gravel, cobbles, boulders flow may become turbulent ∴ Darcy's Law may not be valid		Very good
1			Very good
10^{-1}	Clean gravels		Very good
10^{-2}			Very good
10^{-3}	Clean sands		Good
10^{-4}	Clean sand-gravel mixtures		Good
10^{-5}		Impervious soils modified by the effects of:	Good
10^{-6}	Very fine sands	fissuring	Poor
10^{-7}	Silty sands	desiccation	Poor
10^{-8}	Silts	weathering	Practically impervious
10^{-9}	Stratified clay/silt deposits		Practically impervious
	Unweathered, unfissured, homogeneous clays (Clay content > 20%)		Practically impervious

where C, a, b and c are constants.

None of these correlations is particularly reliable especially compared to *in situ* test results. Determination of the effective size d_{10} can be very inaccurate for gap-graded materials, samples of stratified soils, or where finer particles are lost during sampling as is often the case. The most commonly used correlation is Hazen's relationship for clean (no fines) filter (fairly uniform) sands in a loose condition (near maximum void ratio):

$$k = Cd_{10}^{2} \text{ m/s} \tag{3.7}$$

where:

d_{10} = effective size, mm
C = coefficient, 0.01 to 0.015.

Compared to some *in situ* permeability tests this relationship appears reasonable, see Figure 3.15.

Laboratory tests also show (Cedergren, 1989) that k can vary by as much as one order of magnitude between the loosest and densest states of a soil. For example, a sandy gravel gave:

loose state - $k \approx 10^{-3}$ m/s
dense state - $k \approx 10^{-4}$ m/s

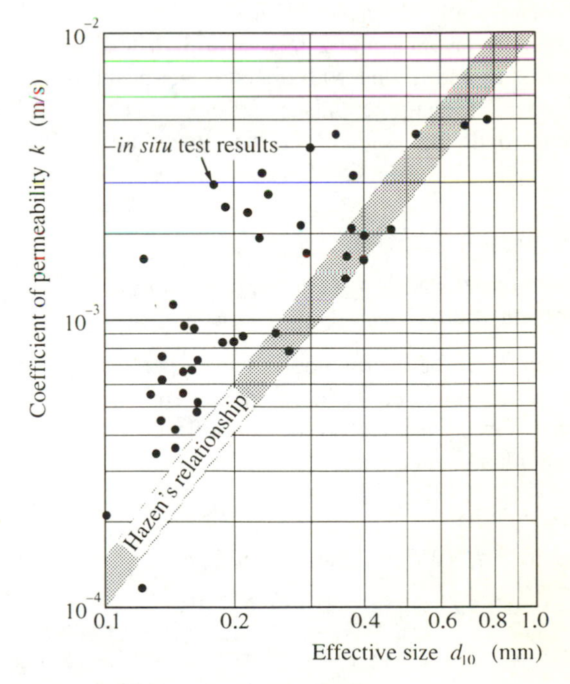

Figure 3.15 *Empirical correlations (From Leonards, 1962)*

Layered soils *(Figure 3.16)*
These are quite common especially where they have been deposited in lakes, estuaries, deltas, flood plains or rivers where layers of clay, silt and sand are deposited either alternately as with laminated clays and varved clays or a thin layer of one soil type may exist within a thicker deposit of another soil type.

The permeabilities of these soil types are several orders of magnitude different so that a layer of sand in a thick clay deposit will greatly increase overall horizontal permeability while a thin clay layer in a thick sand deposit will greatly decrease overall vertical permeability.

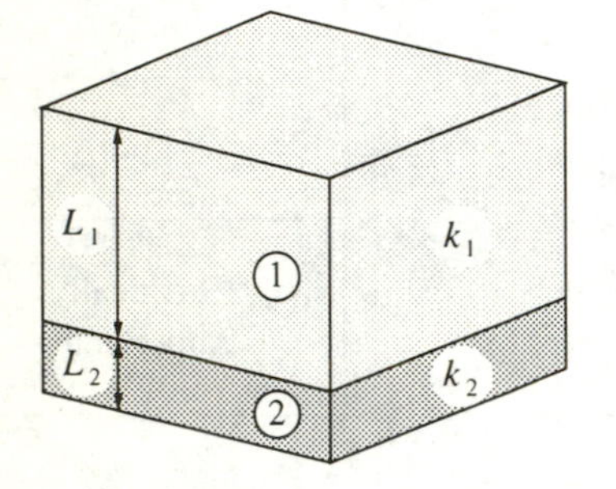

Figure 3.16 *Layered soils*

Layered soils – *horizontal flow*
Consider layers 1 and 2 in Figure 3.16, with thicknesses L_1 and L_2 and with different permeabilities k_1 and k_2. The equipotentials are vertical and equidistant so the hydraulic gradient, i is the same in both layers:

$i_1 = i_2 = i$

Applying Darcy's Law $\quad q = A\,k\,i$

Total horizontal flow $= q_H = 1\,(L_1 + L_2)\,k_H\,i$
per unit width (1 m)

where k_H is the overall coefficient of permeability in the horizontal direction for the layered soil. The overall total flow q_H is equal to the sum of the flows in the individual layers, i.e.

$q_H = q_1 + q_2$

where $\quad q_1 = 1\,L_1 k_1 i_1$ and $q_2 = 1\,L_2 k_2 i_2$

$\therefore (L_1 + L_2)\,k_H\,i = (L_1\,k_1 + L_2\,k_2)i$

and

$$k_H = \frac{L_1 k_1 + L_2 k_2}{L_1 + L_2} \qquad (3.8)$$

or

$$k_H = \frac{\sum_1^n L_i k_i}{\sum_1^n L_i} \qquad (3.9)$$

which is a general expression for several layers.
For horizontal flow $i_1 = i_2$ $\therefore$

$$\frac{q_1}{q_2} = \frac{L_1 k_1}{L_2 k_2} \qquad (3.10)$$

Layered soils – *vertical flow*
For continuity the rate of flow must be the same in each layer, i.e.

$q_1 = q_2 = q_v$

For unit horizontal area (A = 1 m^2) $\quad q_v = 1\,k_v i_v$

where k_v is the overall coefficient of permeability in the vertical direction. The loss in total head over length $L_1 + L_2$ will be equal to the sum of losses in total head in the individual layers, i.e.

$$i_v(L_1 + L_2) = i_1 L_1 + i_2 L_2$$

$$i_v = \frac{i_1 L_1 + i_2 L_2}{L_1 + L_2}$$

$$i_1 = \frac{q_1}{k_1} \quad \text{and} \quad i_2 = \frac{q_2}{k_2}$$

$$k_v = \frac{q_v}{i_v} = \frac{q_v(L_1 + L_2)}{q_1\dfrac{L_1}{k_1} + q_2\dfrac{L_2}{k_2}} = \frac{L_1 + L_2}{\dfrac{L_1}{k_1} + \dfrac{L_2}{k_2}} \qquad (3.11)$$

or

$$k_v = \frac{\sum_1^n L_i}{\sum_1^n \dfrac{L_i}{k_i}} \qquad (3.12)$$

Laboratory test – constant head permeameter *(Figure 3.17)*

The procedure for carrying out this test using a constant head permeameter is described in BS 1377:1990. The test is only suitable for soils having a coefficient of permeability in the range 10^{-2} to 10^{-5} m/s, which applies to clean sand and sand-gravel mixtures with less than 10% fines (silt or clay).

The test is carried out by adjusting the control valve and waiting until the flow, q, through the sample and the hydraulic head loss, h, between the manometer points have reached a steady state. The flow rate, q, and

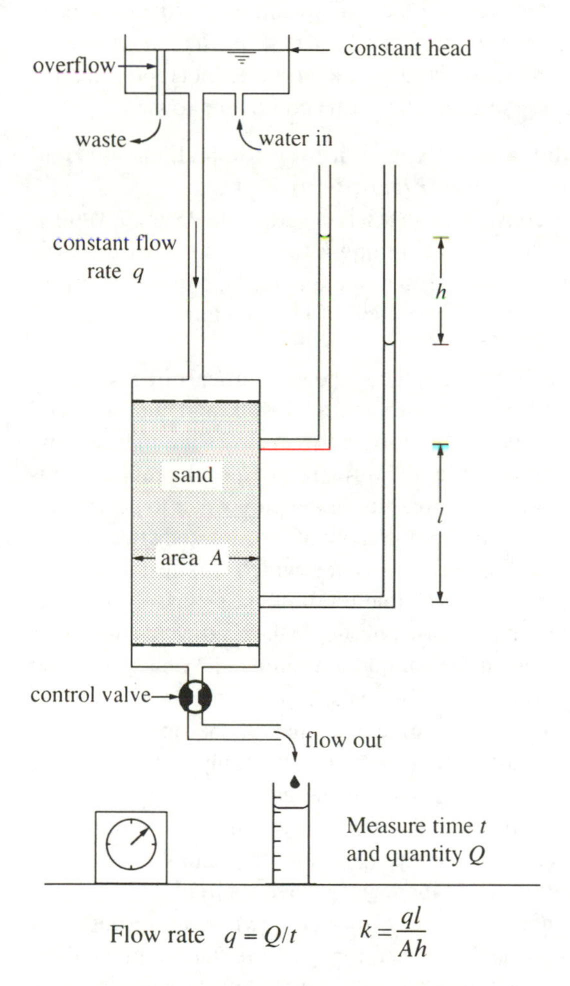

Figure 3.17 *Constant head permeameter*

head loss, h, are measured and the coefficient of permeability is calculated. The test is repeated at different hydraulic gradients by adjusting the control valve to obtain a number of q and h values and determinations of k. An intermediate manometer point is recommended, to ensure that the hydraulic gradient through the sample is uniform.

The permeability values should be corrected for the effect of temperature using the correction factor R_T (see Figure 3.14), and the average value is reported.

The dry mass of the test sample should be obtained and the dry density of the soil determined. The particle density should also be obtained or assumed so that the void ratio can be calculated. It is preferable to carry out the test on samples compacted at a range of densities and to plot k against density or void ratio.

The test is rarely used for natural soil samples since:

- The samples are unrepresentative of *in situ* conditions. The samples are small (75 or 100 mm diameter) and are unlikely to contain sufficient macrostructure (fissures, laminations, root holes etc) to give the appropriate k value.
- One-dimensional (vertical) flow of water *in situ* is an unlikely event on its own whereas the test is constrained to this direction.
- Installation and sealing of 'undisturbed' specimens in the permeameter is difficult and further disturbance is inevitable during sample preparation. This changes the microstructure and macrostructure and tends to reduce flow.
- Leakages around the sides of the sample may give erroneously high results and air bubbles trapped would produce low results.

The test can be suitable for soils when used in their completely disturbed or remoulded states, such as for drainage materials and filters, to confirm that their performance will be adequate. Again, a range of tests at different densities and placement moisture contents would be required to relate k to the compaction characteristics of the materials.

The maximum particle size must be limited to one-twelfth of the permeameter diameter to ensure representative sampling, so for coarse drainage material particles greater than 6 mm must be discarded for a 75 mm diameter cell.

Laboratory test – falling head permeameter *(Figure 3.18)*

This test can be used for soils of low to intermediate permeability but its procedure is not included in BS 1377. Water is allowed to flow down the standpipe and through the sample so that the hydraulic head, h, at any time, t, is the difference between the meniscus level in the standpipe and the overflow level.

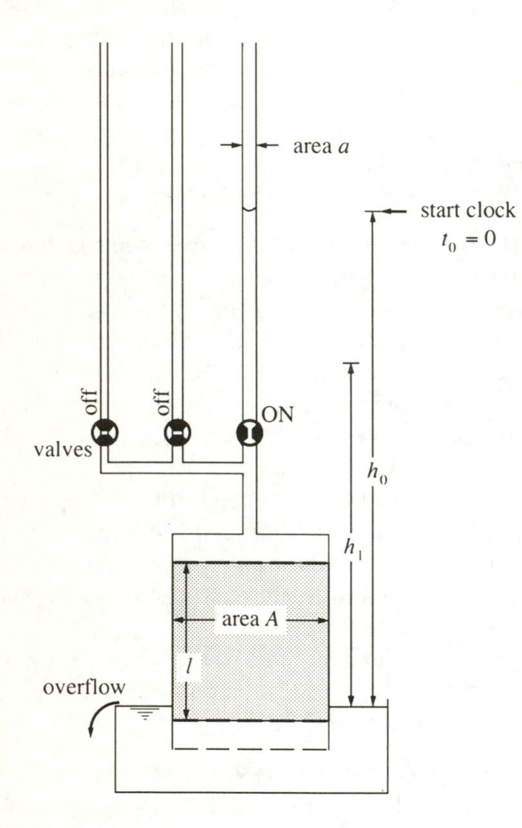

Figure 3.18 *Falling head permeameter*

In the standpipe the water level falls dh in dt and

$$q = -a\frac{dh}{dt}$$

In the sample $q = Aki$

$$i \text{ at any time } t = \frac{h}{l}$$

$$\therefore -\int_{h_0}^{h_1}\frac{dh}{h} = \frac{Ak}{al}\int_{t_0}^{t_1} dt$$

giving

$$\ln\frac{h_1}{h_0} = -\frac{Ak}{al}\left(t_1 - t_0\right)$$

with $h = h_0$ at $t_0 = 0$
and $h = h_1$ at t_1

$$k = 2.3\frac{al\log_{10}\frac{h_0}{h}}{At} \quad (3.13)$$

If a graph of t versus $\log_{10}(h_0/h)$ is plotted, a straight line should be obtained and k can be derived from the gradient.

The main limitations of the test are the potential for leakages around the soil specimen, particularly if an undisturbed specimen is used. The effective stresses to which the soil is subjected *in situ* are not represented in the test so some clay soils could tend to swell.

Laboratory test – hydraulic cell – vertical permeability *(Figure 3.19)*

This constant head test is described in BS 1377:1990 as a means of determining k for an 'undisturbed', large volume (up to 250 mm diameter) and therefore representative sample of soil of low to intermediate permeability.

The sample is subjected to a vertical effective stress representing *in situ* stress levels and the stiff cell provides K_0 conditions. Water flows are induced and measured under the influence of a back pressure. This is an elevated pressure in the pore water to prevent air or gas bubbles coming out of solution and affecting the results. Organic clays for which this test would be appropriate are liable to allow gases dissolved in the pore water to form bubbles in the void system when the stresses on the sample are removed. A back pressure therefore models the *in situ* pore pressure.

With the sample installed in the apparatus it is necessary first to saturate the sample, the cell, the pressure lines and the gauges. The sample is then consolidated under a pressure p_d in order to introduce an effective stress (p_d) into the sample. The inlet pressure p_1 is increased above the outlet pressure p_2 to induce a flow of water downwards through the sample and this difference of pressure is maintained until the flow becomes steady; p_1 must never exceed p_d.

For silty and sandy soils a head difference ($h_1 - h_2$) of only a few centimetres may be necessary to

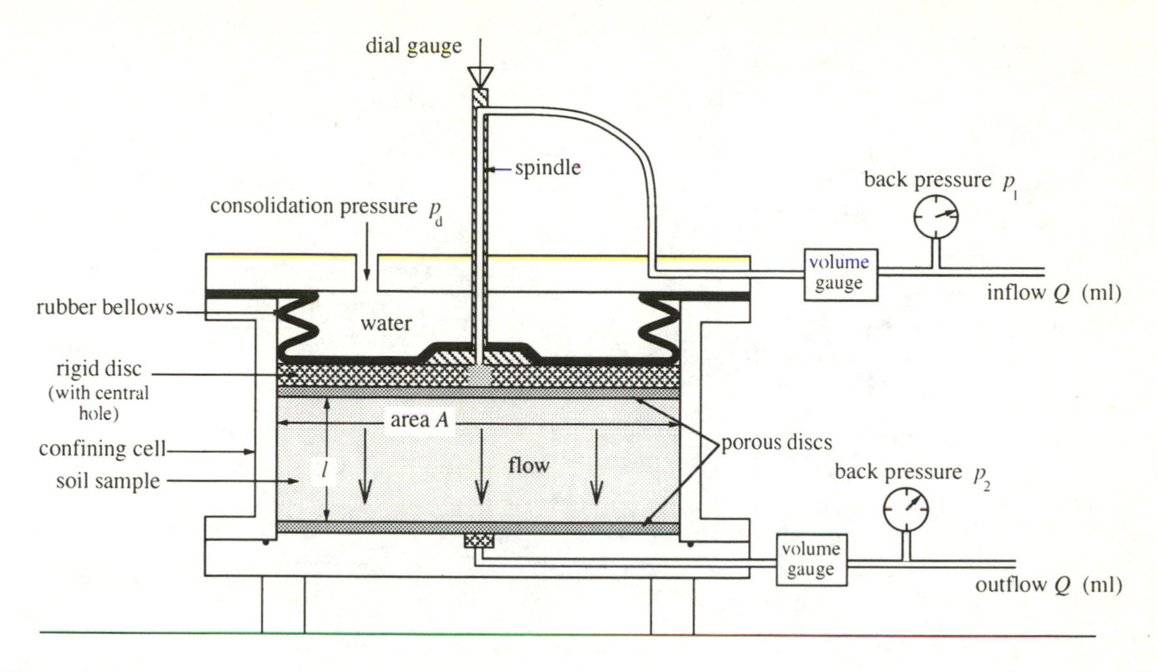

Figure 3.19 *Hydraulic cell – vertical permeability*

produce measurable flow, whereas for clay soils a head difference of up to 2 m or pressure difference of about 20 kN/m^2, may be required, depending on the thickness of the sample and its permeability. When the difference between p_1 and p_2 is small it is best measured using a differential pressure gauge or manometer.

When the rate of flow is high (> 20 ml/minute) there may be a significant pressure or head loss in the system (p_c) due to friction losses in the pressure lines, valves etc. This pressure loss can be determined during the calibration of the apparatus and subtracted from the measured pressure loss.

The expression for the determination of the vertical coefficient of permeability k_v is obtained from:

$$q = k_v A i$$
$$q = Q/t$$

For full saturation and no leakages:

inflow Q = outflow Q

h = head difference across the sample = $h_1 - h_2$

$$h_1 = p_1/\gamma_w \qquad h_2 = p_2/\gamma_w$$

$$k_v = \frac{q}{Ai} = \frac{q}{A}\frac{l\,\gamma_w}{(p_1 - p_2)}R_T \qquad (3.14)$$

Laboratory test – hydraulic cell – horizontal permeability *(Figure 3.20)*

The test is set up in a similar fashion to the previous test but water is prevented from flowing vertically by plugging the central hole in the rigid disc and placing this disc on an impervious layer of latex rubber, similar to the rubber membrane used in the triaxial test. The peripheral porous plastic drain is connected via the rim drain valve to a back pressure system. A central sand drain formed by drilling a hole through the specimen and filling it with sand is connected to the other back pressure system.

Care must be taken with the construction of the sand drain to minimise smear effects and to avoid the drain acting as a hard spot. Smearing can affect the results significantly where there is any anisotropic structure in the soil with a preference for water to flow horizontally such as with laminated or varved clays. For these reasons it is better to place the sand in a loose condition and keep the diameter of the sand drain small in relation to the sample diameter (less than 1:20).

After saturating the specimen and consolidation under the pressure, p_d, the inlet pressure, p_1, is increased above the outlet pressure, p_2, to induce water flow in a radial direction. This flow can either be

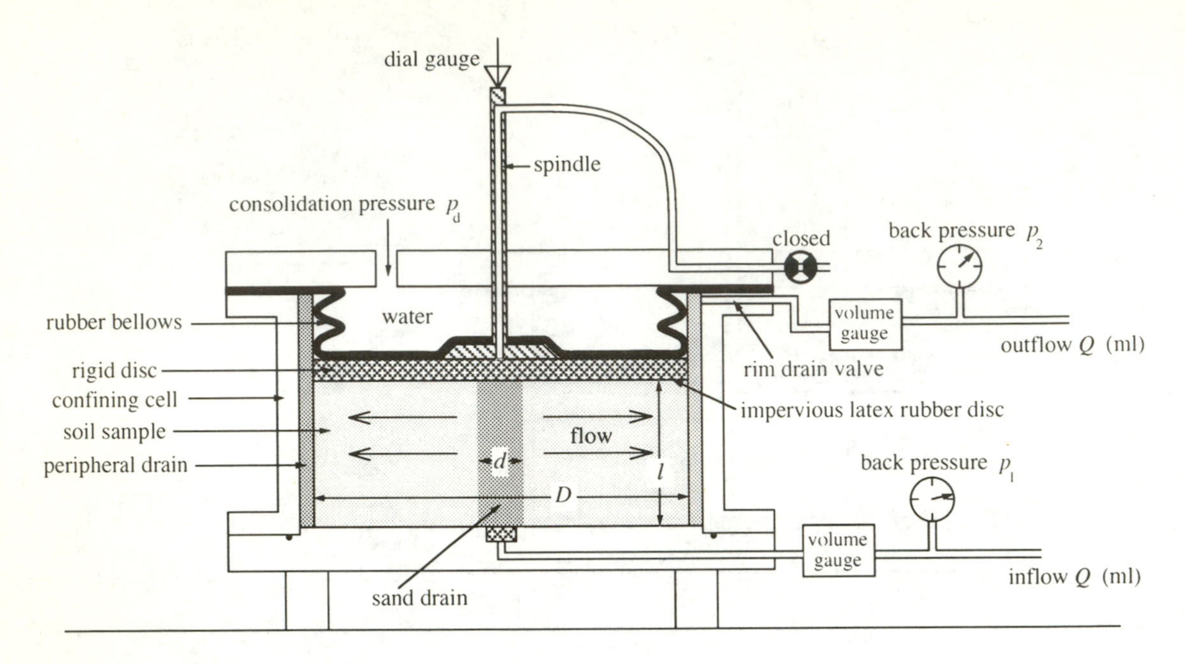

Figure 3.20 *Hydraulic cell - horizontal permeability*

inwards towards the central sand drain or outwards to the peripheral drain and should be measured when a steady flow is achieved.

The expression for the determination of the horizontal coefficient of permeability, k_H, is given by:

$$k_H = \frac{q\gamma_w}{2\pi l(p_1 - p_2)} \ln \frac{D}{d} R_T \qquad (3.15)$$

Borehole tests – open borehole *(Figures 3.21 – 3.23)*
Permeability tests can be carried out in cased or lined boreholes provided the casing extends below the water table. Permeability tests in cased boreholes above the water table can be carried out, but interpretation of the results is dubious since the water flowing out of the borehole must first saturate the voids, and suctions between the pore water and pore air in the voids will affect the hydraulic head.

The test is carried out by providing a head of water inside the borehole above water table level to induce flow into the soil. Either a constant head can be maintained at H_c above the water table by introducing water at a steady flow rate, q, or a falling head test can be carried out by measuring the head of water, H, above the water table at time intervals t as it falls.

Sometimes rising head tests have been used by pumping out water below the water table level but these can be prone to more uncertain results due to potentially unstable conditions (piping) at the bottom of the borehole.

Expressions for determining k are given in Figure 3. 21, Equation 3.16 is for the constant head test and Equation 3.17 is for the falling head test. The variable f is termed a shape factor or intake factor and is related to the condition at the bottom of the borehole. Shape factors for the six cases A – F have been derived from the work of Hvorslev (1951) and are given in Figures 3.22 and 3.23.

These figures show that the condition at the bottom of the borehole has a significant effect, something that drilling and boring techniques cannot easily control. A small amount of soil inside the casing due to insufficient cleaning out, or fines settling, reduces f and it is more likely that only k_v is measured, whereas extending the borehole slightly below the casing increases f and it is more likely that k_H is measured. Thus very different results can be obtained depending on the condition at the bottom of the casing and whether the soil is anisotropic ($k_H > k_v$).

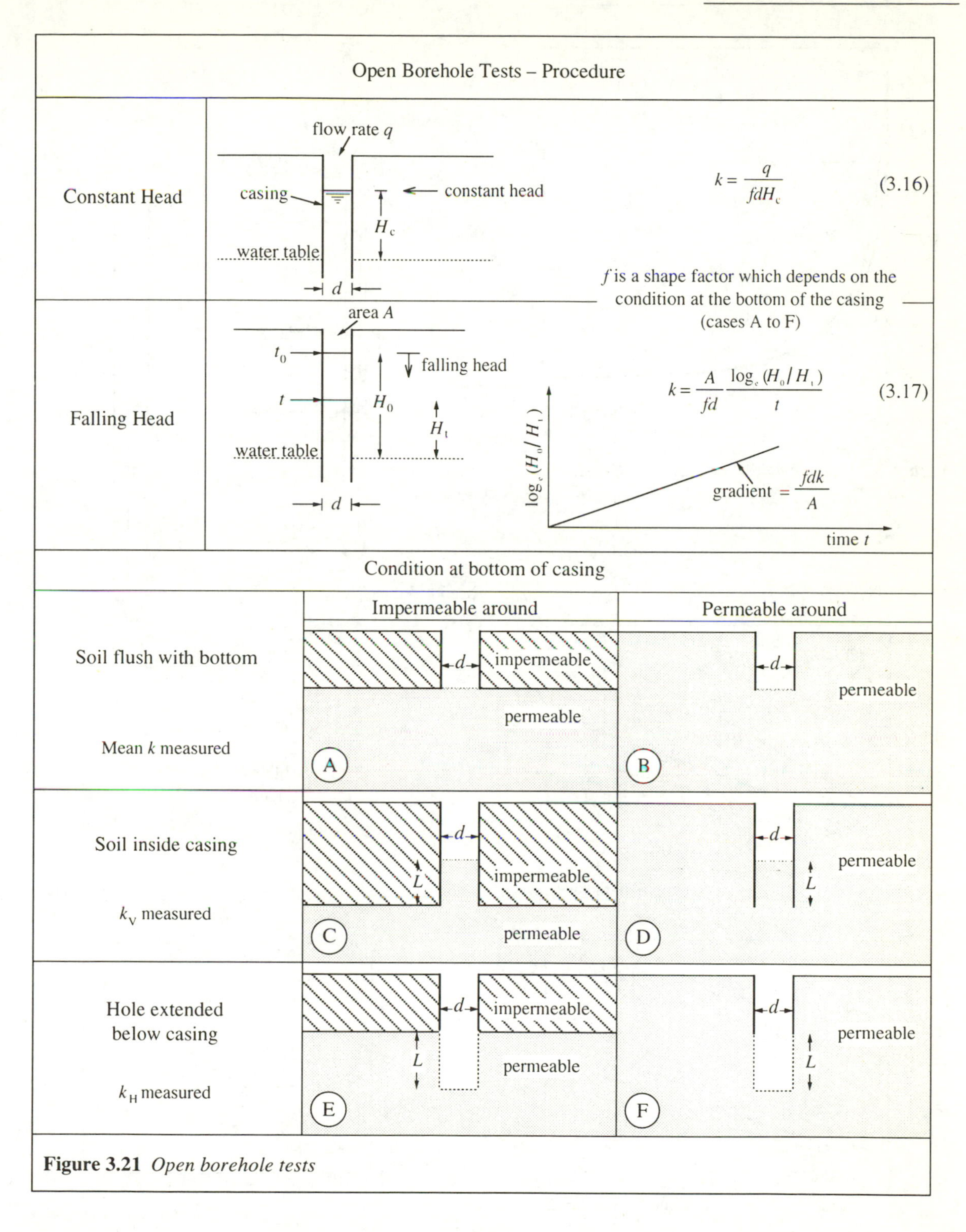

Figure 3.21 *Open borehole tests*

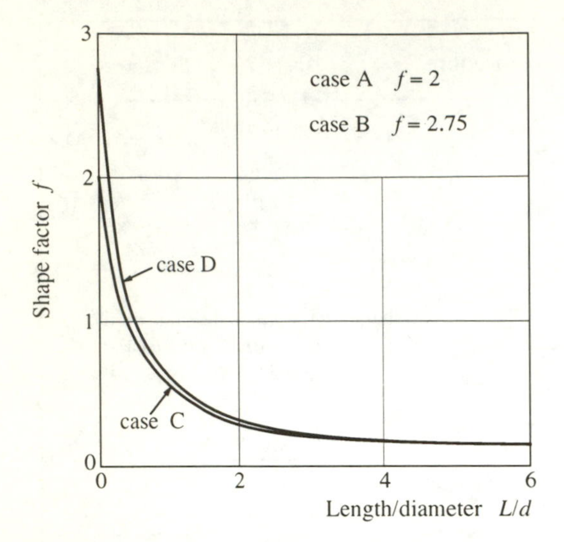

Figure 3.22 *Shape factor for cases A, B, C, D*

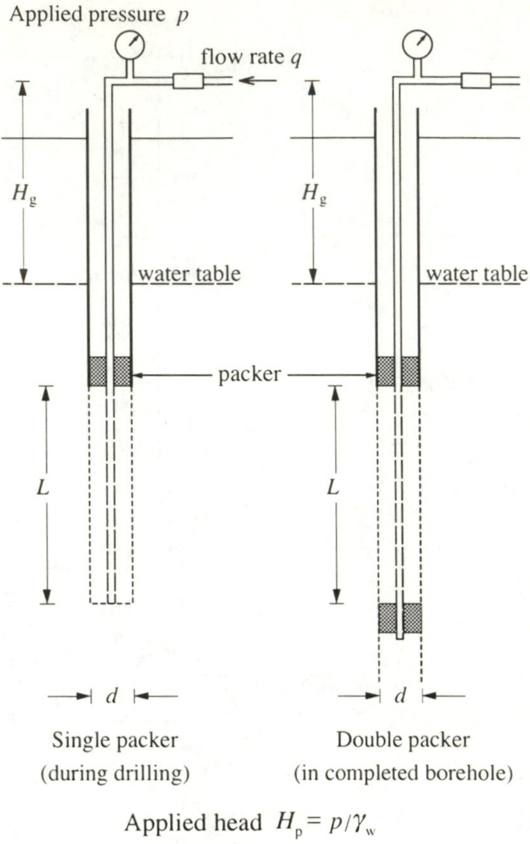

Applied head $H_p = p/\gamma_w$

Gravity head $= H_g$

Total head $H = H_p + H_g$

$$k = \frac{q}{Hfd} \quad (3.18)$$

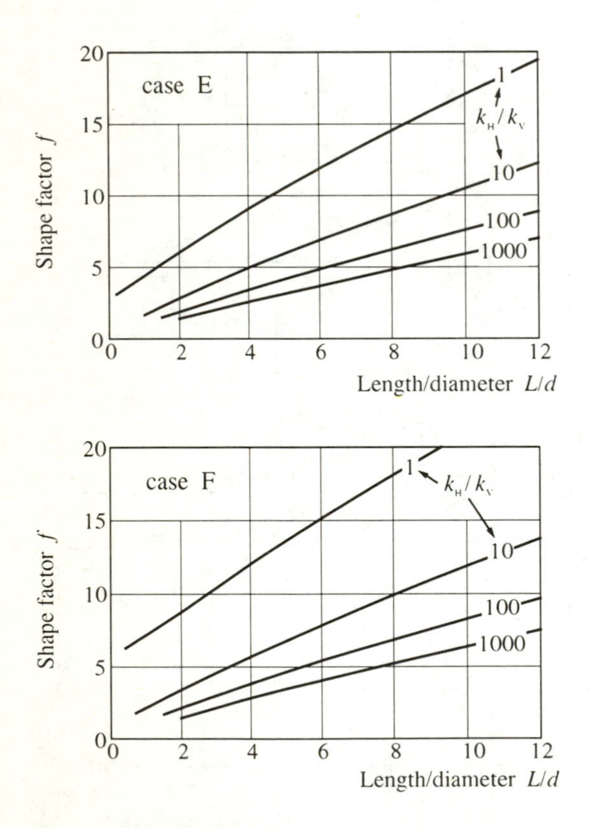

Figure 3.23 *Shape factors – Cases E and F*

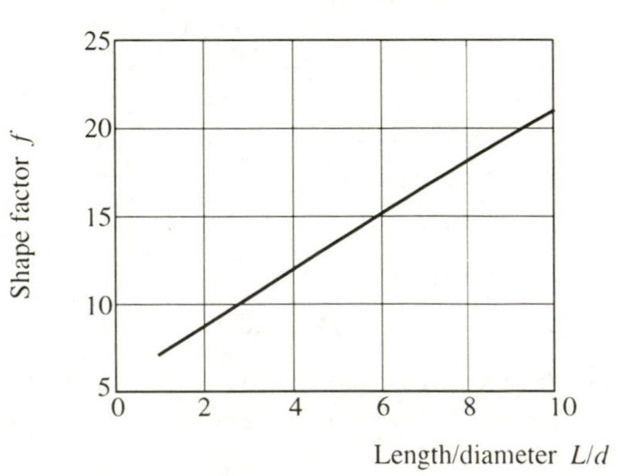

Figure 3.24 *Packer tests*

Borehole tests – packer tests *(Figure 3.24)*
A packer is a rubber bag which is inflated against the sides of the borehole and around a tube to form a seal. The section of borehole beneath the packer can then be tested by applying a head of water, H, and measuring the flow rate, q, to maintain this constant head. The applied head can simply be the gravity head, H_g, of water in the tube above the water table, although it is more common to apply a pressure, p, to increase the water head. The total head must be limited to ensure that hydraulic fracturing does not occur.

The test can be carried out during drilling with a single packer sealed against the inside of the casing, or at selected levels between double packers in a completed borehole. However, with the test carried out below the water table, this requires the soil in the uncased section to be self-supporting so the test is usually restricted to determining the permeability of jointed bedrocks. The sides of the hole must be clean and free from smeared material and the water in the borehole must be clean, so it is common practice to fill the hole with water, surge it, bail it out and fill with clean water.

The test can be carried out above the water table where the gravity head, H_g, will be the head of water in the tube above the mid-point of the tested section but these results must be treated with caution as they will be less accurate. The formula given in Figure 3.24 is also less valid when the thickness of the deposit tested is less than $5L$.

Borehole tests – piezometers *(Figure 3.25)*
In this test a constant head is applied to the sand filter surrounding the piezometer so the dimensions of this zone must be known with reasonable accuracy. The length:diameter ratio of the sand filter should not exceed 5. It is presumed that the sand is much more permeable than the surrounding soil and it has been found (Gibson, 1966) that the permeability of the piezometer tip material itself must be at least ten times more permeable than the surrounding soil, otherwise the test will merely measure the permeability of the ceramic tip. For this reason the test is only suitable for soils of low permeability.

The constant head is maintained by an elevated water container with the tap to the piezometer turned on. This induces flow into the soil surrounding the sand filter and the flow is allowed to continue until a steady

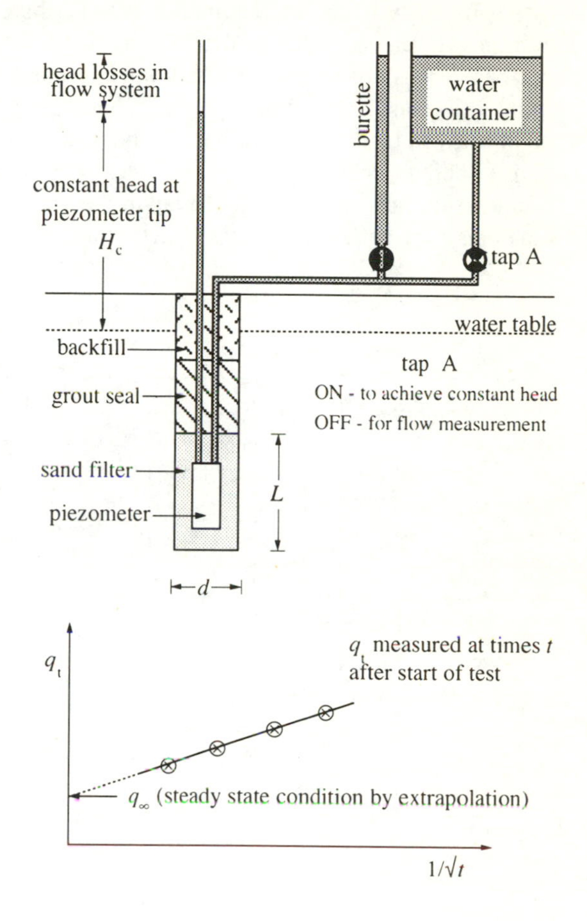

a) Isotropic soil – mean k measured

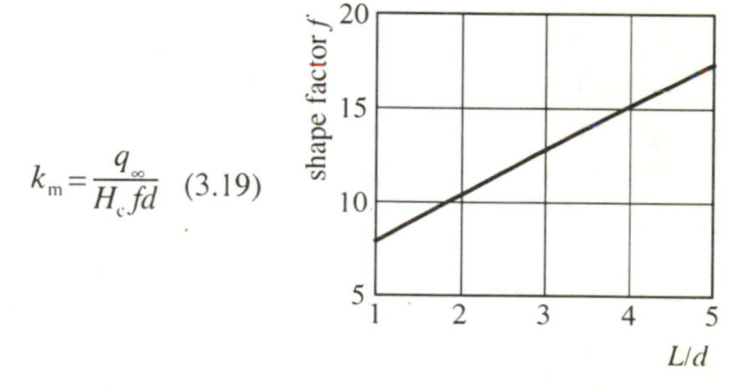

$$k_m = \frac{q_\infty}{H_c f d} \quad (3.19)$$

b) Anisotropic soil – k_{11} measured

$$k_{11} = \frac{q_\infty}{H_c f d} \qquad f = \pi \frac{L}{d} \qquad (3.20)$$

(After Wilkinson, 1968)

Figure 3.25 *Piezometer test*

rate of flow, q_∞, is reached. However, in low permeability soils many hours are required to achieve this condition so q_∞ can be found by determining the flow rate, q_t, at various times t after the start of the test and plotting q_t versus $1/\sqrt{t}$. With sufficient readings a straight line can be extrapolated back to the origin to give q_∞ (when $t = \infty$).

To measure flow at any stage the tap to the water container is turned off. Water flows from the burette to the piezometer tip and the quantity of flow, Q ml, in a given time, is measured. The constant head is maintained by raising the burette so that the meniscus always remains at the same level.

Either two leads to the piezometer tip or two standpipes within the filter zone should be provided, one for flow measurement and one for measuring the head. At higher flow rates head losses in the tubes on the flow side can be significant.

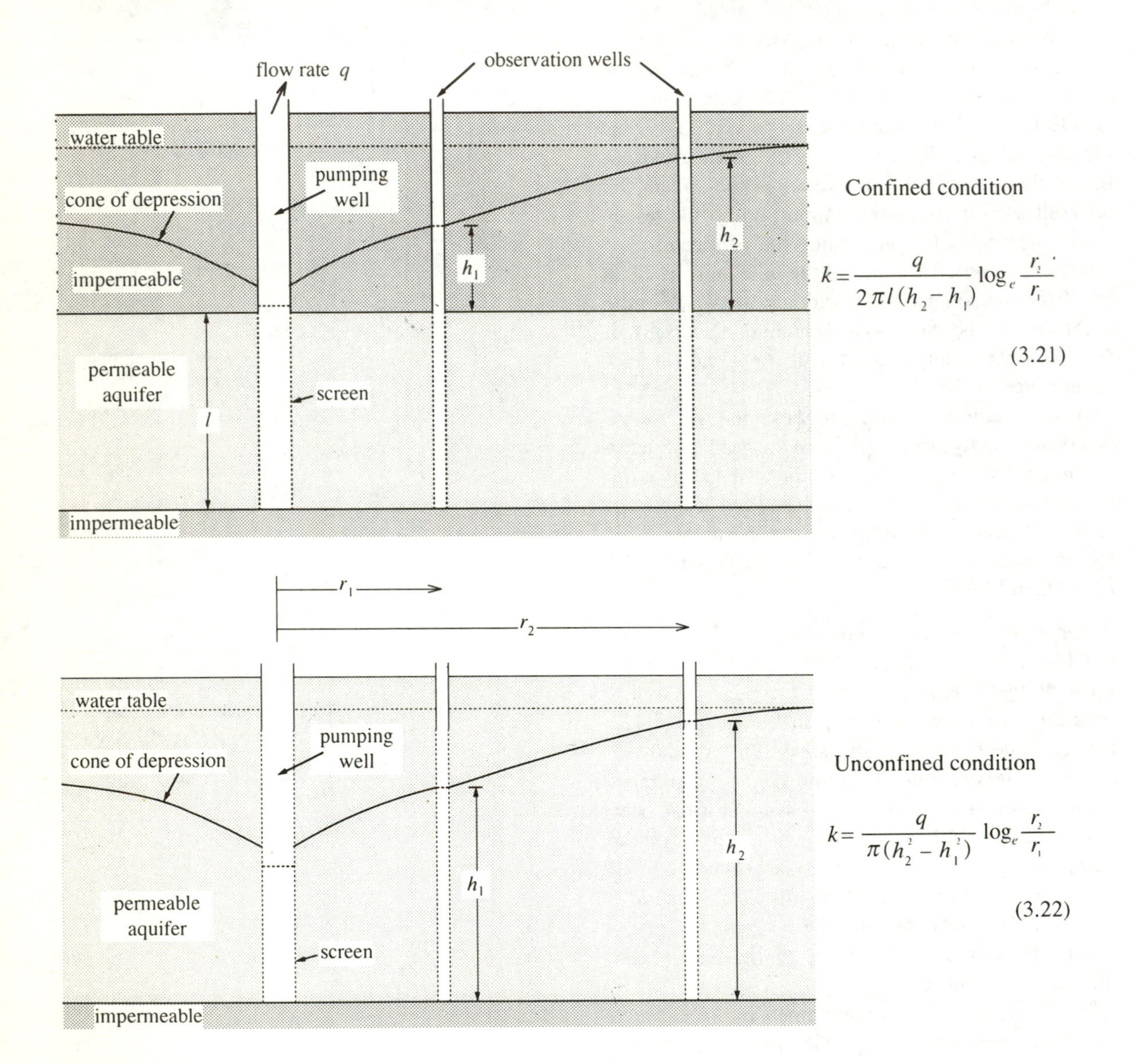

Confined condition

$$k = \frac{q}{2\pi l(h_2 - h_1)} \log_e \frac{r_2}{r_1} \tag{3.21}$$

Unconfined condition

$$k = \frac{q}{\pi(h_2^2 - h_1^2)} \log_e \frac{r_2}{r_1} \tag{3.22}$$

Figure 3.26 *Pumping tests*

The plot of q_t versus $1/\sqrt{t}$ may not always be a good straight line. Curvature can be produced by:

- smearing on the sides of the borehole. This can vary with the different methods of drilling and will be more significant when the soil contains well-defined macrostructure. Smearing is inevitable with the push-in type of piezometer.
- air or gas bubbles in the sand filter. The sand should be de-aired and poured under water.

Pumping tests *(Figure 3.26)*
These tests are described fully in BS 5930:1981. They are carried out by constructing a pumping well to the full thickness of the aquifer and pumping out water from the bottom, using suction pumps for depths less than about 5 m or submersible pumps for greater depths, with a flow meter or notch tank to measure the rate of flow, q.

Pumping will lower the water table and eventually a symmetrical cone of depression will form when a steady state condition has been reached. This occurs when the rate of water pumped out of the well is comparable to the rate of water flow in the aquifer towards the well. This may take several days to achieve for low permeability soils.

The determination of k for the aquifer requires some information on the shape of the cone of depression. For this purpose, four observation wells are installed into the aquifer, in two rows at right angles to each other, and water levels are observed until a steady state has been achieved or the rate of change of drawdown is small. The expression for k then depends on whether the aquifer is confined or unconfined, (see Figure 3.2), and these are given in Figure 3.26.

Seepage theory *(Figure 3.27)*
Seepage can occur in three directions or dimensions, as is the case for flow to a pumping well. The theory considered here is for two-dimensional flow which is applicable for any cross-section through a long structure such as a sheet-pile cofferdam, earth dam or concrete weir.

Flow along the structure (perpendicular to Figure 3.27) cannot occur, but flow can have vertical and horizontal components of discharge velocity, v_x and v_z in the x-z plane (Figure 3.27a). The permeability of the soil is considered initially to be homogeneous, or the same at all locations, and isotropic, the same in all directions, so that $k_H = k_v = k$. This theory only applies to fully saturated soils and considers water flow only below the water table.

Consider an element of soil with dimensions dx and dz in the plane of seepage (Figure 3.27b), and dy perpendicular to the cross-section with flow occurring in the x- z plane. The discharge velocities (or components) are v_x and v_z and the velocity gradients are $\partial v_x/\partial x$ and $\partial v_z/\partial z$, respectively. Assuming the water to be incompressible and the mineral grain structure to be unaffected by seepage, the quantity of water leaving the element must equal the quantity entering it.

Quantity of flow $q = Av$

$$v_x \mathrm{d}y\mathrm{d}z + v_z \mathrm{d}y\mathrm{d}x =$$

$$\left(v_x + \frac{\partial v_x}{\partial x}\mathrm{d}x\right)\mathrm{d}y\mathrm{d}z + \left(v_z + \frac{\partial v_z}{\partial z}\mathrm{d}z\right)\mathrm{d}y\mathrm{d}x$$

Therefore

$$\frac{\partial v_x}{\partial x} + \frac{\partial v_z}{\partial z} = 0 \qquad (3.23)$$

This is the *continuity equation* in two dimensions. The hydraulic gradient components for the element $i_x = \partial h/\partial x$ and $i_z = \partial h/\partial z$ can be inserted into Darcy's Law to give:

$$v_x = ki_x = -k\frac{\partial h}{\partial x}$$

and $$v_z = ki_z = -k\frac{\partial h}{\partial z}$$

where the total head, h, is decreasing in the direction of v_x and v_z.
Two functions:

$\Phi(x,z)$ – potential function
and $\Psi(x,z)$ – flow or stream function

are introduced to give:

$$v_x = -k\frac{\partial h}{\partial x} = \frac{\partial \Phi}{\partial x} = \frac{\partial \Psi}{\partial z} \qquad (3.24a)$$

$$v_z = -k\frac{\partial h}{\partial z} = \frac{\partial \Phi}{\partial z} = -\frac{\partial \Psi}{\partial x} \qquad (3.24b)$$

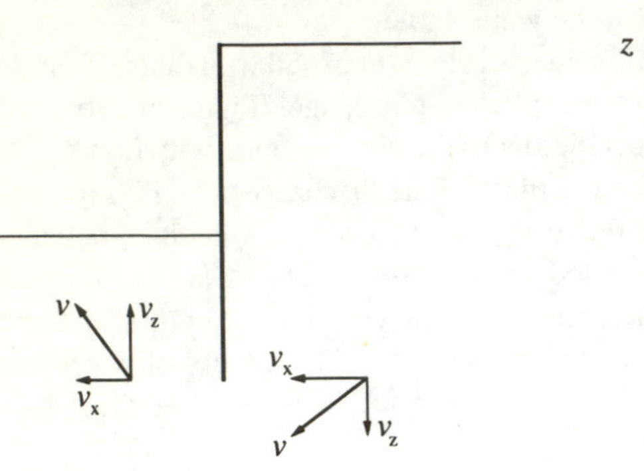

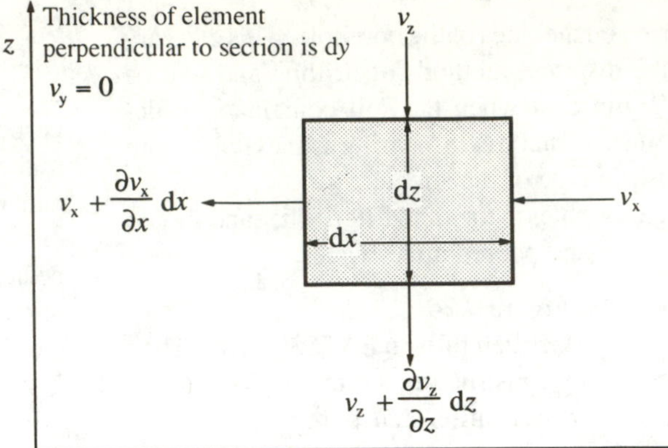

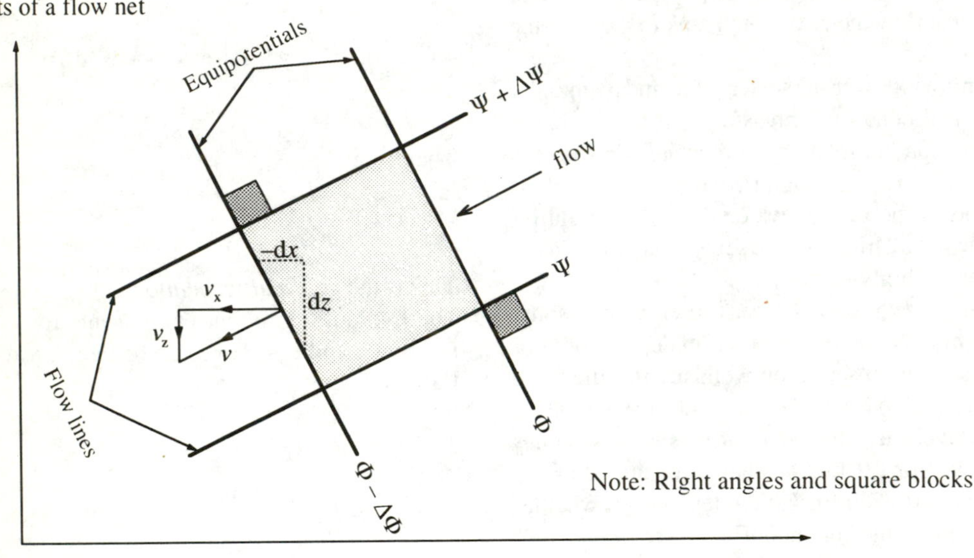

Figure 3.27 *Seepage theory*

From Equation 3.23 it can be seen that:

$$\frac{\partial^2 \Phi}{\partial x^2} + \frac{\partial^2 \Phi}{\partial z^2} = 0$$

and $$\frac{\partial^2 \Psi}{\partial x^2} + \frac{\partial^2 \Psi}{\partial z^2} = 0$$

which are the Laplace equations in two dimensions. Integrating the potential function in Equation 3.24 gives:

$$\Phi(x,z) = -kh(x,z) + c$$

such that

$$\Phi = -kh$$

If this potential function is given a constant value, say Φ_1, then it will form a curve called an equipotential which can be plotted in the x-z plane and along which the total head, h, will be constant.

Differentiating the stream function in Equation 3.24 gives:

$$d\Psi = \frac{\partial \Psi}{\partial x}dx + \frac{\partial \Psi}{\partial z}dz$$

$$= -v_z dx + v_x dz$$

If the stream function is given a constant value, say Ψ_1, then $d\Psi = 0$ and

$$\frac{dz}{dx} = \frac{v_z}{v_x} \quad (3.25)$$

so that the tangent to the stream function curve Ψ_1 plotted in the x-z plane defines the direction of the resultant discharge velocity and this curve, of constant Ψ is called a flow line or streamline.

Differentiating the potential function in Equation 3.24 gives:

$$d\Phi = \frac{\partial \Phi}{\partial x}dx + \frac{\partial \Phi}{\partial z}dz$$

$$= v_x dx + v_z dz$$

and for a particular equipotential where $d\Phi = 0$

$$\frac{dz}{dx} = -\frac{v_x}{v_z} \quad (3.26)$$

Comparing this equation with Equation 3.25 shows that the flow lines intersect the equipotentials at right angles. The equipotentials act as 'contours' of equal potential or total head. Thus the direction of the maximum hydraulic gradient will be at right angles to the equipotentials, and in the direction of the flow lines, with flow occurring in the direction of decreasing values of potential Φ.

It can be shown that if the intervals between equipotentials $\Delta\Phi$ (potential drops) are assumed equal to the intervals between stream lines, $\Delta\Psi$ then they should be plotted on the x-z plane at equal intervals thus forming 'square' shapes and producing a flow net.

Flow nets

When there is a difference in hydraulic head either side of a water-retaining structure such as a dam or sheet-pile wall water will flow beneath and around the structure. This flow can be represented mathematically by the Laplace equations of continuity given above but solutions require complex mathematical procedures.

In seepage problems it has to be accepted that only an estimate of the flows or resulting water pressures can be obtained, since this is the best our determinations of the coefficient of permeability will allow.

A simple approach to the seepage problem can be obtained by representing the Laplace equations in the form of a simple flow net sketched on a cross-section of the problem as carefully as possible following a number of rules.

A flow net consists of two sets of lines:

- ***flow lines***
  These are paths along which water can flow through a cross-section. There are an infinite number of flow lines available but only a few (four to five) need to be selected for an adequate flow net. Drawing too many will complicate the result.
  The intervals between adjacent flow lines (flow channels) represent a constant flow quantity, Δq, so the total seepage flow is given by Δq multiplied by the number of flow channels. The number of flow channels need not be a whole number. This can be useful when plotting a flow net up to an impermeable boundary when only a part flow channel is left. The seepage quantity passing through this remaining part channel is proportional to its width relative to a full channel.

- ***equipotential lines***
  These are lines of equal energy level or equal total head. As the water flows through the pore spaces its energy is dissipated by friction and the equipotential lines act like contours to show how the energy is lost. The intervals between adjacent equipotentials represent a constant difference in total head loss, Δh, and the total head, h, lost around the structure is shared equally among equipotential drops.
  It must be stressed that the head along an equipotential represents total head, not pressure head, see Figure 3.13.
  A flow net is usually used to represent the steady state condition. For example, on impounding a reservoir behind an earth dam, the soil voids must first become saturated before a steady state flow through the dam can develop.

Flow net construction *(Figure 3.28 – 3.31)*
A certain amount of skill is required in drawing a flow net but adequate results can quickly be obtained providing the following rules are observed:

- *right angles*
  Flow lines and equipotential lines must cross at right angles.
- *square blocks*
  The areas formed by intersecting flow lines and equipotential lines must be as near square as possible, i.e. the central dimensions should be equal. A useful test is to visualise whether a circle can be placed inside the block and touch all four sides.
- *impermeable boundaries*
  These are flow lines. Examples are the surface of a clay layer, the vertical surface of sheet piling or the underside of a concrete dam.
- *permeable boundaries*
  Where a permeable soil boundary is in contact with open water as on the upstream face of an earth dam, this boundary will be an equipotential, i.e. the total head is constant on this boundary. This also applies to a horizontal water table within a permeable deposit when flow is occurring vertically downwards, as alongside a sheet piled excavation.
  An example of a flow net sketched (drawn, erased, edited and re-drawn) on a cross-section through a sheet-piled excavation is given in Figure 3.29.
- *entry requirements*
  These apply for the construction of a flow net through an earth dam. The upstream face of the dam is an equipotential so flow lines must intersect at right angles.
- *deflection rule (Figure 3.30)*
  When water flows across a boundary between soils of different permeabilities the flow lines bend, the flow channel width (distance between two flow lines) alters and the distance between equipotentials changes so that the blocks become rectangular. The quantity of flow in both deposits must be the same, i.e. $q = A\,k\,i$.
  When water flows from a soil of high permeability to one of low permeability A and i must increase so the flow channel width increases and the distance, l, between the equipotentials decreases.
  This will apply within a zoned earth dam where water is flowing from the upstream shoulder fill into a central clay core. To the rear of the clay core a downstream shoulder fill is placed which is relatively permeable, or sometimes a chimney drain is placed which is very permeable. Thus when water flows from a soil of low permeability (clay core) to one of higher permeability A and i must decrease so the rectangular blocks elongate to provide narrower flow channels and greater distance between equipotential drops.
- *phreatic surface*
  When water flows through an earth dam or flood bank the upper surface of the flowing water is the upper flow line and the flow is described as

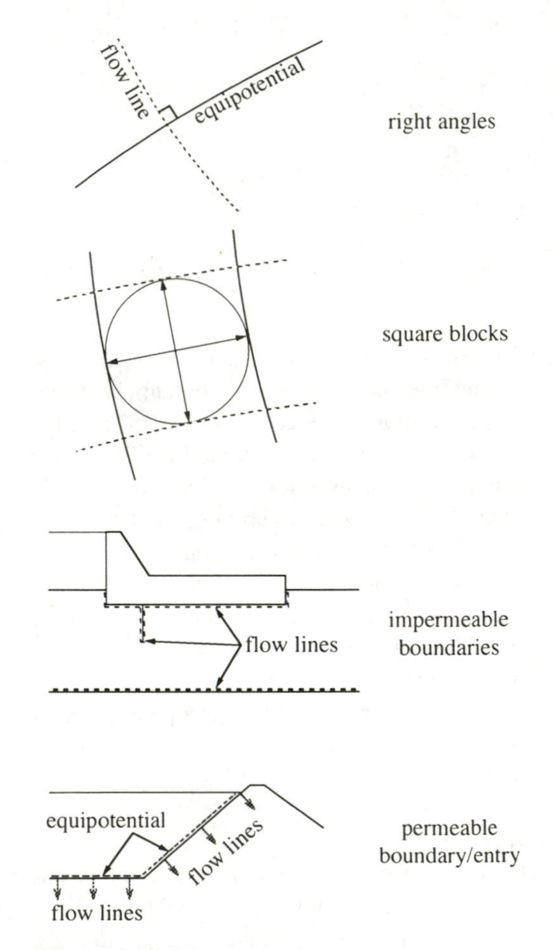

Figure 3.28 *Flow net rules*

unconfined. The location of the phreatic surface is not known, so a construction is adopted, as shown in Figure 3.36, for a homogeneous earth structure.

- *transformed sections (anisotropic soils) (Figure 3.31)*
  Flow nets are constructed as above, on the assumption that permeabilities are equal in both the vertical and horizontal direction, i.e. they are isotropic. However, most natural soils and compacted fills display anisotropic permeabilities. To allow for this, the cross-section is first drawn to a transformed scale, and then the flow net is constructed following the above rules for isotropic conditions.

The transformed section is obtained by keeping the same vertical scale but multiplying the horizontal scale by

$$\sqrt{(k_V/k_H)} \qquad (3.27)$$

Since k_H is usually greater than k_V this means that the horizontal dimensions must be reduced. For example, if $k_H = 9k_V$ all horizontal dimensions are divided by 3. This is illustrated in Worked example 3.9 for seepage beneath a concrete weir, Figure 3.31.

If pore pressures or uplift pressures are required then the flow net produced must be re-drawn to a natural scale and then the flow net will consist of diamond shapes, not squares.

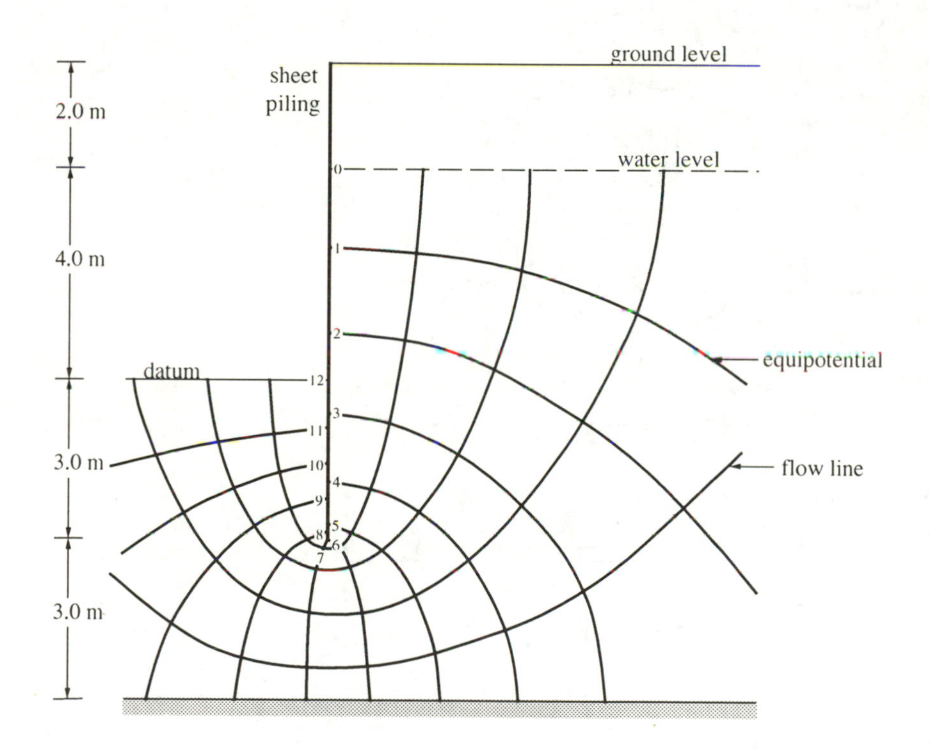

Figure 3.29 *Flow net – sheet piling*

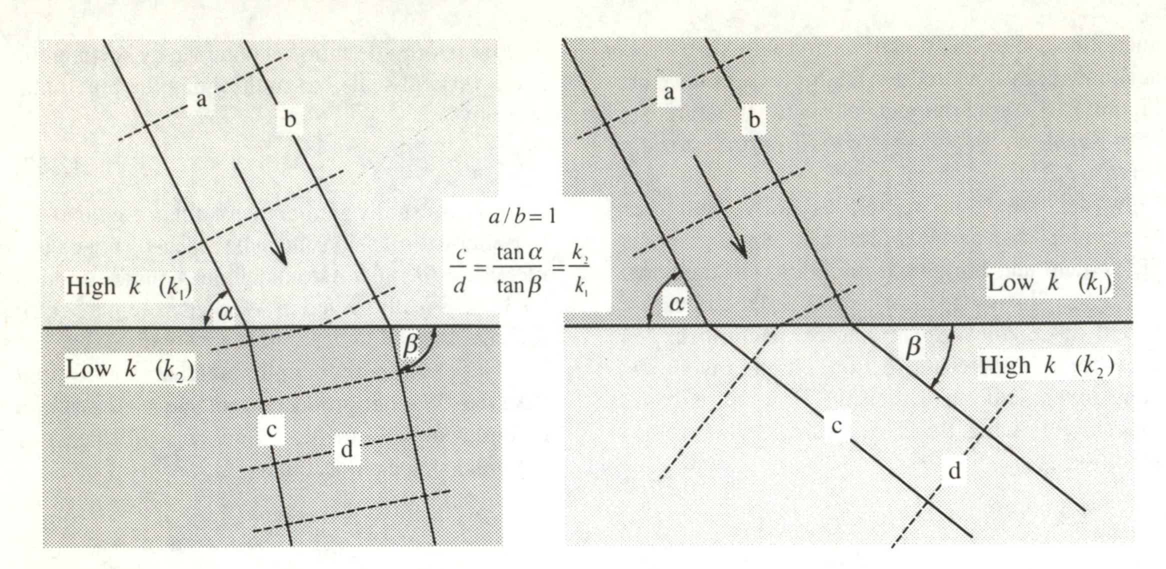

Figure 3.30 *Deflection rule*

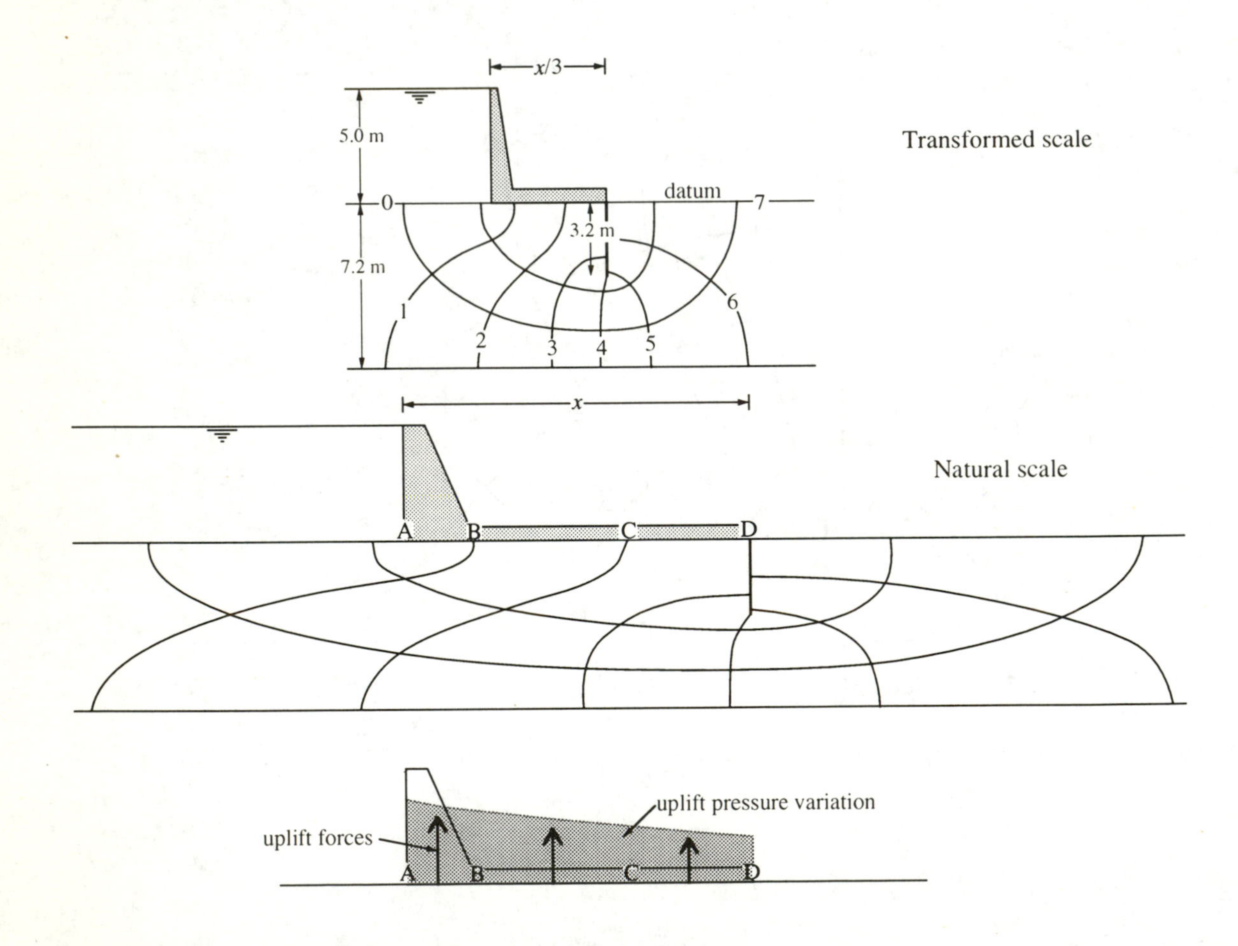

Figure 3.31 *Transformed section – Worked Example 3.9 Seepage beneath a concrete weir*

Seepage quantities

These can be determined from flow nets, such as those in Figures 3.29 and 3.31, using Darcy's Law. Flow is assumed to be two-dimensional so a unit width of the cross-section is considered. The total flow around a structure will then depend on its overall length.

Since the blocks on a flow net are square the width of a flow channel will be equal to its length Δl.

$\therefore$ area $A = 1\ \Delta l$

total flow q $= \Delta q$ in each channel $\times$ no. of channels $= \Delta q \times n_f$

Head loss between equipotentials $= \Delta h$

and $\Delta h = \dfrac{\text{total head lost}}{\text{no. of intervals}} = \dfrac{h}{n_d}$

Hydraulic gradient for any block $= i = \dfrac{\Delta h}{\Delta l}$

The flow in each channel $= \Delta q = Aki$

$$= 1 \times \Delta l k \frac{h}{n_d} \times \frac{1}{\Delta l}$$

$$= \frac{kh}{n_d}$$

The total flow is then given by

$$q = kh\frac{n_f}{n_d} \qquad (3.28)$$

where:

n_f/n_d = shape factor

From Figure 3.31 it can be seen that the shape factor is the same for isotropic and anisotropic permeabilities. For the isotropic case, k is obviously the isotropic k value but when k_H is different from k_v the seepage quantity must be determined using:

$$k = \sqrt{(k_H \times k_v)} \qquad (3.29)$$

Total head, elevation head and pressure head *(Figure 3.32)*

Total head includes position head (or elevation head) and pressure head and represents an energy level. Water will only flow if there is a difference in energy level but it need not flow if there is a difference in pressure level. For example, consider water at two different levels in a lake. There is a difference in pressure but there is no flow between these levels because they have the same energy or total head.

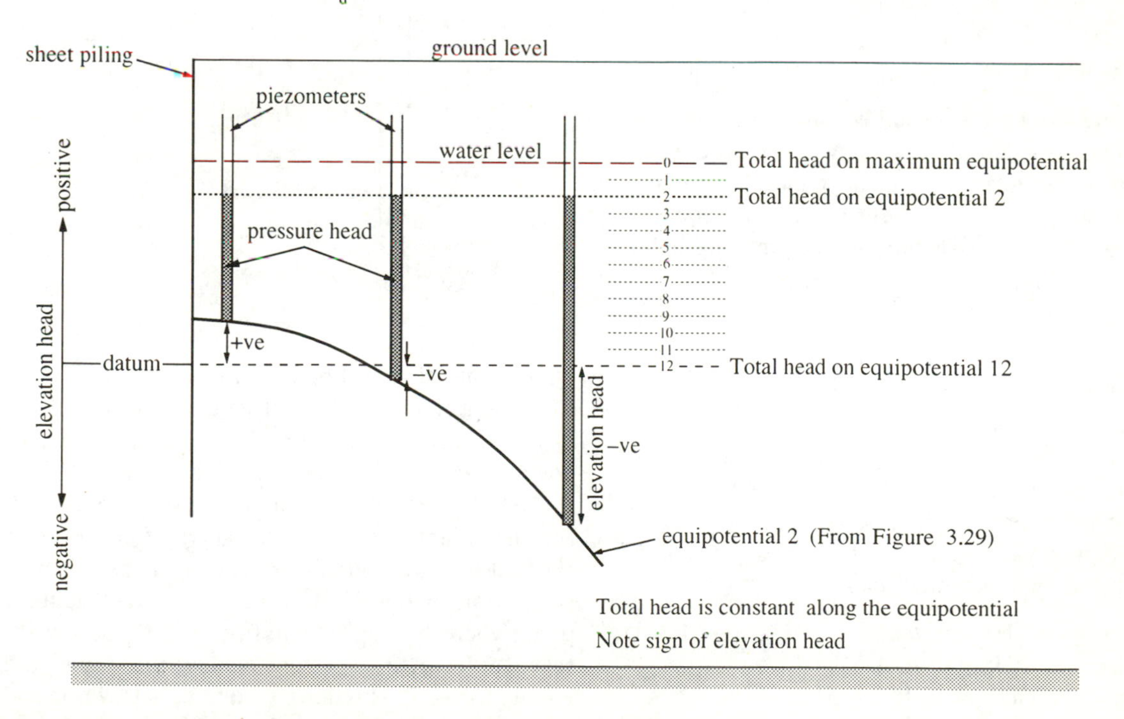

Figure 3.32 *Total head, elevation head and pressure head*

Along an equipotential the total head is constant and can be determined from:

total head on equipotential = total head on maximum equipotential − $\Delta h \times$ number of drops (3.30)

On Figure 3.29 the total head loss occurs over twelve equipotential drops (between thirteen equipotentials) so the total head on the third equipotential (Equipotential 2) is:

$4.0 - 4.0/12 \times 2 = 3.33$ m
or $4.0 \times 10/12 = 3.33$ m

If a standpipe or piezometer is installed at any point on an equipotential then water will rise up the standpipe to the same level since the pressure head, h_P, is given by:

pressure head h_P = total head − elevation head (3.31)

and the total head is constant along the equipotential. This is illustrated in Figure 3.32.

Pore pressure and uplift pressure
Pore water pressure u is then:

$$u = \gamma_w h_P \quad (3.32)$$

where γ_w is the unit weight of water, 9.81 kN/m³.

Since pore water pressure acts equally in all directions (it is hydrostatic), uplift pressure is the pore water pressure at the underside of an impermeable structure. In Worked example 3.9 there is an explanation of transforming a cross-section and determining uplift pressures and forces.

Seepage force *(Figure 3.33)*
As water flows through the voids of a soil it transfers some of its energy to the soil particles, and a force is applied by the flowing water which, in certain circumstances, can be detrimental to stability of the soil and any structure on the soil. This seepage force (and seepage pressure) can be derived by considering a block (Figure 3.33) in a flow net bounded by two flow lines and two equipotential lines:

water force on LHS = $\gamma_w h_1 A$
water force on RHS = $\gamma_w h_2 A$
Area per unit width of section = Δl 1
Volume affected by seepage force = $V = \Delta l^2$ 1

Hydraulic gradient $i = \dfrac{\Delta h}{\Delta l}$

force applied to sand particles = force on LHS − force on RHS
$= \gamma_w h_1 \Delta l\, 1 - \gamma_w h_2 \Delta l\, 1$
$= \gamma_w \Delta h \Delta l\, 1$
$= \gamma_w \dfrac{\Delta h}{\Delta l} \Delta l^2\, 1 = \gamma_w i V$

seepage force J (units of kN) is then:

$$J = \gamma_w i V \quad (3.33)$$

Seepage pressure is:

$$\gamma_w i \quad (3.34)$$

which is the seepage force per unit volume (with units of kN/m³). The term seepage pressure will be misleading since it is not a pressure.

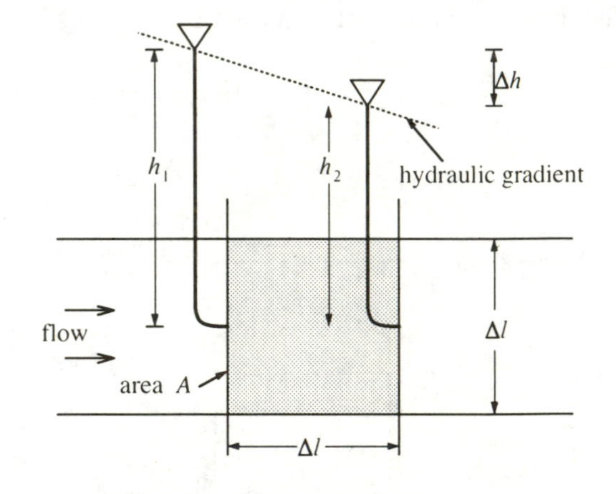

Figure 3.33 *Seepage force*

Quick condition and boiling *(Figure 3.34)*
Consider a block in a flow net at the soil surface on the exit side, such as inside a sheet-pile cofferdam or downstream of a concrete or earth dam.

To the right of the sheet piling the seepage force, J, acts in the same direction as the gravity force, R, in an element of soil, so effective stresses and hence shear strength are increased. However, to the left of the sheet piling where the seepage exits from the soil, the upward seepage force, J, is acting against the downward gravity force, S. This downward force acts within and between the soil grains and represents the effective stresses in the element. Hence

$$S = \gamma_{sub} V$$

However the upward seepage force, J, also acts on the soil grains thereby reducing the effective stresses and shear strength.

The 'quick' condition occurs when $S = J$ and the effective stresses are zero. Then the soil has no strength at all, interlock between the grains is removed and the soil is in a 'quick' or active fluid state. Small upward seepages will be seen as localised 'boiling', carrying sand particles upwards which flow outwards to form small volcano-like mounds.

There is a critical hydraulic gradient, i_c, at which this condition occurs, when:

$$S = J$$

$$\gamma_{sub} V = \gamma_w i_c V$$

$$i_c = \frac{\gamma_{sub}}{\gamma_w} = \frac{G_s - 1}{1 + e} \qquad (3.35)$$

This critical hydraulic gradient will depend, therefore, on the particle density and particle packing. For lightweight particles in a loose condition i_c could be as low as 0.6 but for densely packed quartz grains a value over 1.0 is likely.

A factor of safety against this 'quick' or 'boiling' condition can be obtained by considering the last block of the flow net on the exit side adjacent to the structure. The head loss is Δh and the length of this block is Δl so the exit hydraulic gradient i_e is:

$$i_e = \frac{\Delta h}{\Delta l}$$

and

$$F_{boiling} = \frac{i_c}{i_e} \qquad (3.36)$$

It can be seen that the soil immediately adjacent to a structure is the soil most prone to instability and 'piping' failures typically commence in this region.

Care must also be taken in estimating the exit hydraulic gradient, i_e, since this will be affected by the accuracy with which the flow net is sketched.

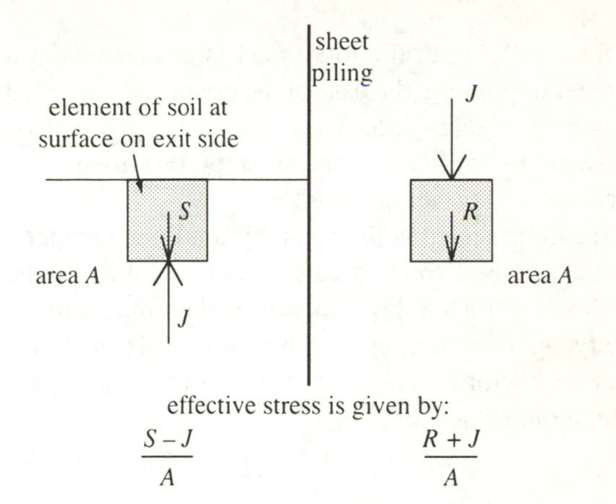

Figure 3.34 *The 'quick' condition and boiling*

Piping adjacent to sheet piling *(Figure 3.35)*
Piping failures are progressive and relentless, starting from
localised 'boiling' at the surface
⇒ soil grains moving apart
⇒ increased permeability
⇒ increased flow
⇒ quick condition
⇒ loss of strength
⇒ catastrophic collapse.

Consider the portion of the flow net on the downstream side of the sheet pile excavation. With the depth of penetration of the sheet piling d it has been shown that the prism of soil of height d and width $d/2$ will be prone to failure due to upward seepage. For this prism:

effective weight of prism $W = \gamma_{sub} \times d \times \frac{d}{2} = \frac{1}{2}\gamma_{sub} d^2$

seepage force on prism $J = \gamma_w i V$
mean total head on base of prism = h_m
total head at top of prism = 0
mean head lost through prism = h_m
mean hydraulic gradient = $h_m/d = i_m$

$$\text{Seepage force } J = \gamma_w i V = \gamma_w \frac{h_m d^2}{d \times 2} = \frac{1}{2}\gamma_w h_m d$$

$$\frac{W}{J} = \frac{\frac{1}{2}\gamma_{sub} d^2}{\frac{1}{2}\gamma_w h_m d} = \frac{\gamma_{sub} d}{\gamma_w h_m}$$

$$F_{\text{piping}} = \frac{i_c}{i_m} \qquad (3.37)$$

From the flow net the total head is determined at a number of points at the base of the prism and plotted as shown in Figure 3.35. The mean head, h_m, can be obtained by estimating the area of this total head variation and dividing by $d/2$.

It can be seen that the mean hydraulic gradient, i_m, for the prism of soil is greater than the exit hydraulic gradient, i_e, for the last block of soil so the factor of safety against 'piping' or 'heaving' will be lower than the factor of safety against 'boiling' and, therefore, more critical.

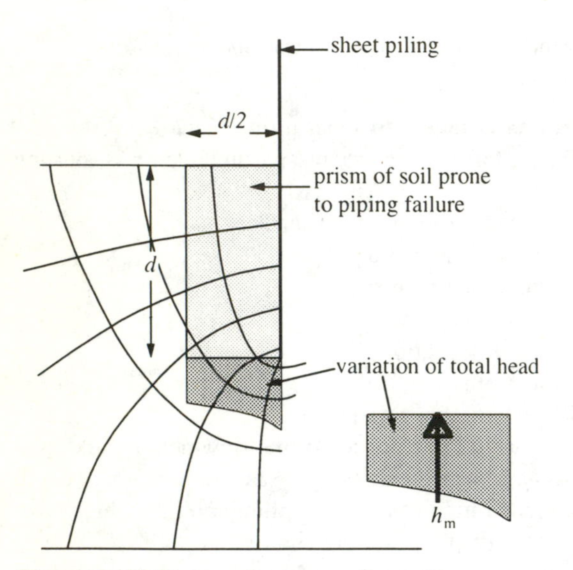

Figure 3.35 *Piping adjacent to sheet piling*

Seepage through earth dams

An earth dam is a mound of soil constructed to retain a fairly permanent reservoir level. The depth of water retained and hence the height of the dam can be considerable (20 – 50 m) and the risk of failure must be very low, considering the catastrophic consequences of failure. Seepages through the dam must be small and the pore pressures produced by seepage inside the dams must not be allowed to produce instability.

For these reasons, earth dams are usually constructed with a composite cross-section with a core of impermeable clay incorporated to minimise seepage losses and filter zones to ensure that seepages are controlled and piping instability prevented. Flow nets through composite cross-sections require skill to construct and can be quite intricate. For more detailed study of this topic the reader is referred to Cedergren (1989).

Seepage through flood banks, levees *(Figure 3.36)*

Levees are mounds of soil placed alongside a river in its flood plain or estuary to prevent flooding of large areas of flat land when river levels are high. A similar form of construction could be an irrigation dam or an earth sea wall. They may be subject to water pressures for only short periods of time, and the risk of collapse may not be as unacceptable as for earth dams especially if the land protected is low-grade agricultural land.

For the above reasons flood banks, irrigation dams and sea walls are constructed in a simple fashion, using local materials, preferably as impermeable as possible. These earth structures are of considerable length so zoning of materials is kept to a minimum to reduce the costs of transporting to site and placing more expensive clay cores, drainage layers and filter materials. The result is usually a simple homogeneous cross-section for which a flow net can be readily drawn.

Water flow through the cross-section can be analysed using seepage theory and a flow net can be constructed using the techniques described above and taking the following into account.

- It is likely that the soil will have different permeabilities in the horizontal and vertical directions, owing to placing and compacting the soil in layers so some degree of transformation of the section should be considered. If care is not taken with the method of embankment construction especially if smooth interfaces between compacted layers are produced then a high horizontal permeability may result leading to the destruction of the embankment by a piping failure.
- The upstream soil surface in contact with the water is an equipotential.
- The flow lines commence from this equipotential at right angles.
- The top flow line or stream line is called the phreatic surface and its location is not known but can be constructed with sufficient accuracy by first drawing a parabola and then applying some adjustments at the entry and exit points.
- The parabola is located by assuming that it passes through point B where BC = 0.3AC and has its focus

at point F. This point F lies either at the downstream toe of a homogeneous embankment or the upper end of the filter drain, if present.

- The property of a parabola is such that any point on it lies at the same distance from the focus F and from the directrix. The directrix DE is a vertical tangent through the point D which lies at the same level as B such that BD = BF. Points on the parabola are then obtained by drawing arcs of varying radius with the compass point at the focus F and intersecting lines parallel to the directrix at distances equal to the radius values. Then EG = FG and FH = JH.
- The entry part of the parabola is corrected in accordance with the foregoing comments, commencing at the point C.
- In a homogeneous earth dam where the parabola cuts the downstream slope above the toe, the phreatic surface is adjusted downwards so that it is tangential to the slope, cutting it at point K such that:

$$\mathrm{FK} = 0.64\mathrm{FL} \quad \text{for } a = 30^\circ$$
$$\mathrm{FK} = 0.68\mathrm{FL} \quad \text{for } a = 60^\circ$$

- The pore pressure or pressure head along the phreatic surface is zero so total head equals elevation head along this surface. The equipotentials will then cut this surface at equal vertical distances which can be marked off and used in the flow net construction. For example, if the height of the phreatic surface is split equally into ten vertical intervals then these will mark the starting points of nine equipotentials.

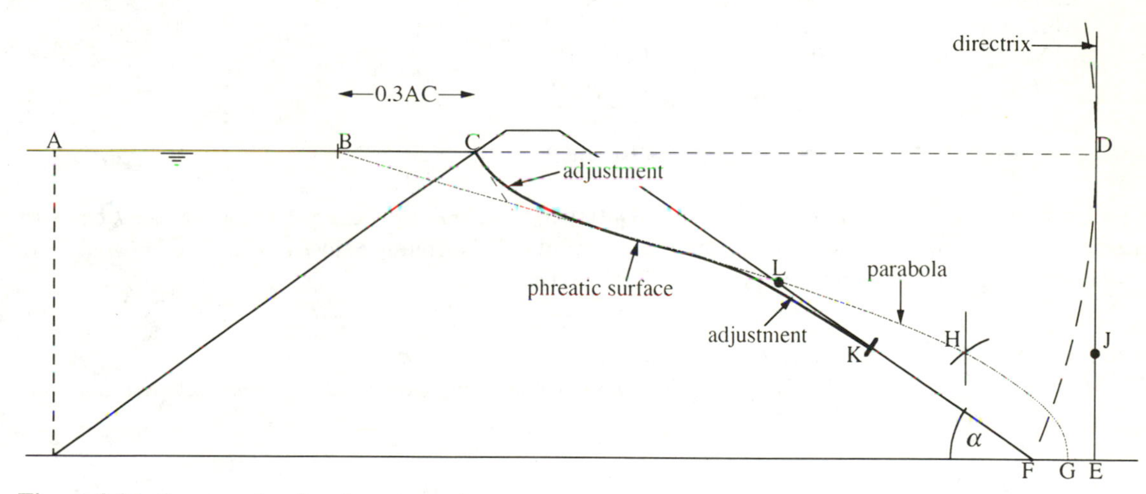

Figure 3.36 *Construction for phreatic surface*

Worked Example 3.1 *Constant head test*

A constant head permeameter was set up containing fine sand with an effective size d_{10} of 0.12mm. The times required to collect 250ml of water were recorded at the following manometer readings. Determine the average value of k and compare this with Hazen's empirical relationship.

Diameter of sample = 100 mm
Distance between manometers, h = 100 mm

Area of sample, $A = \dfrac{\pi \times 100^2}{4} = 7854 \text{ mm}^2$

$Q = 250 \text{ ml} = 250 \times 10^3 \text{ mm}^3$

Times to collect Q	2 min 25s	3 min 15 s	4 min 55 s
Manometer A reading (mm)	157	191	262
Manometer B reading (mm)	24	89	195

$$k = \frac{QL}{Aht} = \frac{250 \times 10^3 \times 100}{7854 \times h \times t}\left[\frac{\text{mm}^3 \times \text{mm}}{\text{mm}^2 \times \text{mm} \times \text{s}}\right] = \frac{3183}{ht}\text{mm/s} = \frac{31.83}{\text{ht}}\text{m/s}$$

h (mm)	t (s)	k (m/s)
133	145	1.65×10^{-4}
102	195	1.60×10^{-4}
67	295	1.61×10^{-4}

$\therefore$ Average $k = 1.62 \times 10^{-4}$ m/s (report as $k = 1.6 \times 10^{-4}$ m/s)

From Hazen's relationship, Equation 3.7, $k = (0.010 - 0.015) \times 0.12^2 = 1.4 - 2.2 \times 10^{-4}$ m/s. Since k can vary considerably these results are quite comparable. The test would give different values of k at different void ratios or densities which would not be reflected in Hazen's relationship.

Worked Example 3.2 *Falling head test*

The following data were obtained during a falling head test on a sample of clayey silt. Determine the average value of the coefficient of permeability k.

diameter of sample = 100 mm
length of sample, l = 100 mm
diameter of standpipe tube = 3 mm

Time after start, t (seconds)	0	15	30	49	70	96
Water level in tube (mm)	1000	900	800	700	600	500

Area of sample, $A = \dfrac{\pi \times 100^2}{4} = 7854 \text{ mm}^2$

Internal area of tube, $a = \dfrac{\pi}{4} \times 3^2 = 7.07 \text{ mm}^2$

Time, t, seconds	15	30	49	70	96
$\text{Log}_{10} h_0/h_t$	0.046	0.097	0.155	0.222	0.301

The gradient of a graph of $\log_{10} h_0/h_t$ versus t gives 3.14×10^{-3} [1/s]

$$k = \frac{3.14\times10^{-3}\times2.3\times7.07\times100}{7854} = 6.5\times10^{-4}\ \text{mm / s} = 6.5\times10^{-7}\ \text{m / s}$$

Worked Example 3.3 *Clay layer in sand*

A sand deposit contains thin (10mm) horizontal layers of clay 1.0 m apart. This is a fairly infrequent spacing. Determine the overall coefficients of permeability in the horizontal and vertical directions given the following:

For the sand $L_s = 1000$ mm $\quad k_s = 1 \times 10^{-3}$ m/s
For the clay $L_c = 10$ mm $\quad k_c = 1 \times 10^{-7}$ m/s

For horizontal flow, the overall coefficient of permeability is given by Equation 3.9. The sequence of layers is consecutive so Equation 3.8 can be used.

$$k_H = \frac{1.0\times10^{-3} + 0.01\times10^{-7}}{1.0+0.01} = 0.99\times10^{-3}\ \text{m / s}$$

$\therefore$ k_H is hardly altered compared to k_s.

Also

$$\frac{q_c}{q_s} = \frac{L_c k_c}{L_s k_s} = \frac{10^{-2}\times10^{-7}}{1.0\times10^{-3}} = 1\times10^{-6}$$

$\therefore$ horizontal flow through the clay layers is negligible.

For vertical flow, use Equation 3.11.

$$k_v = \frac{1.0+0.01}{\dfrac{1.0}{10^{-3}} + \dfrac{0.01}{10^{-7}}} = 1\times10^{-5}\ \text{m/s}$$

$\therefore$ clay thickness is only 1/100 that of the sand but k_v overall is reduced by 100.

Worked Example 3.4 *Sand layer in clay*

A clay deposit contains thin (10 mm) horizontal layers of sand 1.0 m apart. This is a fairly infrequent spacing. Determine the overall coefficients of permeability in the horizontal and vertical directions.

For the sand $L_s = 0.01$ m $\quad k_s = 10^{-3}$ m/s
For the clay $L_c = 1.0$ m $\quad k_c = 10^{-7}$ m/s

For horizontal flow, use Equation 3.8.

$$k_H = \frac{L_s k_s + L_c k_c}{L_s + L_c} = \frac{0.01 \times 10^{-3} + 1.0 \times 10^{-7}}{0.01 + 1.0} = 1.0 \times 10^{-5} \text{ m/s}$$

The sand thickness is 1/100 that of the clay but k_H is increased by 100.

Also

$$\frac{q_s}{q_c} = \frac{10^{-2} \times 10^{-3}}{1.0 \times 10^{-7}} = 100$$

Horizontal flow through the sand is considerable, 100 times that through the clay.

For vertical flow, use Equation 3.11.

$$k_v = \frac{0.01 + 1.0}{\frac{0.01}{10^{-3}} + \frac{1.0}{10^{-7}}} = 1.01 \times 10^{-7} \text{ m/s}$$

k_V is hardly altered compared to k_c.

Worked Example 3.5 *Open borehole test*
A falling head test has been carried out in a borehole sunk below the water table in a uniform deposit of silty fine sand. Details of the test measured as depths below ground level are as follows:

Bottom of casing = 5.7 m
Bottom of borehole = 6.3 m
Water table level = 4.5 m
Initial water level = 1.6 m
Internal diameter = 200 mm
The test observations and calculations are tabulated together for convenience.

Test Observations		Calculations			
time t	Depth to water level	H	H_o/H	$x = \log_e H_o/H$	x/t ($\times 10^3$)
0	1.60	2.90	1.00	0	0
30 s	1.90	2.60	1.12	0.11	3.64
1 min	2.17	2.33	1.24	0.22	3.65
2 min	2.62	1.88	1.54	0.43	3.61
3 min	2.99	1.51	1.92	0.65	3.63
4 min	3.29	1.21	2.40	0.87	3.64
5 min	3.52	0.98	2.96	1.08	3.62
7 min	3.87	0.63	4.60	1.53	3.64
10 min	4.17	0.33	8.79	2.17	3.62
				Ave =	3.63

$\frac{L}{D} = \frac{0.6}{0.2} = 3$ shape factor $f = 10.5$ (Case F) From Figure 3.23

Assume $k_H = k_V$, from Equation 3.17,

$$k = \frac{\pi \times 0.2^2 \times 3.63 \times 10^{-3}}{4 \times 10.5 \times 0.20} = 5.4 \times 10^{-5} \text{ m/s}$$

Worked Example 3.6 *Pumping test in confined aquifer*

Determine the value of k for a confined sand aquifer given the following:

Thickness of overlying clay = 5.5 m

Thickness of sand aquifer = 3.5 m (clay beneath)

Quantity of flow from pumping well = 0.30 m³/min

Observation well	Distance from pumping well, m	Depth to water level, m
1	14	3.4
2	48	2.8

$h_2 - h_1 = 6.2 - 5.6$ or $3.4 - 2.8 = 0.6$ m

From Equation 3.21 on Figure 3.26:

$$k = \frac{0.30 \log_e\left(\frac{48}{14}\right)}{60 \times 2 \times \pi \times 3.5 \times 0.6} = 4.7 \times 10^{-4} \text{ m/s}$$

Worked Example 3.7 *Pumping test in unconfined aquifer*

An unconfined sand aquifer is 9.0 m thick with a water table 1.5 m below ground level and clay underneath. It is required to lower the water table at a point to 3.5 m by pumping from a well, 300 mm diameter, at a distance of 5 m away from the point. The coefficient of permeability of the sand $k = 8 \times 10^{-4}$ m/s. Determine the pumping rate required and the highest level the pump can be placed in the well.

Where drawdowns are a significant proportion of the saturated thickness they must be corrected using the equation:

$$S_c = S_o - \frac{S_o^2}{2h_o}$$

S_c = corrected drawdown

S_0 = observed or required drawdown

h_0 = initial saturated thickness

$S_0 = 3.5 - 1.5 = 2.0$ m $\quad h_o = 9.0 - 1.5 = 7.5$ m

$$S_c = 2.0 - \frac{2.0^2}{2 \times 7.5} = 1.73 \text{ m}$$

$h_1 = 7.5 - 1.73 = 5.77$ m at a distance from the well $r_1 = 5.0$ m

r_0 is the radius when pumping has no influence (at $h = h_0$)

Assume $r_0 = 750$ m (BS 5930: 1981)

From Figure 3.26, rearranging the expression for k, Equation 3.22 gives

$$q = \frac{k\pi\left(h_o^2 - h_1^2\right)}{\log_e \frac{r_o}{r_1}}$$

$$q = \frac{8 \times \pi(7.5^2 - 5.77^2) \times 60}{10^4 \times \log_e\left(\frac{750}{5}\right)} = 0.69\ \text{m}^3/\text{min} = 690\ \text{litres/min}$$

Water level = h_w at the side of the well where $r = r_w$
r_w = effective radius of well = 0.15 × 1.20 = 0.18 m (BS 5930: 1981)
h_0 = 7.5 m at $r = r_0$ = 750 m

$$0.69 = \frac{8 \times \pi(7.5^2 - h_w^{\ 2}) \times 60}{10^4 \times \log_e\left(\frac{750}{0.18}\right)} \qquad \text{giving} \quad h_w = 4.3\ \text{m}$$

Drawdown S_c = 7.5 – 4.3 = 3.2 m
∴ $3.2 = S_o - S_o^2/15$ giving S_o = 4.63 m
Allowing for well losses S_w = 4.63 × 1.33 = 6.2 m
The pumps must be placed at least 6.2 m below the water table or 7.7 m below ground level. This procedure is useful in assessing dewatering requirements.

Worked Example 3.8 *Seepage around a sheet pile cofferdam*
A flow net has been constructed on Figure 3.29 for a cross-section through a sheet pile wall excavation in a uniform sand deposit. Determine the following:
i) the quantity of flow in litres/minute.
ii) the pore pressure distribution on both sides of the sheet piling.
iii) the factors of safety against boiling and heaving.

For the sand $k = 3 \times 10^{-5}$ m/s $G_s = 2.65$ $e = 0.60$
Total head loss h = 4.00 m
Number of equipotential drops n_d = 12
Number of flow channels n_f = 4.5

$$q = \frac{khn_f}{n_d} = \frac{3 \times 4.0 \times 4.5 \times 60 \times 1000}{10^5 \times 12} = 2.7\ \text{litres/min per metre length of wall}$$

Pressure head h_w = total head – elevation head Pore pressure $u = \gamma_w h_w$

Location	Total Head m	Elevation Head m	Pressure Head m	Pore Pressure kN/m²
0	4.0	+4.0	0	0
1	11/12 × 4.0 = 3.67	+2.5	1.17	11.5
2	10/12 × 4.0 = 3.33	+0.9	2.43	23.8
3	9/12 × 4.0 = 3.00	-0.6	3.60	35.3
4	8/12 × 4.0 = 2.67	-1.9	4.57	44.8
5	7/12 × 4.0 = 2.33	-2.6	4.93	48.4
6	6/12 × 4.0 = 2.00	-3.0	5.00	49.0
7	5/12 × 4.0 = 1.67	-3.0	4.67	45.8
8	4/12 × 4.0 = 1.33	-2.8	4.13	40.5
9	3/12 × 4.0 = 1.00	-2.2	3.20	31.4
10	2/12 × 4.0 = 0.67	-1.5	2.17	21.3
11	1/12 × 4.0 = 0.33	-0.9	1.23	12.1
12	0	0	0	0

Factor of safety against boiling $F_b = i_c / i_e$

Critical hydraulic gradient, $i_c = \dfrac{2.65 - 1}{1 + 0.60} = 1.03$

Head loss between equipotentials $\Delta h = \dfrac{4.0}{12} = 0.33$

Δl for last block on exit (11→12) = 0.90 m

$i_e =$ 0.33 / 0.90 = 0.37 $\quad\therefore\ F_b$ = 1.03 / 0.37 = 2.8

Factor of safety against piping $F_h = i_c / i_m \qquad i_m = h_m / d$

$h_m = \dfrac{3.5 \times 4}{12} = 1.17 \qquad i_m = \dfrac{1.17}{3.00} = 0.39$

$F_h = \dfrac{1.03}{0.39} = 2.6$

Worked Example 3.9 *Seepage beneath a concrete weir*
A concrete weir is to be placed on a sand deposit as shown on Figure 3.31. The coefficients of permeability of the sand are $k_H = 9 \times 10^{-5}$ m/s and $k_v = 1 \times 10^{-5}$ m/s. Determine the quantity of flow in litres/minute emanating downstream and the uplift force (upthrust) acting on the underside of the structure.
The soil is anisotropic with respect to permeability so the cross-section must be drawn to a transformed scale by reducing the horizontal dimensions:

$x_T = \sqrt{(k_v / k_H)}\, x = 1/3\, x$

and maintaining the vertical dimensions. A flow net is then drawn on this transformed cross-section (see Figure 3.31), following the rules for construction
Total head loss h = 5.0 m
Overall $k = \sqrt{(k_v \times k_H)} = 1 \times 10^{-5} \times 9 \times 10^{-5} = 3 \times 10^{-5}$ m/s
Number of equipotential drops $n_d = 7$
Number of flow channels $n_f = 3$

$q = \dfrac{khn_f}{n_d} = \dfrac{3 \times 5.0 \times 3 \times 60 \times 1000}{10^5 \times 7} = 3.9$ litres/min per metre length of weir
Elevation head on underside of weir = 0
∴ total head = pressure head

location	total head, m	uplift pressure, kN/m²
A	5.0	49.0
B	6/7 × 5.0 = 4.3	42.2
C	5/7 × 5.0 = 3.6	35.3
D	4.5/7 × 5.0 = 3.2	31.4

Width of weir = 15 m
upthrust A → B = 1/2 (49.0 + 42.2) × 3.0 = 136.8 kN/m
upthrust B → C = 1/2 (42.2 + 35.3) × 6.7 = 259.6 kN/m
upthrust C → D = 1/2 (35.3 + 31.4) × 5.3 = 176.8 kN/m
All are forces per metre length of weir.

Exercises

3.1 A constant head permeameter test has been carried out on a sample of sand. With a head difference of 234 mm, 200 ml of water was collected in 3 minutes 45 seconds. The diameter of the sample is 75 mm and the distance between the manometer points is 100 mm. Determine the coefficient of permeability of the sand.

3.2 A falling head test has been carried out on a sample of soil 120 mm long in a permeameter 100 mm diameter, with a 4 mm diameter standpipe tube attached. The initial head of water in the standpipe was 950 mm and fell to 740 mm after 30 minutes. Determine the coefficient of permeability of the soil.

3.3 A constant head permeameter test has been carried out on the same sand as in Exercise 3.1 but with a layer of silt, 5 mm thick, placed within the sand between the manometer points. With a head difference of 672 mm, 100 ml of water was collected in 12 minutes 25 seconds. Assuming the value of the coefficient of permeability of the sand as obtained from Exercise 3.1 determine the coefficient of permeability of the silt.

3.4 A layer of clay 5.0 m thick overlies a thick deposit of sand. The water table in the sand is sub-artesian at 2.0 m below the top of the clay. The coefficient of permeability of the clay is 5×10^{-8} m/s. A reservoir, 100 m square and 5.0 m deep, is to be impounded above the clay layer. Determine the initial seepage quantity through the clay.

3.5 A constant head test has been carried out in an open cased borehole, 150 mm diameter, in a uniform sand. A flow rate of 2.5 litres/minute was required to maintain the water level inside the casing at 0.5 m below ground level. The water table is at 4.0 m below ground level and the depth of casing and depth of borehole are at 7.5 m below ground level. Determine the coefficient of permeability of the sand.

3.6 A layer of clay, 6.0 m thick overlies a layer of sandy gravel 4.5 m thick which is underlaid by impermeable bedrock. A pumping well has been installed to the bedrock and a steady state rate of pumping of 540 litres per minute established. In two observation wells, 11.0 and 37.0 m away from the pumping well, water levels of 2.65 and 2.20 m, respectively, were recorded. Determine the coefficient of permeability of the sandy gravel.

3.7 A cross-section through a sheet pile wall is shown on Figure 3.37. Pumping from the gravel filter maintains the water level on the downstream side at the base of the filter. The coefficient of permeability of the sand is 3×10^{-5} m/s. Sketch a flow net and determine:
i) the rate of flow per metre of wall
ii) the water pressure variation on both sides of the sheet piling.

3.8 Repeat Exercise 3.7 but with the coefficient of permeability in the vertical direction of $k_v = 3 \times 10^{-5}$ m/s and in the horizontal direction of $k_H = 6 \times 10^{-4}$ m/s.

3.9 From the flow net sketched in Exercise 3.7 determine the factor of safety against:
i) boiling
ii) piping or heaving.

3.10 A cross-section through an earth dam is shown on Figure 3.38. Construct the phreatic surface and sketch a flow net. The coefficient of permeability of the soil forming the dam is 5×10^{-6} m/s. Determine the rate of flow through the dam per metre length.

3.11 From the flow net sketched in Exercise 3.10 determine the variation of pore pressure along the slip surface shown on Figure 3.38.

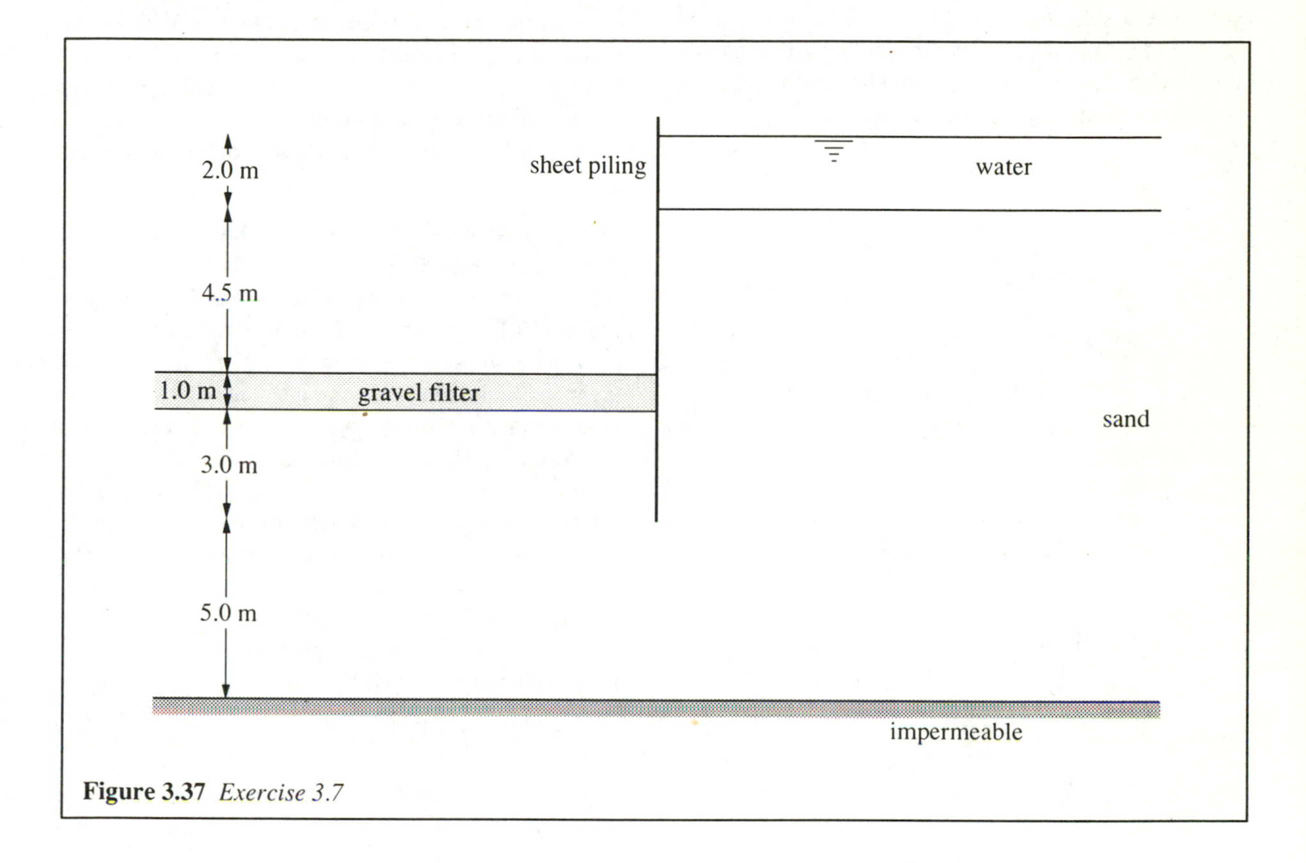

Figure 3.37 *Exercise 3.7*

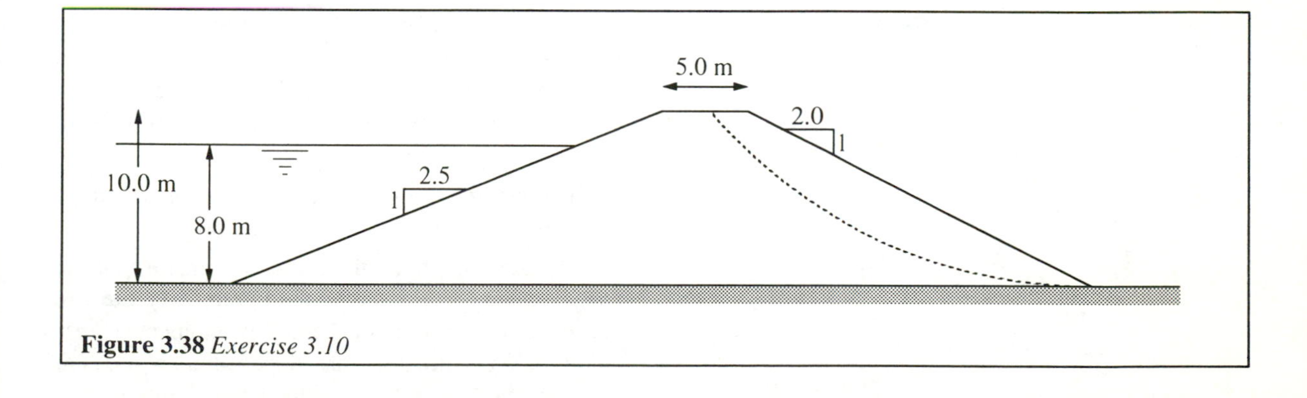

Figure 3.38 *Exercise 3.10*

4 Effective Stress and Pore Pressure

Total stress *(Figure 4.1)*
Total stress is the stress acting on a plane, assuming the soil to be a solid material. For the small soil element shown in Figure 4.1, at a depth z below ground level, the vertical total stress, σ_V, would be the stress acting on the horizontal surface of the element.

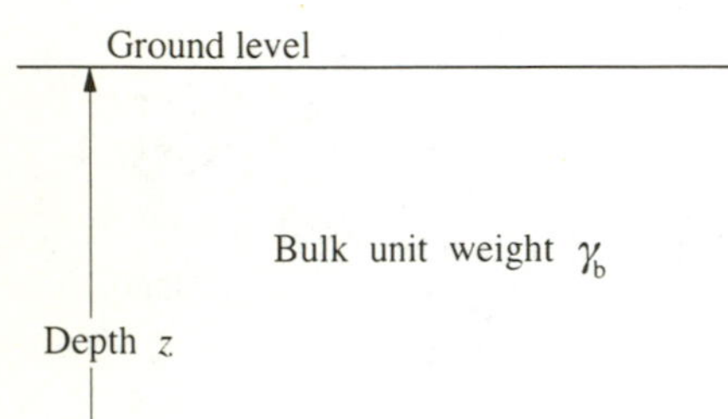

a) Above a water table

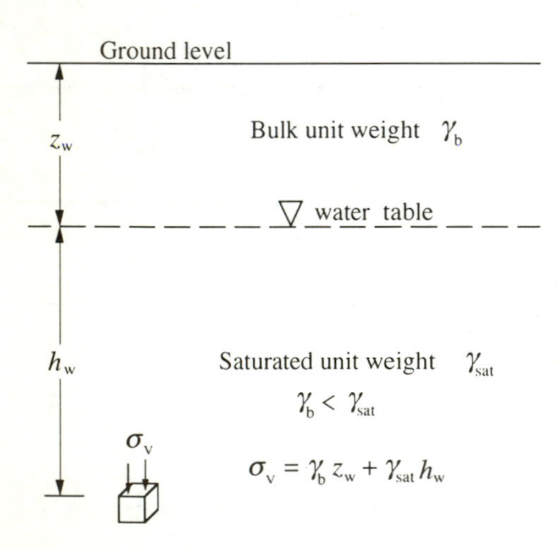

b) Below a water table

Figure 4.1 *Total stress*

This soil element will remain in equilibrium in the ground (no consolidation or shear failure) because of a horizontal total stress, σ_H, acting on the vertical surface of the element. The stresses in soils are not isotropic, i.e. usually σ_V does not equal σ_H. In this chapter the stresses considered are vertical stresses, but it is important to remember that horizontal stresses also act.

Pore pressure below the water table *(Figure 4.2)*
This is the pressure in the water in the void spaces or pores which exist between and around the mineral grains. In the ground, the level at which the water pressure is the same as atmospheric pressure is called the water table. If a tube were inserted below the water table (and no seepage was occurring), then water would rise up the tube to the water table level.

The pores in the soil below this level are fully saturated with water, so this ground is often referred to as the zone of saturation. Water below the water table is called phreatic water so the phreatic surface is another term for the water table.

Only gravitational forces act on the pore water so the pressure within it is given by:

$$u_w = \rho g h_w \quad \text{or} \quad \gamma_w h_w \qquad (4.1)$$

where h_w is the depth below the water table. The pore water pressure, u_w, is hydrostatic, meaning that it has the same value in all directions.

Effective stress *(Figure 4.3)*
The principle of effective stress strictly applies only to fully saturated soils. It equates the total stress which can be imagined to exist external to an element of soil, to the stresses within the two components of the soil, i.e.

- the mineral grain structure (effective stress) and
- the water in the pores (pore water pressure).

The effective stress, therefore, is a measure of the stress existing within the mineral grain structure and since the major soil mechanics phenomena (shear strength, consolidation) are dependent on the soil structure this principle is of fundamental importance.

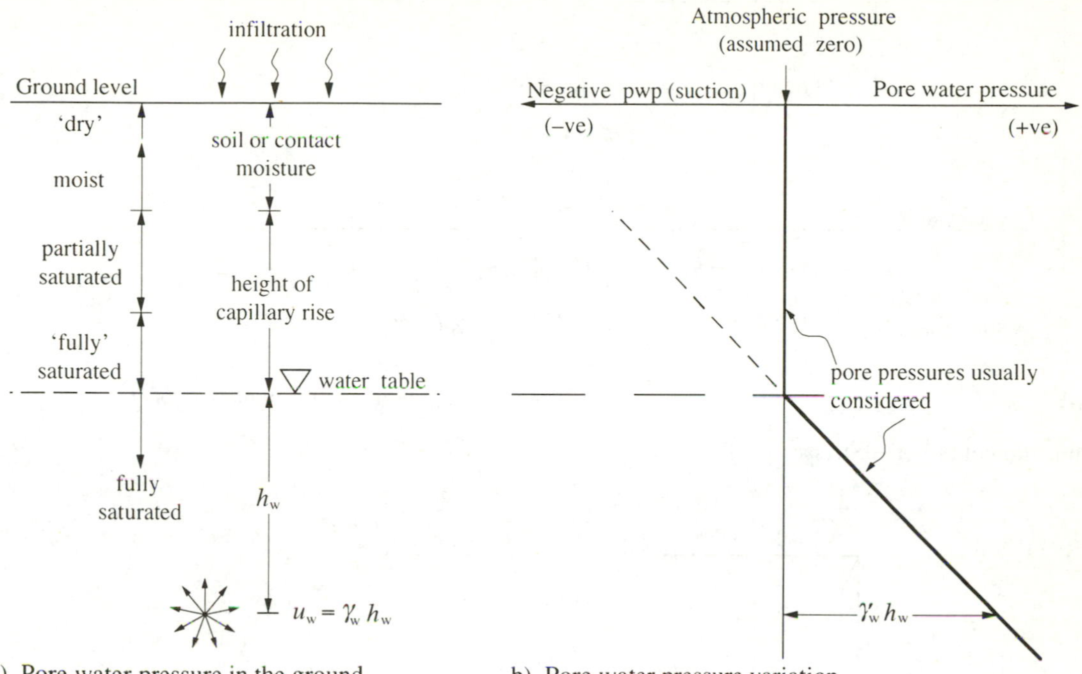

Figure 4.2 *Pore water pressure*

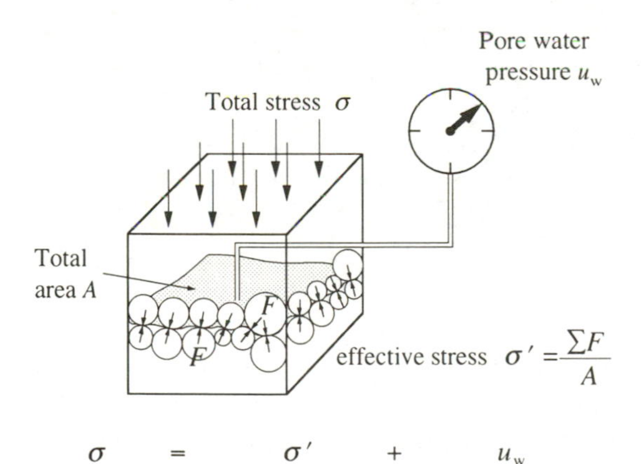

Figure 4.3 *Principle of effective stress*

The effective stress is not the direct contact stress between the soil particles, but represents the contribution provided by the soil structure or soil skeleton to support the total stress.

Effective stress in the ground *(Figure 4.4)*
Above the water table, pore water pressures are usually assumed to be zero (rather than negative) and total stresses then equal effective stresses. Below the water table the effective stress can be obtained from the total stress minus the pore water pressure, or directly by using the submerged unit weight, as shown in Figure 4.4.

Total stress = effective stress + pore water pressure

$$\sigma = \sigma' + u_w \tag{4.2}$$

Stress history
The stresses to which a soil has been subjected, during its formation up to the present time, are referred to as its stress history. During deposition, effective stresses in the soil will increase as more soil particles are placed; and during erosion effective stresses will decrease as soil is removed.

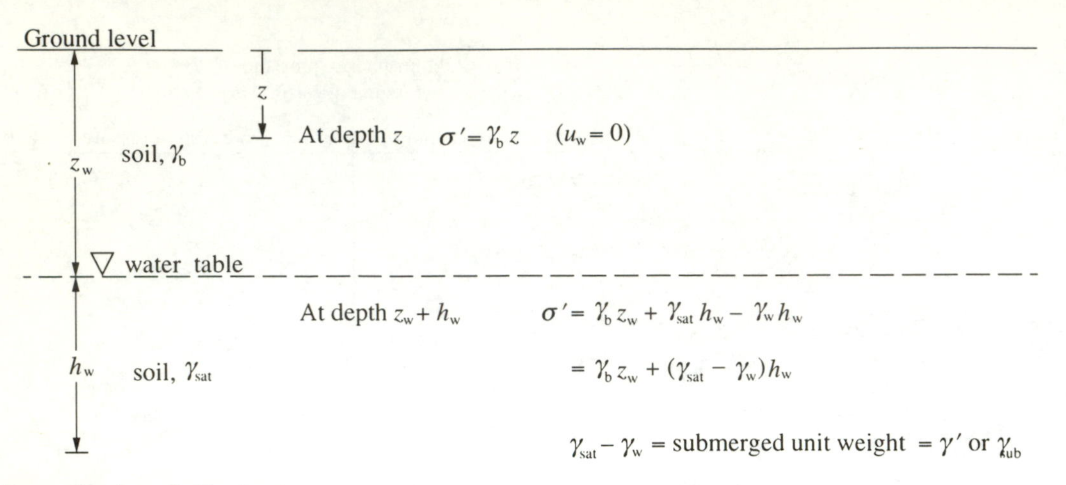

a) Determination of effective stresses

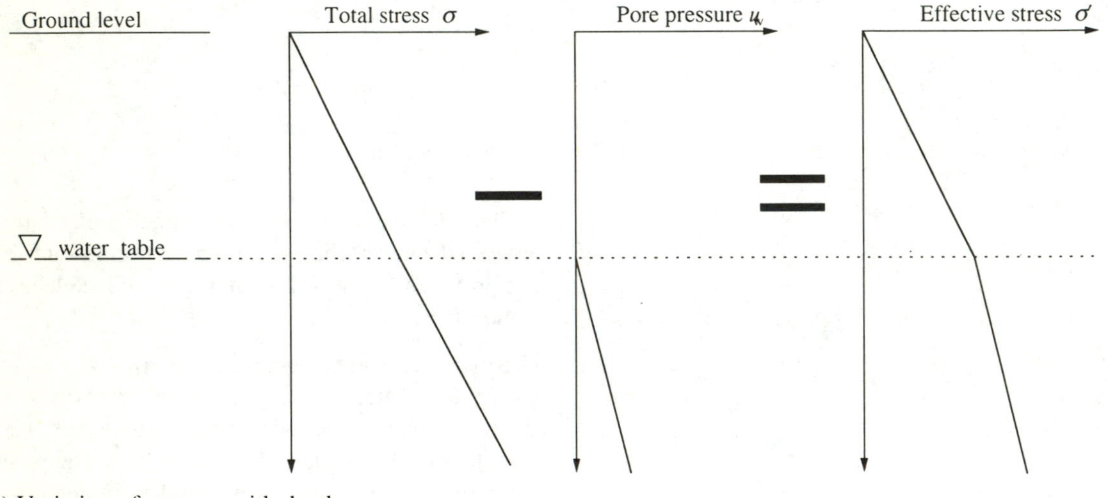

b) Variation of stresses with depth

Figure 4.4 *Effective stress in the ground*

Soils existing in and around river estuaries at the present time may be only centuries or even decades old and have fairly straightforward stress histories of deposition only, whereas older clays such as London Clay are millions of years old and have undergone more complex stress histories, particularly with significant thicknesses removed due to erosion.

Normally consolidated clay *(Figure 4.5)*
This is a clay that has undergone deposition only.

In Figure 4.5 two stages are shown during deposition above a soil element. It is assumed that the water level in the sea or lake remains constant. As deposition occurs, the vertical (and horizontal) effective stress in the soil element increases; due to the process of consolidation the void ratio of the soil is reduced. The graph of void ratio versus stress on a logarithmic scale is usually a straight line. It is denoted as the deposition line or, since there has been no other process involved, the virgin compression curve.

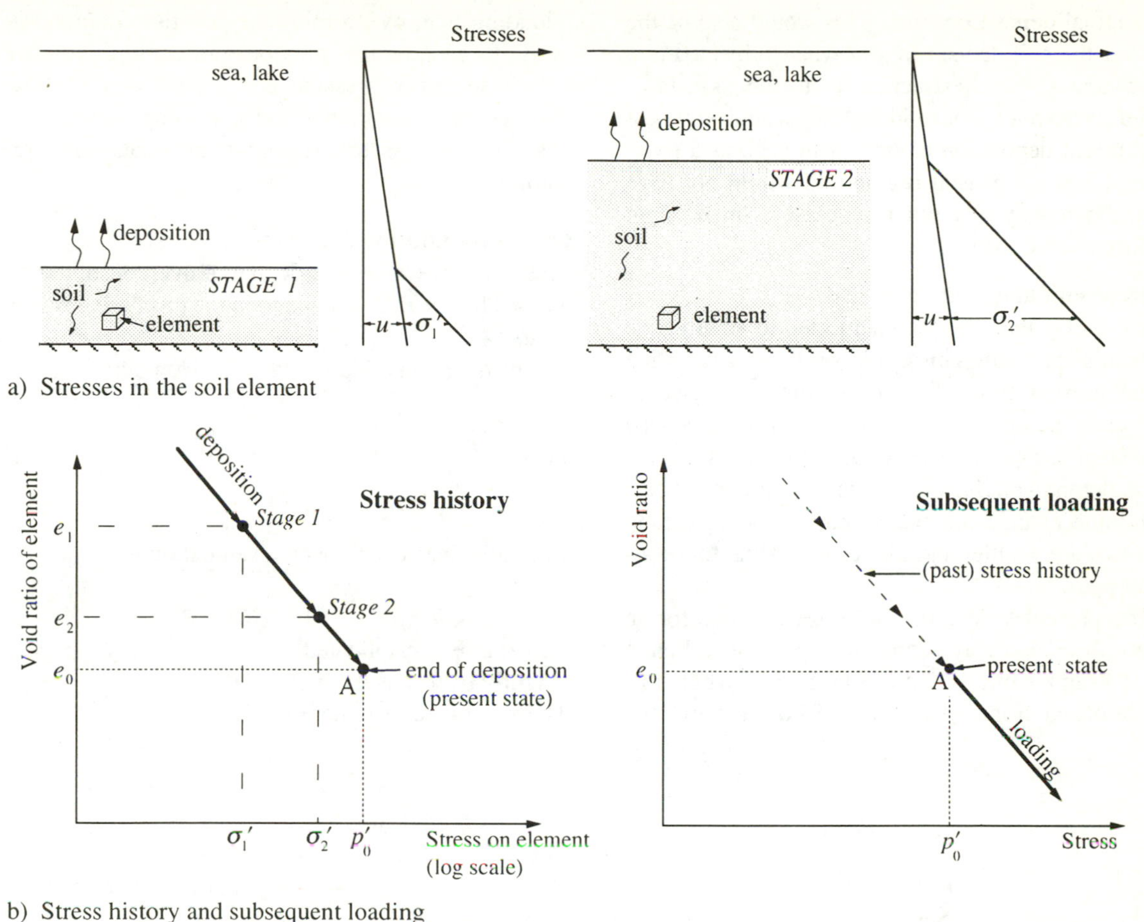

a) Stresses in the soil element

b) Stress history and subsequent loading

Figure 4.5 *Normally consolidated clay*

During deposition the mineral grain structure of the soil element will be adjusting to the changes in void ratio, mostly by structural rearrangement of the particles. This process, where the particles move closer together, is an irreversible one, so that the original particle arrangement could not be recovered even if the stresses were removed.

It is also postulated that some of the volume change of the soil skeleton occurs because of factors such as elastic strains in the particles themselves, and compression of the layers of water molecules attracted around each particle. These volume changes are reversible or recoverable.

At a certain stage, depending on the geological environment, deposition and consolidation will cease and, provided no other geological process occurs, the soil will remain at this state of void ratio and effective stress p_o' (end of deposition). The effective stress at this stage will be the maximum stress ever placed on the soil, p_c', and it will also be the present overburden stress or pressure, p_o'. Thus:

$$p_o' = p_c'$$

If a sample of the soil at the level of the soil element were taken from a borehole with minimal disturbance, placed in a laboratory consolidation apparatus and subjected to a vertical stress, p_o', it would be at point A shown on Figure 4.5. If the pressure on the sample is then increased, the loading line would continue from the original deposition line. This would also be the case for the soil deposit *in situ* when subjected to a stress increase from a structure or an embankment.

A truly normally consolidated clay would only exist in a current deposition environment such as a river, estuary, lake or coastal region and would not have been affected by any other processes which might produce overconsolidation.

Overconsolidated clay *(Figure 4.6)*
This is a clay that has been subjected to an effective stress in its past stress history larger than that existing at the present time. The most common cause of overconsolidation is erosion, but it can be produced by other processes, see below. On removal of stress the soil skeleton swells but only the reversible components of volume change are recovered, so the void ratio increases are smaller and the 'erosion' line follows a flatter path.

The present void ratio and pressure state for an overconsolidated clay is shown as point B in Figure 4.6. If a sample of this soil was taken from a borehole and reloaded from p_o' it would follow a fairly flat reloading line, overcoming the 'elastic' components until it reached the previous maximum pressure (preconsolidation pressure, p_c'). Then it would follow the steeper deposition or virgin compression line, when irreversible structural rearrangements would recommence.

Overconsolidation ratio
This is a measure of the degree of overconsolidation the soil has been subjected to in the past, as shown in Figure 4.6.

The overconsolidation ratio is given by:

$$\text{OCR} = \frac{p_c'}{p_o'} \tag{4.3}$$

Typical values of the overconsolidation ratio are:

Soil type	*OCR*
Normally consolidated	1
Lightly overconsolidated	1.5 – 3
Heavily overconsolidated	> 4

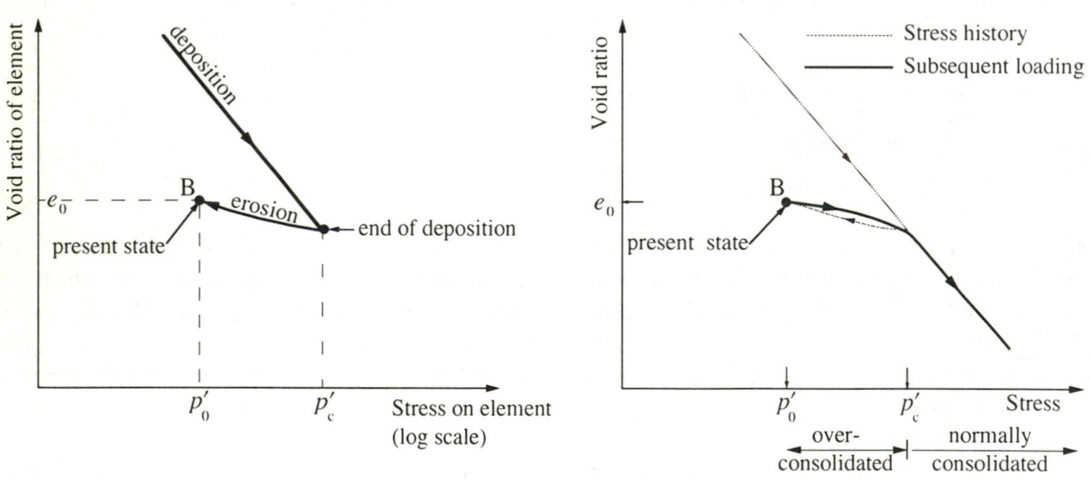

Figure 4.6 *Overconsolidated clay*

Overconsolidation ratio in the ground *(Figure 4.7)*
For a normally consolidated clay the overconsolidation ratio will remain constant with depth and the overconsolidation ration OCR = 1.

For an overconsolidated clay the overconsolidation ratio does not have a constant value. It decreases with depth if due to erosion, as shown in Figure 4.7.

It will also vary according to other processes which the soil may have undergone since deposition. These could include:

a) water table fluctuations
b) desiccation due to emergence, evaporation, vegetation roots
c) physico-chemical processes of cementation, cation exchange, thixotropy, leaching, weathering, etc.
d) delayed compression
e) tectonic forces, ice sheets, sustained seepage forces.

Desiccated crust *(Figure 4.8)*
This is produced in the upper horizons of a normally consolidated clay and is due to water table lowering from *a* to *b* and then back to *c*.

In Figure 4.8 it can be seen that the effective stresses throughout the deposit are increased when the water table drops to level *b*. Suctions above the lowered water table will increase effective stresses and produce further consolidation. At this stage, the soil is still normally consolidated since it has only undergone further loading. If the water table rises to level *c*, the effective stresses decrease and the soil below level *c* then becomes overconsolidated. When this is followed by desiccation and other physico-chemical processes, a stiffer crust at the top of an otherwise very soft clay is produced. This is beneficial for agricultural and building purposes.

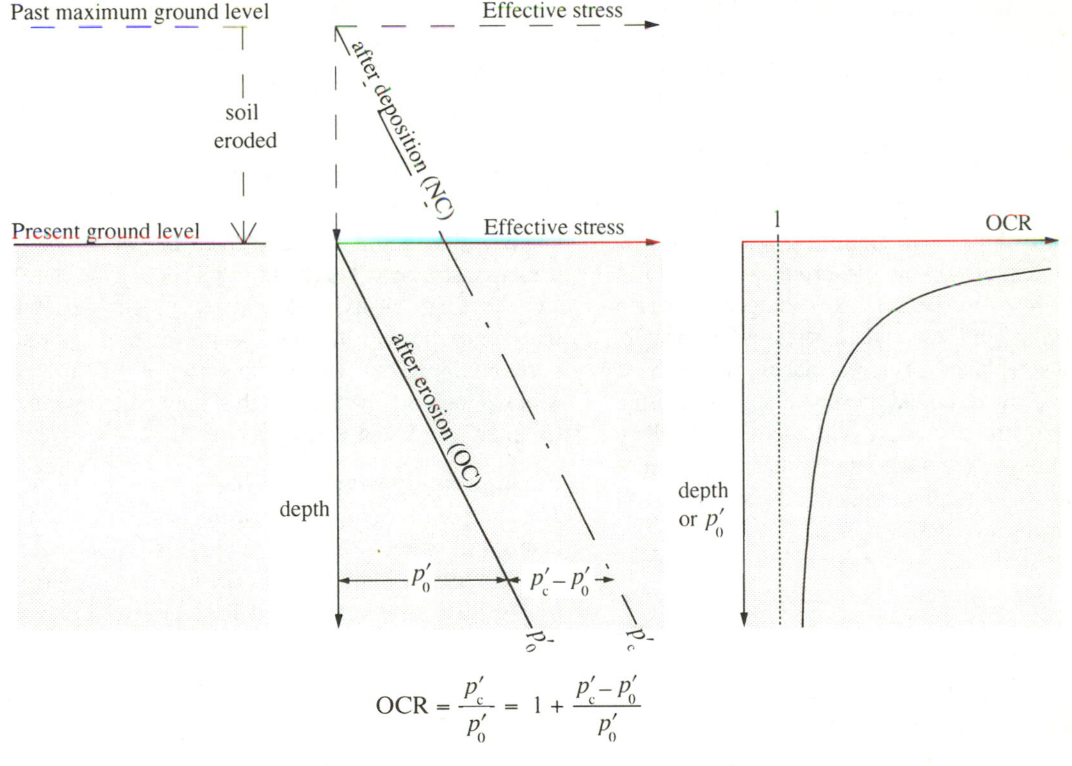

$$\text{OCR} = \frac{p'_c}{p'_0} = 1 + \frac{p'_c - p'_0}{p'_0}$$

$p'_c - p'_0 = \text{constant}$ $\quad\therefore$ OCR decreases as p'_0 increases

Figure 4.7 *Overconsolidation ratio versus depth*

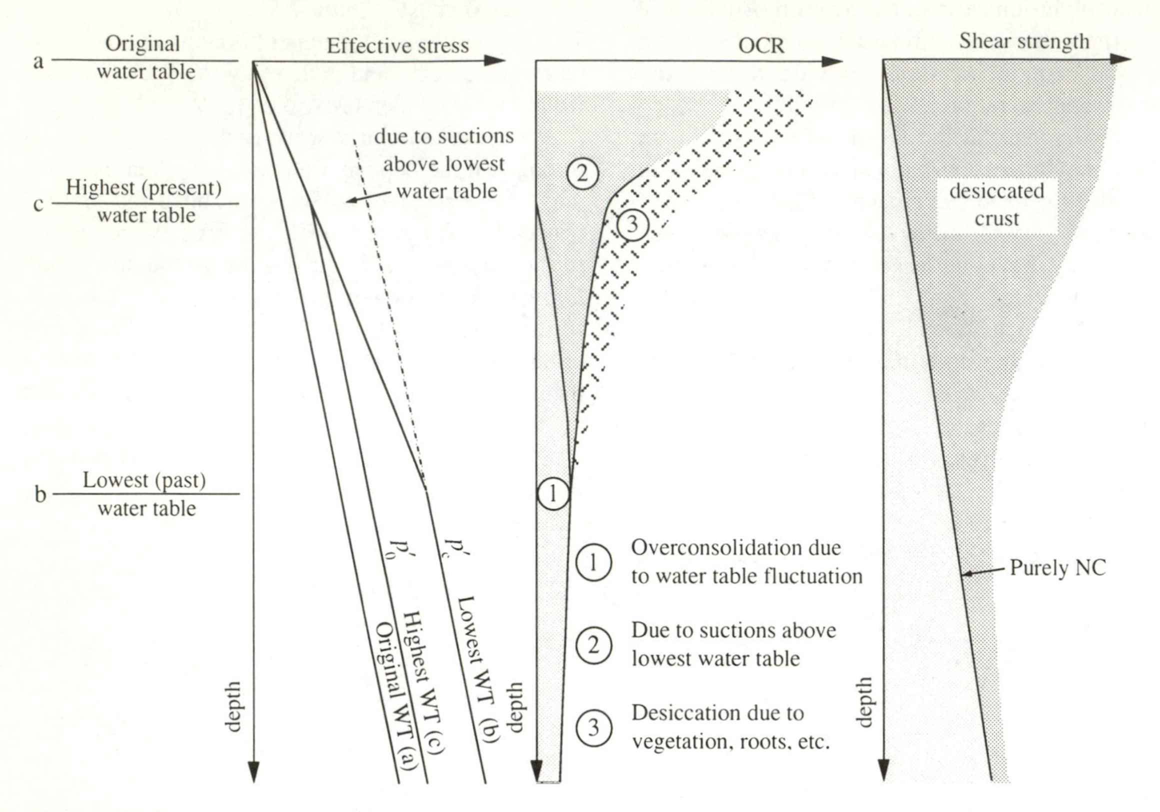

Figure 4.8 *Water table fluctuations and desiccated crust*

Present state of stress in ground *(Figure 4.9)*
For an element of soil in the ground to remain at equilibrium, total stresses on three orthogonal axes are required, σ_1, σ_2 and σ_3. These stresses are usually chosen to act on planes (principal planes) on which the shear stresses are zero, and these stresses are referred to as principal stresses. There will, therefore, be three effective stresses (major, intermediate and minor principal stresses) maintaining equilibrium in the soil structure given as σ_1', σ_2' and σ_3'.

For most studies of soil mechanics it is assumed that $\sigma_2' = \sigma_3'$ (axi-symmetric conditions) and most test procedures are carried out to model this condition. There are situations where $\sigma_2' \neq \sigma_3'$ and plane strain conditions then apply, such as beneath or behind a retaining wall or beneath a slope.

Mohr's circle of stress *(Figure 4.10)*
The Mohr circle is a useful way of determining the shear stress and normal stress acting on a plane at any angle within a soil element given the values of the principal stresses. This method can depict the state of stresses for total as well as effective stresses, with the circles having the same diameter but separated horizontally by the value of pore water pressure, u_w. More detail on the use of this method is given in Chapter 7 on Shear Strength.

Earth pressures at rest *(Figure 4.11)*
The state of a soil *in situ* at equilibrium (no vertical or horizontal strains occurring) is referred to as the 'at-rest' condition. The ratio of horizontal (σ_H') to vertical (σ_V') effective stresses in this state is denoted by K_o, the coefficient of earth pressure at rest.

$$K_o = \sigma_H'/\sigma_V' \tag{4.4}$$

For a normally consolidated clay, the ratio of σ_V' and σ_H' is found to remain constant, so K_o is constant with depth for this type of clay, see Figure 4.11a.

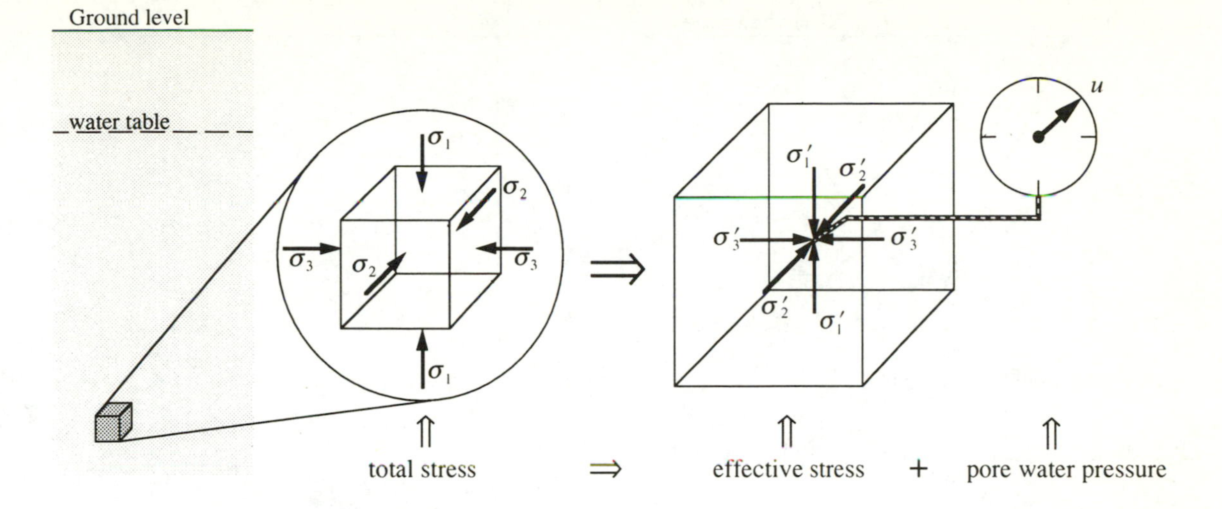

Principal stress	*Total stress*	*Effective stress*
Major	σ_1	$\sigma'_1 = \sigma_1 - u$
Intermediate	σ_2	$\sigma'_2 = \sigma_2 - u$
Minor	σ_3	$\sigma'_3 = \sigma_3 - u$

Figure 4.9 *In situ stresses*

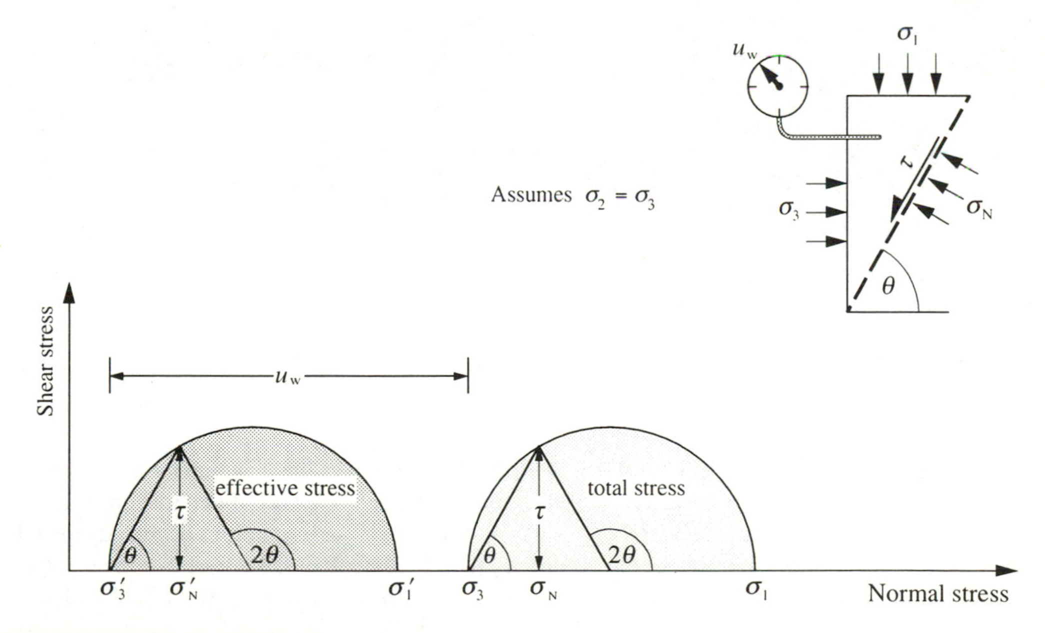

Figure 4.10 *Mohr circle of stress*

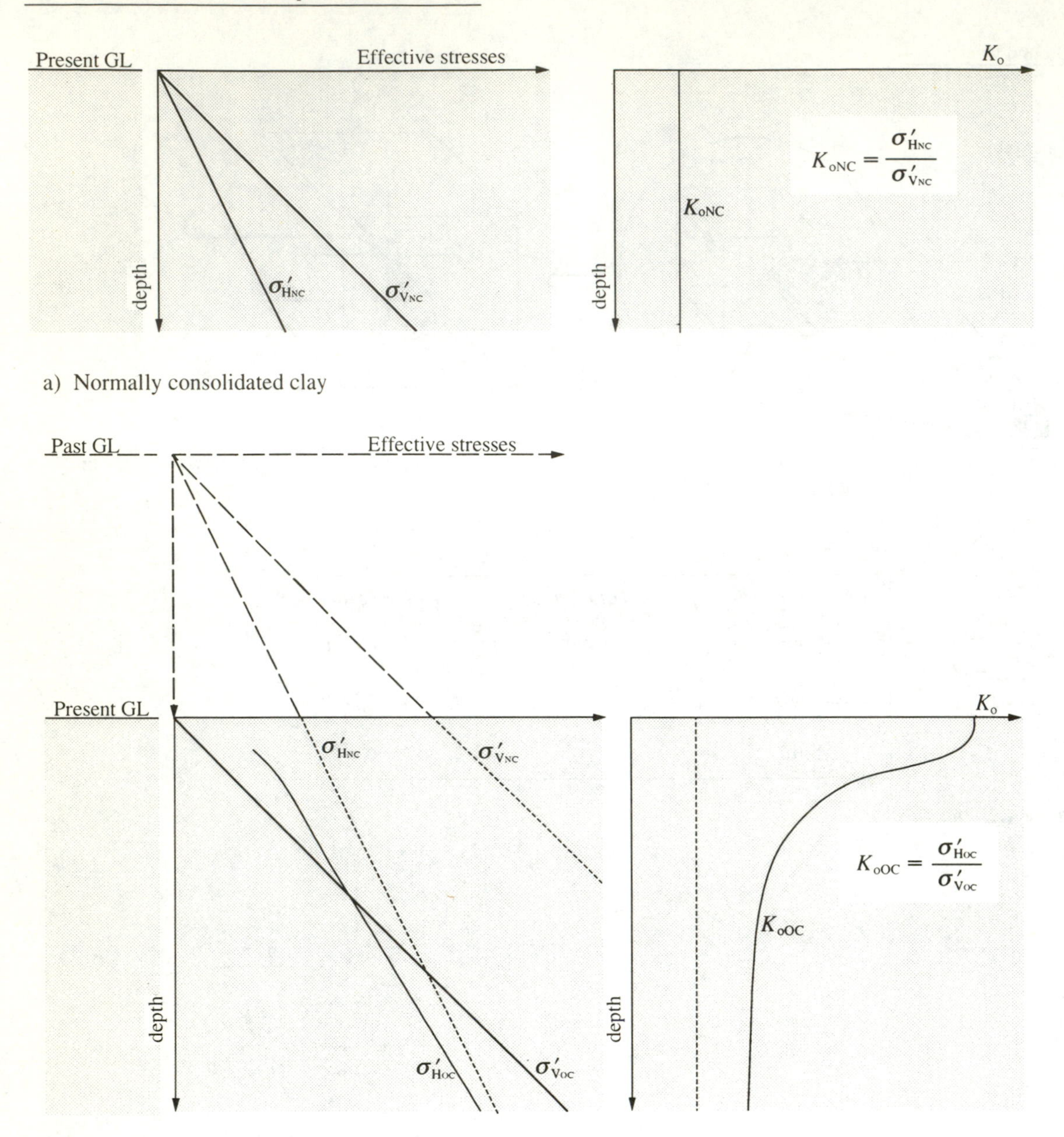

Figure 4.11 *Earth pressures at rest*

For an overconsolidated clay, as the vertical stress is reduced due to erosion the horizontal stresses remain 'locked in', as shown in Figure 4.11b. Then K_o depends on how much the vertical stress is reduced and hence on the overconsolidation ratio. Since the OCR decreases with depth (Figure 4.7) the K_o value increases towards the ground surface. However, there is an upper limit to this value at which passive failure occurs, denoted by the K_p (coefficient of passive earth pressure) value. Some expressions for K_o are given in Table 4.1.

Table 4.1 *Coefficients of earth pressure at rest K_o*

Soil type	Expression	Author
Elastic material	$K_o = \nu/(1 - \nu)$ (ν – Poisson's ratio)	–
Normally consolidated clay	$K_{oNC} = 0.95 - \sin \phi'$ $K_{oNC} = 0.19 + 0.233 \log_{10} I_p$	Brooker and Ireland (1965) Alpan (1967)
Overconsolidated clay	$K_{oOC} = K_{oNC} \times OCR^{\sin\phi'}$	Mayne and Kulwahy (1982)
Normally consolidated sand	$K_{oNC} = 1 - \sin \phi'$	Jaky (1944)
Overconsolidated sand	$K_{oOC} = K_{oNC} \times OCR^{\sin\phi'}$	Mayne and Kulwahy (1982)

Changes in stress due to engineering works *(Figure 4.12)*

A soil element in the ground will remain at equilibrium supporting the total overburden stress, σ, and with a pore water pressure u_w related to the water table level, Point A in Figure 4.12, until a change of stress occurs. The soil will behave differently depending on whether it subsequently undergoes loading or unloading.

Loading

Figure 4.12 represents the consolidation analogy. The 'tap' on the soil element represents the facility for drainage from the soil; the degree of opening of the tap represents the permeability of the soil or the rate at which water can flow out of the soil.

When the engineering works produces an increase in stress (total stress) on the soil element it is assumed that this stress increase is applied quickly, and in relation to the permeability of the soil which is assumed to be low (such as for a clay) it is as though the tap has not yet been turned on so the soil element has to respond in an undrained manner.

When lateral strain in the soil is prevented (one-dimensional condition or K_o consolidation) and water cannot be squeezed out of the soil (undrained conditions) the soil particles cannot rearrange themselves to develop stronger interparticle forces and cannot support the increase in total stress. Then the pore water has to support the total stress entirely and the increase in pore water pressure will be equal to the increase in total stress, i.e. $\Delta u_w = \Delta\sigma$, as in case B in Figure 4.12.

If lateral strains were permitted (three-dimensional conditions) then some particle rearrangement could take place, the interparticle forces could provide some support to the total stress (by an increase in effective stress) and the pore pressure would be lower than the applied total stress, i.e. $\Delta u_w < \Delta\sigma$, see 'Pore water pressure parameters', below.

Drainage will occur from the clay towards the nearest permeable boundary (such as a sand layer) so with the 'tap' now considered open, flow of water will occur from the soil element and the pore pressure will decrease, referred to as dissipation, case C on Figure 4.12. As the pore pressure dissipates and the void volume decreases the soil structure responds, with the particles moving closer together, increasing the interparticle forces and hence the effective stress.

This takes time because of the low permeability of clay soils and the distances the water has to travel to move out of the soil deposit. Eventually, pore pressure throughout the deposit returns to the original steady state pore water pressure, u_w, case D in Figure 4.12. Then the new particle arrangement is in a stronger state to support the change in (external) total stress and the change in effective stress, $\Delta\sigma' = \Delta\sigma$. This process is called consolidation.

Unloading

When the engineering works produces a decrease in total stress on the soil element the soil structure will have a tendency to swell due to the recoverable components of volume change (such as elastic strains of the particles and the bound water layers) but if this stress decrease is applied quickly and the soil is of low permeability the soil will again respond in an undrained manner, i.e. the 'tap' remains closed. This will produce a negative pressure in the pore water but the effective stress will remain unchanged.

As water is drawn towards the soil element the negative pore water pressure change dissipates and increases until it finally reaches the original steady state pore water pressure, u_w, or a new pore pressure value produced by the construction works such as a cutting slope. The change in effective stress will now be a reduction. As the effective stress decreases the particles move further apart and the volume of voids increases, a process referred to as swelling.

At the end of the swelling process, the new effective stress will be less and hence the shear strength will be reduced. Thus when a stress decrease occurs, such as around basements, excavations and cuttings the long-term condition produces the most critical condition.

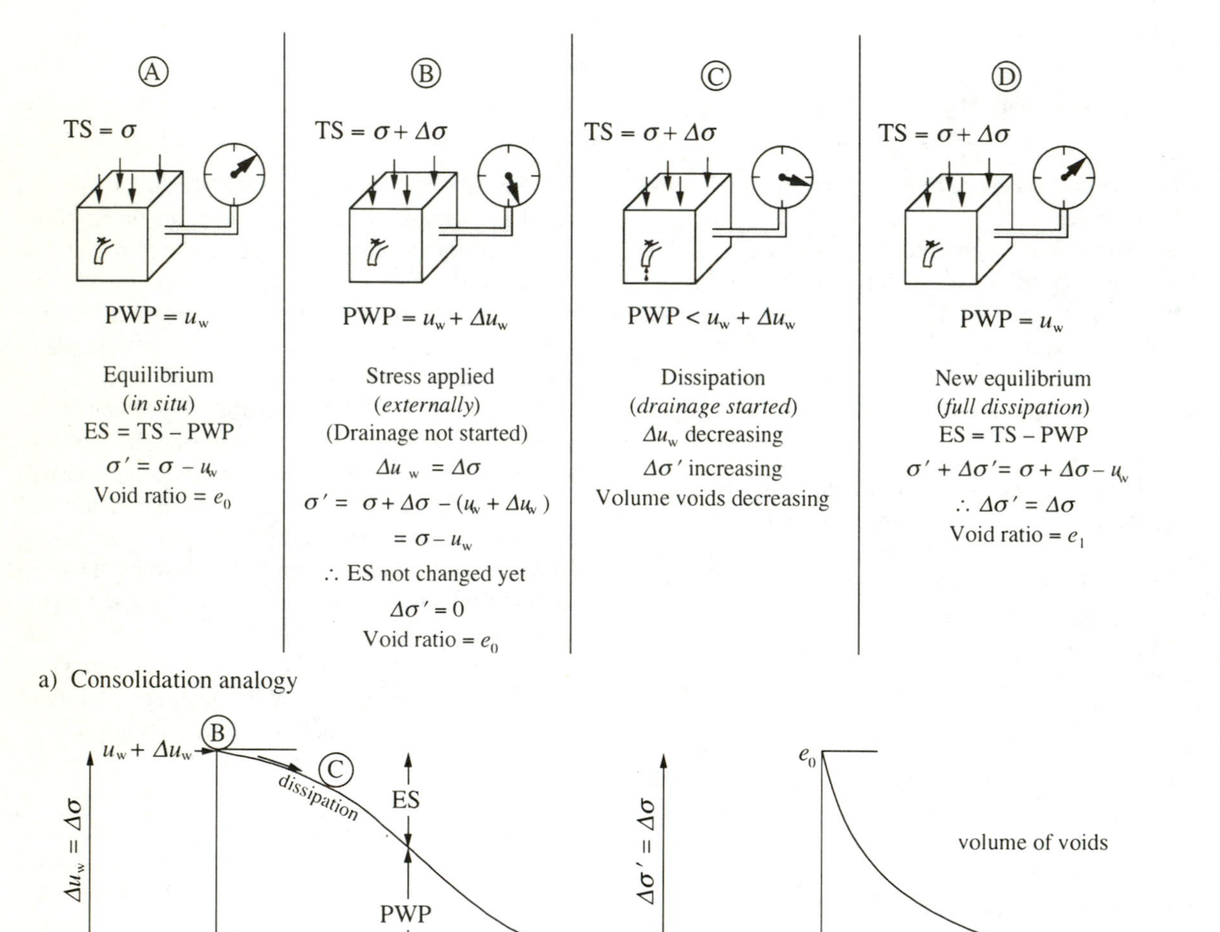

b) Variations with time

Figure 4.12 *Changes in stress*

Pore pressure parameters – theory *(Figure 4.13)*
It would be useful to determine the pore water pressure change (internal) in a soil element when the total stresses (external) are changed. It is assumed that for this purpose the soil behaves in an undrained manner with no water movement to or from the voids.

An explanation for the pore pressure parameters is given in Figure 4.13.

From this various cases can be considered:

1) For fully saturated soils, C_w is the compressibility of the water alone (no air present) and is very small so that B tends to 1 and Equation 4.5 in Figure 4.13 becomes

$$\Delta u = \Delta\sigma_m \tag{4.6}$$

1) The mean total stress $\Delta\sigma_m = 1/3(\Delta\sigma_1 + \Delta\sigma_2 + \Delta\sigma_3)$

$$\Delta\sigma_1' = \Delta\sigma_1 - \Delta u$$
$$\Delta\sigma_2' = \Delta\sigma_2 - \Delta u$$
$$\Delta\sigma_3' = \Delta\sigma_3 - \Delta u$$

2) The mean effective stress

$$\Delta\sigma_m' = 1/3(\Delta\sigma_1' + \Delta\sigma_2' + \Delta\sigma_3')$$
$$= 1/3(\Delta\sigma_1 + \Delta\sigma_2 + \Delta\sigma_3 - 3\Delta u)$$

3) The pore water pressure change (internal) Δu in a soil element due to total stress change (external) is obtained as follows. Undrained conditions (no water entry or exit) are assumed.

4) The soil structure behaves as an elastic isotropic material with compressibility C_s:

$$\frac{\Delta V_s}{V_T} = C_s\Delta\sigma_m' = \frac{1-2\nu'}{E'}\Delta\sigma_m'$$

E' and ν' are the modulus and Poisson's ratio of the soil structure.
The decrease in volume of the soil structure is:

$$-\Delta V_s = V_T\, C_s\, \Delta\sigma_m'$$

5) The relationship between volume change (ΔV_w) of the pore water itself and its pressure (Δu) is linear, i.e.

$$\frac{\Delta V_w}{V_w} = C_w \Delta u$$

where C_w is the compressibility of the pore water. The decrease in volume of the water is:

$$-\Delta V_w = V_w\, C_w\, \Delta u = -n\, V_T\, C_w\, \Delta u$$

6) The soil particles themselves are incompressible.

7) The decrease in volume of water = the decrease in volume of the soil structure:

$$-\Delta V_w = -\Delta V_s$$
$$n\, V_T\, C_w\, \Delta u = V_T\, C_s\, \Delta\sigma_m'$$

$$\Delta u = \frac{C_s}{nC_w}\Delta\sigma_m' \quad \text{in terms of effective stress}$$

$$\Delta u = \frac{C_s}{nC_w}\left[\frac{1}{3}(\Delta\sigma_1 + \Delta\sigma_2 + \Delta\sigma_3) - \Delta u\right]$$

in terms of total stress

8) Rearranging gives:

$$\Delta u = \frac{1}{3}\frac{1}{\left(1+\dfrac{nC_w}{C_s}\right)}(\Delta\sigma_1 + \Delta\sigma_2 + \Delta\sigma_3)$$

9) Putting pore pressure parameter B as:

$$B = \frac{1}{\left(1+\dfrac{nC_w}{C_s}\right)}$$

$$\boxed{\Delta u = B\,\Delta\sigma_m} \tag{4.5}$$

Figure 4.13 *Pore pressure parameters – theory*

2) For axisymmetrical conditions (as in the triaxial test) during isotropic consolidation where only the cell pressure ($\Delta\sigma_3$) is increased:

$$\Delta\sigma_1 = \Delta\sigma_2 = \Delta\sigma_3$$

so $\Delta\sigma_m = \Delta\sigma_3$ and

$$\Delta u = B\Delta\sigma_3 \tag{4.7}$$

B can then be found by applying increments of cell pressure, $\Delta\sigma_3$, and recording the increase in Δu. For partially saturated soils where the degree of saturation S_r is less than 1, B will be less than 1, see below.

3) For axisymmetrical conditions during anisotropic consolidation where lateral or horizontal strain is prevented (see K_o conditions above) as in the ground, an oedometer test or a triaxial test where lateral strain is not permitted it can be shown that for a fully saturated soil:

$$\Delta u = \Delta\sigma_1 \tag{4.8}$$

$\Delta\sigma_1$ is usually the vertical applied stress.

4) In the triaxial compression test where $\Delta\sigma_1 > \Delta\sigma_3$, when the deviator stress ($\Delta\sigma_1 - \Delta\sigma_3$) is applied the general expression can be written:

$$\Delta u = B[{}^1/_3(\Delta\sigma_1 + 2\,\Delta\sigma_3)] \text{ or}$$

$$\Delta u = B[\Delta\sigma_3 + {}^1/_3(\Delta\sigma_1 - \Delta\sigma_3)] \tag{4.9}$$

5) In a triaxial extension test the cell pressure represents the major ($\Delta\sigma_1$) and intermediate ($\Delta\sigma_2$) principal stresses ($\Delta\sigma_1 = \Delta\sigma_2$) and these are greater than the axial stress ($\Delta\sigma_3$) giving:

$$\Delta u = B[{}^1/_3(2\Delta\sigma_1 + \Delta\sigma_3)] \text{ or}$$

$$\Delta u = B[\Delta\sigma_3 + {}^2/_3(\Delta\sigma_1 - \Delta\sigma_3)] \tag{4.10}$$

6) In a plane strain test where $\Delta\sigma_2' = \nu'(\Delta\sigma_1' + \Delta\sigma_3')$ it can be shown (Bishop and Henkel, 1957) that:

$$\Delta u = B[\Delta\sigma_3 + {}^1/_2(\Delta\sigma_1 - \Delta\sigma_3)] \tag{4.11}$$

Pore pressure parameters A and B *(Figures 4.14 and 4.15)*
In reality, soil is not elastic (it is non-linear) nor isotropic so the above equations have been generalised to:

$$\Delta u = B[\Delta\sigma_3 + A(\Delta\sigma_1 - \Delta\sigma_3)] \tag{4.12}$$

after Skempton (1954) where A and B are the pore pressure parameters.

This expression is useful for triaxial tests since it separates the effect of changes of cell pressure ($\Delta\sigma_3$) and changes of deviator stress ($\Delta\sigma_1 - \Delta\sigma_3$) on the changes in pore water pressure.

B decreases as the degree of saturation S_r decreases since an increase in air content increases the overall compressiblity of the pore fluids. A typical relationship is shown in Figure 4.14.

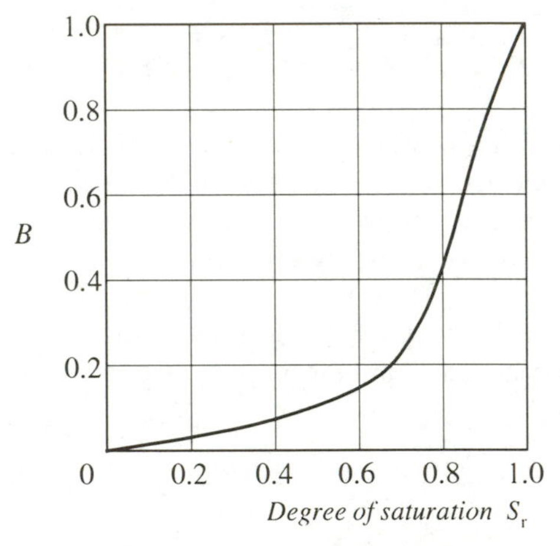

Figure 4.14 *Typical relationship for B*

From above, it can be seen that the parameter A has particular values for certain conditions of the test and stress system. It will also vary during a shear strength test as pore pressures develop with shearing, so the value at failure, A_f, is usually reported. For the triaxial compression test on clays, A_f has been demonstrated to depend on the value of the overconsolidation ratio, see Figure 4.15.

For sands the parameter A will depend on the initial density. For a loose sand A will be high due to contraction of the soil structure during shear. For a dense sand A will be low and probably negative due to dilatancy of the soil structure during shear.

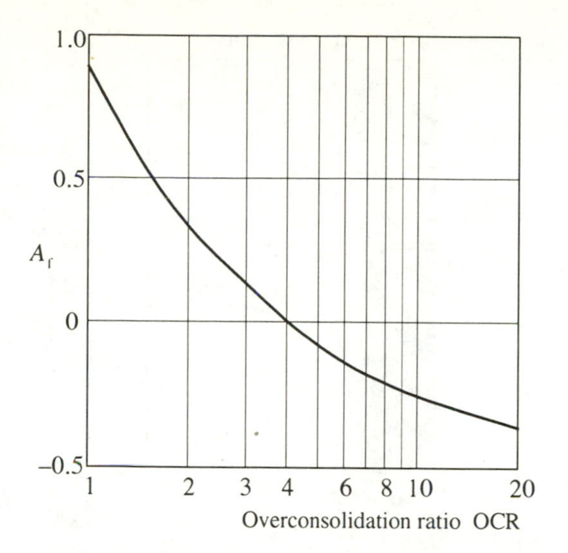

Figure 4.15 *A_f values for overconsolidated clays*

Typical values are:

	A_f
Loose sand	2.0 - 3.0
Medium dense sand	0 - 1.0
Dense sand	−0.3 - 0

Values of A_f should not be used to predict pore water pressure changes at stress levels before failure and since this is the condition required for design, in practice the A_f parameter is of limited value.

Capillary rise above the water table *(Figure 4.16)*
If water did not display the property of surface tension the soil above the water table would be dry, apart from water percolating downwards from the soil surface.

Above the water table the pores may still be completely saturated because water is held within them due to the surface tension of the water. The voids in the soil form an intricate network of continuous channels which decrease in size as the soil particle size decreases and these channels can be imagined as fine capillary tubes. The height to which water will rise in a capillary tube due to surface tension effects increases as the diameter of the capillary tube decreases so this capillary zone will be greater for finer-grained soils.

Soils contain voids (or capillary tubes) of varying sizes. Above the fully saturated capillary zone, h_s, the surface tension cannot hold water in all the voids; then the soil is partially saturated with air filling some voids.

The height, h_c, to which water will rise in a capillary tube is theoretically proportional to the surface tension force, T, and the diameter of the tube d:

$$h_c \propto \frac{T}{d}$$

so the height of the capillary zone will also be affected by the cleanliness of the water; it could be much less for polluted water.

According to Terzaghi and Peck (1967):

$$h_c = \frac{C}{ed_{10}} \qquad (4.13)$$

where:
h_c = maximum height of capillary rise, in mm related to minimum pore size
e = void ratio
d_{10} = effective size, mm
C = constant = 10 – 50 mm^2 (for clean water)

For open gravels capillary rise will be negligible; for clays it will be considerable. Approximate values from the above expression are:

Soil type	h_c *m*
gravel	0.01 – 0.05
sand	0.1 – 1.0
silt	2 – 10
clay	10 – 30

These values apply for the smallest voids in the soil and are therefore the maximum values.

Soils contain a range of void sizes so the zone of complete saturation will be given by the maximum height of capillary rise in the largest voids, h_s. An approximate value for h_s could be obtained by using d_{60} of the soil instead of d_{10} in the above equation for h_c:

$$h_s = \frac{C}{ed_{60}} \qquad (4.14)$$

Above this zone surface tension can no longer support water in the largest voids so the soil then becomes partially saturated.

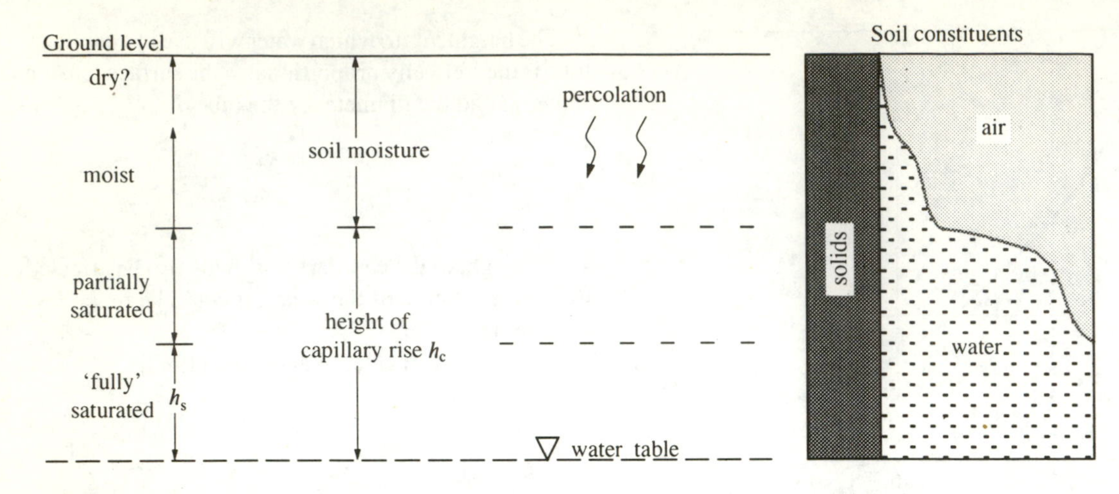

Figure 4.16 *Capillary rise above the water table*

Effective stresses above the water table
(Figure 4.17)
Within the zone of complete saturation the pore water pressures will be negative and can be obtained from:

$$u_w = -\gamma_w z_a \tag{4.15}$$

where z_a is the elevation *above* the water table. Effective stresses will then be given by:

$$\sigma' = \sigma + \gamma_w z_a \tag{4.16}$$

where σ is the total stress obtained in the normal way. In the zone of partial saturation there will be a pressure in the water u_w and in the air voids u_a and the difference between these is defined as suction, $u_a - u_w$.

Bishop *et al* (1960) proposed the following relationship for effective stress in a partially saturated soil:

$$\sigma' = \sigma - u_a + \chi(u_a - u_w) \tag{4.17}$$

where σ is the total stress and χ represents the proportion of a unit cross-sectional area occupied by water. For dry soils $\chi = 0$ and for saturated soils $\chi = 1$.

A reasonable approximation for effective stresses in the partially saturated zone can be obtained assuming $u_a = 0$ and $\chi = S_r$ (%) which gives:

$$\sigma' = \sigma + \gamma_w z_a \frac{S_r}{100} \tag{4.18}$$

This shows that effective stresses above the water table are enhanced and will reduce the instability problems posed by partially saturated soils.

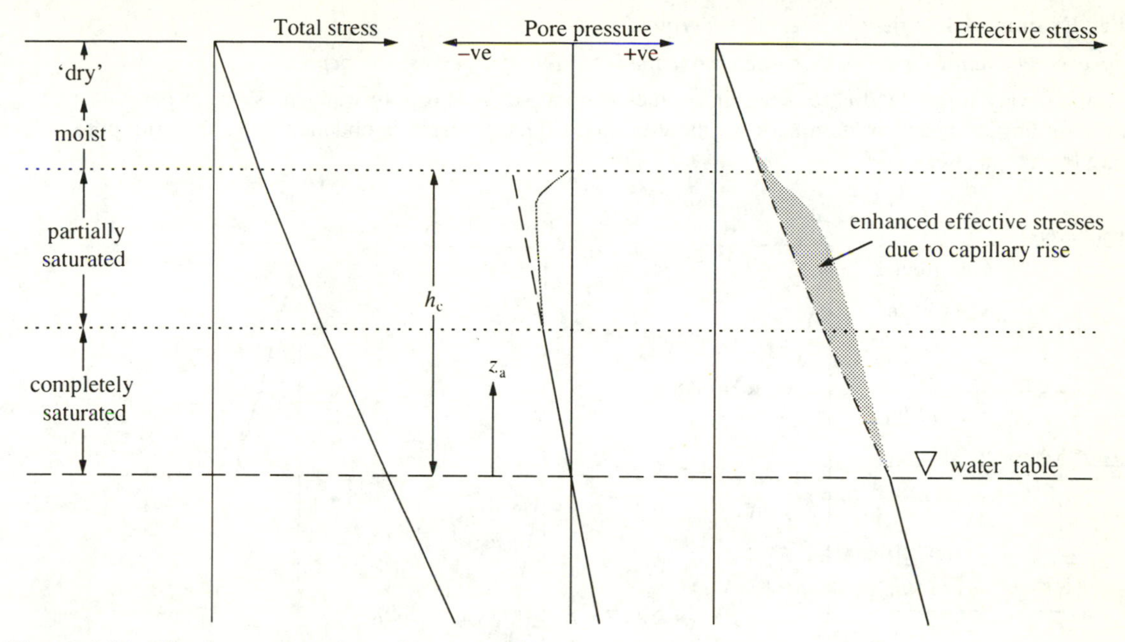

Figure 4.17 *Effective stresses above the water table*

Worked Example 4.1 *Effective stress in the ground*

For the ground conditions given below determine the variation of stresses with depth.

The simplest way to proceed is to determine values of total stress and pore water pressure at particular depths assuming a linear variation between these points. Effective stress is obtained by subtracting pore pressure from total stress.

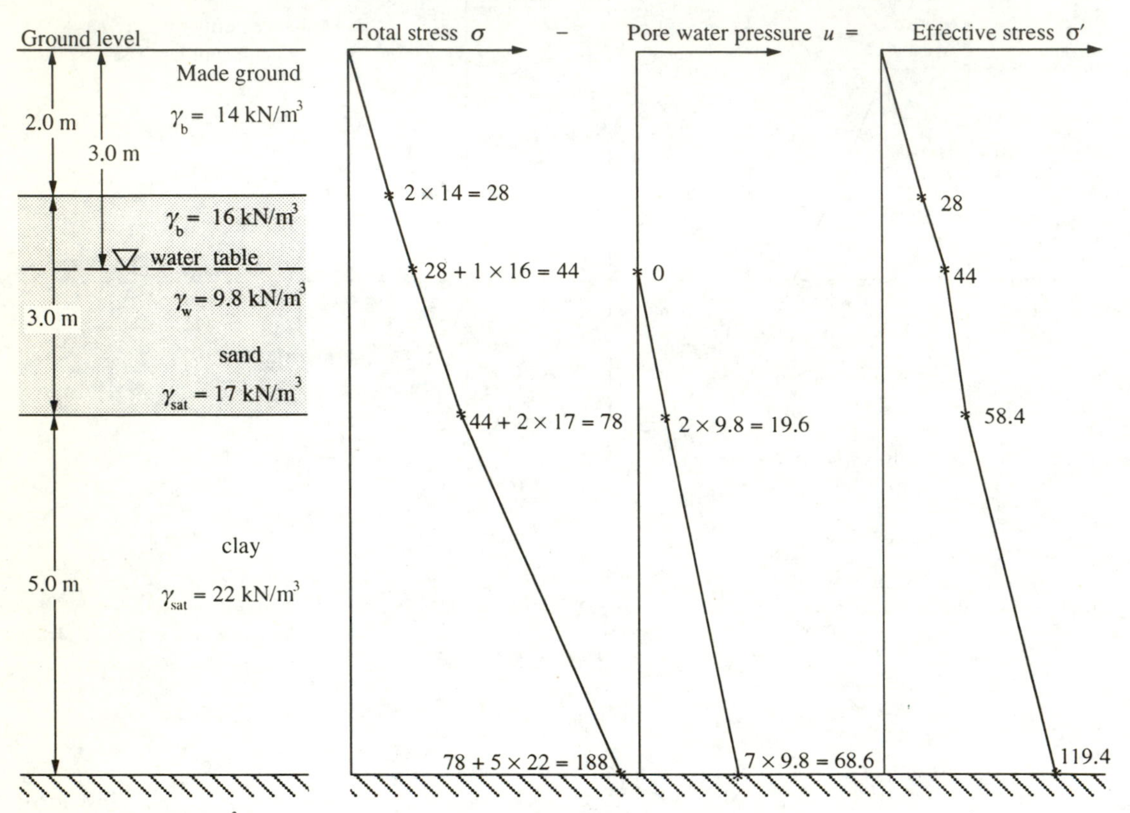

Note: $\gamma_w = 10$ kN/m^3 is often used and gives a sufficiently accurate result

Worked Example 4.2 *Overconsolidation ratio*
A sample of clay has been taken from 6 m below the bed of a river. It is known that the saturated unit weight of the clay γ_{sat} = 21.5 kN/m³ and the overconsolidation ratio of the clay has been found to be OCR = 2.5. Determine how much soil has been removed by erosion.
Assume $\gamma_w = 9.8$ kN/m³
$\gamma_{sub} = 21.5 - 9.8 = 11.7$ kN/m³
The present effective stress or overburden pressure $p_o' = 6 \times 11.7 = 70.2$ kN/m²
From Equation 4.3, the past maximum pressure $p_c' = 70.2 \times 2.5 = 175.5$ kN/m²
Assuming the saturated unit weight of the soil removed to be 21.5 kN/m³ and the original maximum thickness of soil = z then

$z \times 11.7 = 175.5$

$\therefore z = 15.0$ m

Thickness of soil removed = 15.0 – 6.0 = 9.0 m

Worked Example 4.3 K_o *condition*
A sample of clay has been taken from 5 m below ground level with the water table at 1.5 m below ground level. The unit weight of the clay above and below the water table is γ_{sat} = 20.7 kN/m³ and the K_o value has been determined as K_o = 0.85. Determine the total and effective vertical and horizontal stresses at this depth.
Assume $\gamma_w = 9.8$ kN/m³
Pore water pressure $u_w = 3.5 \times 9.8 = 34.3$ kN/m²
Total vertical stress $\sigma_V = 5 \times 20.7 = 103.5$ kN/m²
Effective vertical stress $\sigma_V' = 103.5 - 34.3 = 69.2$ kN/m²
Effective horizontal stress $\sigma_H' = 69.2 \times 0.85 = 58.8$ kN/m² (From Equation 4.4)
Total horizontal stress $\sigma_H = 58.8 + 34.3 = 93.1$ kN/m²

Worked Example 4.4 *Changes in stress*
In a triaxial apparatus a specimen of fully saturated clay has been consolidated, by allowing drainage of pore water from the specimen, under an all-round pressure of 600 kN/m² and a back pressure of 200 kN/m². The drainage tap is then closed and the cell pressure increased to 750 kN/m². Determine the effective stress and pore water pressures before and after increasing the cell pressure.

i) *Before increasing cell pressure*
back pressure = pore water pressure = 200 kN/m²
The consolidation is isotropic (same stresses all-round) so the total stresses are
$\sigma_1 = \sigma_3 = 600$ kN/m²
The effective stresses are
$\sigma_1' = \sigma_3' = 600 - 200 = 400$ kN/m²

ii) *After increasing the cell pressure*
Since the clay is fully saturated, $B = 1$ so the excess pore pressure will be
$\Delta u = \Delta\sigma_3 = 750 - 600 = 150$ kN/m² (From Equation 4.7)
The pore pressure will now be 200 + 150 = 350 kN/m²
The effective stress = 750 – 350 = 400 kN/m². It has remained unchanged in the undrained and fully saturated condition. No consolidation or increase in shear strength due to this increase in cell pressure will take place until the excess pore pressure Δu is allowed to dissipate by drainage from the specimen. An explanation of the use of back pressure is given in Chapter 7 on Shear Strength.

Worked Example 4.5 *Changes in stress in an oedometer*
In an oedometer apparatus a specimen of fully saturated clay has been consolidated under a vertical pressure of 75 kN/m² and is at equilibrium. Determine the effective stress and pore water pressure immediately on increasing the vertical stress to 125 kN/m².
Under the applied pressure of 75 kN/m², the pore pressure will be zero after consolidation is complete.
Total vertical stress, σ_V or $\sigma_1 = 75$ kN/m²
Pore water pressure $u = 0$
Effective vertical stress σ_V' or $\sigma_1' = 75$ kN/m²
The horizontal stress is not known but for this test condition
$\Delta u = \Delta\sigma_1 = 125 - 75 = 50$ kN/m² (From Equation 4.8)
so the pore pressure will rise immediately to 50 kN/m² on increasing the vertical stress to 125 kN/m².
The initial effective stress will be unchanged, i.e.
$\sigma_V' = 125 - 50 = 75$ kN/m²
However, the specimen will immediately commence consolidating with the pore pressure decreasing and the effective stress increasing.

Worked Example 4.6 *Pore pressure parameter B*
In a triaxial test a soil specimen has been consolidated under a cell pressure of 400 kN/m² and a back pressure of 200 kN/m². The drainage tap is then closed, the cell pressure increased to 500 kN/m² and the pore pressure measured as 297 kN/m². Determine the pore pressure parameter B.

The change in pore pressure $\Delta u = 297 - 200 = 97$ kN/m²
The change in total stress $\Delta\sigma_3 = 500 - 400 = 100$ kN/m²
$B = \Delta u/\Delta\sigma_3 = 97/100 = 0.97$ (From Equation 4.7)

Worked Example 4.7 *Pore pressure parameter A*
From the Worked example 4.6, with the specimen remaining under undrained conditions and after the increase in cell pressure to 500 kN/m², an axial load is applied to give a principal stress difference of 645 kN/m² when the pore pressure is measured as 435 kN/m². Determine the pore pressure parameter A at this stage.

From above, $B = 0.97$ $\Delta\sigma_3 = 100$ kN/m²
$\Delta\sigma_1 - \Delta\sigma_3 = 645$ kN/m²
$\Delta u = 435 - 200 = 235$ kN/m²
From Equation 4.12,
$235 = 0.97[100 + A(645)]$ $\therefore A = 0.22$

Exercises

4.1 A river, 5 m deep, flows over a sand deposit. The saturated unit weight of the sand is 18 kN/m^3. At a depth of 5 m below the river bed determine:
a) the total vertical stress
b) the pore water pressure
c) the effective vertical stress.
Assume the unit weight of water is 9.8 kN/m^3.

4.2 In Exercise 4.1 the river level falls to river bed level. Determine the stresses at 5 m below river bed level.

4.3 A layer of clay, 5 m thick, overlies a deposit of sand, 5 m thick, which is underlaid by rock. The water table in the sand is sub-artesian with a level at 2 m below ground level. The saturated unit weights of the clay and sand are 21 and 18 kN/m^3, respectively. Determine the total stress, pore water pressure and effective stress at the top and bottom of the sand.

4.4 In Exercise 4.1 the river dries up and the water table lies at 3 m below ground level. Capillary attraction maintains the soil 1.0 m above the water table in a saturated state. The saturated unit weight of the sand is 18 kN/m^3 and the bulk unit weight (above the saturated zone) is 16.5 kN/m^3. Determine the effective stress at 2 m and 5 m below ground level.

4.5 A layer of sand, 4 m thick, overlies a layer of clay, 5 m thick. The bulk unit weight of the sand is known to be 16.5 kN/m^3; its saturated unit weight is 18 kN/m^3. The saturated unit weight of the clay is 21 kN/m^3. The water table exists initially at 1.0 m below ground level but pumping will permanently and rapidly lower the water table to 3.0 m below ground level. Determine the effective stress at the top, middle and bottom of the clay:
a) before pumping
b) immediately after lowering the water table
c) in the long-term.
Determine the change in effective stress in the clay caused by the pumping.

4.6 In Exercise 4.5, if the unit weight of the sand above and below the water table is assumed to be the same. What is the overall change in effective stress in the clay caused by pumping?

4.7 A clay soil deposited in an estuary and originally normally consolidated has been subjected to water table fluctuations. The lowest water table level was 6 m below ground level and the present water table is at 2 m below ground level. Assume the clay to be fully saturated with a unit weight of 19.5 kN/m^3. Determine the overconsolidation ratio at 2 m, 6 m and 20 m below ground level.

4.8 A pressuremeter test carried out at a depth of 6.0 m below ground level in clay soil has determined the horizontal stress to be 120 kN/m^2. The water table lies at 1.5 m below ground level. Assume the clay to be fully saturated with a unit weight of 20.5 kN/m^3. Determine the coefficient of earth pressure at rest, K_0 at the depth of the test. The angle of shearing resistance of the clay is 25°. Is the clay normally consolidated or overconsolidated?

4.9 The results of the saturation stage of a triaxial test are given below. Determine the pore pressure parameter B at stages a) to f).
Hint: Tabulate the data for stages a) to f) as Δu $\Delta \sigma_3$ B.
All pressures are in kN/m².

	Cell pressure	Back pressure valve	Back pressure	Pore pressure
	0	closed	0	–5
a)	50	closed	–	12
	50	open	40	38
b)	100	closed	–	67
	100	open	90	89
c)	150	closed	–	126
	150	open	140	140
d)	200	closed	–	184
	200	open	190	190
e)	300	closed	–	285
	300	open	290	290
f)	400	closed	–	388

4.10 The results of the shearing stage of a consolidated undrained triaxial test are given below. Determine the pore pressure parameter A for each value.
Cell pressure = 450 kN/m² Back pressure = 300 kN/m²

Deviator stress (kN/m²)	Pore pressure (kN/m²)
0	300
68	315
117	319
146	312
171	301
190	287
198	275

5 Contact Pressure and Stress Distribution

Contact pressure – introduction

A foundation is the interface between a structural load and the ground. The stress q applied by a structure to a foundation is often assumed to be uniform. The actual pressure then applied by the foundation to the soil is a reaction, called the contact pressure p and its distribution beneath the foundation may be far from uniform.

This distribution depends mainly on:

- stiffness of the foundation, i.e.
  flexible ⇒ stiff ⇒ rigid
- compressibility or stiffness of the soil
- loading conditions – uniform or point loading.

Contact pressure – uniform loading *(Figure 5.1)*

The effects of the stiffness of the foundation (flexible or rigid) and the compressibility of the soil (clay or sand) are illustrated in Figure 5.1.

Stiffness of foundation

A flexible foundation has no resistance to deflection and will deform or bend into a dish-shaped profile when stresses are applied. An earth embankment would comprise a flexible structure and foundation.

A stiff foundation provides some resistance to bending and will deform into a flatter dish-shape so that differential settlements are smaller. This forms the basis of design for a raft foundation placed beneath the whole of a structure.

A rigid foundation has infinite stiffness and will not deform or bend, so it moves downwards uniformly. This would apply to a thick, relatively small reinforced concrete pad foundation.

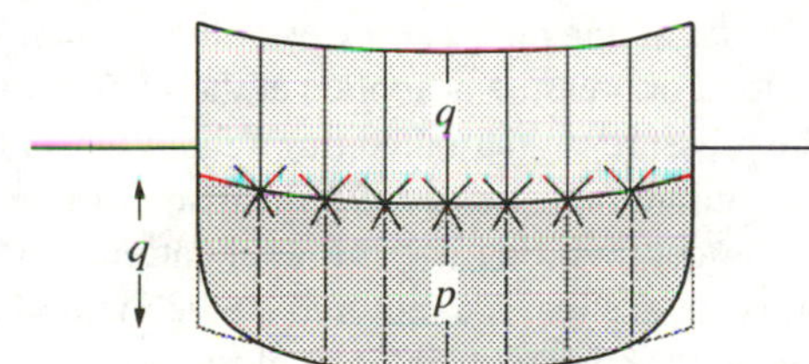

Flexible foundation on clay

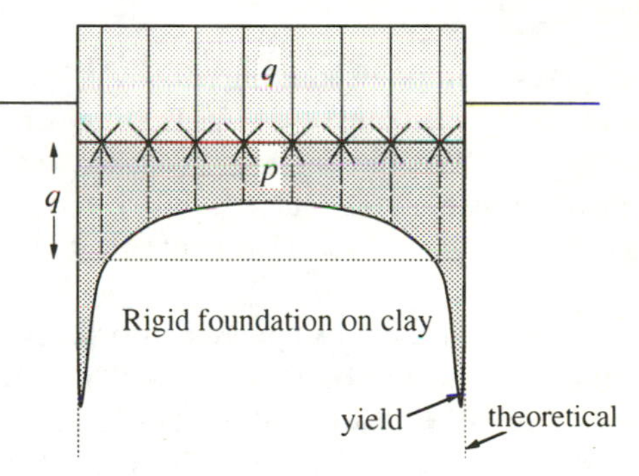

Rigid foundation on clay

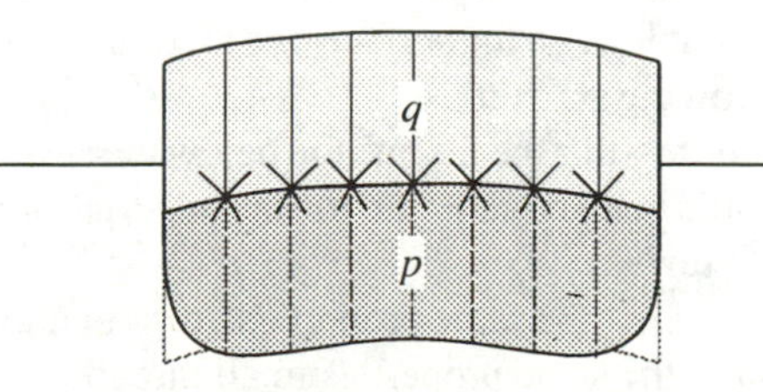

Flexible foundation on sand

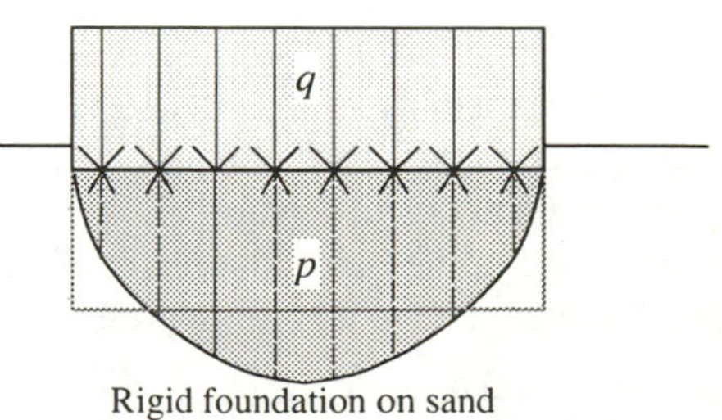

Rigid foundation on sand

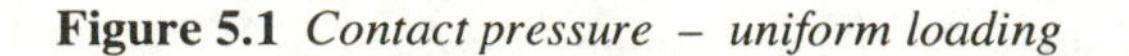

Figure 5.1 *Contact pressure – uniform loading*

Stiffness of soil

The stiffness of a clay will be the same under all parts of the foundation so for a flexible foundation a fairly uniform contact pressure distribution is obtained with a dish-shaped (sagging) settlement profile.

For a rigid foundation, the dish-shaped settlement profile must be flattened out, so beneath the centre of the foundation the contact pressure is reduced, and beneath the edges of the foundation it is increased. Theoretically, the contact pressure increases to a very high value at the edges, although yielding of the soil would occur in practice, leading to some redistribution of stress.

The stiffness of a sand increases as the confining pressures around it increase, so beneath the centre of the foundation the stiffness will be at its greatest, whereas near the edge of the foundation the stiffness of the sand will be smaller.

A flexible foundation will, therefore, produce greater strains at the edges than in the centre, so the settlement profile will be dish-shaped but upside-down (hogging) with a fairly uniform contact pressure.

For a rigid foundation this settlement profile must be flattened out, so the contact pressure beneath the centre would be increased and beneath the edges it would be decreased.

Contact pressure – point loading *(Figure 5.2)*

An analysis for contact pressure beneath a circular raft with a point load, W, at its centre resting on the surface of an incompressible soil (such as clay), has been provided by Borowicka, 1939 (in Poulos and Davis, 1974).

This shows that the contact pressure distribution is non-uniform, irrespective of the stiffness of the raft or foundation. For a flexible foundation, the contact pressure is concentrated beneath the point load, which is to be expected, and for a stiff foundation it is more uniform.

For a rigid foundation, the stresses beneath the edges are very considerably increased, and a pressure distribution similar to the distribution produced by a uniform pressure on a clay (cf. Figure 5.1) is obtained. This suggests that a point load at the centre of a rigid foundation is comparable to a uniform pressure.

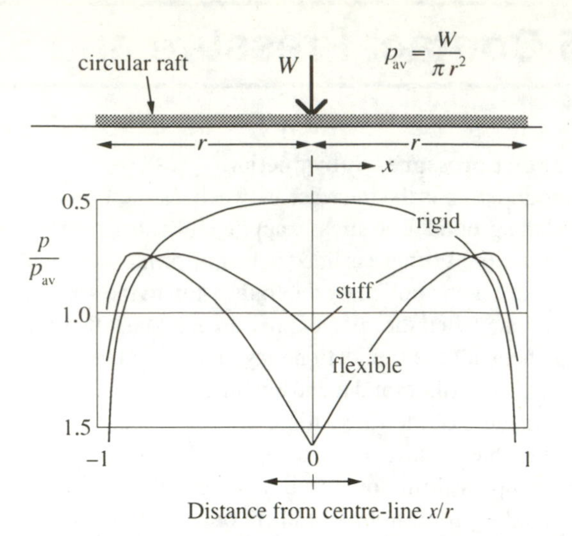

Figure 5.2 *Contact pressure – point loading (After Borowicka, 1939, in Poulos and Davis, 1974)*

Stress distribution – introduction

The stresses due to self-weight of the soil are discussed in Chapter 4. Any element of soil in the ground will be at equilibrium under three normal stresses, σ_x, σ_y and σ_z (or σ_1, σ_2 and σ_3), acting on three orthogonal axes x, y and z. This element will also be subjected to a system of shear stresses acting on the surfaces of the element.

When a load or pressure is applied at the surface of the soil, the pressure is distributed throughout the soil and the original normal stresses and shear stresses are altered. For most civil engineering applications, the changes in vertical stress are required so the methods given below are for increases in vertical stress only. A comprehensive review of solutions for stress distribution is given by Poulos and Davis (1974).

Stresses beneath point load and line load *(Figure 5.3)*

In 1885, Boussinesq published a solution for the stresses beneath a point load on the surface of a material which had the following properties:

- *semi-infinite* – this means infinite below the surface so providing no boundaries to the material apart from the surface
- *homogeneous* – the same properties at all locations
- *isotropic* – the same properties in all directions
- *elastic* – a linear stress-strain relationship.

Expressions for the stresses beneath a point load and line load are given in Figure 5.3.

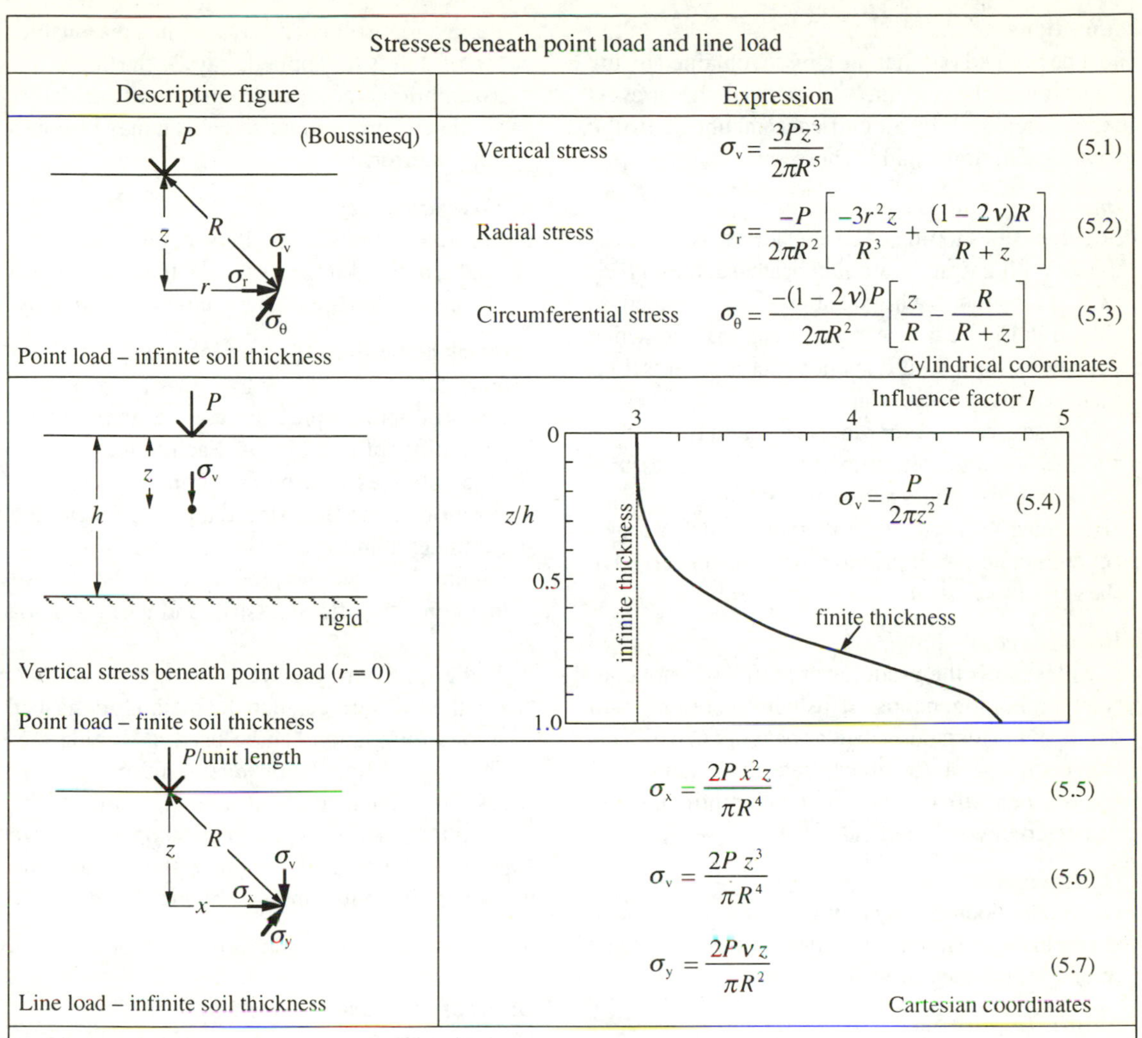

Assumptions

- The soil is homogeneous, isotropic with linear stress-strain (elastic) properties
- The line load is flexible and infinitely long. It is the integration of the point load case.

Notes

- For the finite layer thickness with a rigid stratum beneath the stresses are larger than the infinite thickness case
- Westergaard material – this is the same as the Boussinesq case except that lateral strain is prevented. This is depicted as a homogeneous mass reinforced horizontally with thin but rigid reinforcement and is an extreme case of anisotropy producing stresses which are smaller than for the Boussinesq case. For real soils stresses probably lie between these two cases but Boussinesq produces more conservative values.

Figure 5.3 *Stresses beneath a point load and line load*

Assumptions

It must be pointed out that the stresses obtained by the methods given below may differ from the stresses obtained in real soils by a significant amount, due to the various assumptions made. These are:

- *Infinite layer thickness*
  Soil deposits should not be considered as infinitely thick. With a rigid stratum beneath, such as a bedrock, it is found that higher stresses are obtained, as shown in Figure 5.3. A rigid stratum is one which does not strain, so it does not contribute to settlements or distortions.
  Where several different layers exist on a site, then stresses are often determined for each layer assuming infinite thickness and similar stiffnesses. This is erroneous. At least using methods which adopt a finite thickness will provide for more conservative (larger) stress estimates.
- *Homogeneous*
  For most soils the modulus or stiffness is not constant or homogeneous. It usually increases with depth and is then described as heterogeneous. This has been found to concentrate settlements and stresses beneath a loaded area with minimal stress dispersed beyond the loaded area.
- *Isotropic*
  Overconsolidated clays and rocks can be much stiffer in the horizontal direction than in the vertical direction, i.e. they are anisotropic.
- *Elastic* (*Figure 5.4*)
  The linear elastic assumption will allow stresses to be calculated which far exceed the yield stress of the soil when redistribution of stresses will occur.

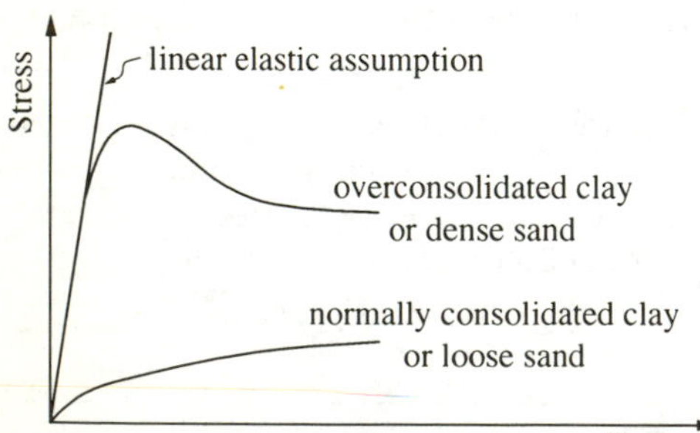

Figure 5.4 *Linear elastic assumption*

For overconsolidated clays when a reasonable factor of safety is applied, say 3, the linear elastic assumption is acceptable but for soft normally consolidated clays and loose sands it may be considerably in error.

- *Foundation depth*
  For foundations at shallow depths assuming the load is applied at the surface of the soil will provide stresses which are on the conservative side.

Stresses beneath uniformly loaded areas *(Figure 5.5)*

A uniform applied pressure can be represented as a large number of point loads. Each of these loads will produce stresses at a point within the soil mass so integration of the Boussinesq equations will give the stress under a uniform pressure.

Figure 5.5 gives expressions for the stresses at points beneath a flexible strip and a circular loaded area.

If the uniform pressure is applied at some depth below the soil surface, then the methods described for a surface foundation can be used, assuming the soil surface to be at foundation level. This may give higher stresses than the theoretical values but considering the assumptions made it is prudent to adopt a conservative approach. The applied pressure q should, however, be the net applied pressure q_{net} obtained from:

$$q_{net} = \text{gross applied pressure} - \text{pressure of soil removed}$$

Bulbs of pressure *(Figure 5.6)*

Lines or contours of equal stress increase can be plotted from the equations available, given in Figure 5.5. Because of their shape they are referred to as bulbs of pressure.

They form the basis of the rule of thumb for depth of site investigation since, if a borehole is sunk to within the $0.2q$ contour, say, stresses can be expected to be small below this depth.

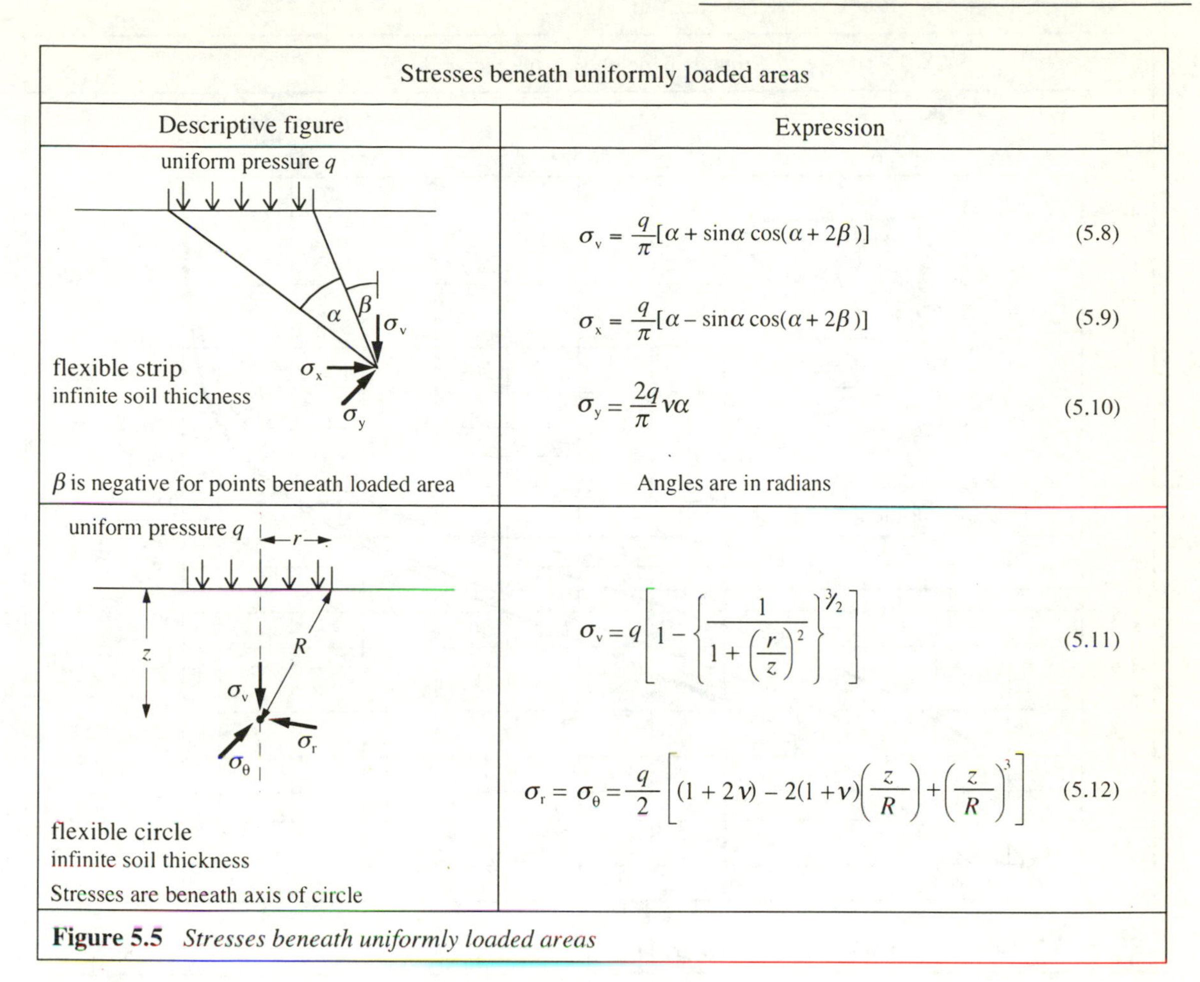

Stresses beneath uniformly loaded areas		
Descriptive figure	**Expression**	
uniform pressure q; flexible strip, infinite soil thickness; β is negative for points beneath loaded area	$\sigma_v = \frac{q}{\pi}[\alpha + \sin\alpha\cos(\alpha + 2\beta)]$	(5.8)
	$\sigma_x = \frac{q}{\pi}[\alpha - \sin\alpha\cos(\alpha + 2\beta)]$	(5.9)
	$\sigma_y = \frac{2q}{\pi}\nu\alpha$	(5.10)
	Angles are in radians	
uniform pressure q; flexible circle, infinite soil thickness; Stresses are beneath axis of circle	$\sigma_v = q\left[1 - \left\{\frac{1}{1+\left(\frac{r}{z}\right)^2}\right\}^{3/2}\right]$	(5.11)
	$\sigma_r = \sigma_\theta = \frac{q}{2}\left[(1+2\nu) - 2(1+\nu)\left(\frac{z}{R}\right) + \left(\frac{z}{R}\right)^3\right]$	(5.12)

Figure 5.5 *Stresses beneath uniformly loaded areas*

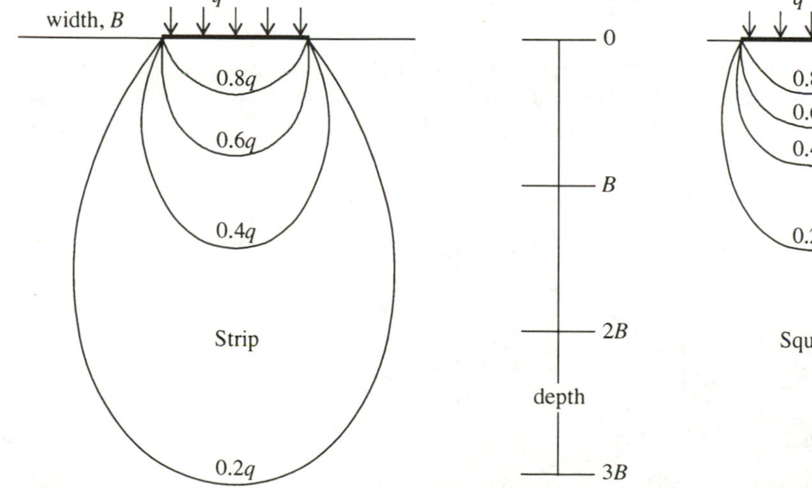

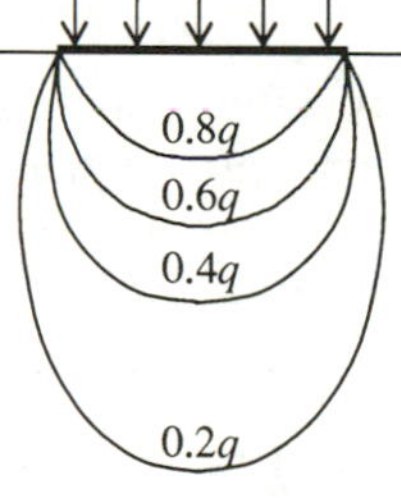

Figure 5.6 *Bulbs of pressure*

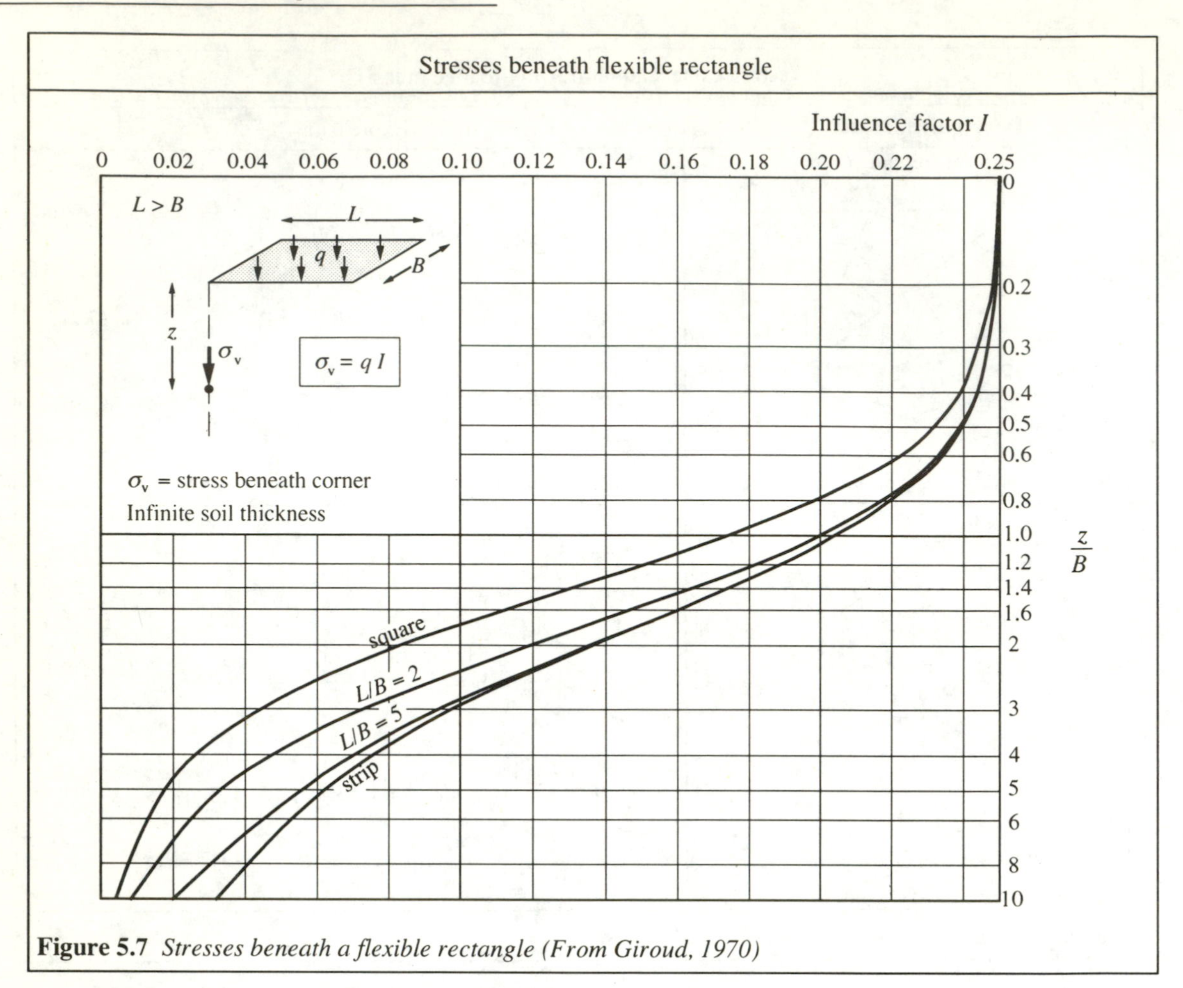

Figure 5.7 *Stresses beneath a flexible rectangle (From Giroud, 1970)*

Principle of superposition

The pressure applied is uniform but the stress distributed in the ground varies beneath the loaded area. The stress distribution methods give the stress at the *corner* of a loaded area so for points other than the corner the principle of superposition should be used. For the stress at the point × split the area into rectangles with their corners at the point ×.

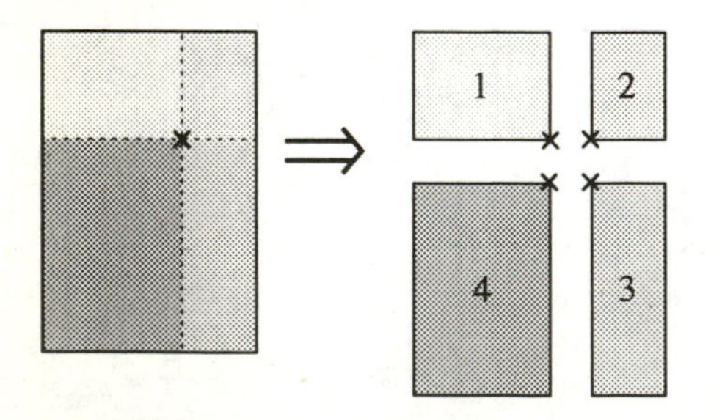

Stress at × = Sum of stresses at the corners of rectangles 1 + 2 + 3 + 4

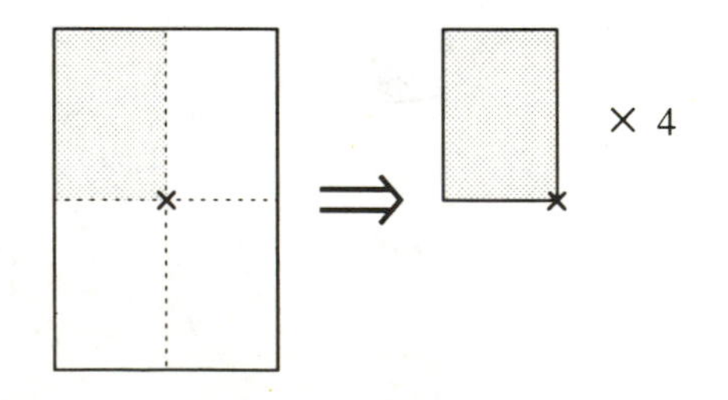

Maximum stress (at centre) = stress at the corner of a quarter foundation, multiplied by 4

Figure 5.8 *Principle of superposition*

Stresses beneath a flexible rectangle *(Figure 5.7)*
The vertical stress, σ_v, at a depth, z, beneath the corner of a flexible rectangle supporting a uniform pressure, q, has been determined using:

$$\sigma_v = qI \tag{5.13}$$

and influence factors I, given by Giroud (1970) are presented in Figure 5.7. They are for an infinite soil thickness. These curves are equivalent to the commonly used charts of Fadum (1948) but are easier to use.

Principle of superposition *(Figure 5.8)*
For stresses beneath points other than the corner of the loaded area the principle of superposition should be used, as described in Figure 5.8.

Stresses beneath flexible area of any shape *(Figure 5.9)*
Newmark (1942) devised charts to obtain the vertical stress at any depth, beneath any point (inside or outside) of an irregular shape. Use of the charts is explained in Figure 5.9.

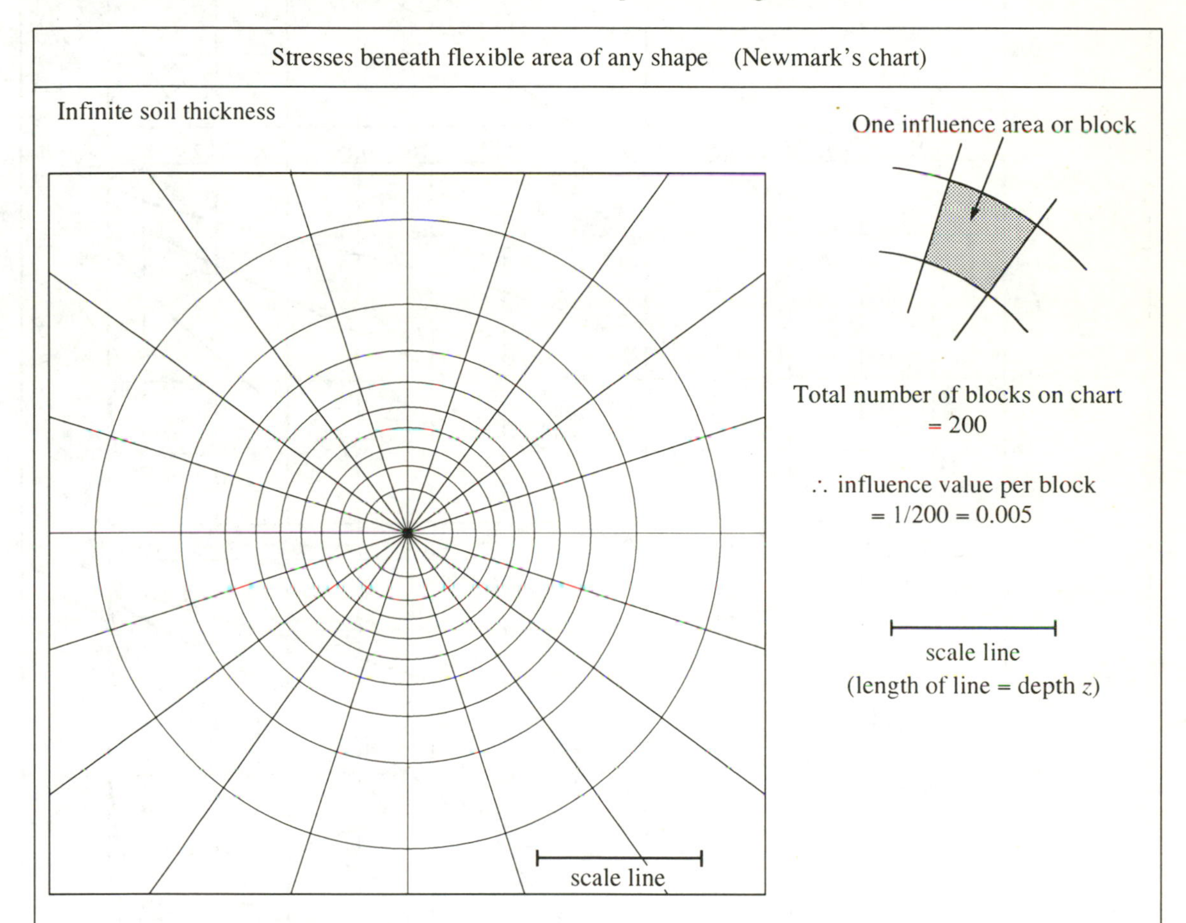

For the vertical stress σ_v at a depth z beneath any point × *on or outside a loaded area:*

1. Draw a plan sketch of the building outline on tracing paper such that the length of the scale line equals the depth z where the stress is required.
2. Place the scale drawing on the chart with the point × at the centre of the chart.
3. Count the number of blocks N covered by the scale drawing. Group together part blocks.
4. The vertical stress at the depth z and beneath the point × is given by $\sigma_v = 0.005\,N\,q$
5. The tracing can then be moved to other locations to obtain the stress beneath other points.

Figure 5.9 *Stress beneath flexible area of any shape (Newmark's chart)*

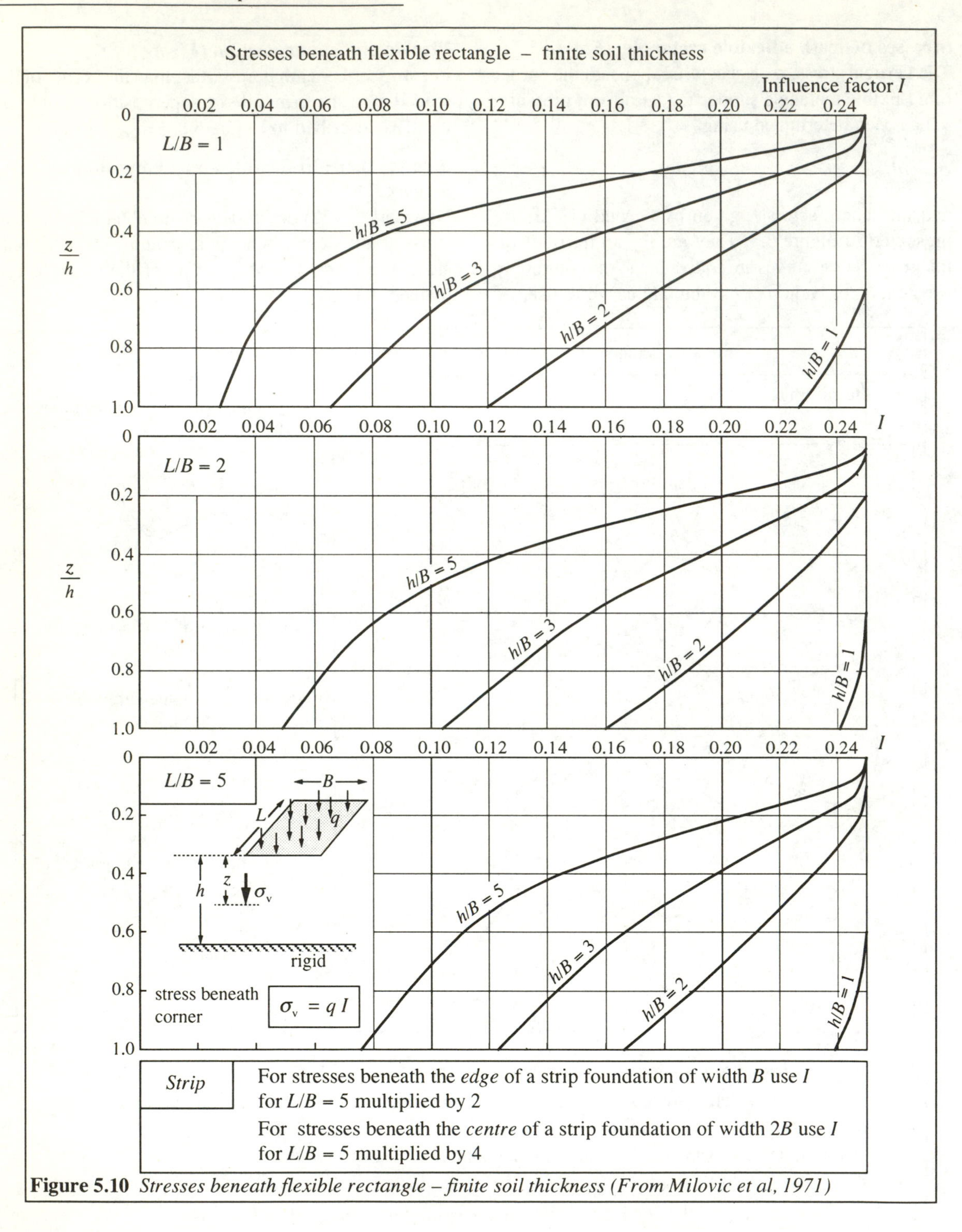

Figure 5.10 *Stresses beneath flexible rectangle – finite soil thickness (From Milovic et al, 1971)*

Stresses beneath flexible rectangle – finite soil thickness *(Figure 5.10)*

Figure 5.7 gives the stresses beneath a flexible rectangle for a deep soil layer (of infinite thickness). Where a rigid underlying deposit exists providing a finite soil thickness, the stresses given by Milovic *et al* (1971) can be used, see Figure 5.10. These give higher stresses than Figure 5.7, particularly in the lower levels of the compressible stratum.

The solutions of Fox (1948a) for a two-layer problem suggest that this 'rigid' stratum need only be about ten times stiffer than the compressible stratum for the finite soil layer case to apply.

Stresses beneath a rigid rectangle *(Figures 5.11 and 5.12)*

Although the stresses within the soil immediately beneath a rigid foundation (contact pressures) vary considerably, see Figure 5.1, the stresses at depth become more uniform as shown in Figure 5.11, taken from Butterfield and Banerjee (1971).

Vertical stresses beneath the centre of a rigid rectangle, given by Butterfield *et al* (1971), are presented in Figure 5.12, with a modification applied for the stresses at shallow depths, as suggested on Figure 5.11. If these stresses are taken as 'mean' stresses beneath a rigid foundation, they will probably be on the safe side.

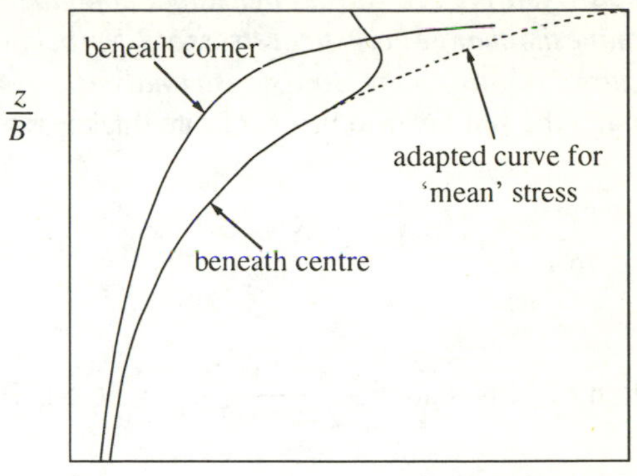

Figure 5.11 *Stresses beneath a rigid rectangle*

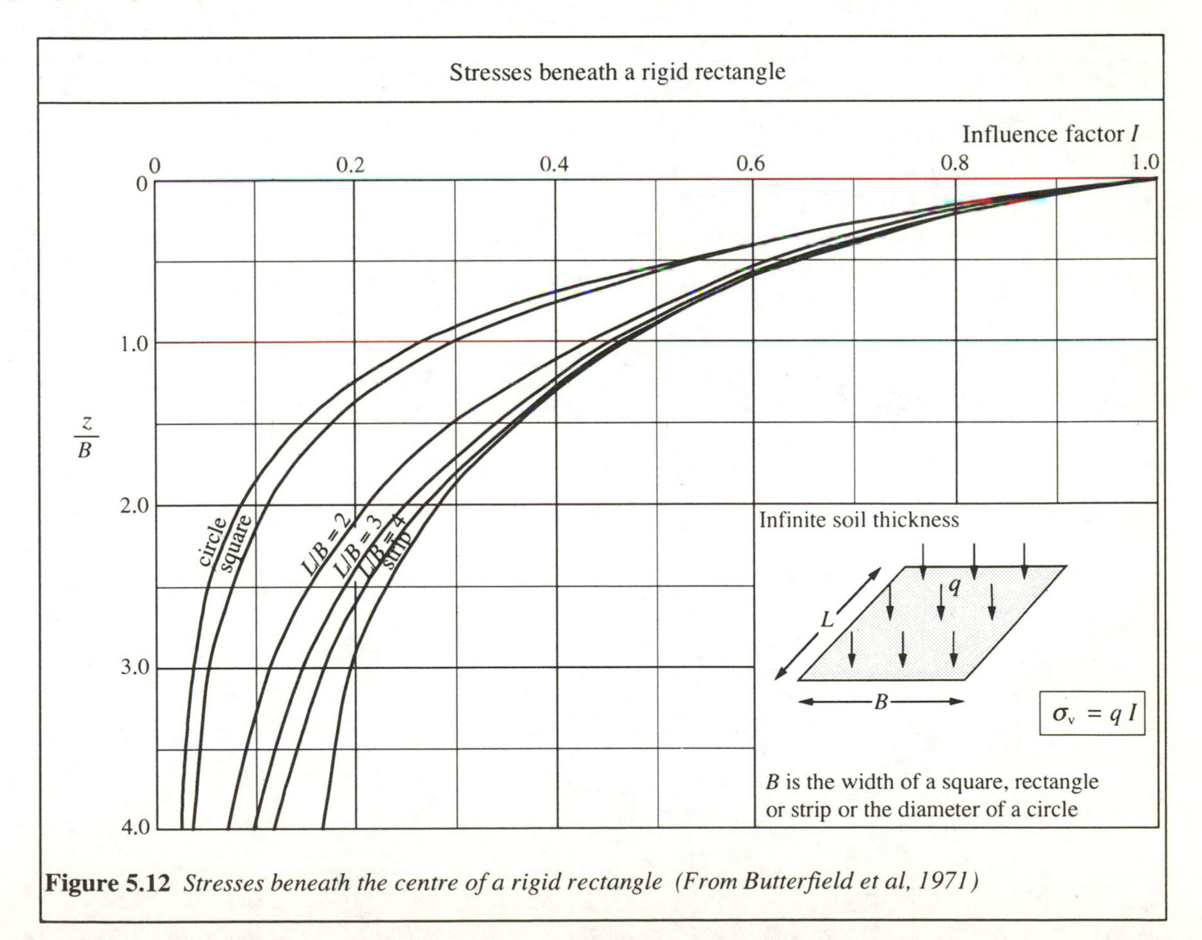

Figure 5.12 *Stresses beneath the centre of a rigid rectangle (From Butterfield et al, 1971)*

Worked Example 5.1 *Stress beneath a point load*

Determine the change in vertical stress at 2.5 m below ground level directly beneath the 870 kN point load, as shown on Figure 5.13, using the Boussinesq analysis.

Assuming the soil layer to be of infinite thickness the vertical stress is given by:

$$\sigma_v = \frac{3Pz^3}{2\pi R^5} \quad \text{(see Figure 5.3)}$$

$$\sigma_v \text{ from 870 kN load } = \frac{3 \times 870 \times 2.5^3}{2 \times \pi \times 2.5^5} = 66.5 \text{ kN/m}^2$$

$$\sigma_v \text{ from 640 kN load } = \frac{3 \times 640 \times 2.5^3}{2 \times \pi \times 4.30^5} = 3.2 \text{ kN/m}^2$$

$$\sigma_v \text{ from 560 kN load } = \frac{3 \times 560 \times 2.5^3}{2 \times \pi \times 5.15^5} = 1.2 \text{ kN/m}^2$$

$$\text{Total} = 70.9 \text{ kN/m}^2$$

Note the influence from each load.

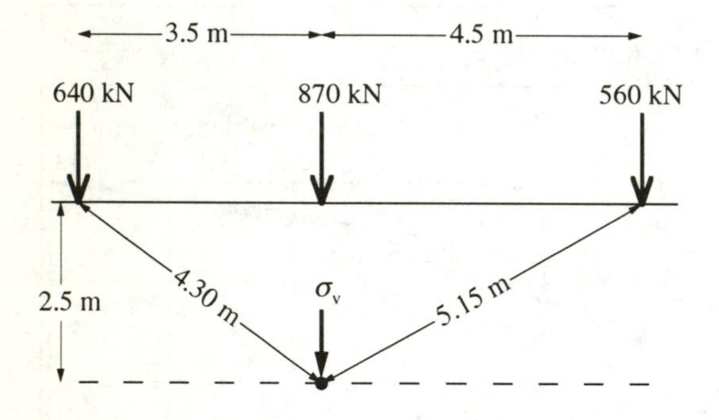

Figure 5.13 *Worked Example 5.1*

Worked Example 5.2 *Stresses beneath a strip load*
A strip foundation, 6 m wide, is uniformly loaded with a pressure of 100 kN/m². Determine the vertical stress distribution beneath and outside the strip at a depth of 3 m below ground level.
Values of α and β have been obtained by first determining ω, see Figure 5.14, for particular values of x, the distance from the centre-line.

x	ω°	β°	α°	α° rads	σ_v kN/m²
0	45.0	–45.0	90.0	1.571	81.8
1.0	53.1	–33.7	86.8	1.515	78.2
2.0	59.0	–18.4	77.4	1.351	66.6
3.0	63.4	0	63.4	1.107	48.0
4.0	66.8	18.4	48.4	0.845	28.9
5.0	69.4	33.7	35.7	0.623	15.6
6.0	71.6	45.0	26.6	0.464	8.4
7.0	73.3	53.1	20.2	0.353	4.7
8.0	74.7	59.0	15.7	0.274	2.8

The variation of stress is plotted on Figure 5.14.

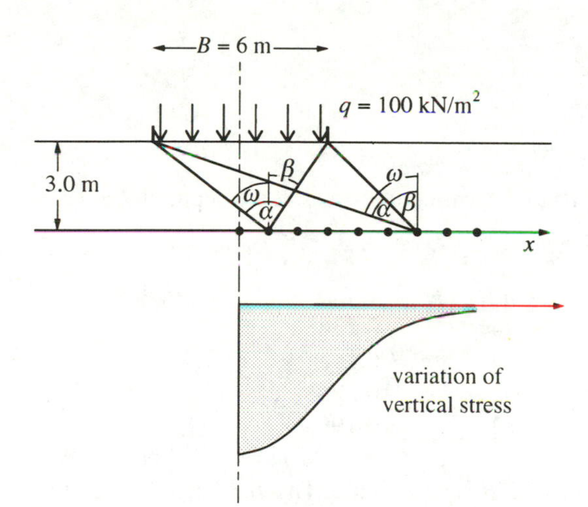

Figure 5.14 *Worked Example 5.2*

Worked Example 5.3 *Stresses beneath a flexible rectangle – infinite thickness*
Determine the vertical stress beneath the point O on a rectangular area, 6 m × 3 m, see Figure 5.15, at a depth of 2 m below ground level when uniformly loaded with a pressure of 50 kN/m².
Using the principle of superposition (Figure 5.8) the stress is required beneath the corner of four smaller areas 1 to 4. The stress values are determined from Figure 5.7.

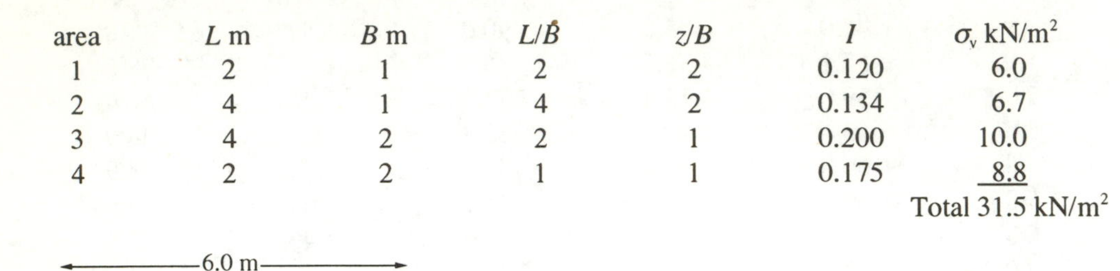

area	L m	B m	L/B	z/B	I	σ_v kN/m²
1	2	1	2	2	0.120	6.0
2	4	1	4	2	0.134	6.7
3	4	2	2	1	0.200	10.0
4	2	2	1	1	0.175	8.8
						Total 31.5 kN/m²

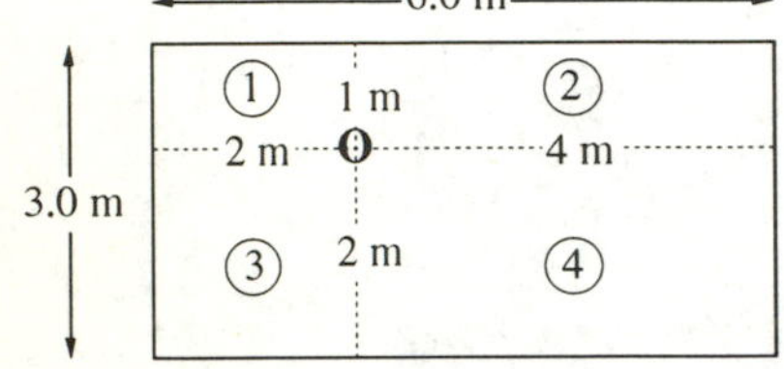

Figure 5.15 *Worked Example 5.3*

Worked Example 5.4 *Stresses beneath a flexible rectangle – finite thickness*
For the same foundation used in Example 5.3, use Figure 5.10 to determine the stress at the depth of 2 m below ground level assuming the layer thickness to be 4 m.
$z/h = 2/4 = 0.5$

area	L m	B m	L/B	h/B	I	σ_v kN/m²
1	2	1	2	4	0.138	6.9
2	4	1	4	4	0.148	7.4
3	4	2	2	2	0.224	11.2
4	2	2	1	2	0.196	9.8
						Total = 35.3 kN/m²

Linear interpolation between the curves for h/B and the charts of L/B is acceptable. This has been adopted for areas 1 and 2.

Worked Example 5.5 *Stresses beneath a rigid rectangle*
Determine the vertical stress beneath a rigid rectangular foundation, 6 m × 3 m, at a depth of 2 m below ground level when uniformly loaded with a pressure of 50 kN/m².
$L/B = 6/3 = 2$
$z/B = 2/3 = 0.67$
From Figure 5.12, $I = 0.55$
vertical stress = $0.55 \times 50 = 27.5$ kN/m²
(beneath centre)

Exercises

5.1 Three point loads of 500 kN lie at ground level in a straight line 4.0 m apart. Determine the increase in vertical stress at 2.0 m below ground level beneath the middle load.

5.2 Three parallel line loads of 500 kN/m lie at ground level 4.0 m apart. Determine the increase in vertical stress at 2.0 m below ground level beneath the middle load.

5.3 A flexible strip foundation 8.0 m wide lies at ground level and applies a uniform pressure of 80 kN/m². Determine the increase in vertical stress at 5.0 m below ground level:
a) beneath the centre of the foundation b) beneath its edge and c) 4.0 m away from the edge.

5.4 A flexible rectangle, which is 12 m long and 6 m wide, lies at ground level and applies a uniform pressure of 105 kN/m². Determine the increase in vertical stress at 6.0 m below ground level:
a) beneath the centre of the foundation and b) beneath its corner.
Assume the soil to be infinitely thick.

5.5 An L-shaped raft as shown in Figure 5.16 lies at ground level and applies a uniform pressure of 80 kN/m². Determine the increase in vertical stress at 8.0 m below ground level:
a) beneath the point A and b) beneath the point B.

5.6 Repeat Exercise 5.4, assuming the soil deposit to be 12 m thick.

5.7 In Exercise 5.4, assume the foundation to be rigid and determine the increase in vertical stress beneath the centre of the foundation.

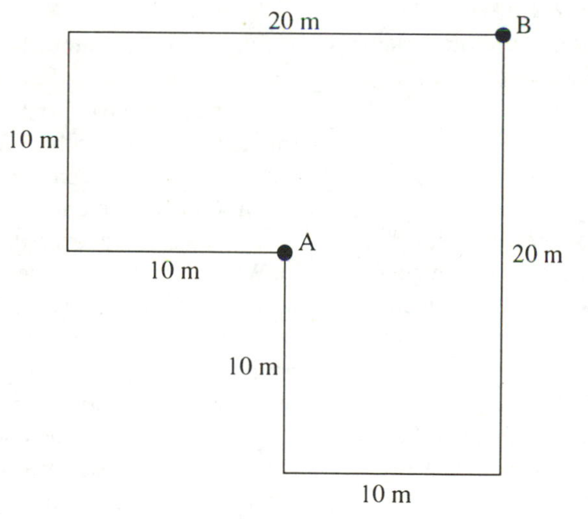

Figure 5.16 *Exercise 5.5*

6 Compressibility and Consolidation

Introduction

Compressibility and consolidation can be distinguished as:

- *compressibility* – volume changes in a soil when subjected to pressure
  ⇒ *amounts* of settlement
- *consolidation* – rate of volume change with time
  ⇒ *time* to produce a given settlement

These must be distinguished from compaction which is the expulsion of air from a soil by applying compaction energy.

Due to the insurmountable problems of obtaining good quality samples of sands from the ground this chapter concentrates on the behaviour of clays.

Soils of this type usually produce large amounts of settlement over a long period of time, after the end of construction. Sands generally produce smaller amounts of settlement in much less time, often during the construction period, so they may be of less concern. Nevertheless, some empirical methods of estimating settlements of structures on sands are given in Chapter 9.

Void ratio/effective stress plot *(Figure 6.1)*

Volume changes in a soil occur because the volume of voids changes. They are defined by the void ratio:

$$\text{void ratio} = \frac{\text{volume of voids}}{\text{volume of solids}} \tag{6.1}$$

and it is assumed that the volume of solids does not alter. The consolidation analogy in Chapter 4 describes the changes which occur in an element of soil when it is subjected to a change in effective stress.

Effective stress is the stress seated in the mineral grain structure or soil skeleton so if it changes the soil skeleton will respond by decreasing or increasing in volume as water is squeezed out of or drawn into the void spaces.

It has been found that there are recoverable and irrecoverable components to the volume changes of the soil skeleton and it is postulated that these are due to:

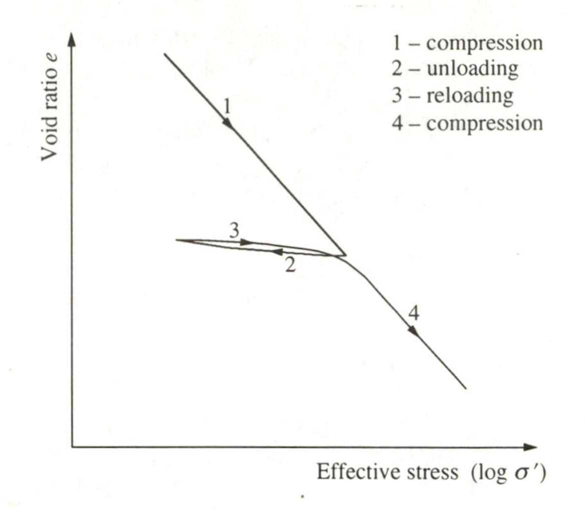

Figure 6.1 *Void ratio – effective stress*

- rearrangement of the soil particles (*a*) – this is permanent or irrecoverable
- elastic strains in the particles (*b*) – this is recoverable
- compression of bound water layers (*c*) – this is recoverable.

All three components will produce volume decrease but only *b* and *c* will allow volume increase.

A typical graph of void ratio versus effective stress (plotted as the logarithm) for a clay soil is shown in Figure 6.1. During compression *a*, *b* and *c* are occurring producing volume decrease, but on unloading only *b* and *c* are recovered.

When the soil is reloaded only *b* and *c* are overcome until the previous maximum pressure is achieved, when particle rearrangement *a* continues. Therefore, particle rearrangement only occurs when the soil lies on the steeper portions, whereas only elastic or recoverable strains occur when on the flatter portion.

The soil is described as normally consolidated when its state exists on the steeper portion (1 and 4) and overconsolidated when it occurs on the flatter portion (2 and 3). This process is related to the formation of a soil, its past deposition and erosion history, as described in Chapter 4.

Settlements of structures placed on or in the soil are related to volume changes, and these will be much smaller for an overconsolidated clay so it is essential to determine the present state of overconsolidation. Overconsolidated clays will generally have lower moisture contents and higher shear strengths.

Reloading curves *(Figure 6.2)*
The past geological and stress history has brought the soil to its present condition. If a sample is taken from the ground and placed in a laboratory consolidation apparatus, it will be possible to obtain a reloading curve by plotting the void ratio produced after the soil has consolidated to a new equilibrium after each change in effective stress.

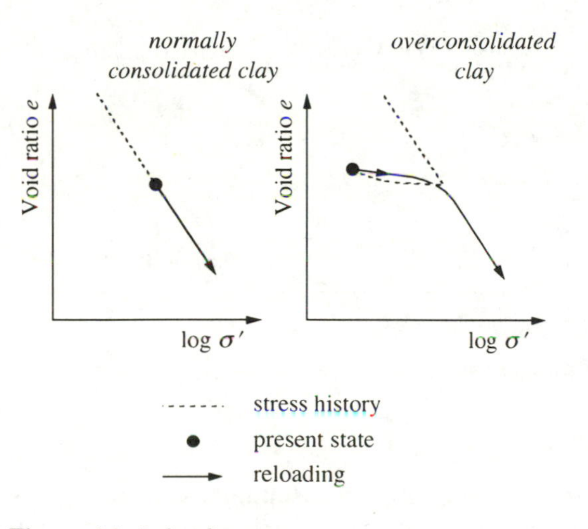

Figure 6.2 *Reloading curves*

The shape of the reloading curve (as well as any available geological information about the site) will help to determine whether the soil is normally consolidated or overconsolidated. If the soil is normally consolidated the reloading curve will continue on the virgin compression line from its present condition and will be a straight line.

An overconsolidated clay will have two portions, one commencing from its present condition up to the preconsolidation pressure followed by a steeper line corresponding to the virgin compression curve.

Preconsolidation pressure p_c' and overconsolidation ratio
The preconsolidation pressure is the previous maximum effective stress to which the soil has been subjected and is described in Chapter 4. It enables an assessment of the degree of overconsolidation using the overconsolidation ratio (OCR):

$$\text{OCR} = \frac{p_c'}{p_o'} \tag{6.2}$$

$$= \frac{\text{previous maximum effective stress}}{\text{present effective stress}}$$

For a normally consolidated clay the present effective stress is also the previous maximum so the OCR = 1. For a heavily overconsolidated clay the OCR may be 4 or more so this type of soil has been subjected to a much greater stress in the past compared to its present condition.

The significance of p_c' for an overconsolidated clay is that, if stresses are kept below this value, settlements can be expected to be small but if the applied stresses due to loading exceed this value, then large settlements will occur.

Casagrande method for p_c' *(Figure 6.3)*
From Figure 6.1 it is clear that p_c' occurs just to the right of the change in slope of the reloading curve of void ratio versus effective stress (plotted as $\log\sigma'$).

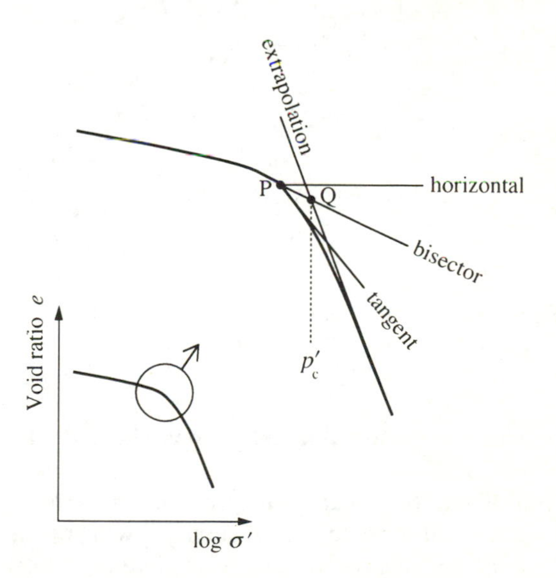

Figure 6.3 *Casagrande construction for p_c'*

Casagrande (1936) proposed a simple graphical method to determine p_c' from the laboratory reloading curve as follows.

1 Locate the point of maximum curvature P.
2 From point P draw a horizontal line and a tangent line and draw a bisector line between them.
3 Extrapolate upwards from the lower straight line to cut the bisector.
4 The point of intersection Q of the extrapolated line and the bisector gives the value of p_c'.

Effect of sampling disturbance *(Figure 6.4)*
Soil disturbance during sampling from the ground will alter and partially destroy the stable arrangement of soil particles and the forces or bonds acting between them. If the soil skeleton is disturbed it is found that a flatter void ratio-effective stress plot is obtained and, since disturbance during sampling from the ground is inevitable, especially for normally consolidated soils, the laboratory test will not truly reflect the *in situ* condition.

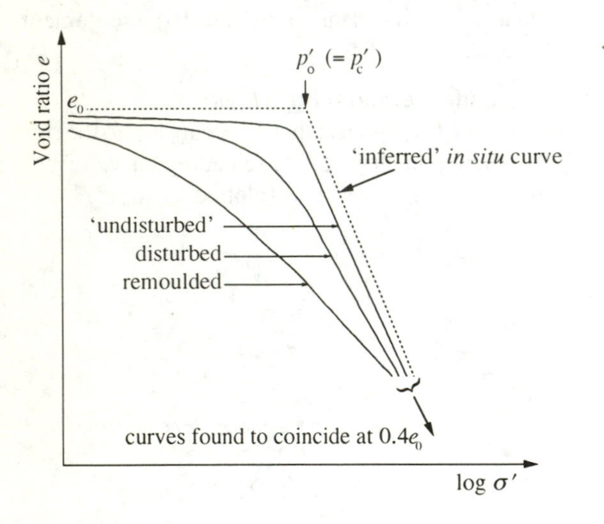

Figure 6.4 *Effect of sampling disturbance on normally consolidated clay*

***In situ* curve for normally consolidated clay** *(Figure 6.5)*
From Figure 6.4 it can be inferred that the *e* versus logσ' relationship for the *in situ* condition will lie to the right of the curve for the least disturbed soil. Schmertmann (1955) found that irrespective of the degree of disturbance all curves coincided at the point of $0.4e_0$ so it is reasonable to assume that the *in-situ* curve also passes through this point.

To plot the *in situ e* versus logσ' curve:

1 Carry out a consolidation test to produce a reloading curve with pressures applied which reduce the void ratio to $0.4e_0$ or at least sufficient values to enable extrapolation to $0.4e_0$.
2 Determine the present void ratio e_0 and effective stress p_0' and plot this as point A on the graph.
3 Plot the point B at $0.4e_0$ on the reloading curve.
4 The *in situ* curve is then plotted as the line A-B.

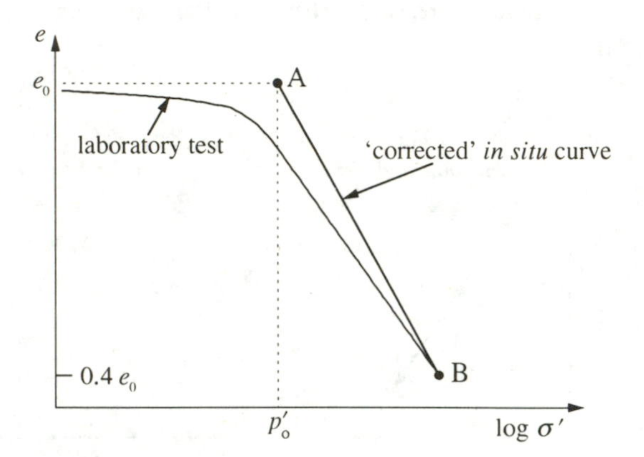

Figure 6.5 *In situ curve – normally consolidated clay*

***In situ* curve for overconsolidated clay** *(Figure 6.6)*
The *in situ* reloading curve will commence at the point (e_0, p_0') and it will have a flatter portion up to p_c', followed by a steeper portion when the soil returns to being normally consolidated. The construction of the *in situ* curve is based on the observations that:

1 All curves irrespective of disturbance pass through the point $0.4e_0$ (Schmertmann, 1955).
2 All unload/reload loops are parallel, irrespective of where they occur.

To plot the *in situ* void ratio versus logσ' curve, it is necessary to carry out a consolidation test to produce a reloading curve, with pressures applied beyond the preconsolidation pressure p_c', followed by an unload/reload loop (BC on Figure 6.6) and then applying pressures which reduce the void ratio to $0.4e_0$ or at least sufficient values to enable extrapolation to $0.4e_0$. The *in situ* curve is then constructed on this reloading plot.

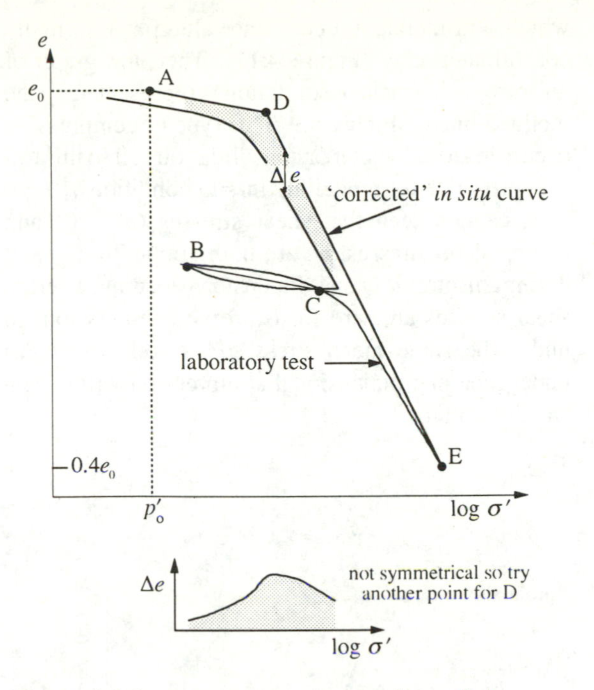

Figure 6.6 *In situ curve – overconsolidated clay*

1. Plot the point A at e_0, p_0' and from A draw a line parallel to the average unload/reload loop BC up to the p_c' value at D. p_c' may be obtained from the Casagrande construction, illustrated on Figure 6.3.
2. Plot the point E at $0.4e_0$ on the reloading curve and draw the line DE to complete the two-limb *in situ* curve.

Alternatively, if p_c' cannot be determined very accurately, choose a reasonable point for D and draw the lines AD and DE. Then Δe is the vertical difference between the trial lines and the laboratory test result and is plotted beneath versus $\log\sigma'$. When the Δe-$\log\sigma'$ curve is symmetrical it is considered that the point D represents the most likely value of p_c'.

Effect of load increments *(Figure 6.7)*

It has been found that the void ratio-$\log\sigma'$ curve can be made more distinct by reducing the increments between applied loads or pressures, especially for normally consolidated clays. This is very important when determining the preconsolidation pressure of lightly overconsolidated clays and the *in situ* curve.

The best quality results will be obtained with the least sample disturbance and small load increments. For some schemes it may be justified to carry out constant rate of strain (CRS) or constant rate of loading (CRL) tests although these would require modified versions of the consolidation test, usually using the Rowe hydraulic cell.

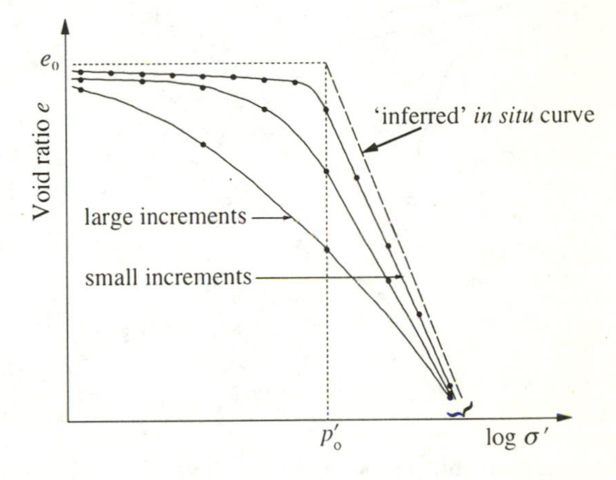

Figure 6.7 *Effect of load increments*

Isotropic compression *(Figure 6.8)*

This occurs when the three principal stresses are equal:

$$\sigma_1 = \sigma_2 = \sigma_3 \tag{6.3}$$

and hence

$$\sigma_1' = \sigma_2' = \sigma_3' \tag{6.4}$$

Since there are no shear stresses applied ($q' = \sigma_1' - \sigma_3' = 0$) it is sufficient to display isotropic compression as a plot of volume change versus mean stress. Here, the parameters used in Critical State Soil Mechanics terminology will be used, namely for volume changes:

$$\text{specific volume,} \quad v = 1 + e \tag{6.5}$$

and for mean stress or pressure p':

$$p' = \tfrac{1}{3}\left(\sigma_1' + \sigma_2' + \sigma_3'\right) \tag{6.6}$$

These can be plotted for both compression and swelling on the flat p'-v plane in Figure 6.8. This type of compression occurs in a triaxial compression apparatus when the cell pressure alone is applied.

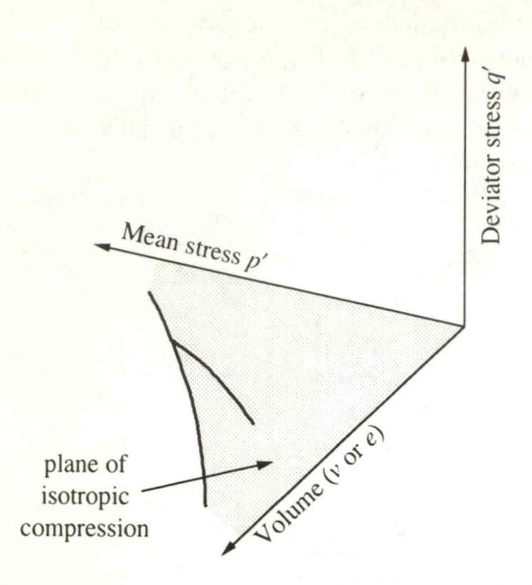

Figure 6.8 *Isotropic compression*

Anisotropic compression *(Figure 6.9)*
This occurs when:

$$\sigma_1' > \sigma_3' \tag{6.7}$$

producing shear stresses represented by the deviator stress q':

$$q' = \sigma_1' - \sigma_3' \tag{6.8}$$

Soil existing *in situ* will already have undergone one-dimensional compression (during deposition or loading) since any element of the soil will not have experienced horizontal strains because of the lateral confinement within the soil deposit. Strains, therefore, only occur in the vertical direction (one-dimensional) and so does pore water flow since it is assumed that there is no escape for it horizontally.

The relationship between vertical and horizontal effective stresses is given by the coefficient of earth pressure at rest K_o:

$$\sigma_H' = K_o \sigma_v' \tag{6.9}$$

so the ratio of q' and p' is given by:

$$\frac{q'}{p'} = \frac{3(\sigma_v' - \sigma_H')}{\sigma_v' + 2\sigma_H'} = \frac{3(1 - K_o)}{1 + 2K_o} \tag{6.10}$$

which will maintain a constant value for a normally consolidated clay (Figure 4.11). Then the graph of anisotropic normal consolidation is represented by the inclined line on Figure 6.9. This type of compression occurs in an oedometer or consolidation cell so this test is appropriate for modelling in-situ conditions.

It can be seen that shear stresses ($q' > 0$) and compression stresses, p', are both applied to the soil during anisotropic or one-dimensional loading, whereas shear stresses are zero for isotropic compression, so under the same mean stress, p', a soil which has undergone one-dimensional compression will have a smaller volume.

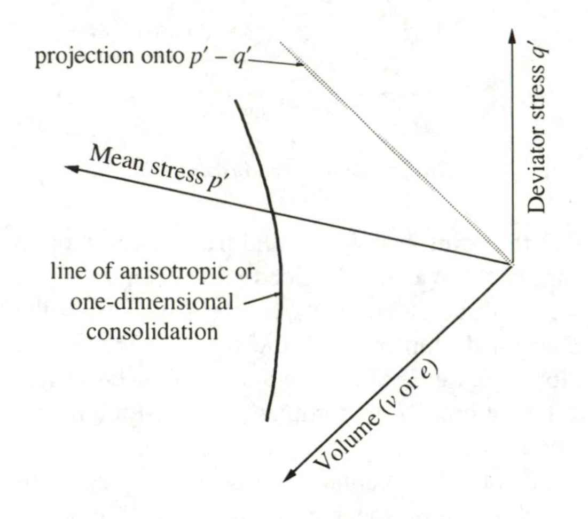

a) Representation in $p' - q' - v$ space

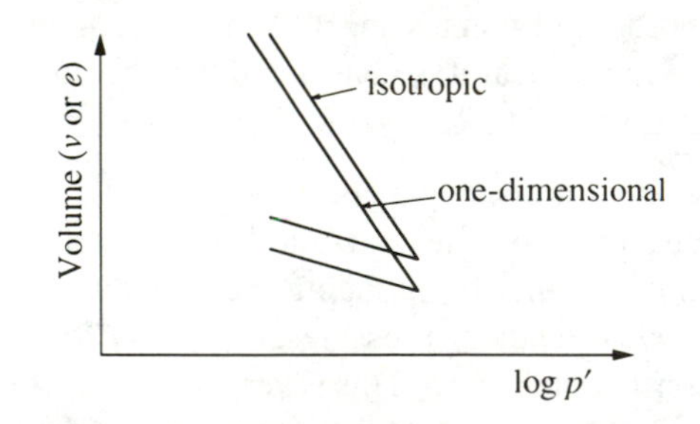

b) Projection onto v – log p' plane

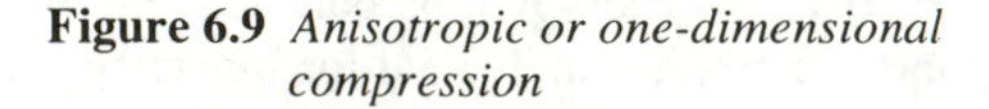

Figure 6.9 *Anisotropic or one-dimensional compression*

Terzaghi theory of one-dimensional consolidation *(Figure 6.10)*

The theory considers the rate at which water is squeezed out of an element of soil and can be used to determine the rates of:

a) volume change of the soil with time
b) settlements at the surface of the soil with time
c) pore pressure dissipation with time.

Several assumptions are made but the tenuous ones are:

1 Compression and flow are one-dimensional (vertical only).
2 Darcy's Law is valid at all hydraulic gradients (deviation may occur at low hydraulic gradients).
3 k and m_v remain constant (whereas they both usually decrease during consolidation).
4 No secondary compression occurs.
5 The load is applied instantaneously and over the whole of the soil layer.

Consider an element of soil in a consolidating layer (Figure 6.10):

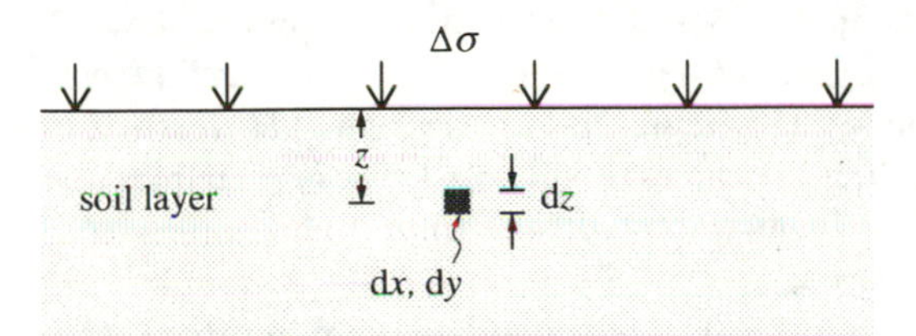

a) Element in a consolidating layer

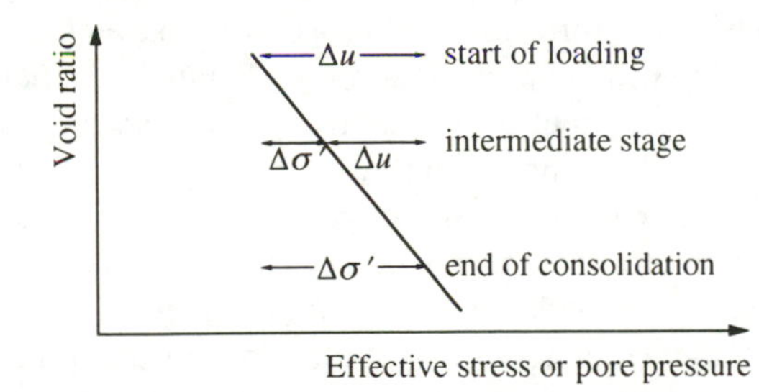

b) Changes in pore pressure and effective stress during consolidation

Figure 6.10 *One-dimensional consolidation*

The hydraulic gradient across the element is:

$$-\frac{\partial h}{\partial z} = \frac{1}{\gamma_w}\frac{\partial^2 u}{\partial z^2}[\partial u = \gamma_w \partial h] \qquad (6.11)$$

The average velocity of water passing through the element, from Darcy's Law:

$$v = -\frac{k}{\gamma_w}\frac{\partial u}{\partial z} \qquad (6.12)$$

The velocity gradient across the element:

$$\frac{\partial v}{\partial z} = -\frac{k}{\gamma_w}\frac{\partial^2 u}{\partial z^2} \qquad (6.13)$$

From the equation of continuity (Equation 3.23), if volume changes in the soil element are occurring the volume change per unit time can be expressed as:

$$\frac{dV}{dt} = \frac{\partial v}{\partial x}dz dx dy \qquad (6.14)$$

$$\frac{dV}{dt} = \frac{-k}{\gamma_w}\frac{\partial^2 u}{\partial z^2}dz dx dy \qquad (6.15)$$

This can now be equated to the volume change of the void space in the element. The total volume of the element = $dx\,dy\,dz$

The proportion of voids in the element is:

$$\frac{e}{1+e_0}dx dy dz \qquad (6.16)$$

The rate of change of void space with respect to time is then:

$$\frac{1}{1+e_0}\frac{\partial e}{\partial t}dx dy dz \qquad (6.17)$$

$$= \left(\frac{1}{1+e_0}\right)\frac{\partial e}{\partial \sigma'}\frac{\partial \sigma'}{\partial t}dx dy dz \qquad (6.18)$$

$$= -m_v\frac{\partial u}{\partial t}dx dy dz \qquad (6.19)$$

Equating 6.15 and 6.19, and dividing by $dx\,dy\,dz$ gives:

$$\frac{\partial u}{\partial t} = \left(\frac{k}{m_v \gamma_w}\right)\frac{\partial^2 u}{\partial z^2} = c_v\frac{\partial^2 u}{\partial z^2} \qquad (6.20)$$

where:
c_v is the coefficient of consolidation
m_v is the coefficient of compressibility
k is the coefficient of permeability

Solution of the consolidation equation
(Figures 6.11 and 6.12)
The basic differential equation of consolidation (Equation 6.20) can be expressed in dimensionless terms as:

$$\frac{dU_v}{dT_v} = \frac{d^2U_v}{dZ^2} \tag{6.21}$$

where:

- $Z = z/d$ – dimensionless depth
- z denotes where the soil element lies within the soil layer
- d is the length of drainage path and represents the *maximum* distance a molecule of water would have to travel to escape from the soil layer. This is illustrated in Figure 6.11.

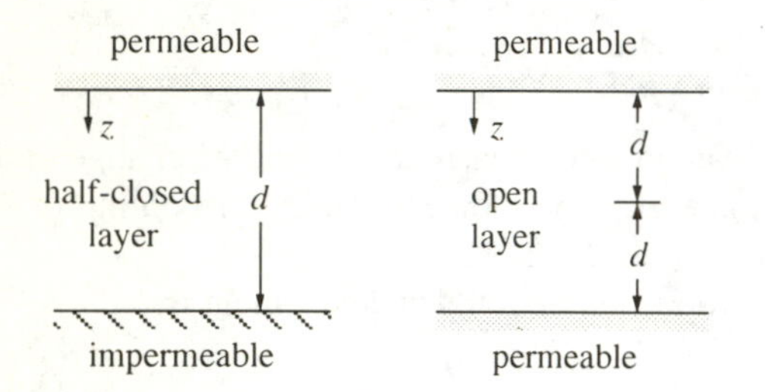

Figure 6.11 *Drainage path length d*

- T_v is a dimensionless time factor:

$$T_v = \frac{c_v t}{d^2} \tag{6.22}$$

c_v is the coefficient of consolidation, as above.
t is the time elapsed after instantaneous loading.

- U_v is the degree of consolidation. It can be represented in three ways. One is the amount of void ratio change which has occurred at the time t compared to the final void ratio change.

$$U_v = \frac{e_0 - e_t}{e_0 - e_f} \tag{6.23}$$

where:
e_0 = initial void ratio
e_t = void ratio at time t
e_f = final void ratio.

It can also be given as the amount of pore water pressure dissipated at the time t compared to the initial pore pressure increase

$$U_v = \frac{\Delta u_i - \Delta u_t}{\Delta u_i} \tag{6.24a}$$

where:
Δu_i = initial excess pore pressure
Δu_t = excess pore pressure at time t
and as the effective stress increase at the time t compared to the final increase in effective stress:

$$U_v = \frac{\Delta\sigma_t'}{\Delta\sigma_f'} \tag{6.24b}$$

where:
$\Delta\sigma_t'$ = increase in effective stress at time t
$\Delta\sigma_f'$ = final increase in effective stress.
The solution to the dimensionless one-dimensional consolidation equation is given as:

$$U_v = 1 - \sum_{m=0}^{m=\infty} \frac{2}{M}\sin(MZ)e^{-M^2T_v} \tag{6.25}$$

where $M = \frac{\pi}{2}(2m+1)$

Although it looks like a daunting expression it merely relates the three parameters U_v, Z and T_v and these are conveniently represented as a graph of U_v versus Z for different T_v values, Figure 6.12. The curved lines refer to constant values of time (or T_v) and so are called isochrones.

From $Z = 0$ to 2 the diagram represents the state or degree of consolidation at any point in an open soil layer whereas half of the diagram, $Z = 0$ to 1 or 1 to 2 would represent a half-closed layer. The diagram also relates to a uniform distribution of excess pore pressure Δu set up within the soil (Case A on Figure 6.13). For triangular distributions (Cases B and C) the isochrones would not be symmetrical.

The line $T_v = 0$ on Figure 6.12 represents the instantaneous loading condition (time $t = 0$) where $U_v = 0$ since consolidation has not yet commenced. Very soon after commencement of the consolidation process, say $T_v = 0.05$, an element of soil adjacent to the permeable boundaries will have been able to fully consolidate (when $U_v = 1.0$) but for a soil element at the middle of the soil layer consolidation will not have started.

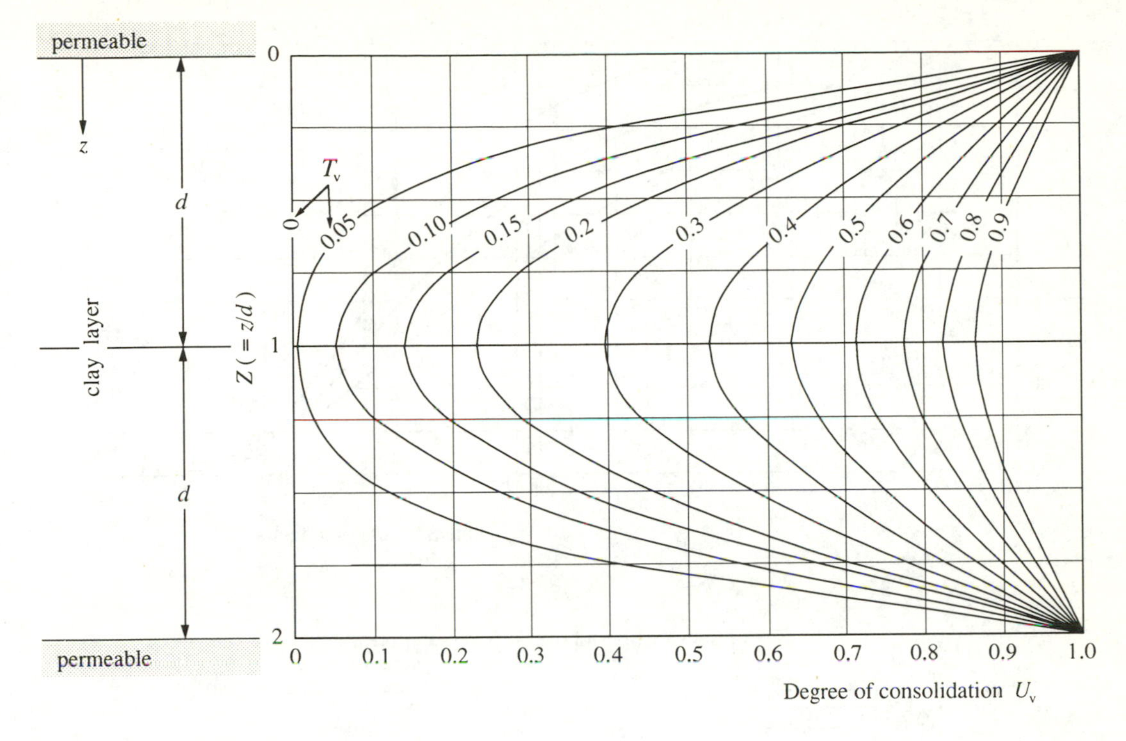

Figure 6.12 *One-dimensional consolidation – Isochrones*

Average degree of consolidation *(Figure 6.13)*
When dealing with settlements of loaded areas placed on the surface of a soil layer the average degree of consolidation at a particular time is required, independent of depth z. This is obtained by integrating with respect to z for a particular T_v value and gives:

$$\overline{U}_v = 1 - \sum_{m=0}^{m=\infty} \frac{2}{M^2} e^{-M^2 T_v} \tag{6.26}$$

which is an expression relating the average degree of consolidation, $\overline{U}_v$ to T_v. The average degree of consolidation is given in terms of settlement:

$$\overline{U}_v = \frac{\rho_t}{\rho_c} \tag{6.27}$$

where
ρ_t = settlement at time t
ρ_c = final consolidation settlement.
There are three relationships between $\overline{U}_v$ and T_v (Curves 1, 2 and 3 on Figure 6.13) depending on the initial variation of excess pore water pressure set up within the soil layer and the permeability of the soil boundaries. The three cases could be produced by:

Case A – wide or extensive applied pressure compared to the thickness of the soil layer, such as an embankment, or general lowering of the water table.
Case B – applied pressure over a small area such as a foundation.
Case C – self-weight of soil forming an embankment.
Values of $\overline{U}_v$ and T_v are given in Table 6.1 and good approximations for the most commonly adopted cases (Curve 1) are:

$$T_v = \frac{\pi}{4}\left(\frac{\overline{U}_v\%}{100}\right)^2 \quad \text{for } \overline{U}_v < 60\% \tag{6.28}$$

and

$$T_v = 1.781 - 0.933\log_{10}\left(100 - \overline{U}_v\%\right) \quad \text{for } \overline{U}_v > 60\% \tag{6.29}$$

Oedometer test *(Figure 6.14)*
The recommended test procedure is described in BS 1377:1990 and a cross-section through the apparatus is shown on Figure 6.14. The test was designed to reproduce one-dimensional consolidation by providing:

1. a rigid confining ring to prevent lateral strains (K_o conditions) and lateral drainage. The inner surface must be smooth and coated with low friction mate-

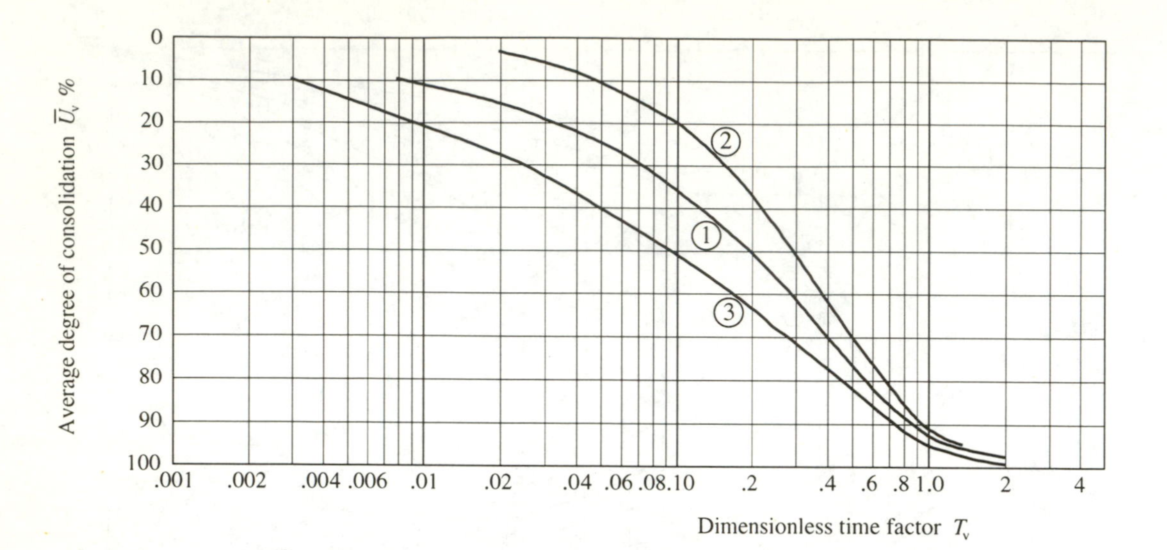

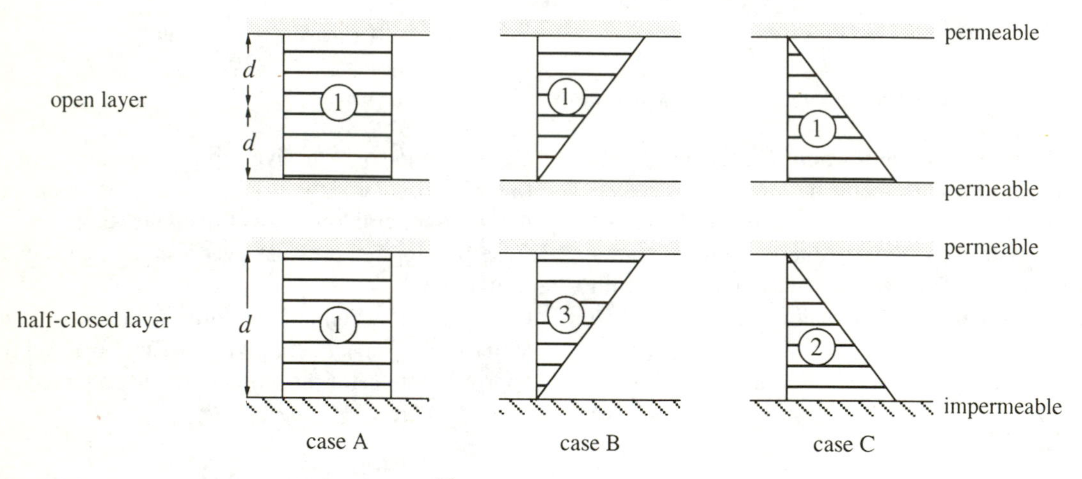

Figure 6.13 *Average degree of consolidation $\overline{U}_v$ versus T_v*

rial or grease to minimise wall friction as the specimen reduces in thickness.

2 porous stones top and bottom to act as permeable boundaries. Coupled with a relatively thin specimen this will provide a small drainage path length and therefore enable consolidation under each increment of applied stress to be completed in a manageable time of 24 hours. A typical specimen height is 19 or 20 mm. Any thicker and the ring could produce excessive friction during loading, excessive disturbance during specimen preparation and increase the time required for consolidation. On the other hand, the specimen should be at least 5 times thicker than the largest particle diameter to avoid edge effects.

3 a rigid loading platen to provide equal settlement.

4 a water container to immerse the specimen and porous stones to ensure full saturation. All of the metallic components must therefore be non-corrodible. The water is added after installation of the specimen and assembling the load frame since some soils may have a tendency to swell in the presence of water and others may settle rapidly due to structural collapse on wetting. This behaviour should be investigated separately.

Table 6.1
Values of T_v

U_v %	T_v		
	Curve 1	Curve 2	Curve 3
10	0.008	0.047	0.003
20	0.031	0.100	0.009
30	0.071	0.158	0.024
40	0.126	0.221	0.048
50	0.196	0.294	0.092
60	0.287	0.383	0.160
70	0.403	0.500	0.271
80	0.567	0.665	0.440
90	0.848	0.940	0.720

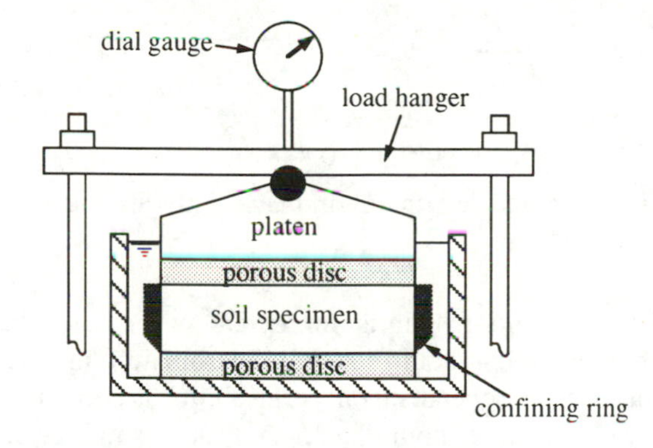

Figure 6.14 *Oedometer test apparatus*

Loading is applied through a loading yoke or load hanger and a counter-balanced lever system with a load ratio of around 10:1 so that a relatively small mass in kg can produce the large stresses required on the specimen.

An initial pressure is applied to return the specimen to near to its in situ effective stress to act as a starting point for reloading or swelling. This stress will depend on the stress history the soil has been subjected to *in situ*.

Each pressure increment requires 24 hours application so to make the test manageable and economical four to six increments of loading and one or two larger unloading increments are normally chosen using the following suggested sequence:

6, 12, 25, 50, 100, 200, 400, 800, 1600, 3200 kN/m^2

although smaller increments may be adopted for samples taken from shallow depths and for small applied stresses from the structure. If the soil is overconsolidated loading should continue into the normal consolidation region so that a measure of the preconsolidation pressure can be obtained.

The change in thickness of the specimen (either compression or swelling) is measured using a dial gauge or a displacement transducer impinging on the loading yoke. Readings must be taken at frequent intervals initially when the specimen is rapidly changing in thickness followed by less frequent intervals. This should give sufficient data when plotting thickness against square root or log time.

Nowadays, it is much more convenient to use displacement transducers attached to an automatic recording device so that continual readings are obtained irrespective of the working hours of the laboratory technicians. These readings can then be plotted to assess whether primary consolidation has been completed when it will be permissible to apply the next increment of pressure. This may occur within the same day for some soils.

From the oedometer test result a number of parameters can be determined:

1 the initial moisture content, bulk density, dry density and degree of saturation
2 void ratio at the end of each pressure increment
3 compression index, C_c for normally consolidated clay
4 coefficient of volume compressibility, m_v, for each loading increment, (not unloading) above p_o'.
5 coefficient of consolidation, c_v, for each loading increment above p_o'.

The initial moisture content w_0% can be obtained from the trimmings around the specimen.

$$\text{bulk density} \quad \rho = \frac{m_0 \times 100}{AH_0} \tag{6.30}$$

where:
m_0 = initial mass of specimen (grams)
A = area of specimen (mm^2)
H_0 = initial height of specimen (mm)

$$\text{dry density} \quad \rho_d = \frac{\rho \times 100}{100 + w_0(\%)} \tag{6.31}$$

$$\text{initial void ratio} \quad e_0 = \frac{\rho_s}{\rho_d} - 1 \tag{6.32}$$

where:
ρ_s = particle density (Mg/m^3)

Initial degree of saturation

$$S_{r0} = \frac{w_0 \rho_s}{e_0} \tag{6.33}$$

Void ratio at the end of each load increment

$$e_f = e_0 - \frac{\Delta H}{H_0}(1 + e_0) \tag{6.34}$$

where, for each particular pressure increment
e_0 = initial void ratio
e_f = final void ratio
H_0 = initial thickness (mm)
ΔH = change in thickness (mm)
The coefficient of volume compressibility m_v with units of m²/kN is given by:

$$m_v = \frac{e_0 - e_f}{1 + e_0} \frac{1}{\Delta\sigma_i} \tag{6.35a}$$

$$\text{or } m_v = \left(\frac{\Delta H_i}{H_i}\right)\frac{1}{\Delta\sigma_i} \tag{6.35b}$$

where:
ΔH_i = change in thickness over the pressure increment considered (mm)
H_i = initial thickness for the pressure increment (mm)
$\Delta\sigma_i$ = pressure increment (kN/m²)
The compression index is given by:

$$C_c = (e_1 - e_2)\log_{10}\frac{\sigma_1'}{\sigma_2'} \tag{6.36}$$

where:
e_1 is the void ratio at the effective stress σ_1' and
e_2 is the void ratio at σ_2'.
C_c represents the gradient of the void ratio-stress graph for normally consolidated clay only.

Coefficient of consolidation, c_v – root time method *(Figure 6.15)*

Taylor (1948) showed that the Terzaghi theory of one-dimensional consolidation gave a straight line when U_v was plotted against the square root of T_v, at least up to U_v = 60%, and that at U_v = 90% the theoretical curve occurred at 1.15 times the extrapolated straight line portion. On Figure 6.15 the length AC = 1.15 × length AB. The process of consolidation relates only to pore water pressure dissipation and is referred to as *primary* consolidation.

The experimental curve often produces a straight line portion when dial gauge readings are plotted against the square root of time but deviates at the beginning and end due to:

1 initial compression produced by bedding of the porous stones, compression of air or gas bubbles which came out of solution following sampling.
2 secondary compression recorded as continuing volume decrease even after all measurable pore pressures have fully dissipated.

In order to apply the consolidation theory to the laboratory test result the 'root time curve-fitting method' can be used as follows.

1 on the dial gauge readings versus square root time plot draw a best fit line through the straightest portion and extrapolate back to a corrected origin R_o to eliminate the initial compression.
2 draw a straight line from R_o at a gradient 1.15 times the gradient of the straight portion of the experimental plot. Where this line cuts the experimental curve is considered to be the point at 90% degree of consolidation.
3 determine the value of $\sqrt{t_{90}}$ and hence t_{90}
4 the coefficient of consolidation, c_v, is then given as:

$$c_v = \frac{T_{v90}\bar{d}^2}{t_{90}} \tag{6.37}$$

where:
$T_{v90} = T_v$ at U_v = 90%, i.e. 0.848
$\bar{d}$ = average length of drainage path for the load increment (= $\bar{H}/2$).

The conventional units for c_v are m²/year so the above equation should be modified from mm²/minute. Since c_v is dependent on permeability a correction for temperature should be applied such as that given in Chapter 3, in Figure 3.14. However, the British Standard states that this correction is not justified in view of the inaccurate and unrepresentative value obtained from the test, see *in situ* c_v values, below.

In practice, the root time method is attempted first because it requires readings over a much shorter time period and less judgement is required. The log-time method is used when a straight line cannot easily be obtained with the root time method.

Coefficient of consolidation, c_v – log-time method *(Figure 6.16)*

This method is due to Casagrande who observed that the graph of U_v versus $\log T_v$ had three portions:

1 an initial portion with a shape very similar to a parabola

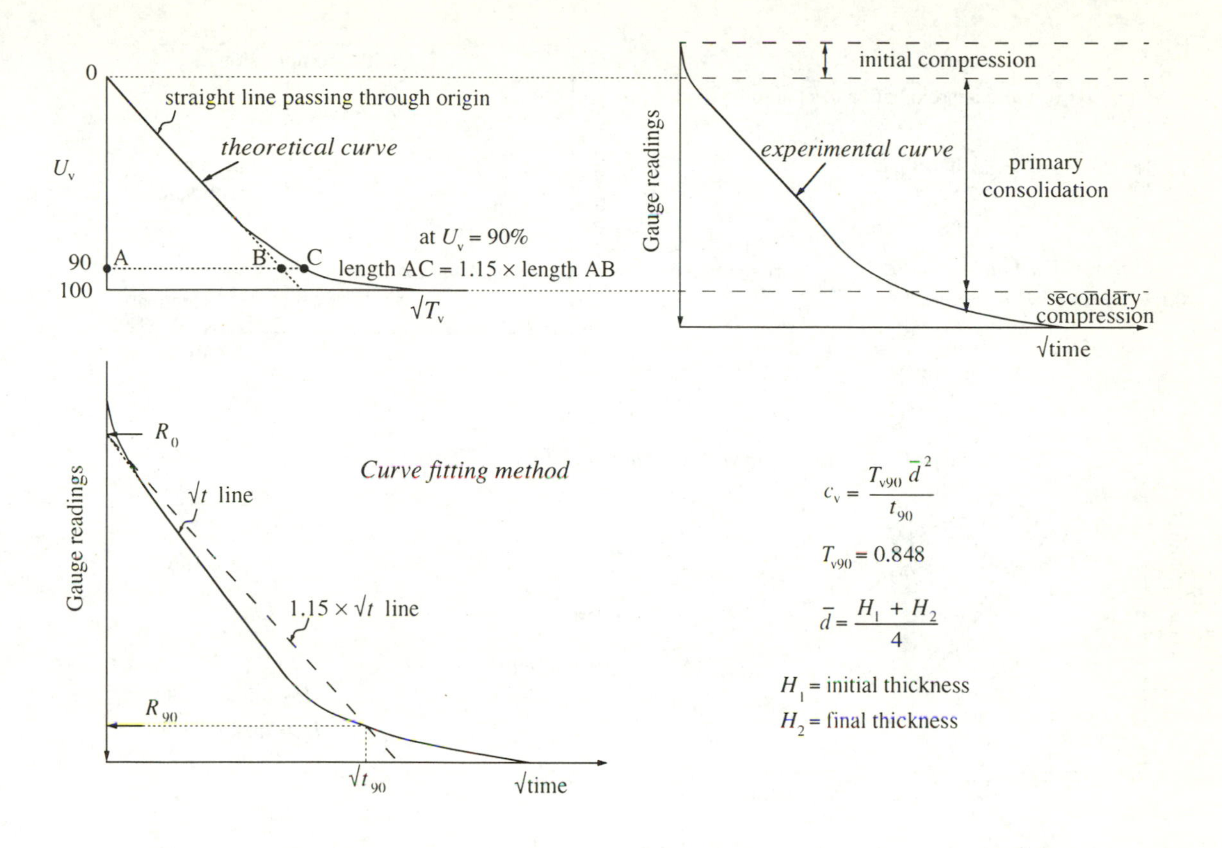

Figure 6.15 *Root time curve fitting method*

2 a linear middle portion
3 a final portion asymptotic to the horizontal.

This relationship applies only to primary consolidation so Casagrande devised two constructions to eliminate the initial compression and secondary compression from the experimental curve of dial gauge readings versus log-time.

Using the properties of a parabola, the initial part is corrected by choosing two points A and B (any two points) at values of time t in the ratio 1:4. The vertical distance between these points D is marked above point A to give the initial reading R_o for primary consolidation. This is the most uncertain part of the method, since the experimental curve is not always parabolic and often the number of data points which can be obtained in the short initial time available is not sufficient to define the curve accurately. The Taylor method gives a more definite initial reading.

For soils which display very little secondary compression, such as heavily overconsolidated clays, the final part of the experimental curve will be a horizontal line which will then define the end reading for primary consolidation R_{100}.

For soils displaying secondary compression, it is commonly found that compressions continue in a straight line on the log graph, so this line is extrapolated back to meet the extrapolated straight middle portion. Where these lines intersect is the final reading for primary consolidation R_{100}. This enables the point R_{50} corresponding to a 50% degree of consolidation to be located mid-way between R_o and R_{100}. The value of t_{50} is obtained and the coefficient of consolidation, c_v is determined from:

$$c_v = \frac{T_{v50}\bar{d}^2}{t_{50}} \qquad (6.38)$$

where

$T_{v50} = T_v$ at U_v = 50%, i.e. 0.196

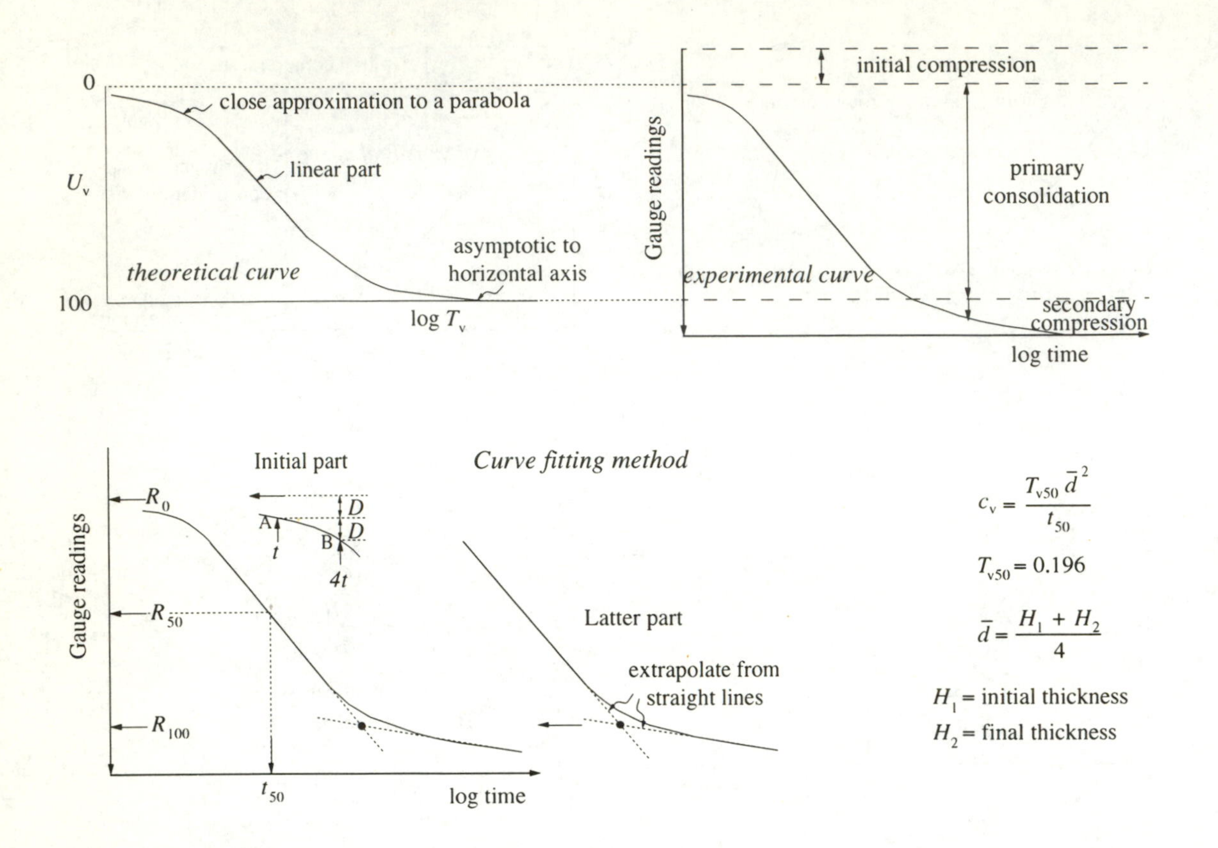

Figure 6.16 *Log time curve fitting method*

In situ c_v values

Calculations for rates of settlement of structures on clay soils, based on laboratory c_v values from the oedometer test, have been found to be grossly in error when compared with the actual observed rates of settlement (Rowe, 1968). These differences are due to the influence of the macro-fabric found in most clay soils but which is not represented in the small scale oedometer test. Macro-fabric or structure in a clay can be:

- horizontal – laminations, layers of peat, silt and fine sand, varves
- vertical – root holes
- inclined – fissures, silt-filled fissures.

This will reduce the time required for consolidation by decreasing the drainage path lengths and increasing the mass permeability of the clay.

To overcome this problem, larger, more representative samples must be tested, for which the Rowe cell was developed, see below. Alternatively, *in situ* permeability tests, such as the constant head piezometer test (Chapter 3), to give an *in situ* k value, can be combined with a laboratory oedometer m_v value to give an *in situ* c_v value:

$$c_{v\,\text{in situ}} = \frac{k_{\text{in situ}}}{m_{v\,\text{lab}}\,\gamma_w} \qquad (6.39)$$

since it has been found that these values are more reliable (Bishop and Al-Dhahir, 1969).

Rowe consolidation cell *(Figure 6.17)*

This test was developed at the University of Manchester (Rowe and Barden, 1966) to improve the quality of data obtained from a consolidation test and to eliminate several of the disadvantages of the oedometer test. The test procedure is detailed in BS 1377: 1990, where the apparatus is described as a hydraulic consolidation cell.

The specimen is enclosed by a cell base, the cell body around and a convoluted rubber membrane above which moves up or down as the specimen changes in thickness. A total stress increment is applied outside the specimen by the constant pressure supply which must be able to compensate for leakages and changes in specimen thickness. Internally, a back pressure may be applied and the pore water pressure can be measured. A more detailed version is illustrated in Figure 3.19.

Drainage from the specimen can be either vertical or radial (Figure 6.17) and as the specimen consolidates the change in thickness is recorded by a dial gauge or a displacement transducer which follows the movement of the hollow drainage spindle.

The main advantages compared to the oedometer are:

1. Large specimens can be tested to ensure that *in situ* macro-fabric is represented in the test. Specimens 250 mm diameter and 100 mm thick are considered to be sufficiently representative to give a good measure of the *in situ* c_v.
2. Pore water pressure can be measured. This is useful to ensure that pore pressures have uniformly increased within the specimen following application of the external total stress increment with the drainage valve closed, i.e. before consolidation commences. During consolidation pore pressure will dissipate and the graph of degree of consolidation U (= pore pressure dissipated/initial pore pressure) versus log-time can be used directly to obtain c_v. In the study of secondary compression, the complete dissipation of pore pressures can be taken as the commencement of this process.

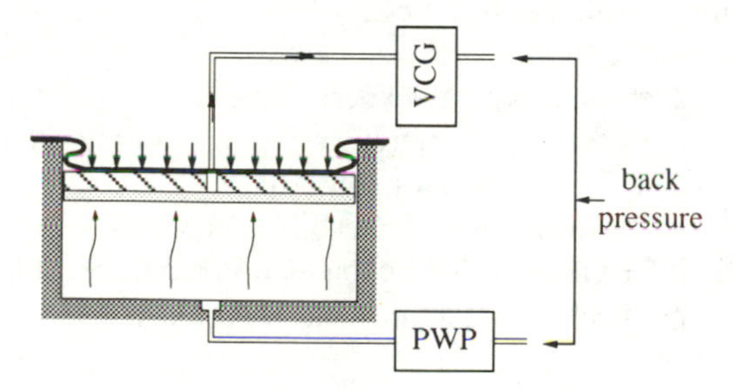

One-way vertical drainage

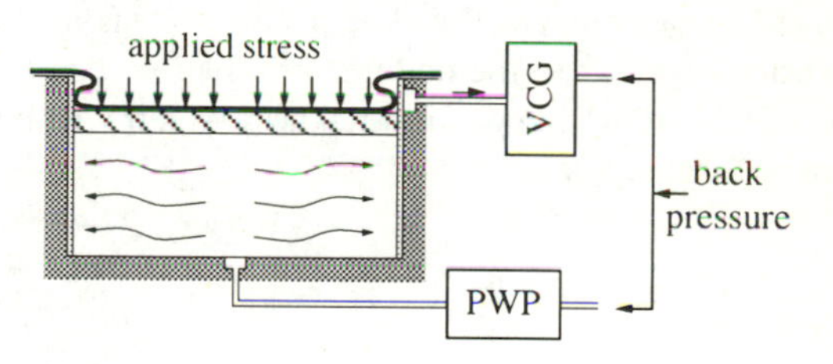

Outward radial drainage

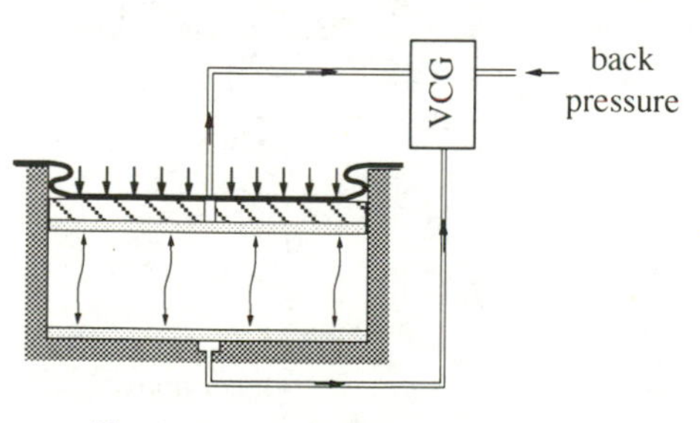

Two-way vertical drainage

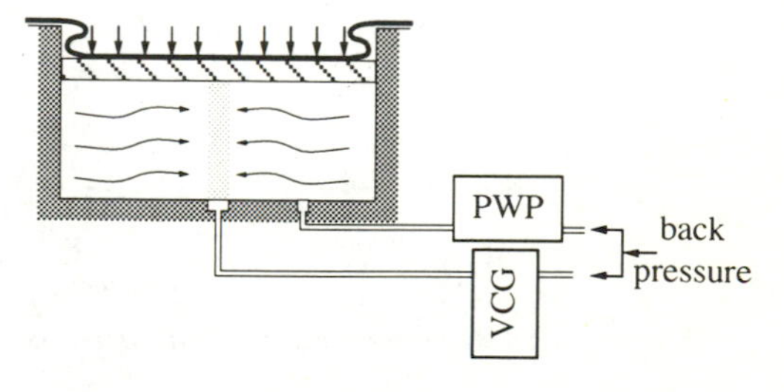

Inward radial drainage

Figure 6.17 *Hydraulic cell drainage conditions*

3 A back pressure can be applied to the pore water in the specimen to ensure full saturation and to prevent gases coming out of solution, particularly for organic soils.
4 The drainage paths can be controlled (Figure 6.17). The drainage can be vertical only (one-way or two-way) or radial only (inwards to a central sand drain or outwards to a peripheral porous lining).
5 A 'free-strain' or 'equal strain' loading condition can be applied. The former is provided by applying the pressure to the specimen directly via the flexible diaphragm, which produces a more uniform pressure distribution of known value (equal to the applied stress) and reduces the effects of side friction.

Flexible loading means that the settlement at the top of the specimen will be non-uniform and, since displacements are measured at the centre of the specimen, the maximum values will be obtained. More appropriate determinations of m_v should be obtained from volume change measurements, although these may not be as accurate as displacement measurements.

'Equal strain' loading is produced by inserting a rigid disc above the specimen to produce conditions similar to the oedometer. Then m_v can be determined from changes in thickness which will be more accurate.

Two- and three-dimensional consolidation *(Figure 6.18)*

One-dimensional consolidation will occur beneath a wide load or where the soil layer thickness is small compared to the width of a load, such as beneath a large fill area or when the water table is lowered generally. Two-dimensional consolidation will occur beneath a strip load and three-dimensional beneath a circle, square or rectangle.

One-dimensional consolidation means no horizontal drainage *or* horizontal strains. With two-dimensional consolidation, vertical drainage and strains will be accompanied by horizontal drainage and strains in *one* horizontal direction only. Three-dimensional consolidation results in horizontal drainage and strains in all directions so it can be seen that the geometrical shape of the load will have a major influence on the rate of settlement.

Conventionally rates of settlement are determined from the one-dimensional theory but Davis and Poulos (1972) have shown that the practical effect of two- and three-dimensional conditions is that the time required for consolidation decreases:

1 as the footing area decreases
2 as the soil layer thickness decreases
3 more for square or circular footings (3-dimensional) than for a strip (2-dimensional).
4 as the horizontal permeability increases ($c_H > c_v$).
5 if the underside of the foundation can be considered permeable, e.g. a thin sand layer exists beneath.

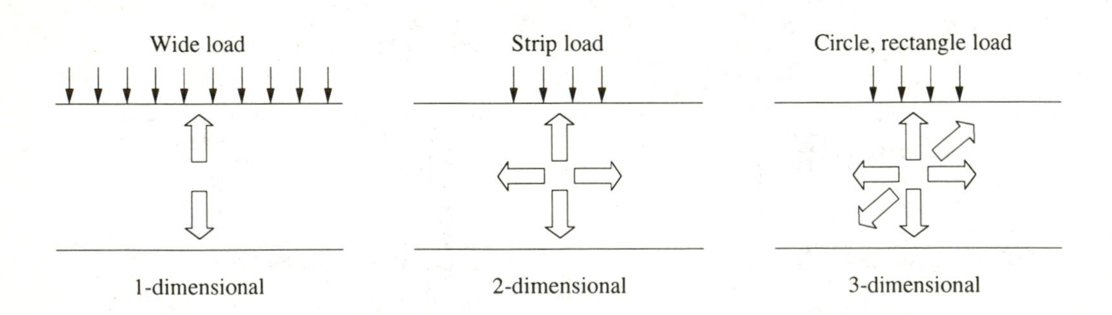

Figure 6.18 *One, two and three-dimensional conditions*

Radial consolidation for vertical drains
(Figure 6.19)

The time for one-dimensional consolidation to take place is inversely proportional to d^2, so for thick layers of clay settlements can take a long time to be completed. By inserting vertical drains at fairly close spacings, much shorter horizontal drainage paths are created, allowing much faster dissipation of pore pressure, removal of pore water and accelerated settlements.

The original 'sand drains' were formed by sinking boreholes 200 – 400 mm diameter through the clay and backfilling with a suitable filter sand. However, these could be slow to construct, produce large amounts of spoil and surface damage and be prone to 'waisting' produced when soft clays squeeze into the borehole on removal of the casing leaving a reduced diameter which could prevent adequate drainage.

Prefabricated drains comprising a continuous filter stocking filled with sand, called sandwicks, provided a more reliable and cost effective solution; since drain diameter is not a particularly important design parameter, smaller diameters of about 70 mm could be used.

Nowadays, the 'band' drain is commonly adopted. This consists of a continuous flat plastic core about 100 mm × 3 mm, corrugated to provide vertical drainage channels and wrapped around by a geotextile fabric to act as a filter. The band drain is installed by attaching one end of the band at the bottom of a rectangular steel mandrel with a clip and driving or pushing the mandrel into the soil to the depth required. As the mandrel is withdrawn, the clip retains the band drain in the soil which will then squeeze around the drain holding it in place.

A solution for radial consolidation has been obtained (Barron, 1948) by considering Terzaghi's equations for three-dimensional consolidation in polar coordinates as:

$$\frac{\partial u}{\partial t} = c_{\mathrm{v}} \frac{\partial^2 u}{\partial z^2} + \left(c_{\mathrm{H}} \frac{\partial^2 u}{\partial r^2} + \frac{1}{r}\frac{\partial u}{\partial r} \right) \tag{6.40}$$

overall drainage = vertical drainage + radial drainage

where $c_{\mathrm{v}} = \dfrac{k_{\mathrm{v}}}{m_{\mathrm{v}}\gamma_{\mathrm{w}}}$ and $c_{\mathrm{H}} = \dfrac{k_{\mathrm{H}}}{m_{\mathrm{v}}\gamma_{\mathrm{w}}}$

The drains of radius r_{d} will have a circular area of influence of radius R (Figure 6.19), with water moving horizontally towards each drain. In practice, drains placed in a square grid pattern will have a square area of influence of size s^2 for a spacing s. This is equated to a circular area of influence such that $s^2 = \pi R^2$ and $R = 0.564s$. For a triangular pattern $R = 0.525s$.

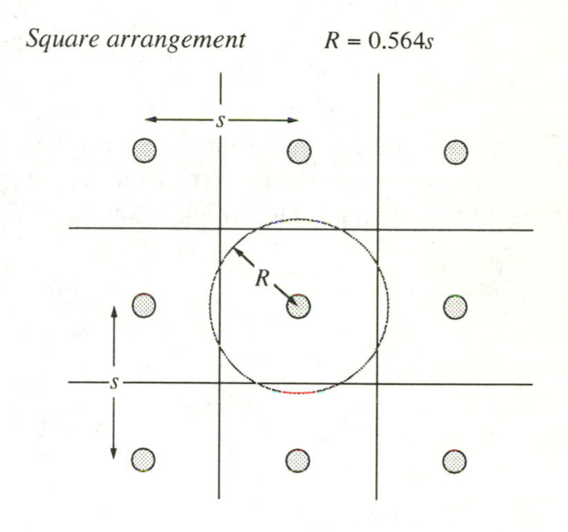

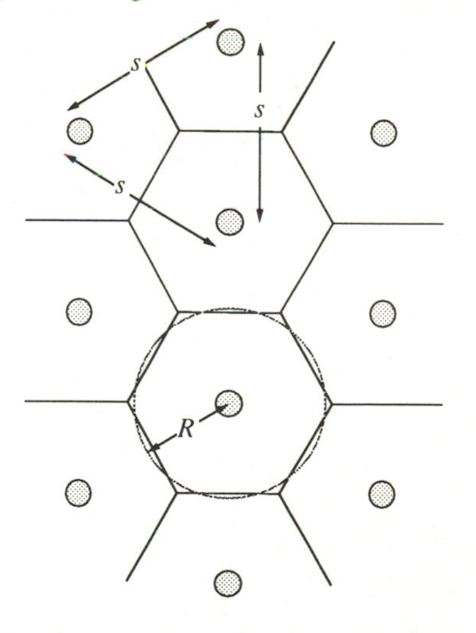

Figure 6.19 *Radius of influence of drains*

The overall degree of consolidation can be obtained from the combination of vertical and radial drainage as:

$$1 - U_c = (1 - U_v)(1 - U_R) \tag{6.41}$$

where:
U_c = overall or combined degree of consolidation
U_v = average U_v for 1-dimensional vertical flow only
U_R = average U_R for radial flow only.
Settlements are now related to the *overall* degree of consolidation as:

$$U_c = \frac{\rho_t}{\rho_c} \tag{6.42}$$

Values of U_v are obtained from Curve 1 of Figure 6.13 or Table 6.1, related to the vertical time factor T_v.

Values of U_R are related to a (radial) time factor T_R:

$$T_R = \frac{c_H t}{4R^2} \tag{6.43}$$

but also to the ratio:

$$n = \frac{R}{r_d} \tag{6.44}$$

and are given in Table 6.2.

For design the size or radius of the drain cannot be varied much, but this does not matter as it is not as critical as the spacing between the drains. It has been found that the time t for consolidation varies approximately as:

$$t \propto s^{2.5} \tag{6.45}$$

so the spacing is critical.

Smearing of the soil during installation will affect the horizontal permeability and hence c_H. To allow for this a reduced value, as low as $c_H = c_v$, could be used or a smaller drain radius of $\frac{1}{2}r_d$ could be used.

It has been found that the cost of a drain installation is inversely proportional to the value of c_v or c_H, so care during the site investigation in obtaining a representative value is essential.

Vertical drains can be ineffective:

1. with thin soil layers where sufficient consolidation may be achieved in one dimension (vertical) alone.
2. if c_H is much larger than c_v. Where macro-fabric will permit sufficient drainage horizontally, drains may not be required.
3. if large secondary compressions are likely. Drains only accelerate pore pressure dissipation or primary consolidation so if:

$$\frac{\rho_c}{\rho_{total}} = \frac{\text{consolidation settlement}}{\text{total settlement}} < 0.25 \tag{6.46}$$

   then drains may not be viable.
4. where they are severed due to large horizontal shear deformations around an incipient slip surface.

Table 6.2

Radial consolidation - Values of T_R

Degree of consolidation U_R %	Time factor T_R										
	$n = 5$	10	15	20	25	30	40	50	60	80	100
10	0.012	0.021	0.026	0.030	0.032	0.035	0.039	0.042	0.044	0.048	0.051
20	0.026	0.044	0.055	0.063	0.069	0.074	0.082	0.088	0.092	0.101	0.107
30	0.042	0.070	0.088	0.101	0.110	0.118	0.131	0.141	0.149	0.162	0.172
40	0.060	0.101	0.125	0.144	0.158	0.170	0.188	0.202	0.214	0.232	0.246
50	0.081	0.137	0.170	0.195	0.214	0.230	0.255	0.274	0.290	0.315	0.334
55	0.094	0.157	0.197	0.225	0.247	0.265	0.294	0.316	0.334	0.363	0.385
60	0.107	0.180	0.226	0.258	0.283	0.304	0.337	0.362	0.383	0.416	0.441
65	0.123	0.207	0.259	0.296	0.325	0.348	0.386	0.415	0.439	0.477	0.506
70	0.137	0.231	0.289	0.330	0.362	0.389	0.431	0.463	0.490	0.532	0.564
75	0.162	0.273	0.342	0.391	0.429	0.460	0.510	0.548	0.579	0.629	0.668
80	0.188	0.317	0.397	0.453	0.498	0.534	0.592	0.636	0.673	0.730	0.775
85	0.222	0.373	0.467	0.534	0.587	0.629	0.697	0.750	0.793	0.861	0.914
90	0.270	0.455	0.567	0.649	0.712	0.764	0.847	0.911	0.963	1.046	1.110
95	0.351	0.590	0.738	0.844	0.926	0.994	1.102	1.185	1.253	1.360	1.444
99	0.539	0.907	1.135	1.298	1.423	1.528	1.693	1.821	1.925	2.091	2.219

Worked Example 6.1 *Compressibility*

The following results were obtained from an oedometer test on a specimen of fully saturated clay. Determine the void ratio at the end of each pressure increment and plot void ratio versus log σ'. *Calculate* m_v *for each loading increment and* C_c *for the normally consolidated portion.*

Initial thickness H_o = 20 mm
Initial moisture content = 24 %
Particle density = 2.70 Mg/m^3

Applied pressure (kN/m^2)	Thickness of specimen (mm)
0	20
25	19.806
50	19.733
100	19.600
200	19.357
400	18.835
800	18.167

Initial void ratio $e_o = m_o\,\rho_s = 0.24 \times 2.70 = 0.648$

From $\dfrac{\Delta H}{H_o} = \dfrac{\Delta e}{1+e_0}$

The void ratio at the end of each pressure increment is given by:

$$e_f = e_0 - \Delta e = e_0 - \frac{\Delta H_f}{H_0}\left(1+e_0\right)$$

where ΔH_f is the change in thickness of the specimen from the initial thickness.
Compare this with Equation 6.34.

$$e_f = 0.648 - \frac{1.648}{20}\Delta H_f$$

Pressure	Thickness (mm)	ΔH_f	Δe	e_f
0	20	0	0	0.648
25	19.806	0.194	0.016	0.632
50	19.733	0.267	0.022	0.626
100	19.600	0.400	0.033	0.615
200	19.357	0.643	0.053	0.595
400	18.835	1.165	0.096	0.552
800	18.167	1.833	0.151	0.497

From Equation 6.35b:

Pressure increment (kN/m^2)	ΔH_i	H_i	$\Delta\sigma_i$	m_v (m^2/MN)
0 – 25	0.194	20	25	0.388
25 – 50	0.073	19.806	25	0.147
50 – 100	0.133	19.733	50	0.135
100 – 200	0.243	19.600	100	0.124
200 – 400	0.522	19.357	200	0.135

The graph of void ratio versus log-pressure or stress (Figure 6.20) shows the soil to be overconsolidated since there are two distinct lines. The value m_v for the first pressure increment is probably unrepresentative due to initial bedding errors and m_v should only be measured for pressure increments above the present overburden pressure. m_v for the pressure increment 200 – 400 kN/m² is larger because it includes volume changes on the normally consolidated limb.

The pressure increment 400 – 800 kN/m² lies on the normally consolidated line so the compression index can be obtained as:

$$C_c = \frac{0.552 - 0.497}{\log_{10}(800 \div 400)} = 0.18$$

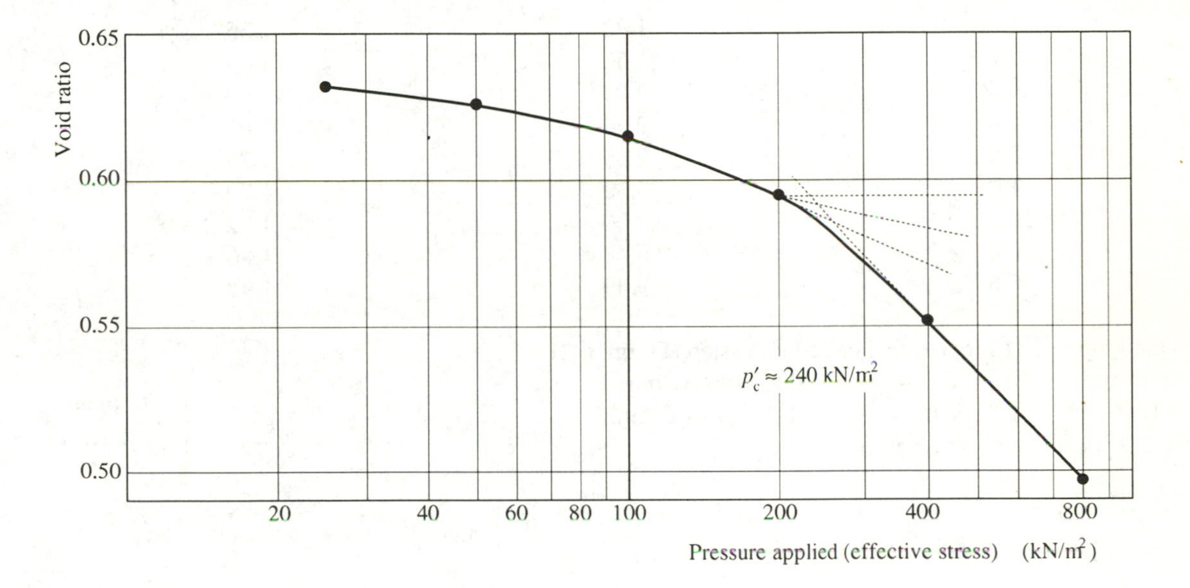

Figure 6.20 Worked Examples 6.1 and 6.2 *Compressibility and the Casagrande construction*

Worked Example 6.2 *Preconsolidation pressure and overconsolidation ratio*
From the graph of void ratio versus logσ' in Example 6.1 (Figure 6.20), use the Casagrande construction to estimate the preconsolidation pressure of the clay. If the specimen of soil was taken from 3 m below ground level in a soil with unit weight of 19 kN/m³ and a water table at ground level determine the overconsolidation ratio.
The accuracy of this method is largely affected by the position chosen for the point of maximum curvature. It is estimated that the preconsolidation pressure is about 240 kN/m².
The present overburden pressure is 3(19 – 9.8) = 27.6 kN/m²

$$\therefore \text{ the overconsolidation ratio OCR} = \frac{240}{27.6} = 8.7$$

This result shows that the soil is heavily overconsolidated.

Worked Example 6.3 *c_v-root time method*

An oedometer test on a specimen of fully saturated stiff clay gave the following results for the pressure increment from 100 to 200 kN/m². The initial thickness of the specimen under no pressure was 19 mm. Determine the values of m_v, c_v and k.

Time from start of loading (minutes)	$\sqrt{t}$	specimen compression (mm)
0	0	0.61
0.25	0.50	0.96
0.50	0.71	1.06
0.75	0.87	1.16
1.00	1.00	1.24
1.50	1.22	1.35
2.25	1.50	1.45
4.00	2.00	1.60
5.0	2.24	1.66
7.0	2.65	1.73
11.0	3.32	1.79
16.0	4.00	1.82
30.0	5.48	1.86
90.0	9.49	1.92

From the graph of $\sqrt{t}$ versus specimen compression (Figure 6.21):

$R_o = 0.69$ mm $\quad \therefore H_o = 19.00 - 0.69 = 18.31$ mm

$R_{100} = 1.92$ mm $\quad \therefore H_f = 19.00 - 1.92 = 17.08$ mm

$$\therefore H_{avc} = \tfrac{1}{2}(18.31 + 17.08) = 17.70 \text{ mm}$$

$$d = \frac{17.70}{2} = 8.85 \text{ mm}$$

From the graph $\sqrt{t_{90}} = 1.85 \quad \therefore t_{90} = 1.85^2 = 3.42$ minutes

From Equation 6.37:

$$c_v = \frac{0.848 \times 8.85^2}{3.42} = 19.4 \text{ mm}^2/\text{minute}$$

$$\Rightarrow c_v = \frac{19.4 \times 60 \times 24 \times 365}{1000 \times 1000} = 10.2 \text{ m}^2/\text{year}$$

$$m_v = \frac{1.92 - 0.61}{19 - 0.61} \times \frac{1}{200 - 100} \times 1000 = 0.71 \text{ m}^2/\text{MN}$$

$$k = c_v m_v \gamma_w = \frac{10.2}{60 \times 60 \times 24 \times 365} \times \frac{0.71}{1000} \times 9.81$$

$= 2.2 \times 10^{-9}$ m/s

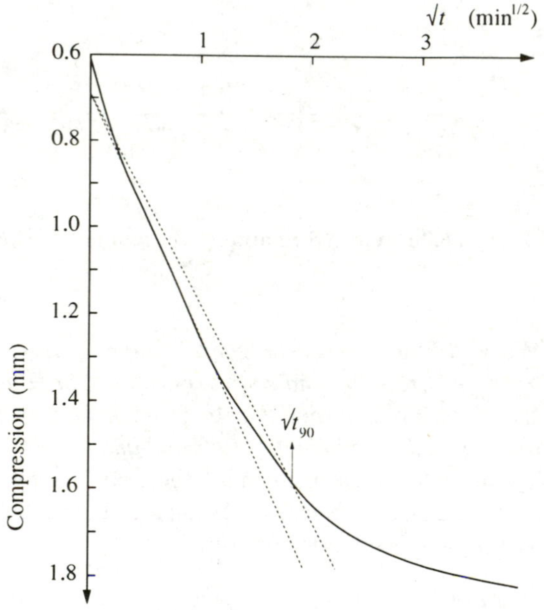

Figure 6.21 Worked Example 6.3
Root time method

Worked Example 6.4 *c_v-log time method*

With the same data in Example 6.3, determine c_v and k using the log time curve fitting method.

From the initial part of the graph of log time versus specimen compression (Figure 6.22):

for times of 0.1 and 0.4 minutes the readings are 0.92 and 1.01 mm.

difference $D = 0.09 \quad \therefore R_o = 0.92 - 0.09 = 0.81$ mm

for times of 0.16 and 0.64 minutes readings are 0.93 and 1.12 mm

difference $D = 0.19 \quad \therefore R_o = 0.93 - 0.19 = 0.74$ mm

Take the average: $R_0 = \frac{1}{2}(0.81 + 0.74) = 0.78$ mm

From Figure 6.22 $R_{100} = 1.79$ mm

$R_{50} = \frac{1}{2}(0.78 + 1.79) = 1.28$ mm $\quad \therefore t_{50} = 1.17$ minutes

$H_o = 19 - 0.78 = 18.22$ mm

$H_f = 19 - 1.79 = 17.21$ mm

$$H_{ave} = \tfrac{1}{2}(18.22 + 17.21) = 17.72 \text{ mm} \quad \therefore d = \frac{17.72}{2} = 8.86 \text{ mm}$$

From Equation 6.38:

$$c_v = \frac{0.196 \times 8.86^2}{1.17} = 13.2 \text{ mm}^2 / \text{minute}$$

$$\Rightarrow c_v = \frac{13.2 \times 60 \times 24 \times 365}{1000 \times 1000} = 6.9 \text{ m}^2 / \text{year}$$

From before, $m_v = 0.71$ m²/MN

$$k = c_v m_v \gamma_w = \frac{6.9}{60 \times 60 \times 24 \times 365} \times \frac{0.71}{1000} \times 9.81 = 1.5 \times 10^{-9} \text{ m/s}$$

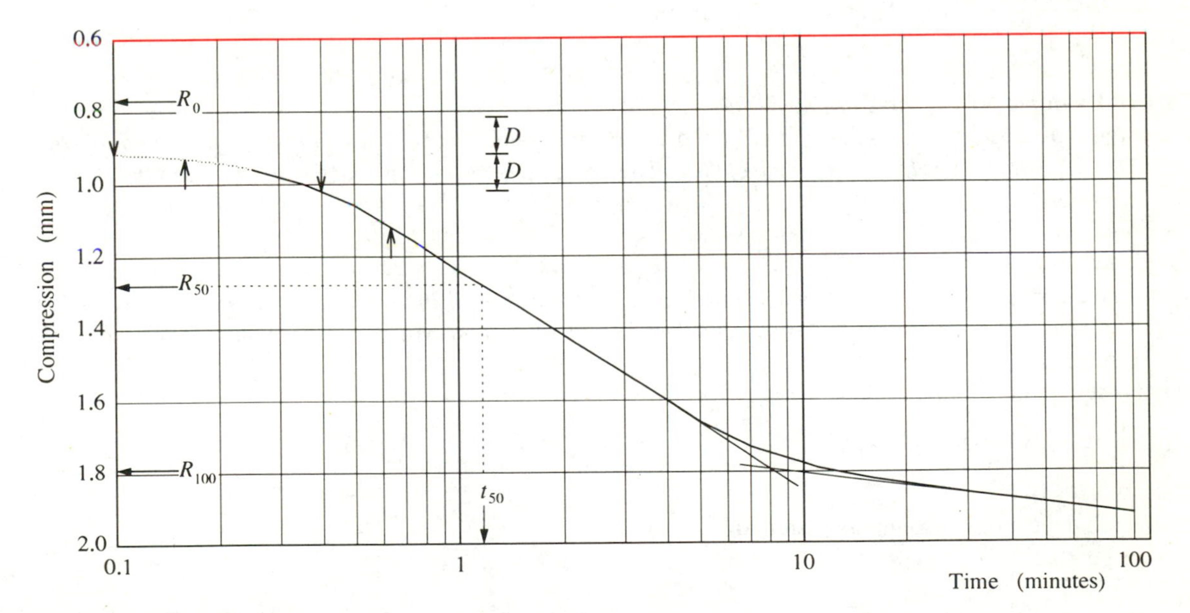

Figure 6.22 Worked Example 6.4 *Log time method*

Worked Example 6.5 *Time required for consolidation settlement*

An embankment for a highway, 5 m high above existing ground level, is to be placed on a deposit of soft clay, which is 8 m thick with m_v = 0.50 m²/MN and c_v = 10 m²/year. If the final road pavement on top of the layer can tolerate 50 mm settlement after it is constructed calculate how soon the pavement can be placed.

The bulk density of the embankment fill is 2200 kg/m³. Assume permeable boundaries top and bottom of the clay, the load is placed instantaneously and $\rho_{oed} = \rho_c$.

The amount of consolidation settlement is given by:

$$\rho_c = m_v \Delta\sigma H$$

For 5 m height of fill $\Delta\sigma = 5 \times 2200 \times \frac{9.81}{1000} = 107.9\ \mathrm{kN/m^2}$

$$\therefore \rho_c = \frac{0.50}{1000} \times 107.9 \times 8.0 = 0.432\ \mathrm{mm}$$

If only 5 m of fill was placed the top of the embankment would be 5.00 – 0.432 = 4.57 m above existing ground level. Since it is necessary for the top of the embankment to be 5.0 m above existing ground level the amount of fill to be placed must be 5.0 + ρ_c m. Then:

$$\rho_c = \frac{0.50}{1000}(5 + \rho_c) \times 2200 \times \frac{9.81}{1000} \times 8$$

giving ρ_c = 0.472 mm

Check : for 5.472 m of fill:

$$\rho_c = \frac{0.50}{1000} \times 5.472 \times 2200 \times \frac{9.81}{1000} \times 8 = 0.472\ \mathrm{m}$$

At the time required to place the pavement the degree of consolidation must be

$$U_v = \frac{472 - 50}{472} = 0.894$$

From Equation 6.29:

T_v = 0.824

$$t = \frac{T_v d^2}{c_v} = \frac{0.824 \times 4^2}{10} = 1.32\ \text{years} \quad \text{or } 15.8\ \text{months}$$

Worked Example 6.6 *Vertical drains*

An embankment is to be constructed up to 7 m above existing ground level above a clay deposit, 15 m thick, over a period of 1.5 months. The bulk unit weight of the embankment fill is 21.5 kN/m² and the properties of the clay are:

m_v = 0.25 m²/MN

c_v = 2.5 m²/year

c_H = 5.5 m²/year

Determine the spacing required for band drains with a cross-section 112 mm × 6 mm installed in a square grid pattern to meet the requirement that only a further 50 mm settlement will occur 2.5 months after the end of construction.

Actual height of embankment fill = 7.0 + ρ_c

$$\therefore \rho_c = \frac{0.25}{1000}(7.0 + \rho_c) \times 21.5 \times 15$$

giving ρ_c = 0.614 m

The overall degree of consolidation required

$$U_c = \frac{614 - 50}{614} = 0.919$$

at an 'instantaneous' time of $\frac{1.5}{2} + 2.5 = 3.25$ months

assuming the load to be placed instantaneously halfway through the construction period.

For vertical (one-dimensional) consolidation, from Equation 6.22:

$$T_v = \frac{2.5 \times 3.25}{7.5^2 \times 12} = 0.012 \quad \text{giving } U_v = 0.124$$

For radial consolidation, from Equation 6.41

$1 - 0.919 = (1 - 0.124)\,(1 - U_R)$ giving $U_R = 0.908$

The 'radius' of the band drain is obtained assuming equivalent perimeters:

$2\,\pi\, r_d = 2(6 + 112)$

$\therefore\ r_d = 37.6$ mm

The radius of influence of each drain $R = 0.564s$

From Equation 6.44:

$$n = \frac{R}{r_d} = \frac{0.564s}{0.0376} = 15.0s \quad \therefore \text{spacing } s = \frac{n}{15.0}$$

The radial time factor T_R, from Equation 6.43

$$= \frac{5.5 \times 3.25}{4 \times (0.564s)^2 \times 12} = \frac{1.17}{s^2}$$

$$\therefore \text{spacing } s = \sqrt{\frac{1.17}{T_R}}$$

Since U_R is a function of both T_R and n it is necessary to find for $U_R = 0.908$ when:

$$\frac{n}{15.0} = \sqrt{\frac{1.17}{T_R}} \quad \text{i.e. spacing in terms of } n = \text{spacing in terms of } T_R$$

Interpolating from Table 6.2 values of n and T_R are obtained when $U_R = 0.908$

n	T_R	$s = \frac{n}{15.0}$	$s = \sqrt{\frac{1.17}{T_R}}$
10	0.477	0.67	1.57
15	0.594	1.00	1.40
20	0.680	1.33	1.31 ⇐
25	0.746	1.67	1.25

These values show that a spacing of about 1.3 m would be required.

Exercises

6.1 The results of an oedometer test on a sample of fully saturated clay are given below. At the end of the test the moisture content was determined to be 27.3% and the particle density was 2.70 Mg/m^3. Determine values of void ratio at the end of each pressure increment and plot them against the logarithm of pressure.

Pressure (kN/m^2)	Thickness (mm)
0	19.000
25	18.959
50	18.918
100	18.836
200	18.457
400	17.946
800	17.444
200	17.526
25	17.669
0	17.782

6.2 From the graph of void ratio versus log pressure in Exercise 6.1 estimate the preconsolidation pressure. If the sample of clay was taken from 5 m below ground level, at a site where the water table exists at ground level, determine the overconsolidation ratio of the clay at this depth. The saturated unit weight of the clay is 21.2 kN/m^3.

6.3 In Exercise 6.1 determine the coefficient of compressibility, m_v in m^2/MN for each pressure increment (loading only). From the graph of void ratio versus log pressure determine the coefficient of compressibility for the pressure increments $p_o' + 50$ and $p_o' + 100$ kN/m^2.

6.4 For the pressure increments on the normally consolidated limb of the pressure-void ratio curve in Exercise 6.1, determine the compression index, C_c.

6.5 The results of an oedometer test on a sample of fully saturated clay are given below for the pressure increment from 200 to 400 kN/m^2. Using the root time curve fitting method, determine the coefficient of consolidation, c_v. Determine the coefficient of compressibility, m_v and derive the coefficient of permeability, k.

Time (minutes)	Thickness (mm)	Time (minutes)	Thickness (mm)
0	18.500	6	17.930
0.25	18.341	9	17.861
0.50	18.282	12	17.822
1	18.223	16	17.794
2	18.135	25	17.771
3	18.073	36	17.759
4	18.017		

6.6 A clay layer, 10 m thick, with sand beneath is to be loaded with a wide layer of fill, 2.5 m thick, and with unit weight 20 kN/m^3. The coefficient of compressibility of the clay decreases with depth z in the form $m_v = 0.24 - 0.02z$. Determine the settlement due to consolidation of the clay.

6.7 A clay layer, 6 m thick with a water table at 2m below ground level, is to be loaded over a wide area, with a pressure of 50 kN/m^2. The coefficient of consolidation of the clay layer is c_v = 12 m^2/year. Assuming the load to be placed instantaneously determine the pore water pressure after 6 months:
(a) at the middle of the clay layer, assuming permeable strata above and below
(b) at the bottom of the clay layer, assuming a permeable stratum at the top and an impermeable stratum beneath the clay.
Assume the bulk unit weight of the clay is 21KN/m^3.

6.8 A 6 m layer of sand overlies a 5 m thick layer of clay with a sand deposit beneath. The water table is to be lowered permanently from 1 m below ground level to 4 m below ground level by pumping from the sand over a period of 6 weeks. Determine the settlement due to consolidation of the clay layer 6 months from the start of pumping. For the clay,
m_v = 0.45 m^2/MN and c_v = 5.5 m^2/year.

6.9 In Exercise 6.8, determine the time from the start of pumping to achieve 90% of the final consolidation settlement.

6.10 A motorway embankment is to be constructed over a layer of soft compressible clay, 10 m thick. At the time required to place the road pavement on top of the embankment it has been calculated that the average degree of consolidation is only 30% without the installation of a sand drain arrangement assuming one-dimensional consolidation only.
To reduce the settlements which would occur after pavement construction it is proposed to install sand drains through the soft clay. Determine the overall degree of consolidation which will be achieved when the pavement is constructed, if sand drains, 200 mm diameter, are installed at 3.5 m centres on a square grid pattern.
Assume that sand exists above and below the clay layer and the coefficients of horizontal and vertical consolidation are in the ratio $c_H = 3.5\ c_v$.

6.11 An embankment is to be constructed up to 6 m above a layer of clay, 13 m thick, over a period of 1 month. The properties of the clay are :
m_v = 0.30 m^2/MN
c_v = 2.5 m^2/year
c_H = 7.0 m^2/year.
The bulk unit weight of the embankment fill is 20.5 kN/m^3. It is required to construct the road pavement on top of the embankment two months after the end of construction but the pavement can only tolerate a settlement of 50 mm. Using vertical band drains with a rectangular cross-section of 8 mm thick and 70 mm wide installed in a triangular grid pattern determine the spacing of these drains to meet the settlement requirement.
Assume the settlement within the embankment itself is minimal.

7 Shear Strength

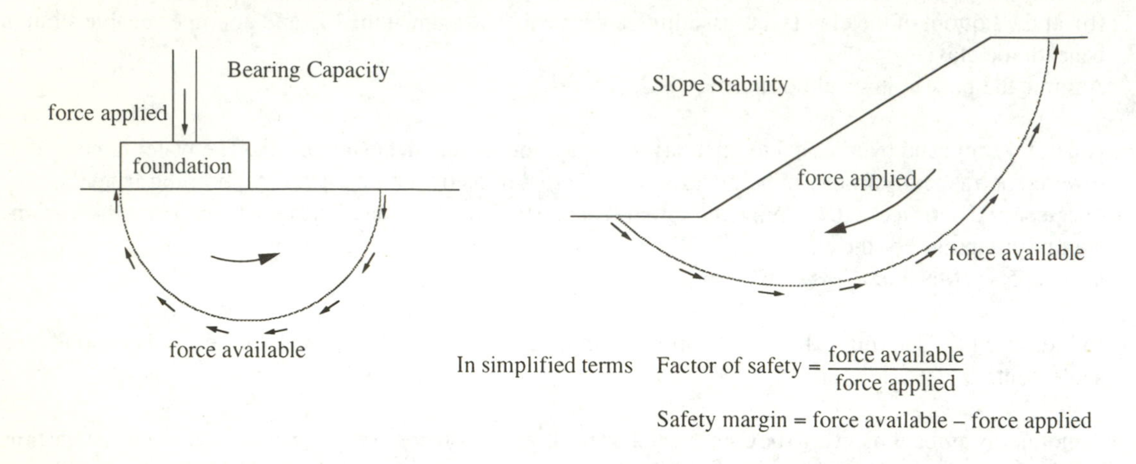

Figure 7.1 *Typical failure modes*

Introduction *(Figure 7.1)*
In soils, failure occurs as a result of mobilising the maximum shear stress the soil can sustain, so an understanding of shear strength is fundamental to the behaviour of a soil mass. The shear strength of the soil, allied with a particular method of analysis, will determine the ultimate (failure) load which can be applied on a foundation resting on soil, or the ultimate force required to cause failure of a soil mass forming a slope.

These values permit the determination of the factor of safety, which is the ratio of the ultimate loads or forces available to the load or force applied. The factor of safety is a measure of the degree or intensity of loading and the designer must be satisifed that there is an adequate value of this ratio to ensure that the earth structure will remain stable. Assessment of the appropriate soil shear strength value is, therefore, of prime importance for a limit state design approach.

Effect of strain *(Figure 7.2)*
Shear stresses in a soil mass are only produced when shear strains can occur, so placing a foundation load on a soil mobilises the available shear strength (or a part of it). Similarly, for a slope, gravity forces mobilise the shear strength available within the slope, although it is usually assumed (for analytical simplicity) that shear failure occurs along a simple, single slip surface.

Strictly speaking, it is the shear strain, γ, which produces shear stress, τ (or *vice versa*), but this is not easy to determine, *in situ* or in laboratory tests, so direct strains, ε or just displacement δL, have to be used.

A typical stress-strain curve is shown in Figure 7.2. As shear stress is applied the soil structure distorts. Initially, this distortion is proportional to the stress applied and if the stress is removed the distortions are recovered. These distortions are probably associated with small rotations at the numerous particle contacts and some elastic compression of the particles themselves. The soil in this region is said to behave in an elastic manner.

At a certain stress level, depending on the soil type, the soil structure will deform in a plastic manner by rearrangement of the particle locations and strains from this point (yield) will comprise both elastic and plastic components but the plastic strains will not be recovered on removal of the stress since the soil particles have moved into a new arrangement.

As the soil is strained or 'worked' further additional shear stress can be sustained due to a process described as work-hardening.

In dense sands and stiff overconsolidated clays, this would be due to expansion of the mineral grain structure (dilatancy) as more stress is required to achieve further strain. After reaching a peak shear stress these

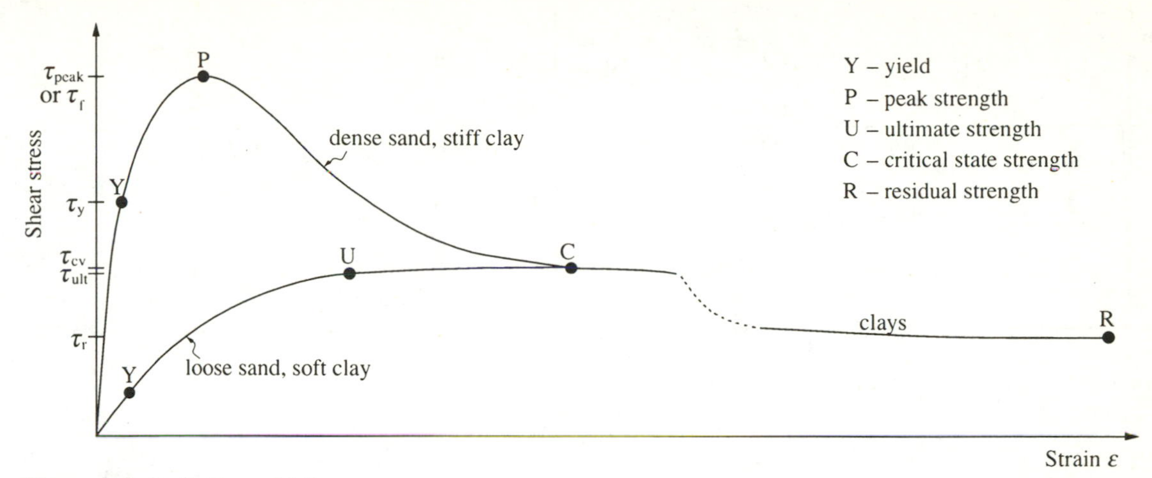

Figure 7.2 *Definition of failure*

soils typically display work-softening since strains beyond this value are being applied to a soil structure which has been weakened. This phenomenon would also apply to a soil which has developed cementation or chemical bonding since deposition. A reduction of strength beyond a peak value is referred to as brittleness; some materials such as sensitive clays and 'collapsing' soils will suddenly crush or collapse due to breakdown of interparticle bonds.

For loose sands and soft normally consolidated clays work-hardening would commence at much lower stress and strain levels since their relatively open mineral grain structures will be contracting as shearing occurs making the structure progressively more able to support more stress. These materials do not tend to display a peak value followed by work-softening or brittleness.

Figure 7.2 shows that choosing the point at which the soil has 'failed' requires a definition. This could be:

1 ***yield*** – although not the maximum shear stress available, if the soil is stressed any further beyond the point Y on Figure 7.2 the strains and movements of the earth structure (foundation, slope etc.) will be so large that they could be considered to have failed. The quantity τ_y represents a yield stress.
2 ***peak shear strength*** – this is the maximum shear stress which can be sustained. It may be dangerous to rely on this value for some brittle soils, due to the rapid loss of strength if the soil is strained beyond this point.
3 ***ultimate strength*** – for loose sands and soft clays work-hardening may continue to increase the shear stress that can be sustained, even at very large strains so a maximum strain limit must be imposed, usually related to the performance of the earth structure, say 10–20% strain, point U on Figure 7.2.
4 ***critical state strength*** – this is sometimes referred to as the ultimate strength. After a considerable amount of shear strain, a soil will achieve a constant volume state (by the soil structure expanding or contracting) and it will continue to shear at this constant volume without change in volume or void ratio. These shear strains must be uniform throughout the soil and not localised. It is sometimes referred to as the constant volume strength (ϕ_{cv}).
5 ***residual strength*** – this is also sometimes referred to as ultimate strength. After a considerable amount of strain on a single slip zone or surface (point R on Figure 7.2) the particles either side of this surface will rearrange to produce a more parallel orientation and this will produce the lowest possible or residual strength. This strength is very important in the reactivation of old landslides and is obviously more significant for platy minerals (e.g. clays).

Friction between the surfaces of particles where they are in contact with each other is related to the effective stress and is represented by the angle of interparticle friction, ϕ_μ. This friction forms the basis for the above observed shear strengths the differences being produced by rolling friction and dilatancy (Rowe, 1962).

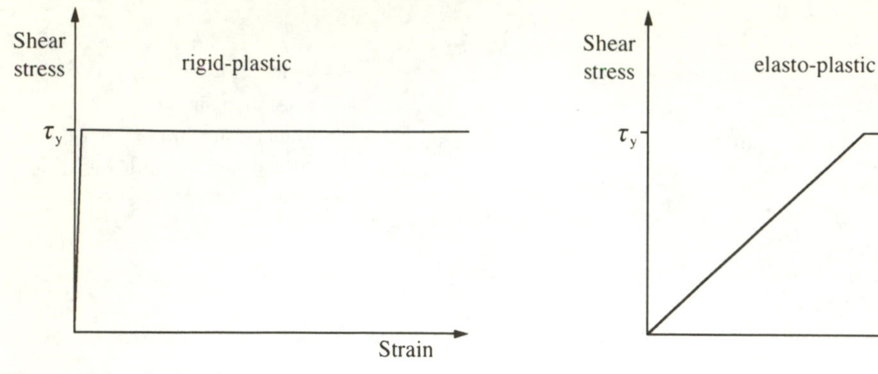

Figure 7.3 *Idealised stress-strain relationships*

Idealised stress-strain relationships *(Figure 7.3)*
These are adopted to assist the methods of analysis used in Soil Mechanics. The most common of these is the rigid-plastic model although the elasto-plastic form is a more realistic relationship.

Failure criterion *(Figure 7.4)*
The Mohr - Coulomb relationship is the most appropriate strength criterion adopted in Soil Mechanics. It simply relates the shear stress at failure (shear strength, τ_f) on a failure plane or slip surface to the normal effective stress σ_N' acting on that plane:

$$\tau_f = c' + \sigma_N' \tan \phi' \qquad (7.1)$$

A different relationship is obtained, depending on the definition of failure adopted, e.g. peak, critical state or residual, as shown in Figure 7.4, and the drainage conditions that are applicable, see 'Effects of drainage' below.

Failure of soil in the ground *(Figure 7.5)*
Failure will be produced by changing the natural *in situ* stress state which would lie below the failure criterion to a state coinciding with the criterion. This will occur when the *in situ* Mohr circle (Figure 4.10) is enlarged or moved (see 'Stress paths' below) and becomes tangential with the failure criterion. A theoretical plane of failure or slip surface at an angle θ to the major principal plane (σ_1' plane in this case) is then postulated. For more brittle soils, a single shear plane or zone would form, whereas for softer plastic soils uniform shear would develop.

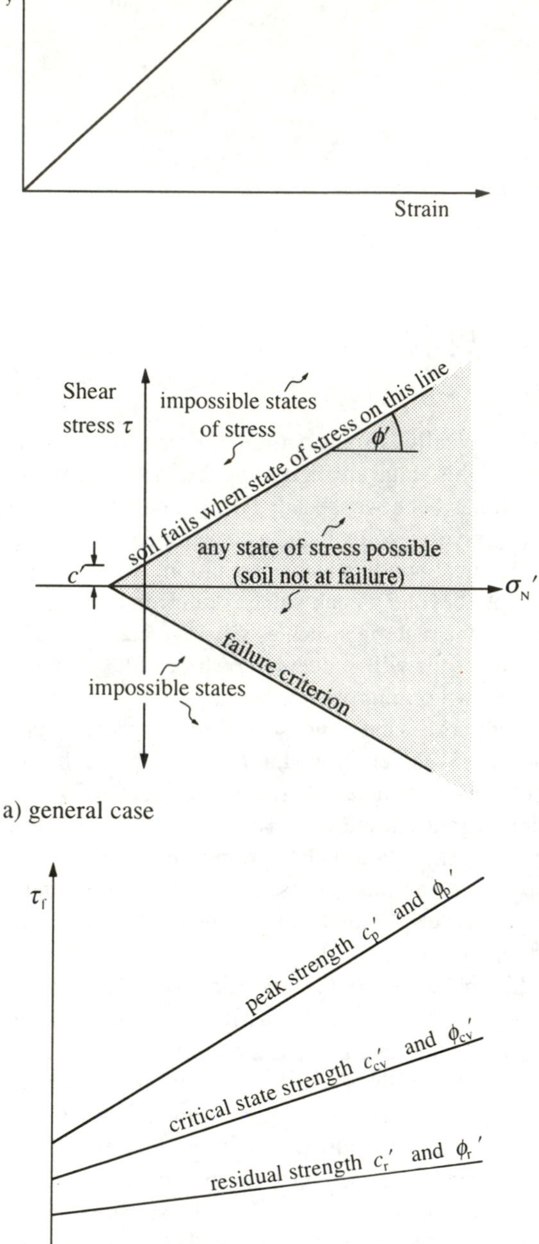

Figure 7.4 *Mohr-Coulomb failure condition*

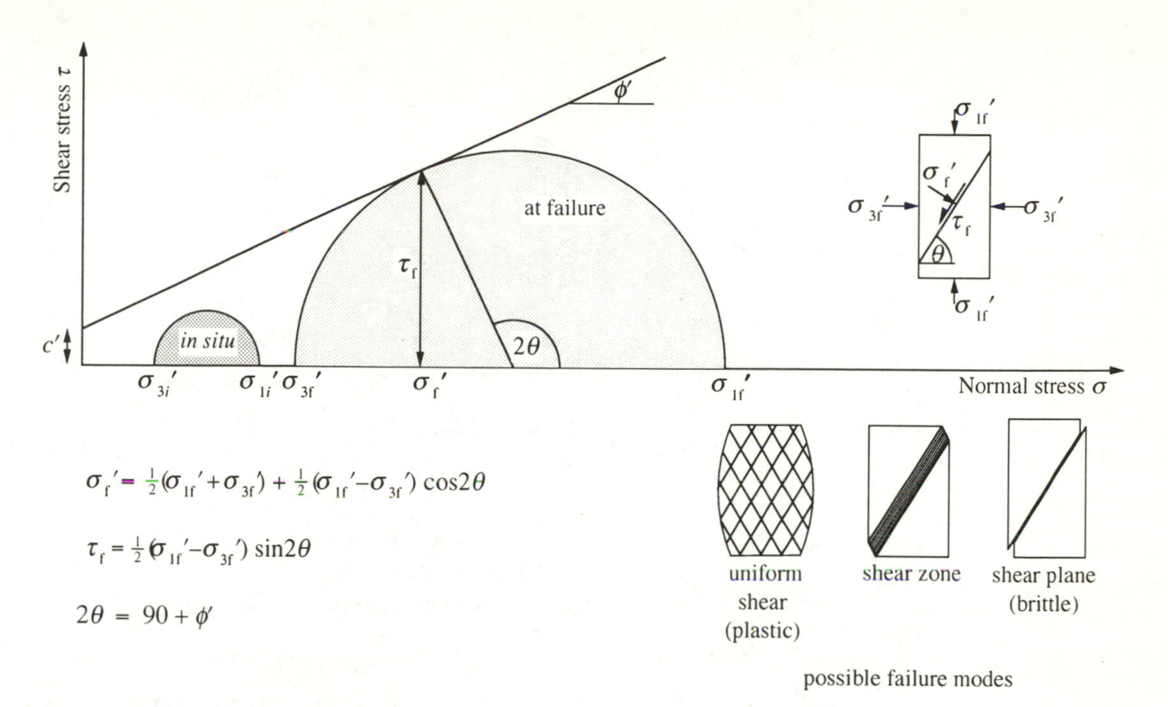

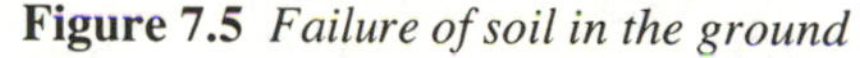

Figure 7.5 *Failure of soil in the ground*

Stress paths *(Figures 7.6 and 7.7)*
The behaviour of a soil will depend on the initial (*in situ*) and final (failure) states as described above but will also depend on the route taken between these states. These routes are referred to as stress paths and may be plotted as either effective stresses or total stresses. For plane strain conditions they are plotted as t' and s' or t and s and for axial symmetry conditions they are plotted as q' and p' or q and p (as shown in Figure 7.6).

The values t' and s' refer to the apex of a Mohr circle of stress while q' and p' are the deviator stress and mean stress, respectively. In both cases, the total and effective stress paths are separated horizontally by an amount equal to u_w, the pore water pressure since:

$$t' = t \quad \text{and } s' = s - u_w \tag{7.2}$$

and

$$q' = q \quad \text{and } p' = p - u_w \tag{7.3}$$

Plane strain conditions ($\sigma_2 \neq \sigma_3$) are obtained beneath a long foundation or a long slope, where the strain along the length of the foundation or slope is assumed to be zero and two-dimensional conditions apply. Axial symmetry means that stresses around one axis, usually the vertical axis, are symmetrical, i.e. $\sigma_2 = \sigma_3$. This situation would exist beneath the centre of a circular foundation and in the triaxial apparatus where three dimensional conditions apply.

A soil element will take a different stress path to failure from the *in situ* equilibrium state depending on the type of loading (Figure 7.7).

Effects of drainage *(Figures 7.8 and 7.9)*
It has been shown in Chapter 4 that, when stresses (loading or unloading) are applied to the ground, the immediate response of the soil is for all of this stress to be supported by the pore water with a consequent change in pore water pressure Δu_w (an excess above the static pore water pressure already existing in the pores). The terms 'drained' or 'undrained' are used in Soil Mechanics to denote whether dissipation of this pore water pressure change can occur or not.

Dissipation means the return of the altered pore water pressure to its original static value, and can be a decrease if pore pressures were raised above the static

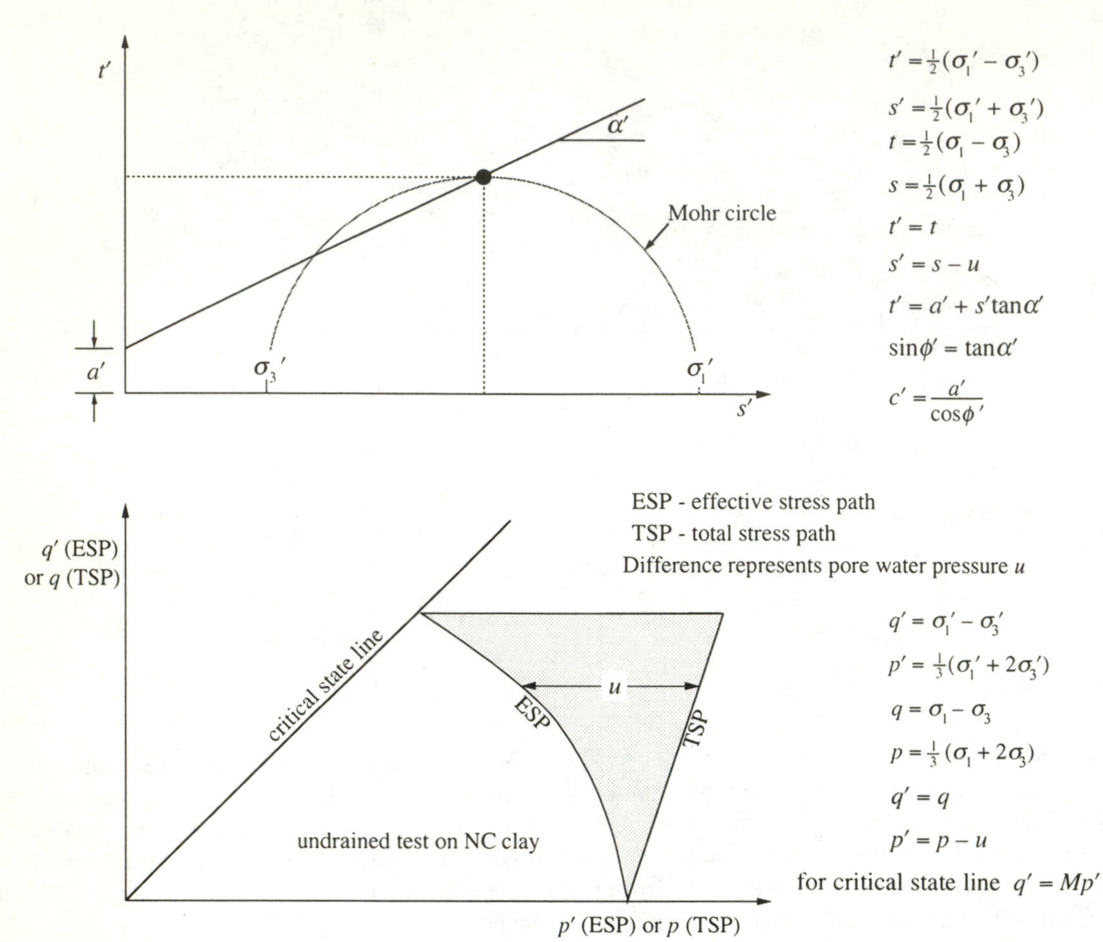

Figure 7.6 *Stress paths*

value (e.g. beneath a foundation) or an increase if they were lowered below the static, e.g. beneath a basement or cutting. The rate of dissipation will largely depend on the permeability of the soil, the proximity of permeable boundaries for water to be forced towards and the time allowed for dissipation in relation to the rate of loading or unloading.

This aspect of 'drainage' is different from the flow of free water under the force of gravity towards a drain or sump.

Thus the terms used are:

- *undrained*
  Dissipation of the excess pore water pressure Δu_w is prevented. This condition is produced when a soil of low permeability such as a clay is loaded quickly.
- *drained* (fully)
  Dissipation of any excess pore water pressure is permitted fully at all times so that effectively there is no measurable excess pore water pressure, i.e. $\Delta u_w = 0$. This condition is produced when a soil of high permeability such as sand is loaded slowly.
- *drained* (partially)
  An excess pore water pressure develops to a certain extent due to loading, but not fully since dissipation is proceeding at the same time and reducing this pore water pressure. This is probably the situation in many engineering applications but the assumptions of fully drained or fully undrained conditions are adopted for simplicity in applying test results and analytical procedures.

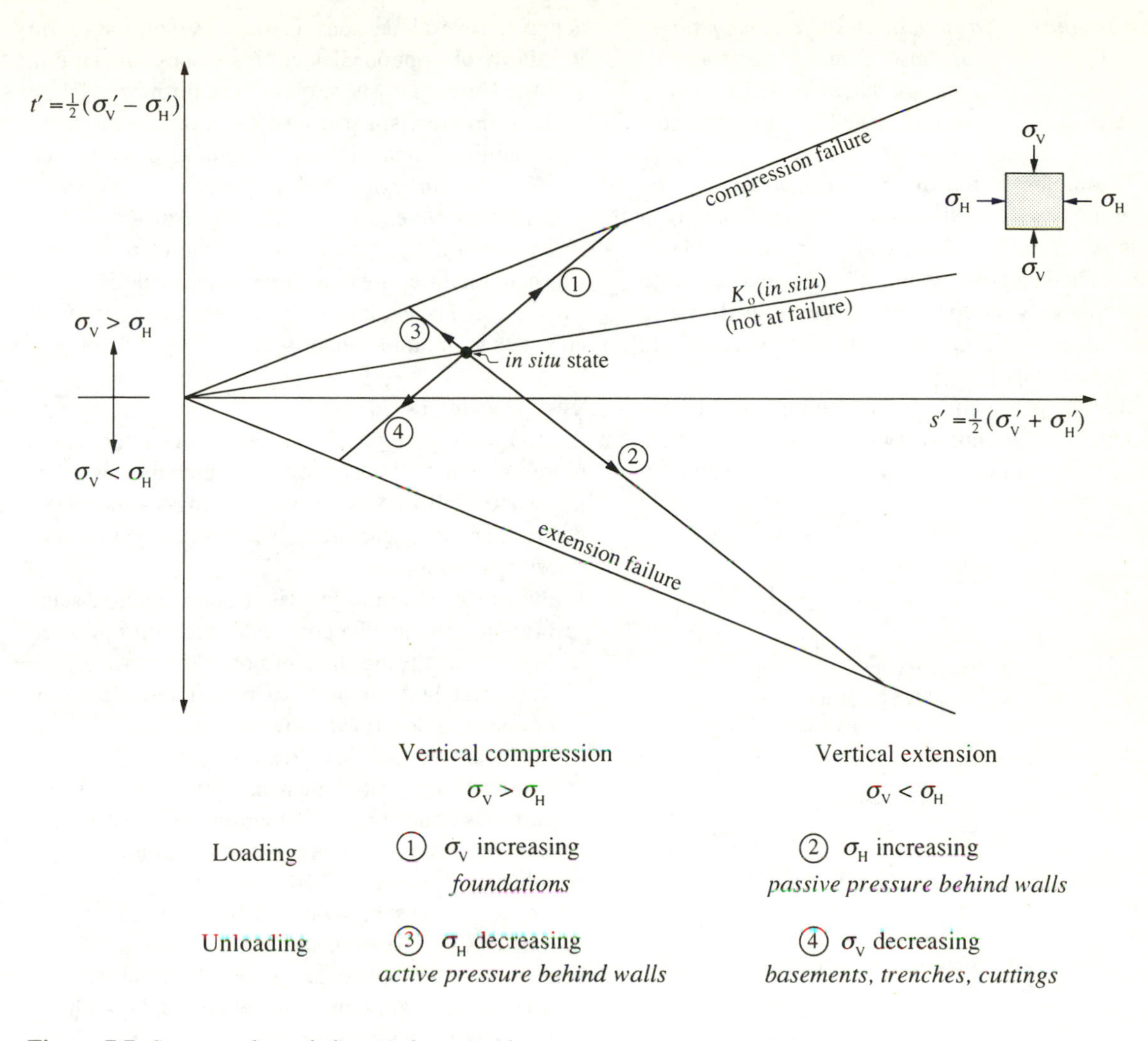

Figure 7.7 *Stress paths to failure in the ground*

Engineering works change the total stresses, $\Delta\sigma_1$ and $\Delta\sigma_3$ or $\Delta\sigma_v$ and $\Delta\sigma_H$, in the ground in various ways, see Figure 7.7. The excess pore water pressure Δu_w, produced by these changes, can be considered as being made up from two components. The first is a consolidation or mean stress change occurring before or during shear and the second is a deviator stress change during shear.

For a fully saturated soil ($B = 1$) the pore pressure parameter expression (Equation 4.12) can be rearranged to give:

$$\Delta u_w = \underbrace{\tfrac{1}{3}\left(\Delta\sigma_1 + 2\Delta\sigma_3\right)}_{\substack{\text{due to mean stress} \\ \text{(consolidation stage)}}} + \underbrace{\left(A - \tfrac{1}{3}\right)\left(\Delta\sigma_1 - \Delta\sigma_3\right)}_{\substack{\text{due to deviator stress} \\ \text{(shear stage)}}} \qquad (7.4)$$

and these stages may occur either separately or concurrently.

The various engineering works will stress the ground in different ways and three common test procedures have been devised to model or represent these applications, as follows.

Test procedure	*Consolidation stage*	*Shear stage*
UU	unconsolidated	undrained
CD	consolidated	drained
CU	consolidated	undrained

Some examples of the drainage behaviour and the appropriate types of test for sands and clays are given in Figures 7.8 and 7.9. It is obvious that unconsolidated and undrained mean no volume change or no moisture content change while consolidated and drained involve volume changes during both the consolidation and shear stages, respectively.

Where the permeability of the soil is low the consolidated drained condition (which is the most critical for unloading situations) will require a long time to achieve, several decades in the case of long-term instability of London Clay cutting slopes (Skempton, 1964). However, where clays contain macro-fabric such as fissures, silt partings, etc. making the mass permeability much higher the critical consolidated drained case can be obtained very soon after unloading, within hours in the case of trench or trial pit sides. This risk is often unappreciated and is one of the largest causes of fatalities in excavations on construction sites. This can only be avoided by the immediate insertion of adequate temporary supports.

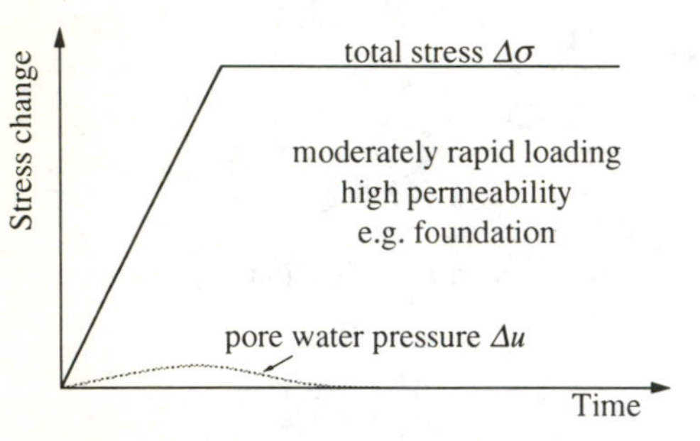

full dissipation at all times ($\Delta u \Rightarrow 0$) during both consolidation and shear stages

CD test

usually (slow) direct shear test

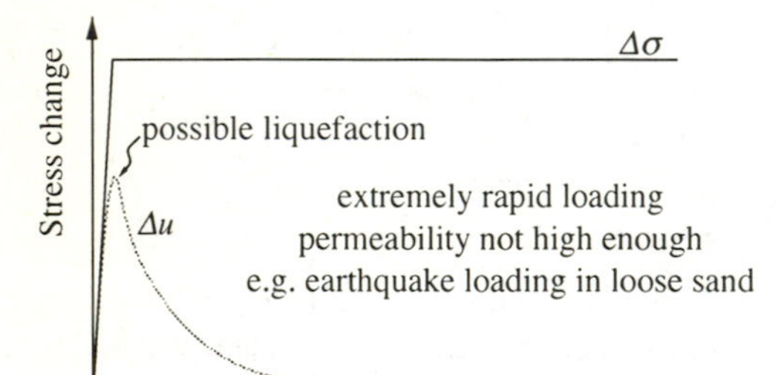

pore water pressure increased due to rapid cyclic shear loading and collapse of loose structure
none or only partial consolidation
none or only partial dissipation during shear

UU test

undrained cyclic shear test

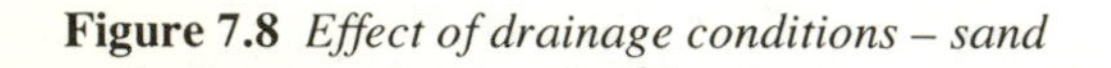

Figure 7.8 *Effect of drainage conditions – sand*

Test procedures *(Figure 7.10)*

A laboratory test on a soil sample is intended to represent or model the conditions which the engineering works will impose on the soil mass. The major determinants of shear strength controlled by the laboratory apparatus are:

1 the mode of drainage – is a consolidation stage provided and are the pore water pressures allowed to dissipate during shear or not? This aspect of the test procedure attempts to model the effects of drainage as described above.

2 the means of applying stress changes.

With regards to the latter, there are three ways in which the strain is controlled by the engineering works, the triaxial, plane strain and direct shear conditions. These are illustrated in Figure 7.10.

The triaxial system relates to the condition of axial symmetry where $\sigma_2 = \sigma_3$ and $\varepsilon_2 = \varepsilon_3$ and would apply beneath the centre of square or circular foundations.

The plane strain condition relates to two-dimensional shearing where strain in the intermediate direction is prevented, i.e. $\varepsilon_2 = 0$ and $\sigma_2 \neq \sigma_3$. This condition would apply beneath a long foundation or long slope.

The direct shear condition relates to shearing on a slip plane or a narrow shear zone where the soil is strained in a fairly pure shear manner.

There are also three ways in which the stresses are changed, by compression, extension and direct shear. With the first two, the shear stress is applied indirectly by changes in the principal stresses (σ_V and σ_H) while the latter occurs where the shear stress is applied directly. Where the engineering works increases the vertical stress above the horizontal stress the stress change is referred to as compression and extension occurs where the horizontal stress exceeds the vertical stress change. This is illustrated in Figure 7.10.

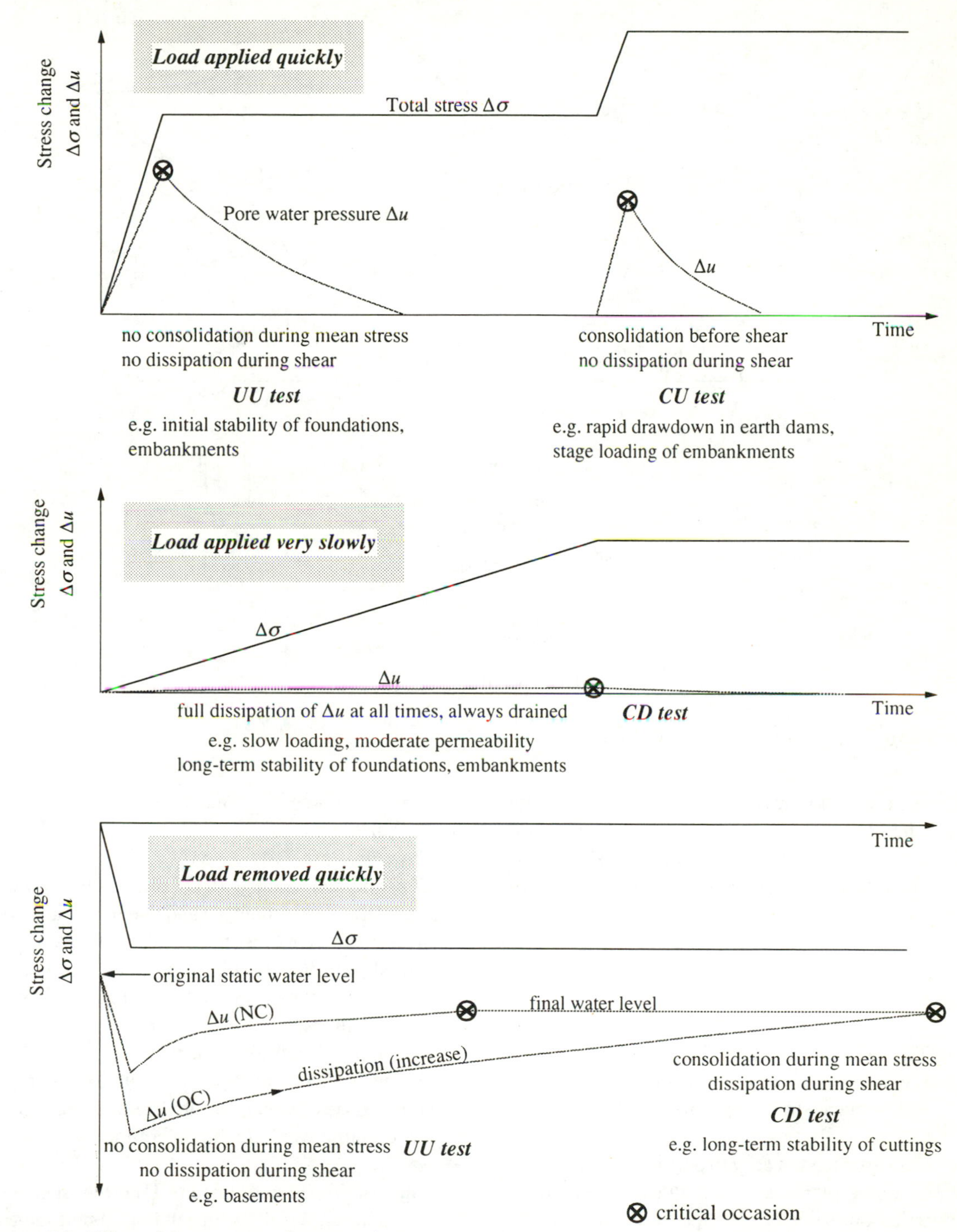

Figure 7.9 *Effect of drainage conditions – clay*

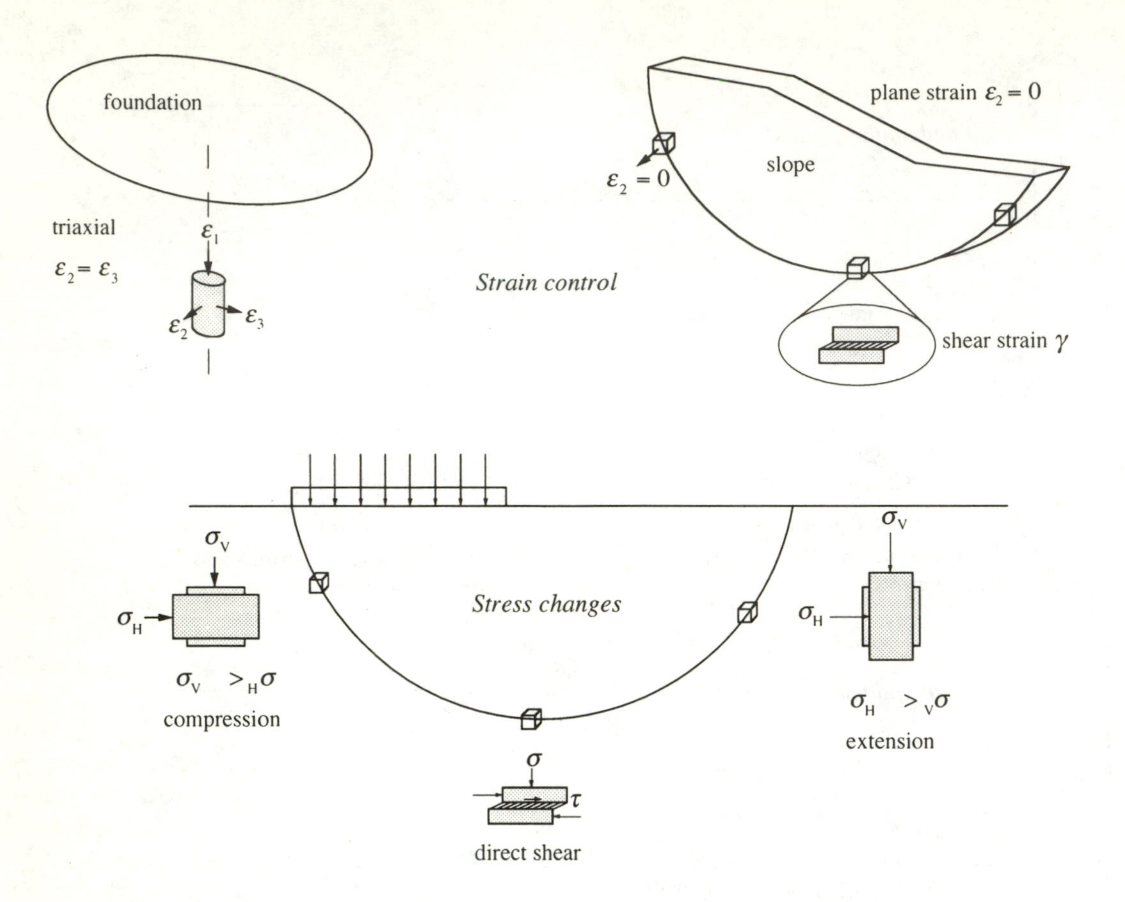

Figure 7.10 *Strain control and stress changes*

These strain and stress conditions are important since it has been found that the strength of a soil differs according to the strain control and stress change and test procedures have been designated accordingly:

DS – direct shear
UC – uniaxial compression
TC – triaxial compression
TE – triaxial extension
PSC – plane strain compression
PSE – plane strain extension.

Plane strain tests are not easy to perform and remain in the laboratories of research institutes. Research has been carried out to determine the different strengths obtained by these procedures and, fortunately, these differences are not particularly large.

In most commercial soils laboratories the strength tests available are:

- *shear box test*
  Consolidated drained (CD) direct shear (DS) on reconstituted samples of sand to determine ϕ' peak or ultimate. This test may also be used on undisturbed samples of clay for peak drained and undrained strength and on cut plane samples for residual strength.
- *unconfined compression test*
  Unconsolidated undrained (UU) uniaxial compression (UC) test on reconstituted, remoulded or undisturbed samples of clay to determine c_u
- *vane test*
  Unconsolidated undrained (UU) direct shear (DS) test in boreholes in the ground or on remoulded or undisturbed samples of clay in the laboratory to determine c_u

- *triaxial test*
  UU, CU or CD test on remoulded or undisturbed samples of clay to determine c_u, c' or ϕ' and occasionally CU or CD tests on sand to determine ϕ'
- *ring shear test*
  Consolidated drained (CD) direct shear (DS) test on remoulded samples of clay to determine the residual strength, c_r' and ϕ_r'.

Shear strength of sand

The following section discusses the behaviour of sand as found from re-constituted specimens in the laboratory. The virtually insurmountable problems of stress relief and fabric changes following sampling from the ground make the study of sand *in situ* very difficult.

Emphasis has, therefore, been placed on the use of *in situ* testing techniques to assess the state of the sand. These can only give an indirect measure of the relevant parameters although correlations have been developed empirically, for example between ϕ' and SPT '*N*' value.

Stress-strain behaviour *(Figure 7.11)*
When a sand particle arrangement is confined laterally strains can only occur in the vertical direction, such as in an oedometer test. As the vertical stress is increased small groups or arrays of particles in a loose state will collapse into the surrounding voids producing volumetric strains and providing a more tightly packed arrangement. The contact stresses between particles increase and each particle becomes more fixed in place because it is given less freedom of movement, a phenomenon known as 'locking'. Thus the vertical stress can increase with less increase of strain and the stress -strain curve is concave upwards (Figure 7.11).

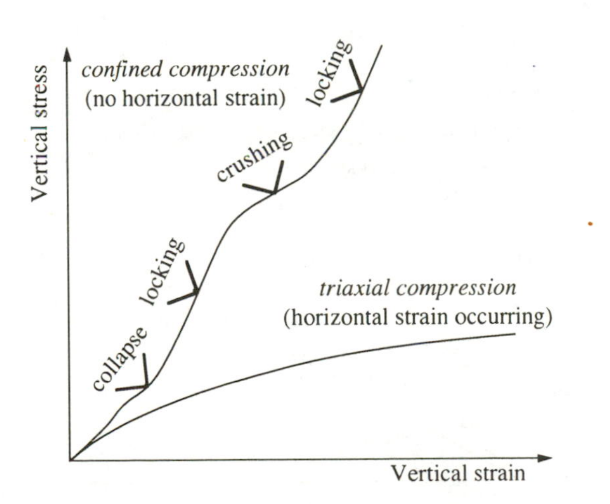

Figure 7.11 *Stress-strain behaviour of sand*

The term 'stiffness' represents the gradient of this curve so it can be seen that the soil is becoming stiffer. At very high stress levels the particles will begin to crush and fracture (yielding) allowing vertical strains to increase. An increase in the numbers of particles then produces further particle contacts, reduces the average contact stress between particles and causes the stiffness to continue increasing due to further 'locking'.

The stress levels at which structural collapse, locking and yielding occur will depend on the initial overall density and the inherent strength or crushability of the particles.

As the vertical stress increases the horizontal stress increases. The horizontal stress need not be as large as the vertical stress because part of the latter will be supported by the shearing resistance of the sand. The ratio of horizontal stress to vertical stress is given by K_o, the 'at rest' or no lateral strain condition (Equation 4.4). An established relationship (Jaky,1944):

$$K_0 = 1 - \sin \phi' \tag{7.5}$$

(see Table 4.1) shows that as the angle of internal friction ϕ' increases the horizontal stress decreases.

With triaxial compression as the vertical stress increases the horizontal stress does not increase, it is usually kept constant, allowing some of the sand particles to move horizontally which, in turn, allows the particles above to move downwards under the vertical stress so producing larger vertical strains. When the sand is initially loose, large vertical strains can occur for a given horizontal strain since there are void spaces for the grains to move into and there will be a net volume reduction or contraction.

In a dense sand, where the particles are in close contact with each other, a vertical strain will be accompanied by a large horizontal strain as the particles are forced outwards under the low confinement. This results in a net volume increase, referred to as 'dilatancy'.

If the pores are full of water and dissipation is prevented (undrained condition), this volume increase will cause a reduction in the pore water pressure, which will increase effective stresses and, in turn, increase the shear strength.

Dilatancy can be demonstrated simply by placing a small pat of wet silt or sand in the hand and squeezing it. As the soil structure expands, the water is drawn into the soil giving it a matt surface and a greater strength. If the soil pat is then jolted or shaken the soil structure contracts and water oozes out of the soil giving it a shiny surface and a weak consistency.

Shear box test *(Figure 7.12)*

When a direct measure of the shear strength of a granular soil is desired, a shear box test can be carried out, but it must be remembered that the results obtained will be for reconstituted specimens with densities and particle arrangements different to those found *in situ* and allowance should be made for this.

Nowadays, the use of the shear box test tends to be for the investigation of the shear strength properties of the more unusual granular materials where correlations are not available or are unreliable. These include crushable sands (e.g. calcareous, vesicular), granular fills (fragmented rock particles, both soft and hard) waste materials (e.g. colliery spoil) and the shear strength of interfaces between two construction materials, e.g. steel and sand (steel piles) or plastic and clay (geomembrane and clay liner).

Other applications which have utilised this test are the quick, undrained strength of clay and cut-plane or reversal tests (returning the split specimen to its starting position) for the determination of drained residual strength, but these have generally been superseded by the triaxial and ring shear tests, respectively.

The basic principle of the test and typical results are illustrated in Figure 7.12. The standard shear box tests a 60 mm square specimen, although apparatus for 100 mm and 300 mm specimens is available.

The test is strain controlled in that the shear force is applied at a constant rate of strain. Different rates of strain are available so that undrained (quick) or drained (slow) conditions can be assumed in a clay specimen. These rates may vary from 1 mm/minute, requiring about 10 minutes to conclude a quick undrained test in a clay to less than 0.001 mm/minute, requiring several days to perform a test to ensure that drained conditions

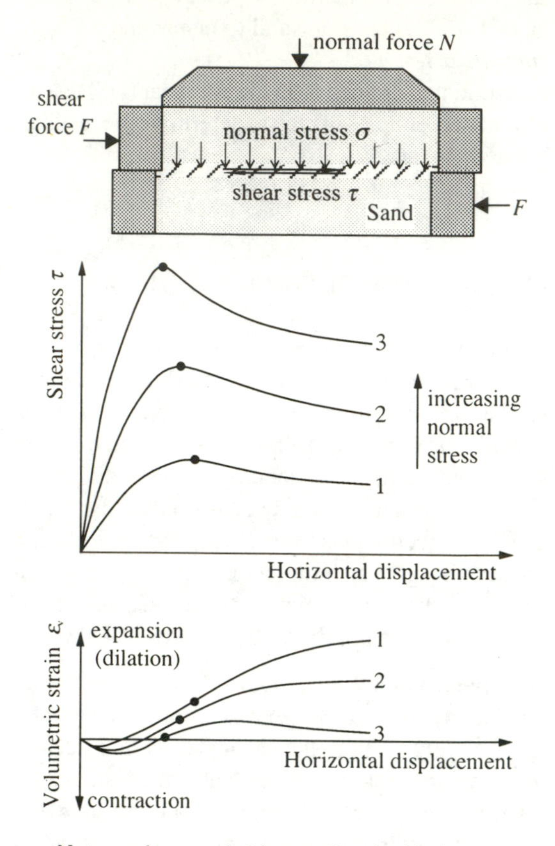

Note: peak strength • is associated with the maximum rate of dilation • given by

$$-\frac{d\varepsilon_v}{d\varepsilon_1}$$

where

ε_v = volumetric strain (negative for expansion)

ε_1 = major principal strain

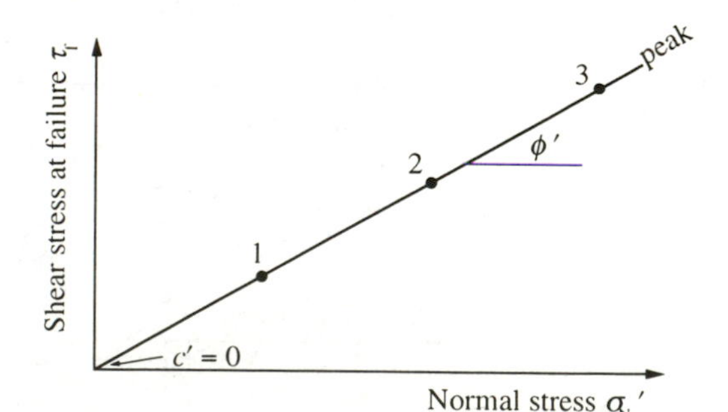

Figure 7.12 *Shear box test on dense sand*

apply in a fairly low permeability clay. For a clean free-draining sand, the 'quick' test is appropriate to ensure drained conditions but where the permeability is reduced by the presence of fines a slower rate of strain will be required.

For granular soils the test is fast and simple and, therefore, relatively inexpensive. The test is also appropriate in that shear planes or thin shear zones are often encountered in failed soil in nature. However, there are significant disadvantages which include:

1. the poor, uncertain control of drainage conditions and the inability to measure pore water pressures
2. the distribution of stresses on the shear plane are non-uniform and complex with stress concentrations at the sample boundaries
3. the specimen is forced to fail along a predetermined plane which may not be the weakest zone
4. as the shear stress is applied the planes on which the principal stresses act will rotate which will not accurately model the in-situ loading conditions.

Effect of packing and particle nature *(Figures 7.13 and 7.14)*

The shear strength (as given by the angle of shearing resistance ϕ') of a granular soil is obtained from two components:

- friction between the grains at particle contacts
- interlocking of the particles.

The former is denoted by the interparticle friction angle ϕ_μ which depends on the mineral type. For hard quartz grains this value has been found (Rowe, 1962) to be a function of grain size (Figure 7.13) and can be considered as the absolute lowest possible shear strength of the soil, see 'Constant volume condition' below. For other weaker or platy minerals, a lower interparticle friction angle can be expected, e.g. ϕ_μ for mica will be less than 15°.

In order to shear a mass of dense sand it will be necessary to move the particles up and over each other. This will require shear stresses in addition to those required to overcome interparticle friction. The additional shear stress will be larger for a greater degree of interlock.

Shearing a dense sand will, therefore, entail volume increases and greater shear strength (Figure 7.14). However, as the peak shear strength approaches the particles continue to move further apart and are unable to maintain the same degree of interlock so the shear strength after the peak decreases (strain-softening).

A loose sand will contain groups of particles which can collapse on shearing so a volume decrease is likely but as the particles move closer together there is a tendency for greater interlock and the shear stress is continually increasing (strain-hardening).

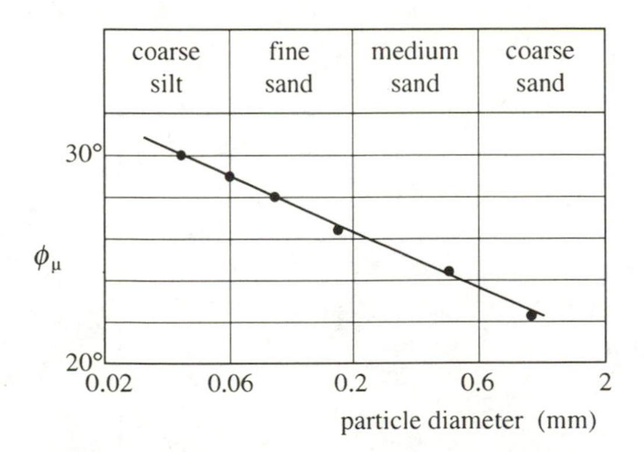

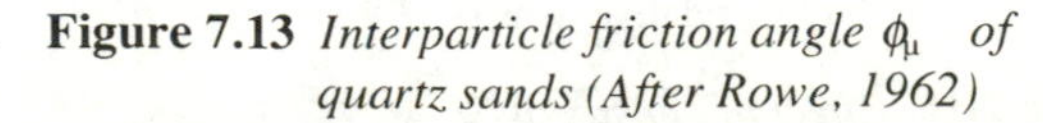

Figure 7.13 *Interparticle friction angle ϕ_μ of quartz sands (After Rowe, 1962)*

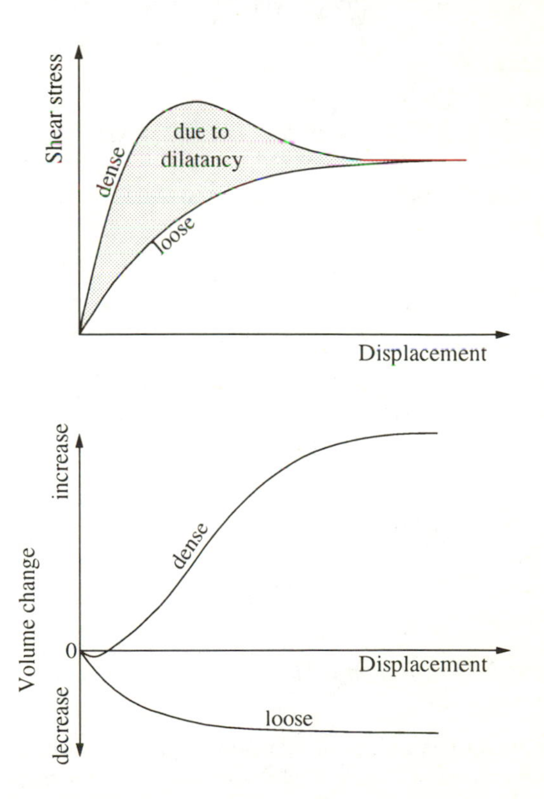

Figure 7.14 *Shear strength – effect of packing*

The angle of shearing resistance ϕ' of a granular soil is generally found to increase with:

- *increasing particle size*
  However, larger rock particles can contain more natural fractures than well-worn sands and with larger contact stresses because of the smaller number of particles present, crushing is much more likely to occur. This could cause additional settlements in granular backfills and rockfill structures.
- *increasing angularity*
  Angular particles may give a ϕ' value up to 5º more than rounded particles but these too could be more prone to crushing at high stresses.
- *increasing uniformity coefficient*
  More well-graded soils will produce better interlock and have more interparticle contacts so that the shear strength is increased (by up to 5º) and the risk of crushing is reduced.
- *stronger mineral particles*
  Weak particles are prone to crushing which reduces the ϕ' value, especially at higher confining stresses.

Constant volume condition *(Figures 7.15 and 7.16)*

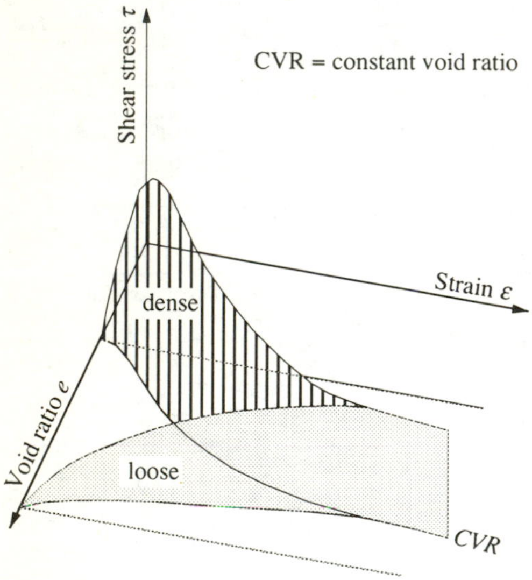

Figure 7.15 *Constant volume condition*

Figure 7.15 is a three-dimensional representation of the two parts of Figure 7.14 and shows that if the tests are continued to large strains, the void ratios will eventually coincide at the constant volume or constant void ratio (CVR) condition. The initially loose sand is increasing its strength as it densifies and the dense sand is losing its ability to support shear stress as it loosens until a common shear strength is obtained at the constant volume condition.

The constant volume strength can be used to define an angle of shearing resistance, ϕ_{cv} (Figure 7.4b) which can be adopted as the lowest possible value for a dense granular soil if there is some doubt that strains may exceed those at the peak value. The value of ϕ_{cv}' is always greater than ϕ_r' because there will still be some interlocking even when the constant volume condition is attained.

The effect of density either side of the constant volume condition on the angle of shearing resistance, ϕ', is illustrated in Figure 7.16.

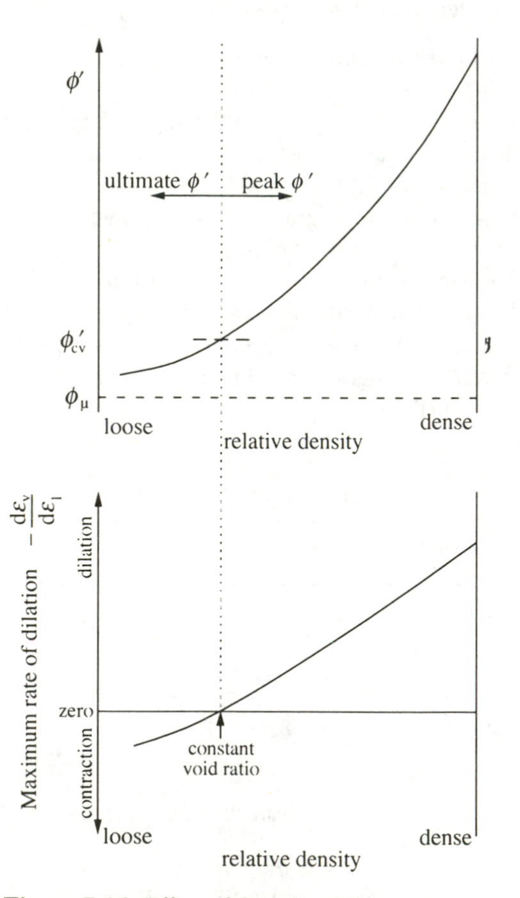

Figure 7.16 *Effect of density on ϕ'*

Shear strength of clay

Effect of sampling

Sampling clay soils from the ground is much more successful than sampling of sands because of:

- the low permeability of the clay preventing free drainage of the pore water. Removal of the *in situ* stresses following sampling transfers to a decrease in the pore water pressure (suction) which maintains the effective stress state and holds the soil structure together.
- the bonds between the mineral grain particles.

Various laboratory test procedures have been developed successfully for clay soils, in the knowledge that they would resemble the *in situ* condition.

The shear strength of a soil can only be provided by the resistance to shearing of the soil structure, the water in the pores having no shear strength at all. To obtain the shear strength of a clay soil as it exists *in situ* its structure must not be altered before it is sheared in the laboratory apparatus. To achieve this it is necessary to:

- take samples in a manner which produces the least disturbance such as using thin walled sampling tubes, laboratory specimens the same size as the sample and careful hand trimming
- avoid moisture content changes after sampling due to drying out or wetting up
- adopt a test procedure which controls water leaving or entering the specimen, i.e. ensure undrained or drained conditions apply.

Undrained cohesion, c_u

This parameter is determined for a clay in its *in situ* state ensuring no moisture content change, since c_u is uniquely related to the moisture content. The unconsolidated undrained (UU) test provides a good measure of *in situ* shear strength and is appropriate for methods of analysis where the rate of loading is fast enough to prevent pore pressure dissipation (drainage) such as with bearing capacity of foundations and piles, trench stability. The tests carried out are:

- unconfined compression test
- vane test
- quick undrained triaxial test

Unconfined compression test

In this test the soil is taken to failure by increasing the axial load only, with no surrounding confining stresses. The test is carried out on cylindrical specimens, usually 38 mm diameter and 76 mm long but larger diameters (100 mm) can be accommodated in larger compression machines.

The autographic apparatus (BS 1377:1990) is most commonly used for this test using 38 mm diameter specimens. The apparatus is portable, self-contained and hand-operated so it lends itself to use for on-site determination of clay strengths. The apparatus can be easily adapted to provide greater accuracy and sensitivity of results when clays of different strengths are tested and automatic adjustment for area corrections as the sample changes shape by barrelling is included to give directly the unconfined compressive strength.

The undrained shear strength of the clay is:

$$c_u = \tfrac{1}{2} \times \text{compressive strength} \tag{7.6}$$

The test has the advantage of being fast, simple, compact and, therefore, inexpensive. However, there are limitations which include the following.

1. The specimen must be fully saturated, otherwise compression of any air voids and expulsion of air will produce an increased strength and excessive movement is recorded on the autograph. This should be borne in mind when testing compacted clays particularly if they are at or below optimum moisture content
2. If the specimen contains any macro-fabric such as fissures, silt partings, varves, gravel particles or defects such as cracks or air voids, then premature failure may result because of these inherent weaknesses.
3. If the specimen has a low clay content then premature failure is likely since it will have poor cohesion.
4. The drainage conditions are not controlled. The test must not be carried out too slowly, otherwise undrained shear conditions may not exist.

Vane test *(Figure 7.17)*

This test consists of rotating a cruciform-shaped vane in a soil and producing a direct shear test on a cylindrical plane surface formed by the vane during rotation. The torque is applied at a constant rate of rotation and is measured by a spring balance or calibrated spring. The test provides a direct measure of the shear strength on the cylindrical failure surface.

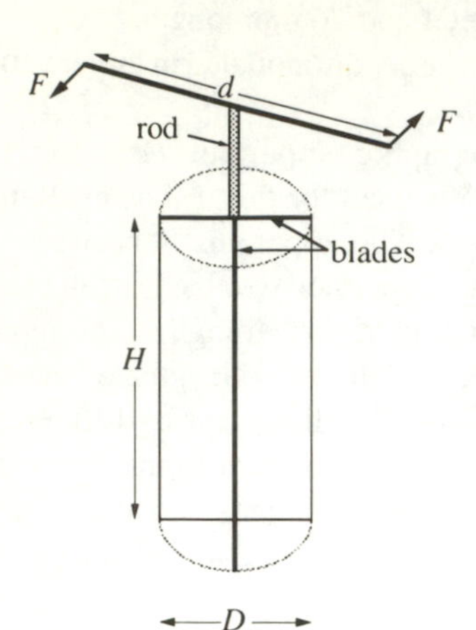

Torque applied $= T = F \times d$

Resisting torque

1) from surface of cylinder

$$\pi D H c_u \frac{D}{2}$$

2) from circular end areas

$$\frac{2\pi D^2}{4}\,\frac{2}{3}\,\frac{D}{2}\,c_u$$

$$\therefore\ T = \frac{\pi D^2}{4}\left(2H + \frac{2D}{3}\right) c_u$$

$$\text{for } \frac{H}{D} = 2 \quad T = \frac{7\pi D^3}{6} c_u$$

Undisturbed strength c_u from T_u (or F_u)

Remoulded strength c_r from T_r (or F_r)
(after several turns of vane)

$$\text{Sensitivity} = \frac{c_u}{c_r} = \frac{F_u}{F_r}$$

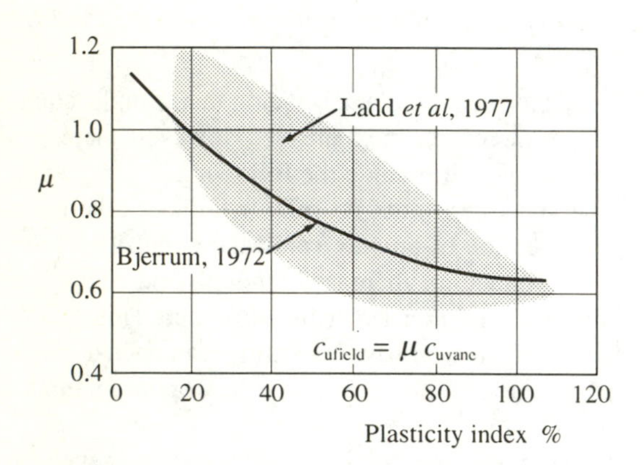

Figure 7.17 *Vane test*

No consolidation is permitted before shearing, and the soil is sheared quickly so the test is presumed to be unconsolidated undrained (UU) and, therefore, only suitable for clay soils. In sands or clays with sand layers, dilatancy during undrained or partially drained shear produces very high torque requirements and the test provides inappropriate shearing conditions for these soil types.

The test was devised for use in the field (BS 1377:1990) carried out at the bottom of a borehole. The vanes are typically 50 – 75 mm diameter and 100 – 150 mm long (height: diameter ratio of 2) with an area ratio less than 15% to minimise disturbance when pushed into the soil. The latter is the ratio of the cross-sectional area of the blades themselves to the area of the circle they produce. The determination of c_u is derived in Figure 7.17.

Smaller versions exist such as the laboratory vane (12.7 mm diameter × 12.7 mm high) used to determine the undrained shear strength of soft clays in tube samples and the hand vane (19 mm or 33 mm diameter). This is very useful as a laboratory vane, but also in the field such as in trial pits and for control of compacted clay fill strength in embankments and earth dams.

The test has the advantage of being fast and simple and is, therefore, inexpensive. It is also useful for obtaining an *in situ* undrained shear strength profile with depth for soft clays and clays which are difficult to sample such as sensitive clays. For these clays, thinner blades should be used to minimise disturbance.

However, there are disadvantages which include:

1 Uncertain control of drainage; undrained conditions are assumed but may not occur if permeable macro-fabric (silt layers) exists.

2 A fairly fast rate of rotation is adopted to provide undrained conditions but the strength can be overestimated if the rate of strain is too high.
3 Uncertainty of shear stress distribution on the cylindrical shear plane.
4 The shear surface tested differs from the field loading condition especially if the clay strength is anisotropic.
5 The results are affected by macro-fabric effects such as fissures, stones, silt partings, fibrous inclusions, roots.
6 The major disadvantage is that the test results show little correlation with the strengths (c_{ufield}) back-analysed from embankments placed on soft clays and which have failed. The vane usually overestimates the strength obtainable in situ and although a correction factor μ has been proposed by Bjerrum (1972) the scatter of the available data, shown in Figure 7.17, demonstrates its uncertain value (Ladd *et al*, 1977).

Triaxial test *(Figure 7.18)*
This apparatus was first developed in the 1930s and has largely replaced the direct shear test in commercial laboratories. The test consists of applying shear stresses within a cylindrical sample of soil by changing the principal stresses σ_1 and σ_3, the most common procedure being to keep the triaxial cell pressure σ_3 constant and increasing the axial or vertical stress σ_1 until failure is achieved. The essential features of the apparatus are illustrated in Figure 7.18.

Specimen sizes in the UK are standardised at 38 mm and 100 mm diameter with a height:diameter ratio of 2:1 to ensure that the middle section of the specimen is free to shear. If this ratio is less than 2:1 then shear stresses at the ends of the sample in contact with the platens will affect the results by constraining the failure planes.

The specimen is surrounded by a rubber membrane to prevent the cell fluid (water) entering the specimen and altering its moisture content. For weaker soil specimens a correction to account for the restraint provided by the membrane should be applied. This correction will be small if the specimen deforms into a barrel-shape and can often be ignored particularly for higher strength soils but if a single plane develops the membrane restraint can be significant.

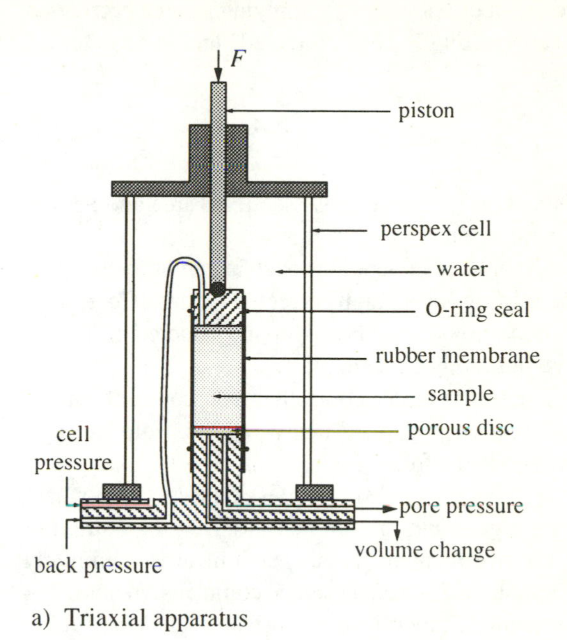

a) Triaxial apparatus

σ_1 $\frac{F}{A}$ σ_3

deviator stress

$$\therefore \frac{F}{A} = \sigma_1 - \sigma_3$$

specimen

b) Determination of deviator stress

Figure 7.18 *Triaxial test*

The axial stress is applied by a motorised drive which raises the specimen (and cell) against a piston reacting on a load frame. A proving ring between the piston and load frame measures the force, *F*, from which the principal stress difference or deviator stress, $\sigma_1 - \sigma_3$, is calculated, as shown in Figure 7.18.

As the stress is applied the specimen often becomes barrel-shaped so the vertical stress in the middle of the specimen must be determined from the force measured and this increased area by applying an area correction to each reading. The 'corrected' area A is obtained from:

$$A = \frac{A_0}{1-\varepsilon} \tag{7.7}$$

where A_0 is the initial cross-sectional area and ε is the vertical strain.

A more convenient approach is to plot the proving ring readings on a graph corrected for this effect. This method should not, however, be adopted if failure develops along a single plane.

The test is strain controlled as a constant rate of compression is applied and a rate of strain must be chosen for the following reasons.

1 In quick undrained compression tests (UU) where pore pressures are not measured a rate of strain of 2% of specimen length per minute is commonly adopted so that a test can be completed in about ten minutes. Impermeable end platens and the rubber membrane ensure undrained conditions.
2 In a consolidated drained test (CD) the shear stresses (via the axial stress) must be applied slowly so that excess pore pressure which may develop in the middle of the specimen can dissipate to ensure fully drained conditions throughout. The rate of strain and hence time to failure will be determined by the permeability of the clay. For many clays several days will be required to reach failure.
3 In a consolidated undrained test (CU) where during the undrained shearing stage the pore water pressure is measured at the base of the specimen sufficient time must be allowed to ensure that the pore pressure produced in the middle of the specimen is the same as at the base (equalisation) where it is measured.

For both the CD and CU tests the consolidation stage and the time to failure in shear can be reduced considerably by providing filter paper strip drains around the specimen and porous discs at the ends of the specimen permitting radial drainage or radial equalisation.

The major advantage of the triaxial test is that drainage conditions can be controlled so that conditions pertaining in the field can be modelled in the laboratory, see 'Effects of drainage', above.

Triaxial unconsolidated undrained test (UU) *(Figure 7.19)*

The unconfined compression test and the vane test are versions of this test approach but they have their limitations, see above. The UU test or quick-undrained test in the triaxial apparatus is one of the most common tests carried out in practice to determine the undrained shear strength of a clay at its *in situ* natural moisture content.

Its main application is in the design of shallow and pile foundations and in assessing initial stability of embankments on soft clays and suitability of clay fill for earthworks construction.

Standard procedure is to prepare three 38 mm diameter specimens from a 100 mm diameter 'undisturbed' sample (U100) and to test each specimen under a different confining pressure. The use of solid end

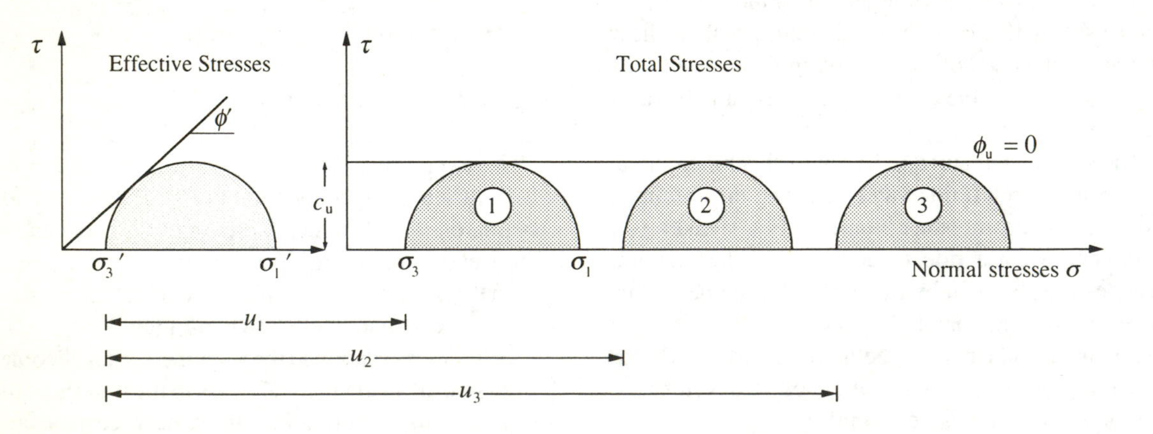

Figure 7.19 *Unconsolidated undrained (UU) triaxial test*

platens and rubber membrane ensure that no consolidation stage is permitted and that the specimen is undrained during shearing. If the specimens are fully saturated, of the same moisture content and have similar soil structure, then similar shear strength (c_u) values should be obtained.

Provided the specimens are fully saturated ($B = 1$), the application of the confining pressure in the cell will simply mean that the pore water pressure in each specimen is increased by the amount of confining pressure. No change in effective stresses occurs, so the Mohr circle at failure in effective stress terms will be the same and the shear strength measured will be the same irrespective of confining pressure (Figure 7.19).

The Mohr-Coulomb envelope for this test is, therefore, represented in total stress terms with c_u being the radius of the Mohr circle at failure $1/2(\sigma_1 - \sigma_3)$. All of these circles will have the same radius so the Mohr-Coulomb envelope is a horizontal straight line with intercept c_u and gradient $\phi_u = 0°$. Thus there is an apparent cohesion but no apparent friction.

For clays even of the same moisture content different strengths may be obtained due to the presence of macro-fabric such as inclined joints in fissured clays or gravel particles in glacial clays. Difficulties in preparing smaller diameter specimens from a U100 may also be experienced, particularly with stones present.

It is usually recommended that the test then be carried out on 100 mm diameter specimens (i.e. straight out of the U100) to avoid further disturbance and to represent the *in situ* mass fabric better. However, there is insufficient material in a U100 to prepare three separate specimens so either testing at one cell pressure (single stage) or testing the same specimen at three different cell pressures (multi-stage) can be adopted. The former will only give one Mohr circle which may be considered a disadvantage but if the material can be assumed to be fully saturated then this should not be so.

The **multi-stage (UU) test** *(Figure 7.20)* consists of applying the shear stresses under the first confining pressure at a somewhat slower rate to enable more readings to be taken until the load - strain plot starts to flatten. At this point the cell pressure is increased and shearing is continued for the second stage. It is then increased again for the third stage. The load-strain plots are extrapolated to a similar strain value (such as 20%) and the deviator stresses calculated from these loads for each confining pressure. The test is only

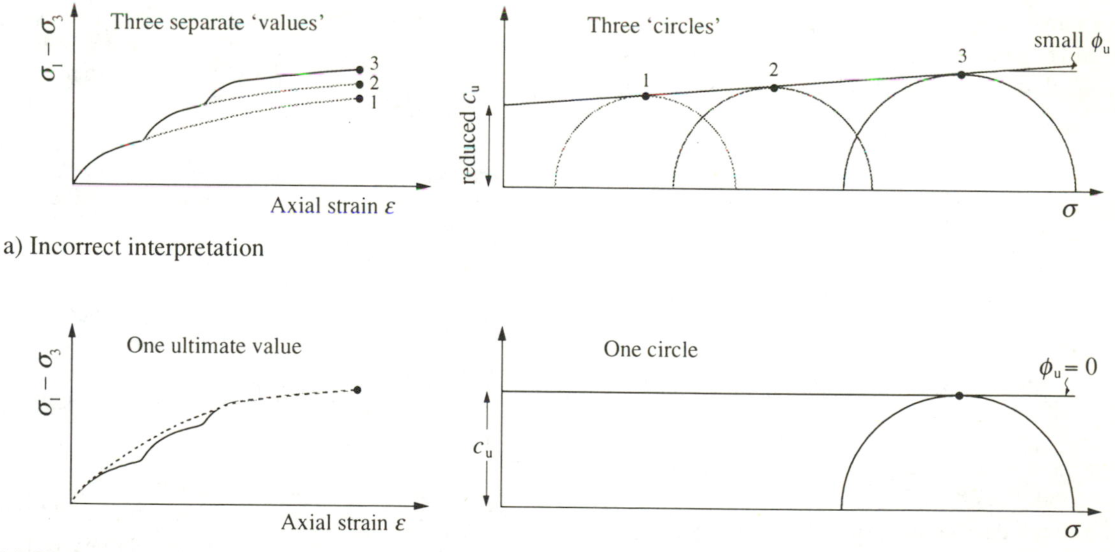

Figure 7.20 *Multi-stage (UU) triaxial test*

really suitable for plastic soils requiring large strains to failure and with no pronounced peak shear strength, as is the case for many normally consolidated clays.

If the soil is tested at constant moisture content, it should strictly only have one failure strength, irrespective of cell pressure. The 'steps' in the stress-strain curve may be caused by an initial increase in stiffness when the cell pressure is increased. There is a tendency to imagine three different Mohr circles when really the stress-strain plot is continuous. Thus results are often reported with three different Mohr circles giving a reduced cohesion intercept c_u and with a ϕ_u value greater than zero, up to 10º which is erroneous. Using these small ϕ_u values in a shallow foundation or immediate slope stability analysis can lead to a dangerous over-estimate of stability.

Effect of clay content and mineralogy

Although the undrained shear strength of a clay is related to its moisture content, decreasing as the moisture content increases, it also depends on the mineralogy of the clay (see Figure 12.5 in Chapter 12) and the amount of clay present (see Figure 12.6).

Partially saturated clays *(Figure 7.21)*

If the triaxial test is carried out on a partly saturated sample, Equation 4.5 shows that the application of the cell pressure σ_3 produces a pore water pressure smaller than the value of σ_3 ($B < 1$), so there is an increase in effective stress, with the soil structure compressing slightly and becoming stronger, so when the sample is sheared the deviator stress at failure is increased.

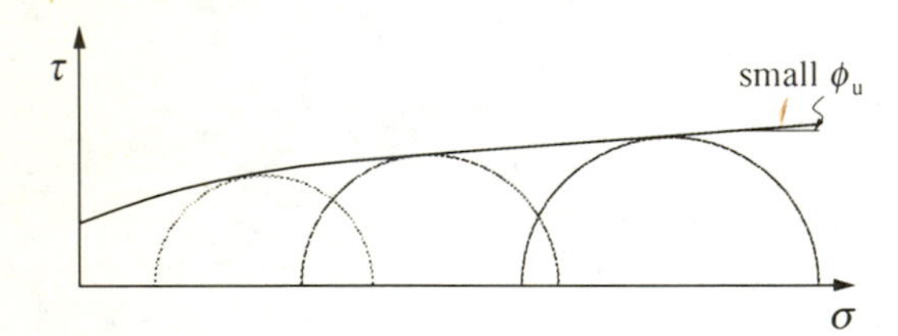

Figure 7.21 *Unconsolidated undrained triaxial test on partially saturated clay*

When these Mohr circles at failure are plotted in terms of total stresses, an inclined or curved Mohr-Coulomb envelope is obtained and the results then tend to be reported as an intercept c_u and an angle of $\phi_u > 0$. Again, care should be exercised in using these values in say a bearing capacity analysis since the state of partial saturation may be erroneous (due to the sample drying) or it may be transient (sampling carried out above a fluctuating water table or during a dry season).

A check on the degree of saturation (from moisture content and bulk density values) should be made to confirm the state of the specimen and hence its relevance to *in situ* conditions.

Fissured clays *(Figure 7.22)*

Fissures exist in most overconsolidated clays producing planes of weakness. The strength on these fissures is described as the fissure strength and is much lower (as low as 10%) than the intact strength of the clay between the fissures. The fissures, therefore, dictate

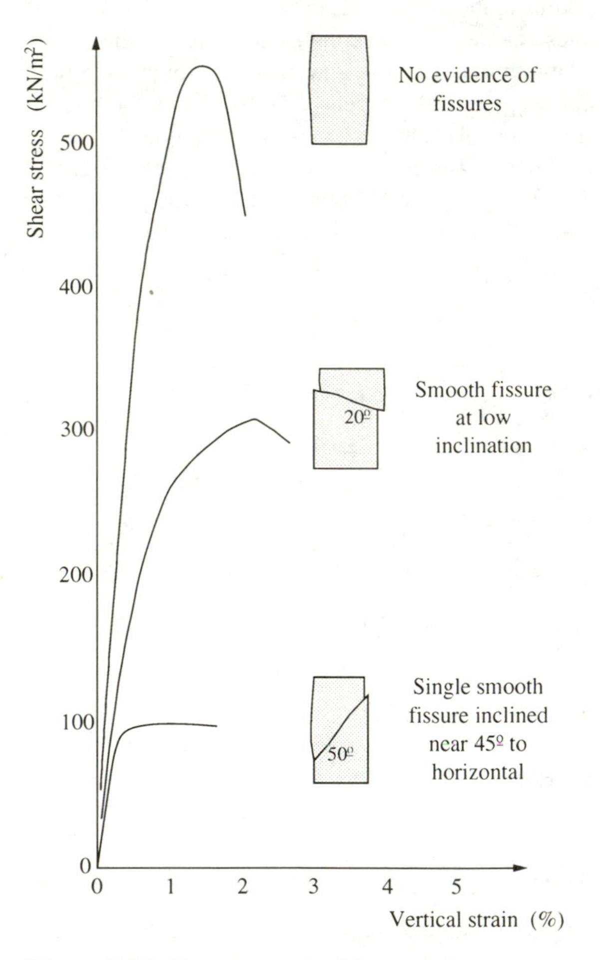

Figure 7.22 *Shear strength of fissured clays (From Marsland, 1971)*

the strength en masse of the clay so that the mass strength or design strength should be nearer the fissure strength but not necessarily equal to it.

The spacing between fissures usually increases with depth from typically 10 – 50 mm in a highly fissured upper zone to more than 100 mm at depth. It has been found that to represent the mass strength of the clay, specimens with a diameter at least equal to the fissure spacing are required for testing.

Some tests reported by Marsland (1971) are reproduced in Figure 7.22 and show the effect of the presence and inclination of fissures in 38 mm diameter specimens. If several of these results are then plotted at the depth of sampling a wide scatter is obtained. Testing larger samples (100 mm diameter) should produce less scatter and be nearer to the mass strengths although there may still be discrepancies at greater depths where fissure spacing increases.

If the samples are not tested soon after sampling, opening and extension of the fissures and small microcracks will cause a reduction of strength which can be significant even within hours after sampling. The samples should also be tested under confining pressures at least comparable to the in situ stress, otherwise reduced strengths may be recorded because the fissures have not been entirely re-closed.

Variation with depth *(Figures 7.23 and 7.24)*

It can be seen, from the discussion in Chapter 4 on the compressibility of a clay soil, that for a normally consolidated clay as depth increases the effective stress increases and the void ratio decreases.

For a truly normally consolidated clay the soil at the surface (present ground level) will be of a mud consistency with a moisture content close to its liquid limit. Assuming a constant compression index C_c with depth, the variation of moisture content for low and high plasticity clays has been plotted in Figures 7.23 and 7.24. The undrained shear strength, c_u, of a normally consolidated clay is related to moisture content and has been found by Skempton (1953) to increase linearly with effective stress p_o':

$$\frac{c_u}{p_0'} = 0.11 + 0.0037 I_p \qquad (7.8)$$

where I_p is the plasticity index.

For a lightly overconsolidated clay, say with 5 m of soil removed due to erosion, the clay at all depths will swell and its moisture content will increase, but to a lesser extent, determined by the swelling index C_s. This is the gradient of the void ratio-log effective stress curve on the unloading/reloading line. Assuming the swelling index to be $0.25 \times C_c$ and that the water table remains at ground level, a moisture content profile has been plotted in Figure 7.23.

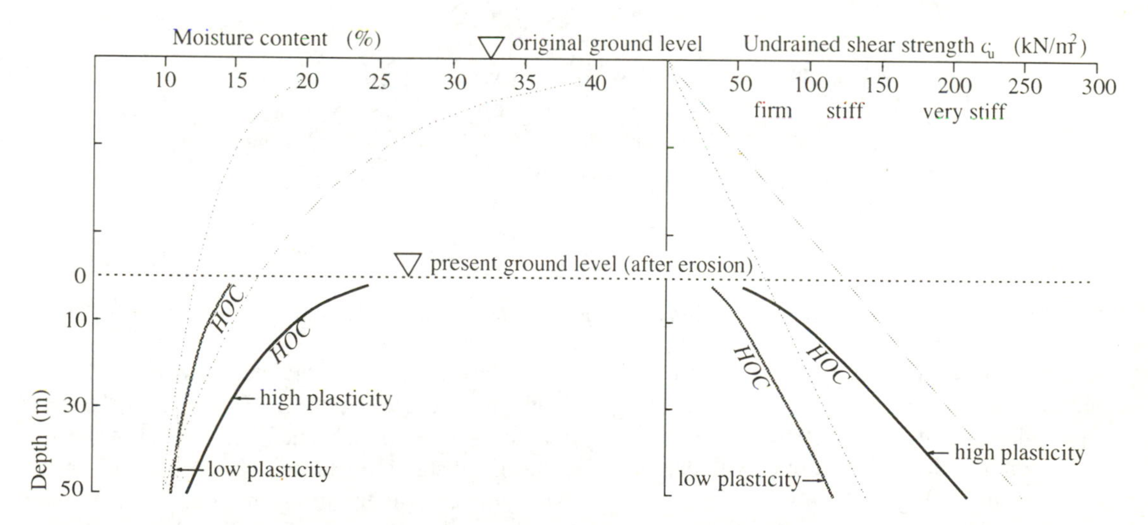

Figure 7.24 *Strength variation with depth – heavily overconsolidated clays*

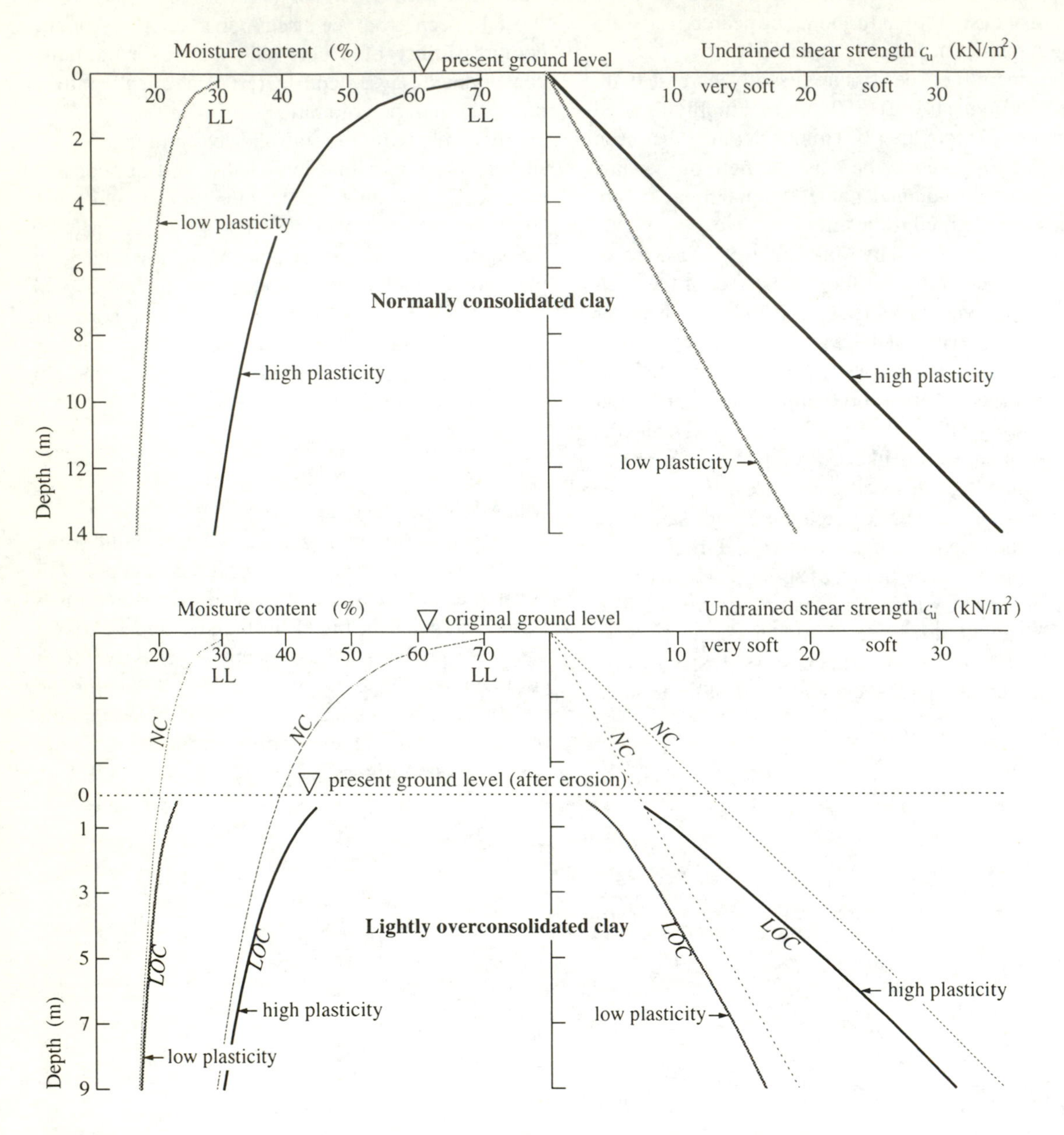

Figure 7.23 *Strength variation with depth – normally and lightly overconsolidated clays*

The undrained shear strength of a lightly overconsolidated clay is the same as a normally consolidated clay, provided their moisture content is the same. With these assumptions the variation of c_u with depth for both clays is plotted in Figure 7.23. A site comprising lightly overconsolidated clay has shear strengths greater than one of normally consolidated clay.

For a heavily overconsolidated clay, say with 50 m of soil removed due to erosion (Figure 7.24), the shear strengths will be significant largely because the soil has been substantially normally consolidated in the past. Note the higher moisture contents and lower shear strengths which are possible close to present ground level.

The above does not include allowances for the effects of desiccation, cementation, secondary compressions and other processes.

Shear strength of clay – frictional characteristics *(Figure 7.25)*
It cannot be stressed enough that the shear strength of clay is determined by the effective stresses and the frictional characteristics both of which reside in the mineral grain structure. As effective stresses increase, the shear strength increases and clay minerals with lower plasticity tend to give greater frictional characteristics, represented by a higher ϕ' value (Figure 7.25).

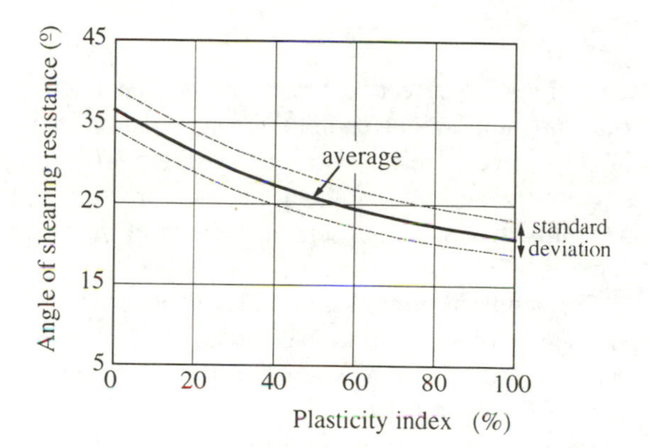

Figure 7.25 *Angle of shearing resistance of clays (From NAVFAC, 1971)*

Test procedures
The Mohr-Coulomb envelope is represented by the angle of shearing resistance ϕ' and a small cohesion intercept c' plotted in effective stress terms. To increase the effective stress in a clay specimen it is necessary to increase the total stress around it and allow the pore water pressure increase to dissipate so each specimen must be consolidated to a different effective stress state before shearing.

The shearing stage is then carried out by changing the external total stresses on the specimen with either:

1 *undrained conditions*
No volume changes are permitted. The pore water pressure must then be measured to obtain effective stresses. This test procedure is described as consolidated undrained (CU).

or

2 *drained conditions*
Volume changes are permitted and shearing is carried out at a rate slow enough to ensure no pore water pressure changes occur so that effective stresses then equal the total stresses. The test procedure is described as consolidated drained (CD).

The stress history dictates the behaviour during shear in that the mineral grain structure:

- of *normally consolidated clays* contracts in a drained test resulting in reducing volumes. It attempts to contract in an undrained test but cannot since volume changes are prevented, resulting in increased pore water pressures.
- of *overconsolidated clays* expands or dilates in a drained test resulting in increased volumes. It attempts to expand in an undrained test resulting in decreased pore water pressures.

These effects are illustrated in Figures 7.26 and 7.27 which show that normally consolidated and overconsolidated clays of the same moisture content give the same strength when tested undrained but when tested drained the normally consolidated clay gives a much greater strength since it is becoming stronger during contraction whereas the overconsolidated clay is becoming weaker during expansion.

Triaxial consolidated undrained test (CU) *(Figure 7.26)*
The procedure carried out for this test relates to a number of field applications, see Effects of drainage. It has two main applications:

1 for the undrained shear strength (c_u) in total stress terms following a consolidation stage. This gives the relationship between c_u and confining stress σ_3 or overburden pressure p_o'.

2 for the effective stress parameters c' and ϕ'. For this pore water pressure measurements are taken during the shear stage to determine effective stresses at failure.

The test consists of three stages:

- saturation
- consolidation
- compression (shear).

For the accurate measurement of pore water pressures, it is essential that air is prevented from entering the pressure system so all air voids must be eliminated to produce a saturated specimen. The *B* value is then

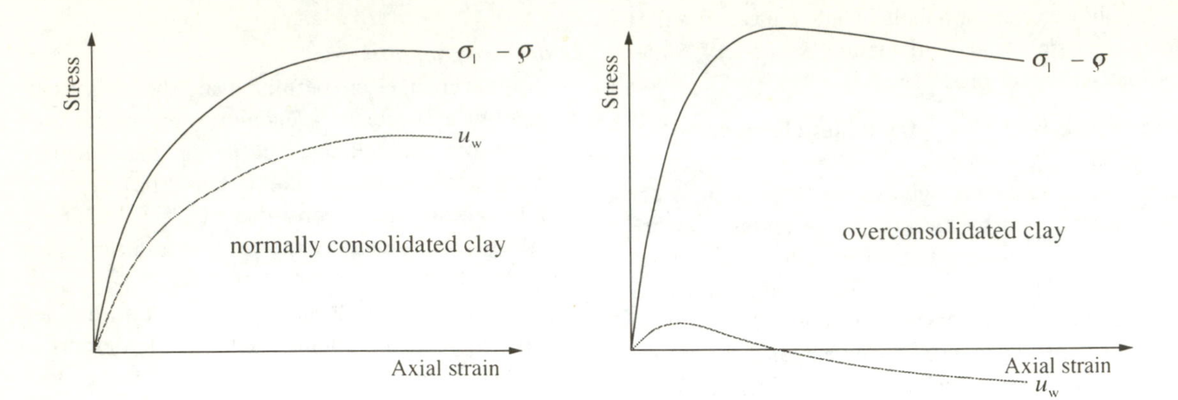

Figure 7.26 *Consolidated undrained test on clay*

virtually unity and changes of confining pressure $\Delta\sigma_3$ are reflected by equal changes in pore water pressure Δu_w.

Saturation is produced by increasing the pore water pressure in the specimen to remove the air. This is achieved by increasing the back pressure to the pore fluid in increments while, at the same time, increasing the cell pressure in increments so that changes in the effective stress are kept to a minimum. During this process small amounts of water will move into the specimen to achieve full saturation.

The cell pressure increments must be larger than the back pressure increments, by about 10 kN/m^2, so that effective stresses are always positive. At the end of the saturation stage the cell pressure and back pressure will be quite large values but their difference will be small. The specimen is now ready to commence the consolidation stage.

The consolidation stage is commenced by increasing the cell pressure and keeping the back pressure constant. With drainage prevented (valve closed), this induces an increase in pore pressure in the specimen which should be equal to the increase in cell pressure, if B is close to unity. An example of the determination of the pore pressure parameter B is given in Worked example 4.6.

The difference between this pore pressure and the back pressure is the excess pore pressure to be dissipated during consolidation, once the drainage valve is opened. During this stage the pore pressure then dissipates (reduces) to the back pressure value and water is squeezed out of the specimen. At the same time the effective stress in the specimen is increasing until, at the end of the consolidation stage, the effective stress is given by the difference between the cell pressure and the back pressure and is the effective stress present at the start of the shear stage.

The sample volume change and pore pressure dissipation are plotted against time (square root or log) and when 95% dissipation is reached the consolidation stage can be terminated.

The volume change versus root time curve will give a measure of the permeability of the soil and, using methods derived by Bishop and Henkel (1962), the time to failure for a particular test (full drainage in the drained shear test or full equalisation in the undrained shear test) can be determined and hence the rate of strain. The specimen is now ready to commence the shear stage.

With the drainage valve closed to prevent movement of water into or out of the specimen (undrained condition) the axial stress σ_1 is increased while keeping the cell pressure σ_3 constant. This produces shear stresses in the specimen which give rise to changes in the pore water pressure and these are measured, see Figure 7.26. Thus the effective stresses σ_1' and σ_3' at failure (for the Mohr circle) and at any time during the shear stage (for stress paths) are obtained.

Three specimens should be tested at different cell pressures (three different effective stresses at the start of the shearing stage). One specimen should be tested at the *in situ* effective stress level with the others at two, three or four times this value.

The main advantage of the CU test compared to the CD test is that the shear stage is much faster, about eight times faster if side drains are provided and sixteen times faster without side drains (Head, 1986). The stress paths from a CU test are also more informative.

Triaxial consolidated drained test (CD)
(Figure 7.27)
This test is carried out with the same saturation and consolidation stages as for the CU test, and then ensuring that no pore pressures develop during the shearing stage by allowing full drainage. Thus total stresses applied to the specimen will also be the effective stresses inside the specimen. The rate of strain must be slow enough to ensure full equalisation.

The main application for the test is to determine the effective stress parameters of the soil, c_d and ϕ_d, see Exercise 7.9. The main disadvantage is that the shear stage of the test takes much longer. However, for soils which are virtually fully saturated the saturation stage is not so important and drainage to an open burette can be permitted, eliminating the need for a back pressure system.

The axial stress σ_1 is increased while keeping the cell pressure σ_3 constant. This produces shear stresses in the specimen which give rise to volume changes which are measured. These volume changes will involve specimen reductions for a contractant soil and increases for a dilatant soil. The total stresses σ_1 and σ_3 at failure are also the effective stresses σ_1' and σ_3' so the Mohr circle can be plotted directly.

Critical state theory

This model of soil behaviour was developed at Cambridge University by Roscoe, Schofield and Wroth (Roscoe et al, 1958). It has been described by Schofield and Wroth (1968) and simply explained by Atkinson and Bransby (1978).

The theory provides a unified model for the behaviour of saturated remoulded clay by observing changes in:

- the mean stress or consolidation stress p'
- the shear stress or deviatior stress q'
- the volume change v, specific volume.

These are observed on a three-dimensional plot, $p' - q' - v$, Figure 7.28, where:

$$p' = \tfrac{1}{3}(\sigma_1' + \sigma_2' + \sigma_3') \tag{7.9}$$

$$q' = \frac{1}{\sqrt{2}}\left[(\sigma_1' - \sigma_2')^2 + (\sigma_2' - \sigma_3')^2 + (\sigma_3' - \sigma_1')^2\right]^{\frac{1}{2}} \tag{7.10}$$

$$v = 1 + e \tag{7.11}$$

For the common case of axial symmetry (oedometer, triaxial tests) where $\sigma_2' = \sigma_3'$:

$$p' = \tfrac{1}{3}(\sigma_1' + 2\sigma_3') \tag{7.12}$$

$$q' = \sigma_1' - \sigma_3' \tag{7.13}$$

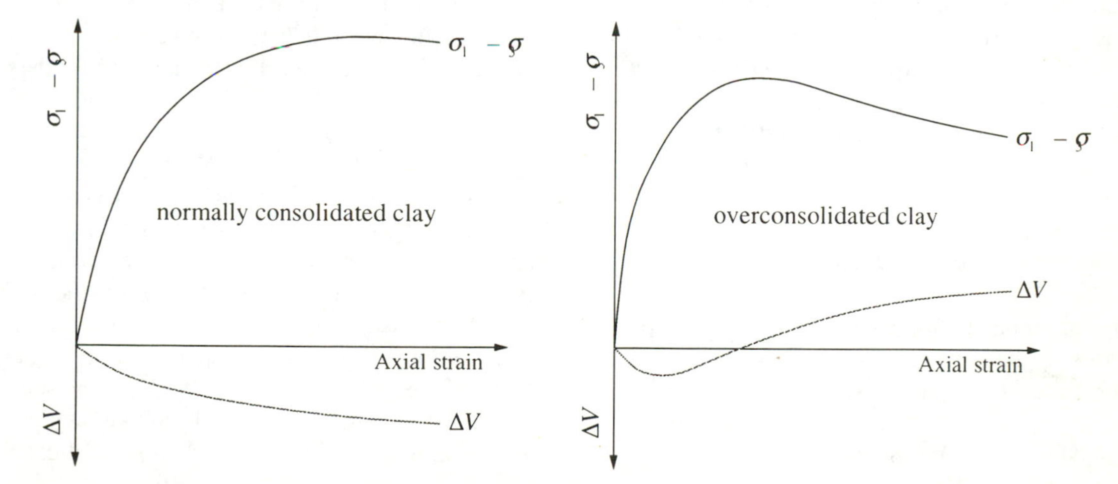

Figure 7.27 *Consolidated drained test on clay*

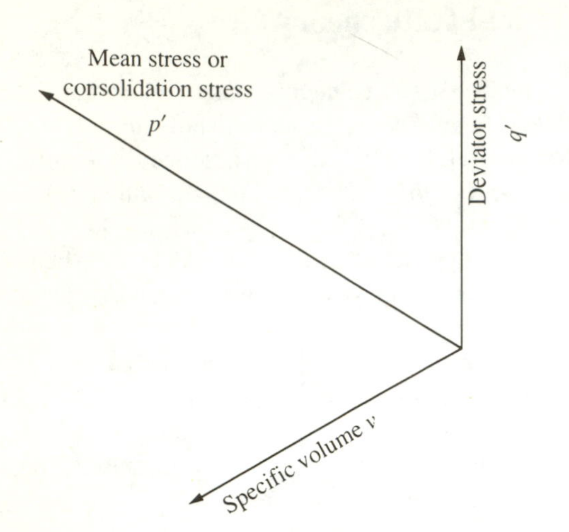

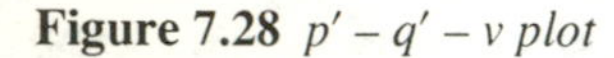
Figure 7.28 *$p' - q' - v$ plot*

The theory postulates that there is a state boundary surface on the $p' - q' - v$ graph which separates states in which a sample can exist and states in which a sample could not exist. As a summary at this stage the complete state boundary surface is shown in Figure 7.29 and the various components which make up this surface are described separately below.

The surface is assumed to be a mirror image for negative values of q'. The state of a soil can only reach the state boundary surface by loading, unloading takes the state beneath the surface. Strains are not represented on the three-dimensional plot but it is assumed that deformations are elastic and recoverable while the state remains beneath the surface but when loading takes the soil to the boundary surface the soil yields and when the state moves across the boundary surface strains are plastic and irrecoverable.

The state boundary surface is a curved surface in all directions but when idealised it may be seen to be made up of:

- ***isotropic normal consolidation line (ICL)*** *(Figure 7.30)*
  Isotropic compression means that the principal stresses applied are equal such as in the triaxial compression test when only the cell pressure is applied ($\sigma_1' = \sigma_2' = \sigma_3'$). No shear stresses are applied so $q' = 0$ and the state exists on the p'-v plane, see Figure 6.8.

  Normal compression or consolidation means that the consolidation stress p¢ is increasing and producing plastic strains leading to a permanent reduction of the void space so v decreases. Plastic straining is occurring by structural rearrangement of the soil skeleton and the state of the soil is moving along part of the state boundary surface. As shown in Chapter 6, when this line is plotted on a log scale a straight line is often obtained so the relationship for the ICL is given by:

$$v = N - \lambda \ln p' \qquad (7.14)$$

  If at any point (such as A on Figure 7.30) the stress is reduced the state will not return back up the ICL since there has been irrecoverable strains. Only the elastic strains are recovered and the state then moves away from the state boundary surface back along the swelling line (ISL).
  If the soil is then loaded again the state will follow the swelling line along the recompression line (IRL) until reaching the state boundary surface (at the ICL) when it will move along the ICL with further plastic strains. The swelling and recompression lines are represented by:

$$v = v_\kappa - \kappa \ln p' \qquad (7.15)$$

  Soils for which the state exists on the ICL part of the state boundary surface are described as normally consolidated and loading will immediately and continuously cause plastic strains, while soils existing on a swelling line are described as overconsolidated and loading will produce only elastic strains until the state reaches the boundary surface.

- ***K_o normal consolidation line (K_oCL)*** *(Figure 7.31)*
  Although not an essential part of the state boundary surface, this line is of fundamental importance, since it represents the condition of the soil in the ground. During deposition and erosion the soil exists in a state of no horizontal strain (known as the K_o condition). The graph is represented in Figure 7.31, where the line ABCD (K_oCL) represents normal compression and CE is a swelling line (K_oSL) on which the soil is in an overconsolidated state.

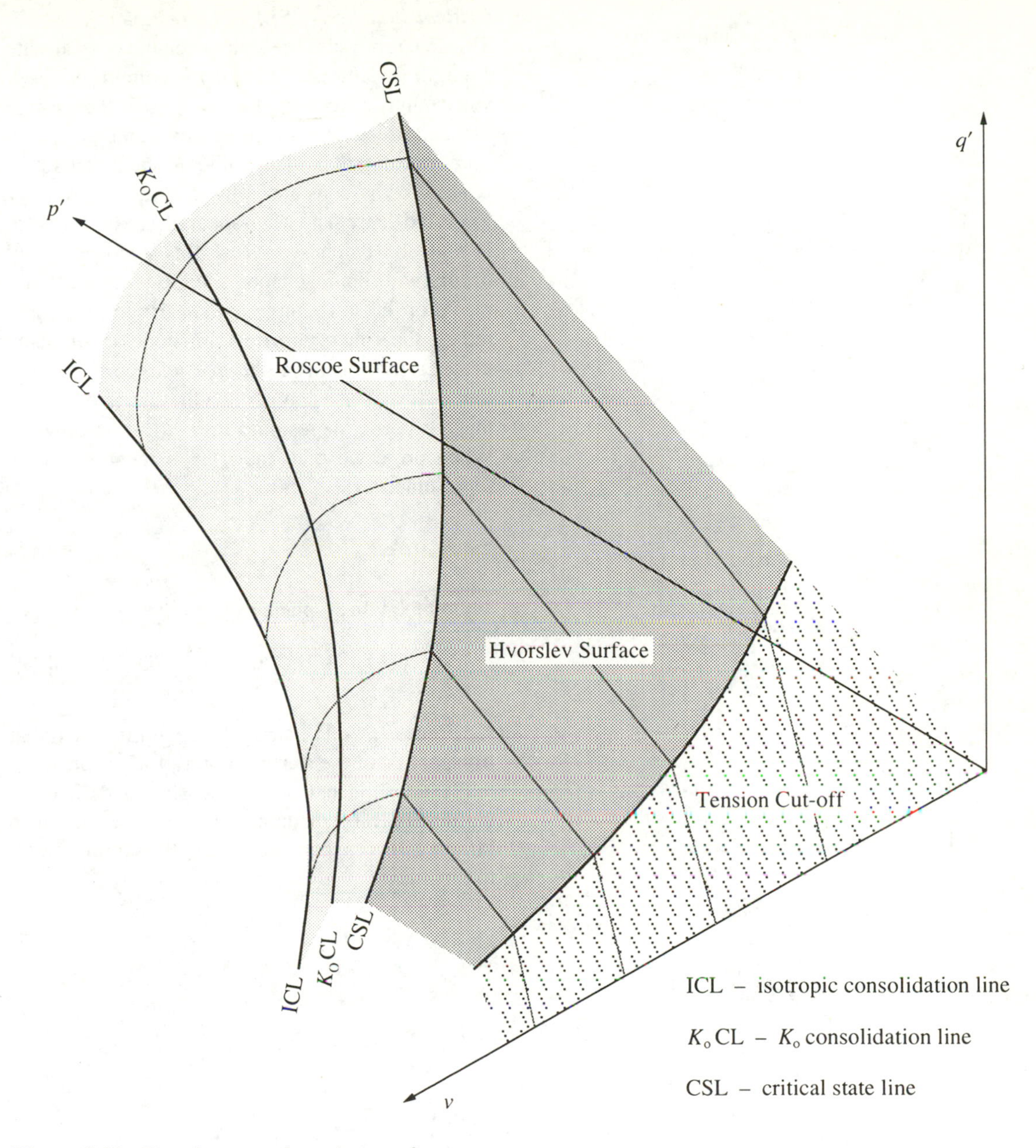

Figure 7.29 *Complete state boundary surface*

Elastic and plastic strains occur on ABCD but only elastic strains occur on CE and EB. The equation for the K_o normal compression line (K_oCL) is given by:

$$v = N_0 - \lambda \ln p' \tag{7.16}$$

and the equation for the swelling line (K_oSL) is given by:

$$v = v_{\kappa 0} - \kappa \ln p' \tag{7.17}$$

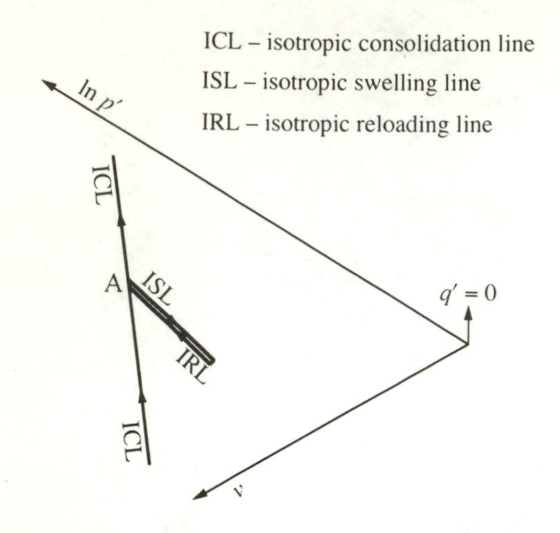

Figure 7.30 *Isotropic consolidation and swelling lines*

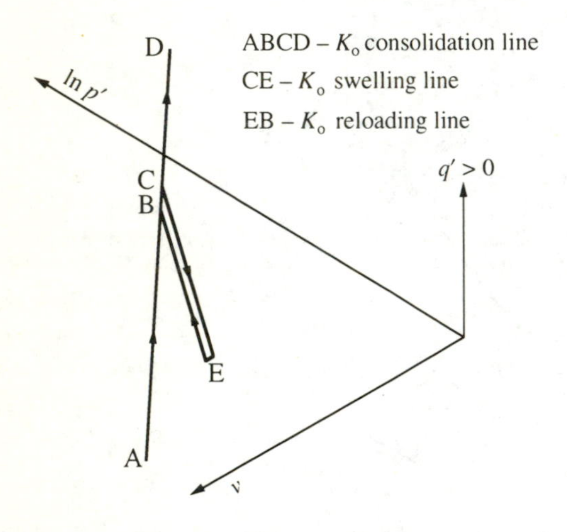

Figure 7.31 K_o *consolidation and swelling lines*

- ***Critical state line (CSL)*** *(Figure 7.32)*
  This line represents the state when the critical state strength (or ultimate strength) is being mobilised, when failure has occurred and with no further change in mean stress, deviator stress or volume (p', q' and v are constant) the soil continues to shear. This state is akin to the constant volume state achieved by loose and dense sand when sheared, see Figure 7.15. It must be emphasised that this condition only occurs with homogeneous shearing when all elements of the soil sample are shearing at the same rate. If shearing occurs on a thin zone or slip plane then rearrangement of clay particles semi-parallel to the slip surface may reduce the strength only on that surface to the residual strength, see below.
  When projected onto the q'-p' plane this line is represented by:

$$\pm q' = Mp' \tag{7.18}$$

and on the v-ln p' plane by:

$$v = \Gamma - \lambda \ln p' \tag{7.19}$$

see Figure 7.32. The significance of this line is that irrespective of the starting point of a sample (normally consolidated or overconsolidated) and regardless of test path, undrained or drained, ultimate failure will occur once the state of the sample moves to this line.

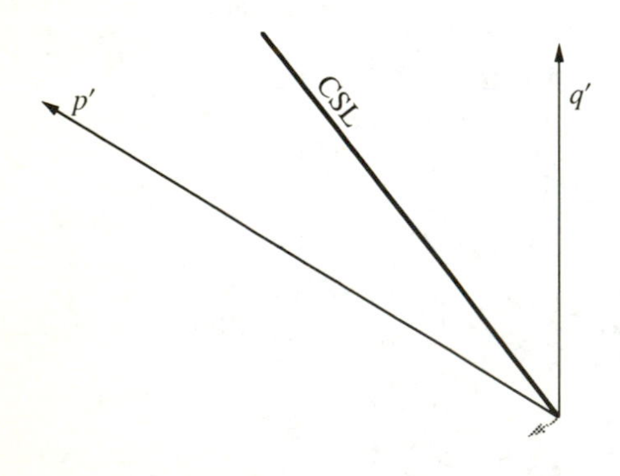

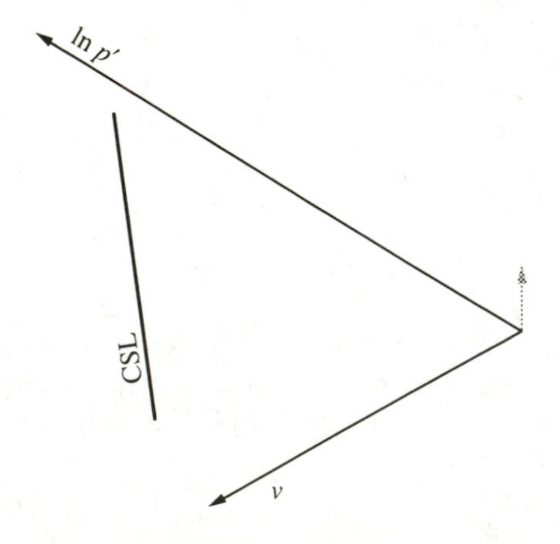

Figure 7.32 *Critical state line projections*

- ***Roscoe surface*** *(Figures 7.33–7.37)*
  This is the part of the state boundary surface between the ICL and the CSL. Tests which commence on the ICL (isotropically normally consolidated) follow the Roscoe surface between the ICL and the CSL, irrespective of whether the test is undrained or drained.

An undrained test must exist on a constant volume plane (Figure 7.33), so there is an undrained plane for each value of v. The path followed by an undrained test starting from the ICL follows the Roscoe surface on the undrained plane up to the CSL.

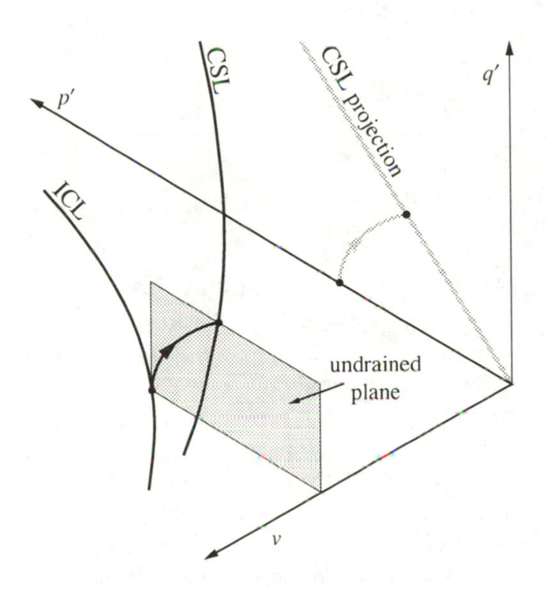

Figure 7.33 *Stress path for an undrained test*

For a drained test it can be shown that $dq/dp = 3$ so all drained planes must be parallel to the v - axis and have a slope of 1 in 3 on the q' - p' projection (Figure 7.34). Samples starting from the ICL will have a tendency to contract or compress during shear so the test will follow the Roscoe surface on the drained plane by reducing in volume.

It has been found that the shape of the undrained test path starting from the ICL is the same for all values of p_e', the initial isotropic stress. Normalising the q'-p' curves by using q'/p_e' and p'/p_e' instead, one curve is obtained for undrained tests carried out at all values of v, as shown in Figure 7.35. The same

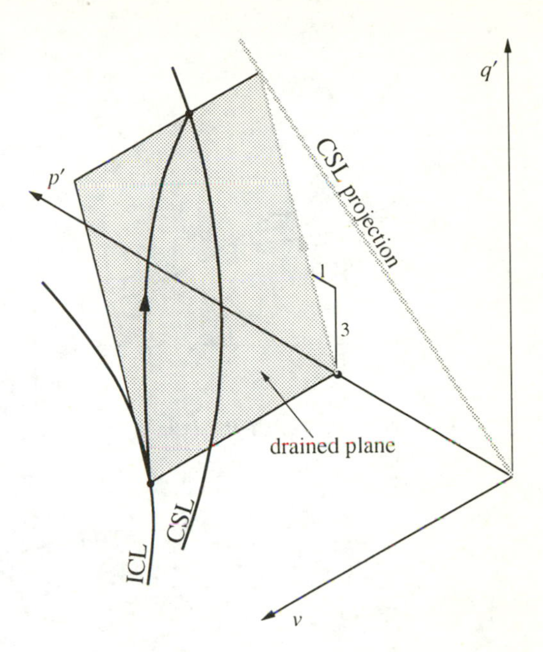

Figure 7.34 *Stress path for a drained test*

curve is obtained for a drained test provided the value of p_e' associated with each value of specific volume as the sample compresses is used.

Soils which are lightly overconsolidated with different overconsolidation ratios can be obtained with the same moisture content and, therefore, lying on the same undrained plane, as shown in Figure 7.36. The paths for undrained tests will commence beneath the Roscoe surface and rise almost vertically until they reach the Roscoe surface where they move along the surface until reaching the critical state (Figure 7.37).

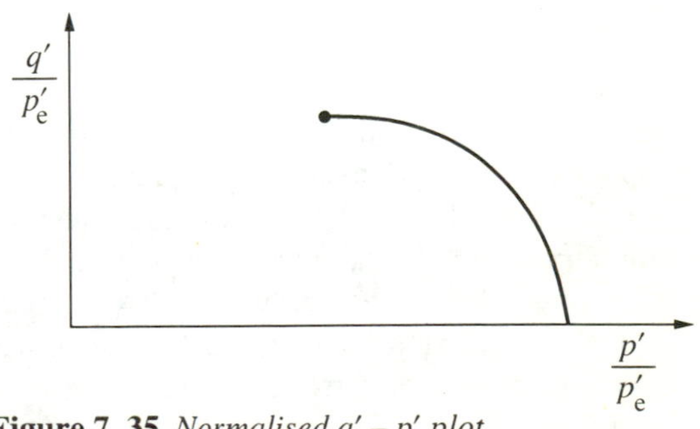

Figure 7. 35 *Normalised q' – p' plot*

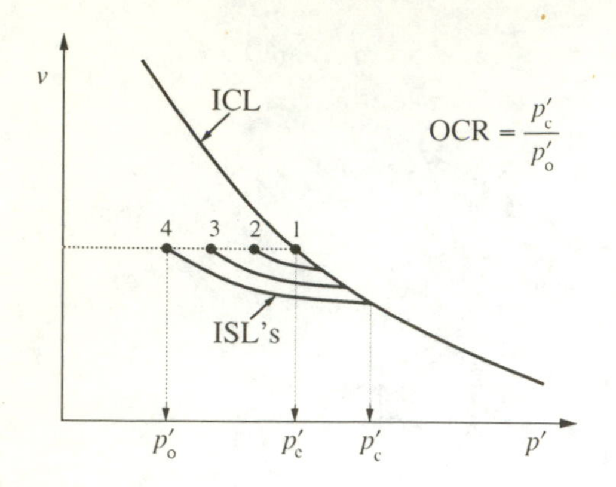

Figure 7.36 *Lightly overconsolidated clays*

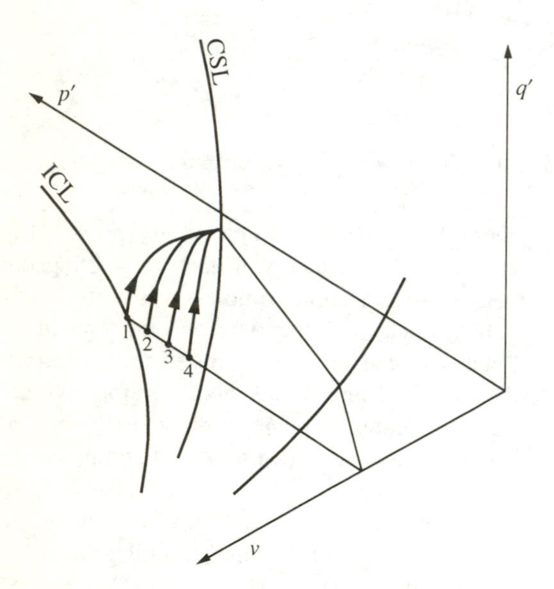

Figure 7.37 *Undrained test paths for lightly overconsolidated clays*

- ***tension cut-off** (Figure 7.38)*
  When the mean effective stress in a soil is zero it is implicit that no shear stress can be applied, i.e. when $p' = 0, q' = 0$ so the state boundary surface must pass through this point. The highest value of the ratio q'/p' corresponds to the case when $\sigma_3' = 0$. For a triaxial compression test by increasing the vertical effective stress σ_1' the tensile strength will be mobilised. Then:

$$q' = \sigma_1' \qquad p' = \tfrac{1}{3}\sigma_1'$$

so

$$q' = 3p' \tag{7.20}$$

giving an upper limit to the state boundary surface referred to as the tension cut-off, as in Figure 7.38.

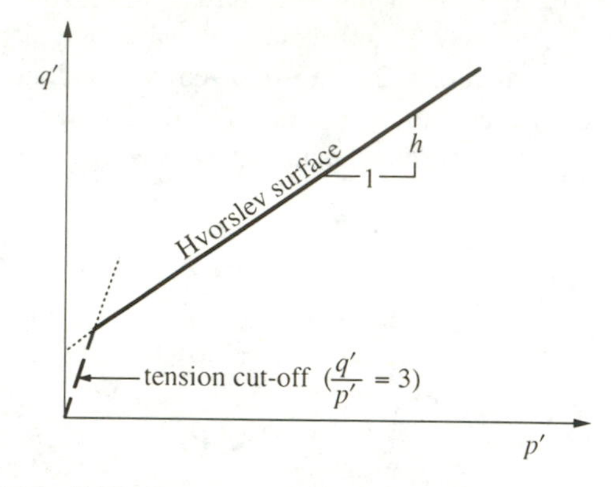

Figure 7.38 *Tension cut-off and the Hvorslev surface*

- ***Hvorslev surface** (Figures 7.38–7.40)*
  This surface intersects with the Roscoe surface at the critical state line. It represents the state boundary surface for heavily overconsolidated samples. For these soils, with different overconsolidation ratios following isotropic consolidation and swelling (Figure 7.39), the paths for undrained tests (all on an undrained, constant v plane) commence beneath the Hvorslev surface, rise near vertically until reaching the Hvorslev surface and then move along the surface until reaching the critical state.
  The path for a drained test will lie on a drained plane (Figure 7.40), commencing beneath the Hvorslev surface rising up the drained plane and then down the Hvorslev surface to the critical state line. Note that the maximum stress or peak strength q' occurs before the critical state is reached with volume expansion and work-softening occurring during the reduction in strength from the peak.
  With overconsolidated clays mobilising the peak strength often produces thin slip zones or planes in the sample with only the material in this zone moving down the Hvorslev surface and softening but with little softening of the remainder of the sample.

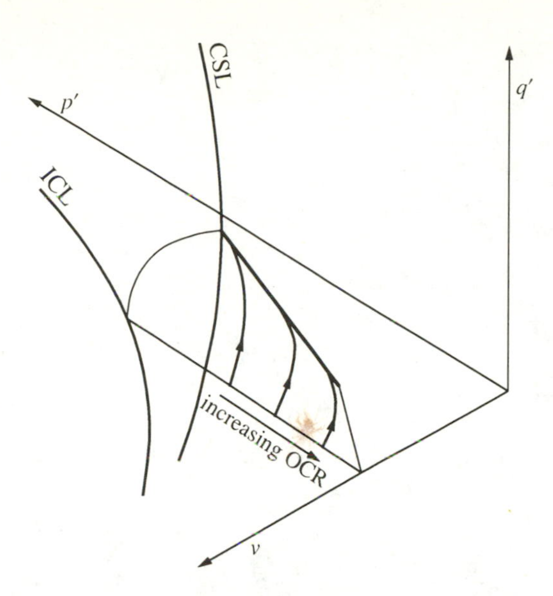

Figure 7.39 *Undrained test paths for heavily overconsolidated clays*

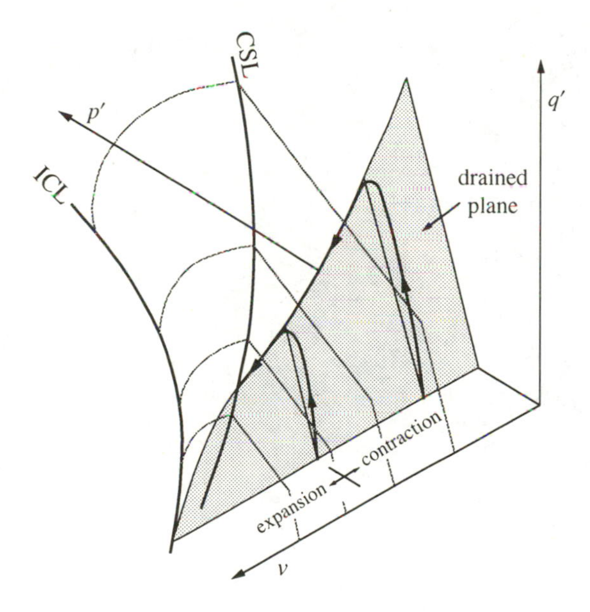

Figure 7.40 *Drained test paths for heavily overconsolidated clays*

Assuming for one value of specific volume v the Hvorslev surface is a straight line with a slope h the equation for this surface is:

$$q' = (M - h)\exp\frac{\Gamma - v}{\lambda} + hp' \qquad (7.21)$$

Residual strength

Although the critical state strength is often referred to as the ultimate strength this condition is achieved with homogeneous shearing, i.e. all of the sample is undergoing the same shear strain and these strains are not excessively large.

It has been observed, particularly from the study of old landslips (Skempton, 1964) where significant straining has occurred on thin shear surfaces that the operative shear strength on these surfaces was much lower than the critical state strength. For example, the residual ϕ_r' value for London Clay can be as low 10º whereas at the critical state ϕ_{cv}' is greater than 20º.

It is essential, therefore, to identify the presence, or otherwise, of pre-existing slip surfaces in a clay soil on a sloping site. Small changes in surface topography or pore pressure conditions can reactivate an ancient landslip with catastrophic consequences.

Residual strength is attained when large shear strains have occurred on a thin zone or plane of sliding in a clay soil where the clay particles have been rearranged to produce a strong preferred orientation in the direction of the slip surface. Lupini *et al* (1981) recognised three modes of residual shear behaviour, as follows.

- *Turbulent*
  This occurs where behaviour is dominated by rotund particles. For soils dominated by platy particles with high interparticle friction this mode may also occur. In this mode energy is dissipated by particle rolling and translation. No preferred particle orientation occurs and residual strength still remains high so that ϕ_r' can be taken as ϕ_{cv}'.
- *Sliding*
  When behaviour is dominated by platy, low friction particles sliding occurs on a shear surface with strongly oriented particles and the strength is low. ϕ_r' depends mainly on the mineralogy, coefficient of interparticle friction, μ and pore water chemistry.
- *Transitional*
  This involves turbulent and sliding behaviour in different parts of a shear zone.

The residual shear strength can be obtained using a ring shear apparatus (Bishop *et al*, 1971 and Bromhead, 1978). A ring-shaped thin sample of remoulded soil is sheared in a direct shear manner, by rotating the upper half of the sample above the lower half with sufficient strain until a slip surface is formed on which the lowest strength is measured from the torsion applied. The residual strength τ_r is related to the normal stress σ_N' applied:

$$\tau_r = \sigma_N' \tan \phi_r' \tag{7.22}$$

although for many soils, the graph of τ_r versus σ_N' shows a small cohesion intercept c_r' or a curvature of the graph, so that the stress range applicable to the site condition must be used for the determination of ϕ_r'.

In general, if the clay content is 40 – 50% or more or the plasticity index is 30 – 40% or more then the ϕ_r' value can be expected to be lower than 15º (Lupini *et al*, 1981).

Worked Example 7.1 *Shear box test*

The following results were obtained from a shear box test on a sample of dense sand. Determine the shear strength parameters for peak and ultimate strengths.

Normal load (N)	105	203	294
Shear load (N) at peak	95	183	265
Shear load (N) at ultimate	65	127	184

(N = newtons)

Since the areas for both the normal stress and shear stress are the same it is not necessary to determine stress values, the shear loads can be plotted directly against the normal loads, as on Figure 7.41. Assuming $c' = 0$ the results give:

peak $\phi' = 42^{\circ}$

ultimate $\phi' = 32^{\circ}$

Drained conditions are assumed so total stresses = effective stresses. The ultimate ϕ' could be taken as the value of ϕ_{cv}'.

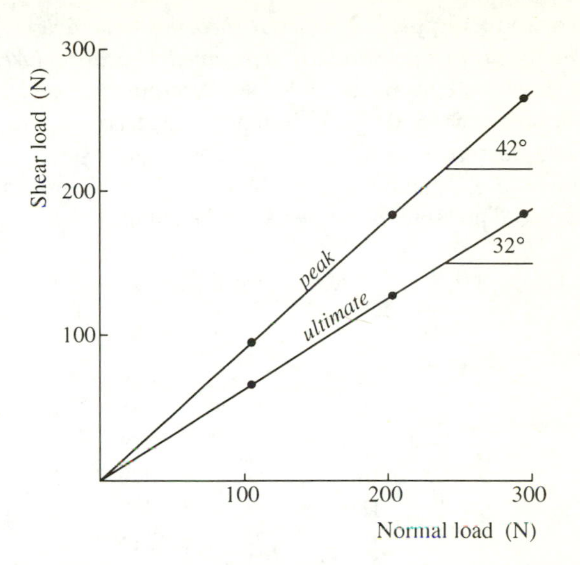

Figure 7.41 Worked Example 7.1 *Shear box test*

Worked Example 7.2 *Triaxial saturation – coefficient B*

The following are the results of the saturation stage of a triaxial test. Determine the value of B for each stage.

An explanation of the pore pressure parameters is given in Chapter 4. The parameter B is applied to the triaxial test to ensure the state of full saturation.

By increasing the cell pressure and the back pressure in a triaxial specimen of clay the voids are eventually filled with water by water entering the specimen from the back pressure system and air dissolving in the water.

The procedure is first to apply a cell pressure and to record the pore water pressure. Then the back pressure (in the pore water) is increased and the pore pressure recorded when it virtually equals the back pressure applied. The back pressure must be about 10 kN/m^2 less than the cell pressure to ensure positive effective stress in the soil.

B is determined from Equation 4.7 and when it reaches about 0.97 the specimen may be assumed to be sufficiently fully saturated.

cell pressure (kN/m^2)	back pressure (kN/m^2)	pore pressure (kN/m^2)	Δu (kN/m^2)	$\Delta\sigma_3$ (kN/m^2)	B
0	0	-4			
50	–	7	11	50	0.22
50	40	39	–	–	–
100	–	62	23	50	0.46
100	90	89	–	–	–
150	–	126	37	50	0.74
150	140	139	–	–	–
200	–	182	43	50	0.86
200	190	190	–	–	–
300	–	285	95	100	0.95
300	290	290	–	–	–
400	–	388	98	100	0.98

Worked Example 7.3 *Triaxial shearing – coefficient A*

The results of a consolidated undrained triaxial compression test with pore pressure measurements on a sample of saturated clay are given below. Determine the variation of the pore pressure parameter A during the test.

cell pressure = 600 kN/m^2 throughout the test

back pressure = 400 kN/m^2 at the start of the test

effective cell pressure = 200 kN/m^2 = effective stress at the start of the test

The pore pressure parameter A can be obtained from Equation 4.12 assuming $B = 1$ and $\Delta\sigma_3 = 0$.

$\Delta\sigma_1 - \Delta\sigma_3$	u	Δu	A
0	400	0	–
58	419	19	0.33
104	441	41	0.39
140	463	63	0.45
158	479	79	0.50
180	499	99	0.55
192	515	115	0.60

Worked Example 7.4 *Consolidated undrained triaxial test*

The results of a consolidated undrained triaxial compression test on a sample of fully saturated clay are given below. Each specimen has been consolidated to a back pressure of 200 kN/m^2.

Parameter (kN/m^2)	Specimen 1	Specimen 2	Specimen 3
cell pressure	300	400	600
deviator stress at failure	326	416	635
pore pressure at failure	146	206	280

Calculate the required stresses and plot them on a Mohr cirle diagram to obtain the effective stress shear strength parameters for the clay.

At failure:

Parameter	Specimen 1	Specimen 2	Specimen 3
σ_3	300	400	600
u_f	146	206	280
(b) σ_3'	154	194	320
$\sigma_1 - \sigma_3$	326	416	635
(c) $\sigma_3' + \frac{1}{2}(\sigma_1' - \sigma_3')$	317	402	638

(c) represents the centre of the Mohr circle and (b) the point of circumscription, see Figure 7.42.

The pore pressure parameter at failure, A_f for each specimen can be obtained from Equation 4.12 assuming that $B = 1$ and $\Delta\sigma_3 = 0$.

Parameter	Specimen 1	Specimen 2	Specimen 3
Δu	–54	6	80
$\Delta\sigma_1 - \Delta\sigma_3$	326	416	635
A_f	–0.17	0.01	0.13

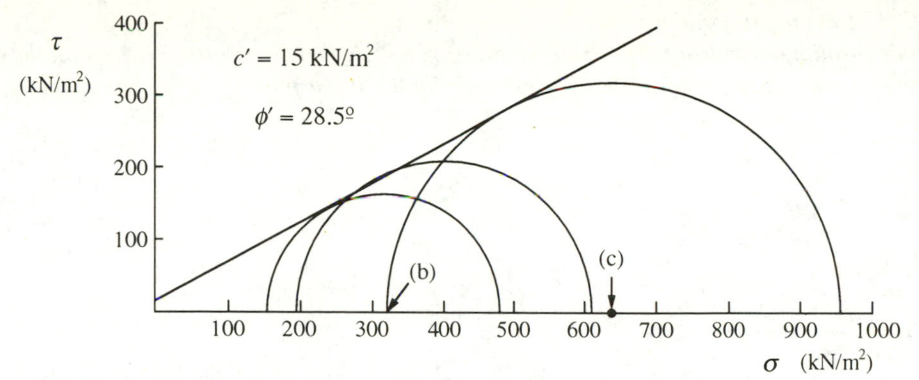

Figure 7.42 Worked Example 7.4 *Consolidated undrained triaxial test*

Worked Example 7.5 *Stress paths t and s*

From the results of the consolidated undrained triaxial compression test given in Example 7.3 determine the values of t and s (total stresses) and t′ and s′ (effective stresses) and plot the stress paths.

$s = \sigma_3 + t$ $\qquad t' = t$ $\qquad s' = s - u$

$\sigma_1 - \sigma_3$	u	t	s	t'	s'
0	400	0	600	0	200
58	419	29	629	29	210
104	441	52	652	52	211
140	463	70	670	70	207
158	479	79	679	79	200
180	499	90	690	90	191
192	515	96	696	96	181

The total and effective stress paths are plotted on Figure 7.43.

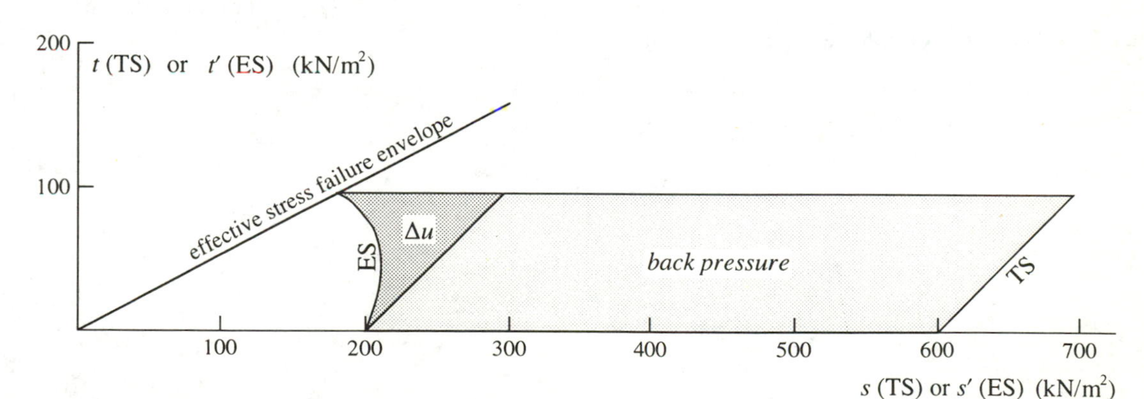

Figure 7.43 Worked Example 7.5 *Stress paths t and s*

Worked Example 7.6 *Stress paths q and p*

From the results of the consolidated undrained triaxial compression test given in Example 7.3 determine the values of q and p (total stresses) and q′ and p′ (effective stresses) and plot the stress paths.

$q = q' = \sigma_1' - \sigma_3'$

$p = \frac{1}{3}q + \sigma_3$

$p' = p - u$

$\sigma_3 = 600\ \text{kN/m}^2$

u	q	p	q'	p'
400	0	600	0	200
419	58	619	58	200
441	104	635	104	194
463	140	647	140	184
479	158	653	158	174
499	180	660	180	161
515	192	664	192	149

The total and effective stress paths are plotted on Figure 7.44.

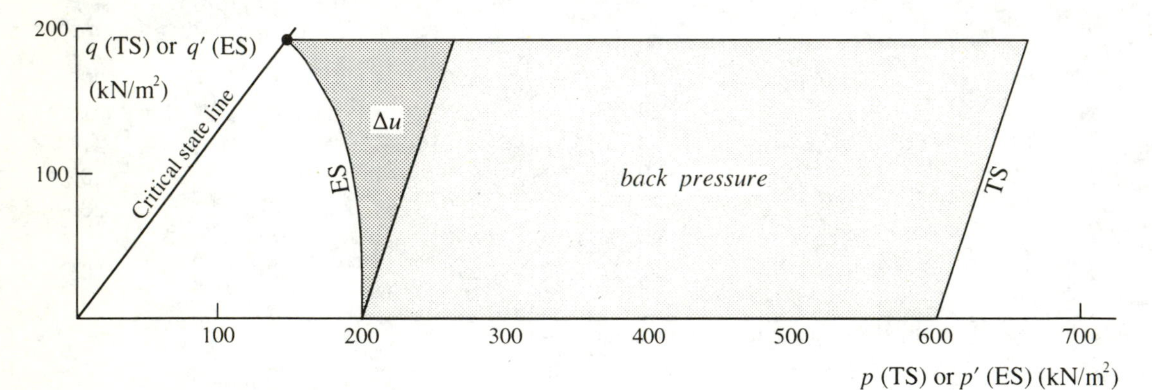

Figure 7.44 Worked Example 7.6 *Stress paths q and p*

Worked Example 7.7 *Critical state approach to a drained test*

A sample of clay has been isotropically normally consolidated to a stress $p_o' = 200$ kN/m^2 and a final void ratio of 0.92 and is sheared in a drained triaxial compression test. For the soil constants given below calculate q' and p' at failure and the final specific volume and volumetric strain at failure.

$\lambda = 0.16$ $\Gamma = 2.76$ $M = 0.89$

From Figure 7.34 the specimen state commences on the ICL and during shearing it rises up the Roscoe surface and over a drained plane to failure at the CSL.

The gradient of the drained plane in p' - q' space is 3 so p' at failure is given by:

$$p_f' = p_0' + \tfrac{1}{3} q_f'$$

From Equation 7.18 $q_f' = M\, p_f'$

so $p_f' = p_0' + \tfrac{1}{3} M p_f'$

giving $p_f' = \dfrac{p_0'}{1 - \frac{1}{3}M} = \dfrac{200}{1 - \frac{1}{3} \times 0.89} = 284 \text{ kN/m}^2$

$\therefore\ q_f' = 0.89 \times 284 = 253 \text{ kN/m}^2$

Equation 7.19 gives the specific volume for the critical state line so the final specific volume is:

$v_f = 2.76 - 0.16 \ln 284 = 1.856$

The change in volume during shear is

$\Delta v = 1.920 - 1.856 = 0.064$

so the volumetric strain is

$$\frac{\Delta v}{v} = \frac{0.064}{1.920} \times 100 = 3.3\%$$

Exercises

7.1 The normal total stress acting on a plane within a soil is 185 kN/m^2 and the effective stress shear strength parameters are:

peak $c_p' = 5$ kN/m^2 $\phi_p' = 29°$ residual $c_r' = 0$ $\phi_r' = 12°$

When the pore water pressure in the soil is 90 kN/m^2 determine the maximum shear strength which can be obtained on the plane and the shear strength after considerable strain.

7.2 The vertical total stress acting on an element of soil in the ground is 125 kN/m^2 and the pore water pressure is 32 kN/m^2. The effective stress shear strength parameters of the soil are $c' = 8$ kN/m^2 and $\phi' = 28°$. The horizontal total stress is measured to be 97 kN/m^2. Confirm that the soil element is not at failure. (Hint: consider simply the apex of the Mohr circle for the *in situ* condition and compare this with the available shear strength, see Figure 7.5.)

7.3 In Exercise 7.2, using the expressions given on Figure 7.5, determine the lowest horizontal effective stress the soil can sustain before failure occurs. Assume the vertical stress remains constant.
The horizontal stress at failure is referred to as the active pressure and a simpler method for its determination is given in Chapter 11.

7.4 From the results of the consolidated undrained test given in Exercise 4.10 determine the stresses:
(a) σ_1, σ_1', σ_3'
(b) the stress path values s' and t' and p' and q
Plot the stress path s' versus t' and determine the effective stress shear strength parameter ϕ'.

7.5 A shear box test carried out at a slow rate of strain on 'undisturbed' specimens of a cemented sand gave the following results at failure.

Normal stress (kN/m^2)	40	80	120
Shear stress (kN/m^2)	47.5	80.0	112.0

Determine the effective stress shear strength parameters of the soil.

7.6 On a potential plane of sliding in a mass of this cemented sand the normal stress is estimated to be 100 kN/m^2 and the shear stress is 64 kN/m^2. Determine the factor of safety against failure.

7.7 The results of unconsolidated undrained (UU) tests on three similar specimens of a fully saturated clay soil at failure are given below. Determine the deviator stress and shear strength for each specimen.

	cell pressure (kN/m^2)	Δl (mm)	Axial force (N)
(a)	100	11.4	331
(b)	200	9.6	309
(c)	400	8.5	312

Initial length = 76 mm Initial diameter = 38 mm

7.8 The results of consolidated undrained (CU) tests on three similar specimens of a fully saturated clay soil at failure are given below.

	cell pressure (kN/m^2)	deviator stress (kN/m^2)	pore pressure (kN/m^2)
(i)	350	143	326
(ii)	400	208	354
(iii)	500	312	418

(a) Determine the effective stresses σ_1' and σ_3' for each specimen.
(b) Plot the Mohr circles and determine the effective stress shear strength parameters c' and ϕ'.
(c) Plot the stress path points s' and t', determine the values a and α and derive c' and ϕ'.
The back pressure used for all three specimens was 300 kN/m^2.

7.9 The results of consolidated drained (CD) tests on three similar specimens of a fully saturated clay soil at failure are given below.
(a) Determine the effective stresses σ_1' and σ_3' for each specimen.
(b) Plot the Mohr circles and determine the effective stress shear strength parameters c' and ϕ'.
(c) Plot the stress path points s' and t', determine the values a and α and derive c' and ϕ'.

	cell pressure (kN/m^2)	axial force, (N)	Δl (mm)	ΔV (ml)
(i)	100	300	7.75	2.8
(ii)	200	545	8.97	4.4
(iii)	300	787	10.34	5.9

$l_0 = 76$ mm $\quad A_0 = 1140$ mm^2 $\quad V_0 = 86.65$ ml

Area $A = A_0\,(1 - \varepsilon_v)/(1 - \varepsilon_a)$ $\quad \varepsilon_v = \Delta V/V_0$ $\quad \varepsilon_a = \Delta l/l_0$

7.10 Specimens of fully saturated normally consolidated clay have been isotropically consolidated to 150 and 300 kN/m^2 with void ratios of 0.75 and 0.65, respectively. For the isotropic normal consolidation line (ICL) determine the values of N and λ for this clay.

7.11 The specimen consolidated to 300 kN/m^2 in Exercise 7.10 has been sheared to failure under drained conditions with the deviator stress at failure (at the critical state) of 368 kN/m^2. Determine the value of M for this clay.

7.12 What would be the deviator stress at failure in a drained test for the specimen consolidated to 150 kN/m^2 in Exercise 7.10?

7.13 In Exercises 7.11 and 7.12 determine the final void ratios for the specimens that are consolidated to 150 and 300 kN/m^2. $\Gamma = 2.37$.

7.14 Specimens of the clay in Exercise 7.10 are sheared to failure under undrained conditions. Determine the deviator stress and the pore pressure at failure for the specimens that are consolidated initially to 150 and 300 kN/m^2. $\Gamma = 2.37$.

8 Shallow Foundations – Stability

Shallow Foundations

Introduction

The term 'shallow foundations' refers to spread foundations as opposed to pile foundations.

They are usually placed at shallow depths such as a depth less than their width i.e. $D < B$. Alternatively, they may be considered as being constructed within the reach of normal excavation plant, i.e. D less than about 3 m where D/B may be up to 3.

If they are open below ground level they would be treated as basement foundations or buoyant foundations.

Spread foundations

This description is often used to show that these foundations convert a localised line load or point load from the structure into a pressure by spreading it over the area of the foundation. The pressure then applied by the foundation must be supported by the soil. For line loading on a strip foundation:

$$\text{width } B = \frac{\text{wall line load (kN/m)}}{\text{applied bearing pressure}} \tag{8.1}$$

For point loading on a pad foundation:
area $A =$ Length $L \times$ width B

$$= \frac{\text{column load (kN)}}{\text{applied bearing pressure}} \tag{8.2}$$

Raft foundations are of necessity about the same size as the whole building itself and are designed to spread both line loads and point loads as much as possible over their whole area.

Design requirements *(Figure 8.1)*

The purposes of a foundation are to ensure:

- ***stability*** of the structure supported.
  Failure could occur due to inadequate bearing capacity, overturning or sliding. There must be an adequate factor of safety against these occurrences. The applied bearing pressure must not exceed the safe bearing capacity, q_s, which is the ultimate bearing capacity, q_{ult}, divided by an appropriate factor of safety F (Figure 8.1). It is then assumed that the foundation and the structure supported will not fail due to the mechanisms considered. This represents an ultimate limit state approach.
- ***settlements*** do not exceed the tolerable limits of the structure supported by the foundation.
  The applied bearing pressure must not exceed the allowable bearing pressure, q_a otherwise settlements greater than the allowable or tolerable settlement of the structure will be produced (Figure 8.1). Settlement analyses are considered separately in Chapter 9.
  If the settlements are greater than the allowable settlements then the foundation will have exceeded the serviceability limit state and can be considered to have 'failed'. Thus design of foundations is more strictly based on the allowable bearing pressure rather than the safe bearing capacity.

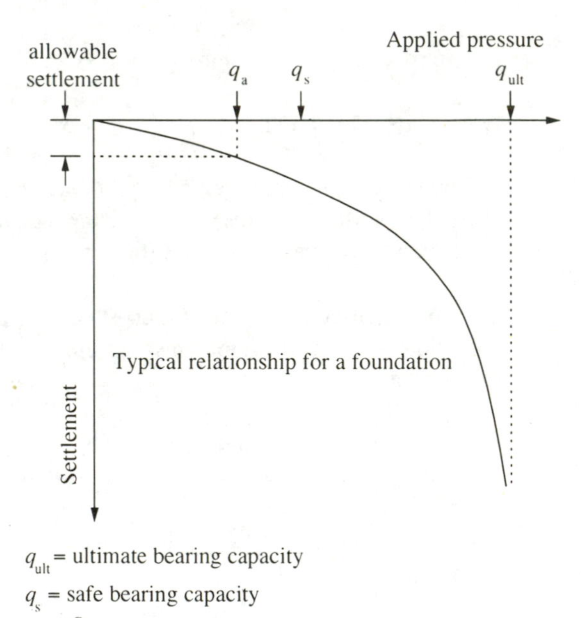

q_{ult} = ultimate bearing capacity
q_s = safe bearing capacity
$= \frac{q_{ult}}{F}$ F = factor of safety
q_a = allowable bearing pressure
(related to allowable settlement of structure)

Figure 8.1 *Bearing pressures*

- ***other ground movements*** do not adversely affect the structure.

  Many of these ground movements are associated with the depth of the foundation, discussed below, while others such as mining subsidence, earthquakes, vibrations will require special measures.

Types of foundation

The type and shape of the loading generally determines the shape of a foundation, e.g. strip foundation beneath a wall, pad foundation beneath a column, raft foundation beneath the whole of a building. Nearly all modern foundations will be constructed of concrete and all but the simplest will contain steel reinforcement, to enable loads to be spread over wider areas.

Strip foundations *(Figure 8.2)*

Strip foundations are designed to provide sufficient width, either as plain concrete for lightly loaded walls such as domestic properties, or as reinforced concrete for heavier retaining walls.

For plain concrete foundations adequate strength in the concrete can be achieved if the thickness, T, is greater than the projection, P, from the face of the wall.

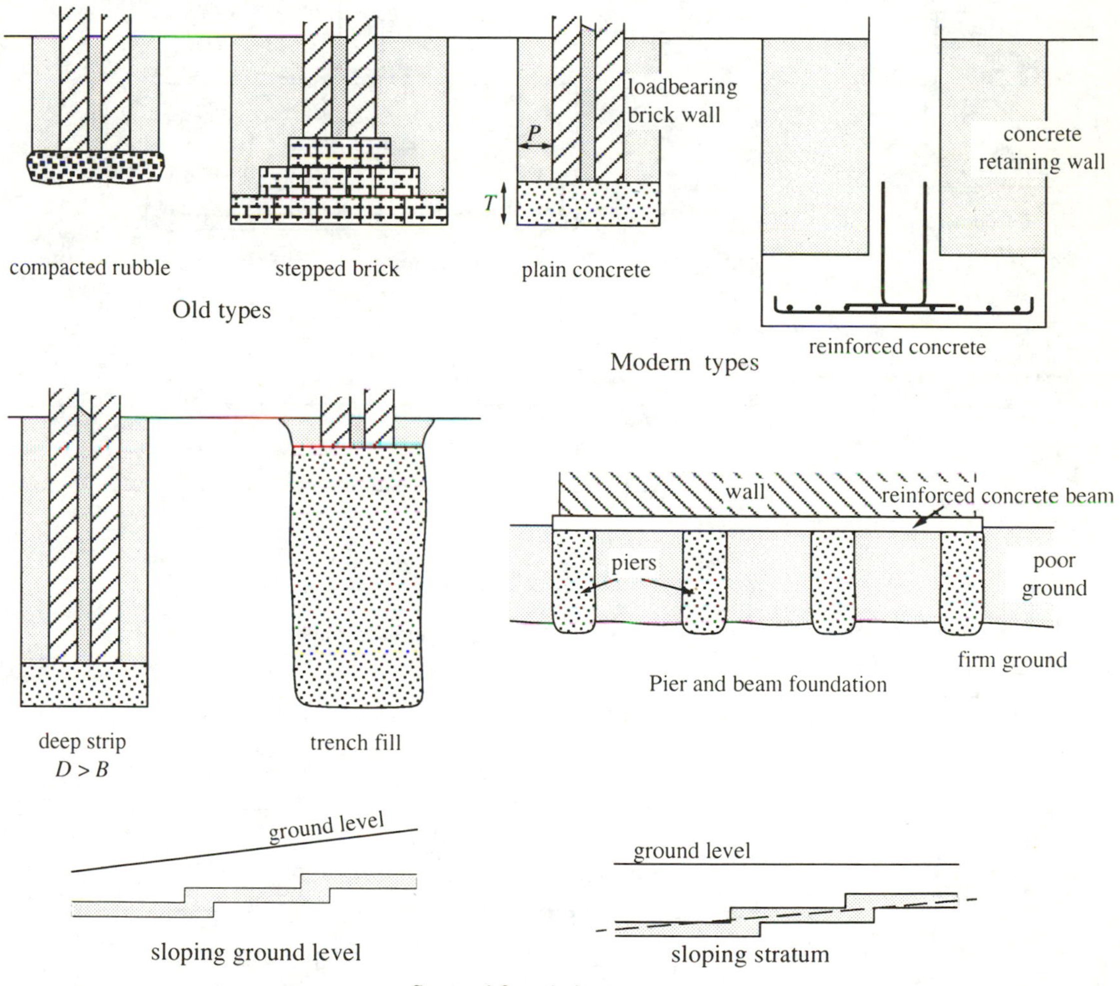

Figure 8.2 *Strip foundations*

However, Equation 8.1 shows that for heavy loads or weak ground the width, B, and hence the thickness, T, must be large. To limit the depth of foundation and the amount of concrete required plain concrete foundations are restricted to narrow widths.

The efficiency of reinforcement in concrete means that wide foundations at shallow depths can be provided economically using reinforced concrete, with transverse reinforcement at the base of the foundation resisting the bending stresses.

These foundations often traverse variable ground or weak spots, so it is advisable to provide at least nominal reinforcement longitudinally to increase bending resistance or stiffness along the length of the wall. This will reduce differential settlements and minimise the risk of cracking in the wall.

Deep strip foundations are difficult to construct, since they require placing brickwork in a narrow confined trench, so the trench fill foundation is often preferred. This may be more expensive in concrete but

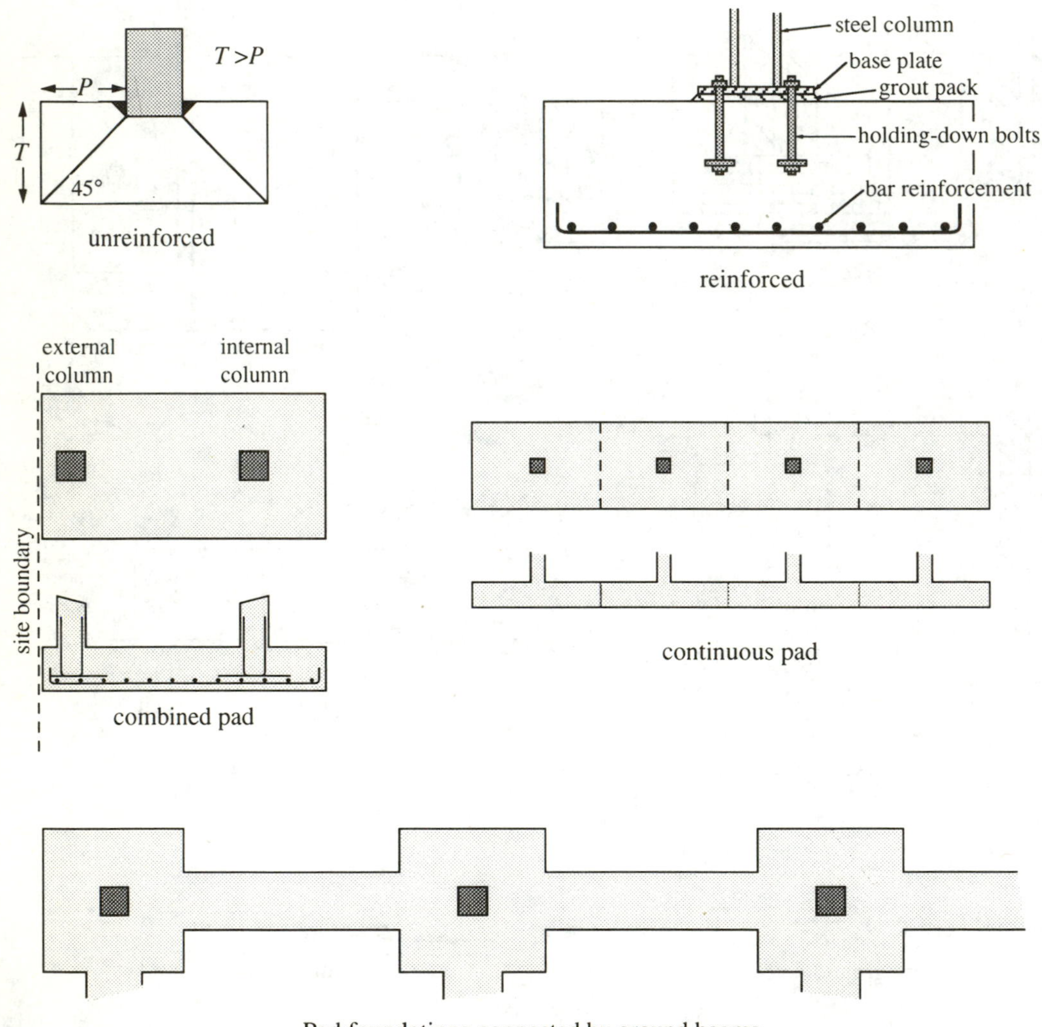

Figure 8.3 *Pad foundations*

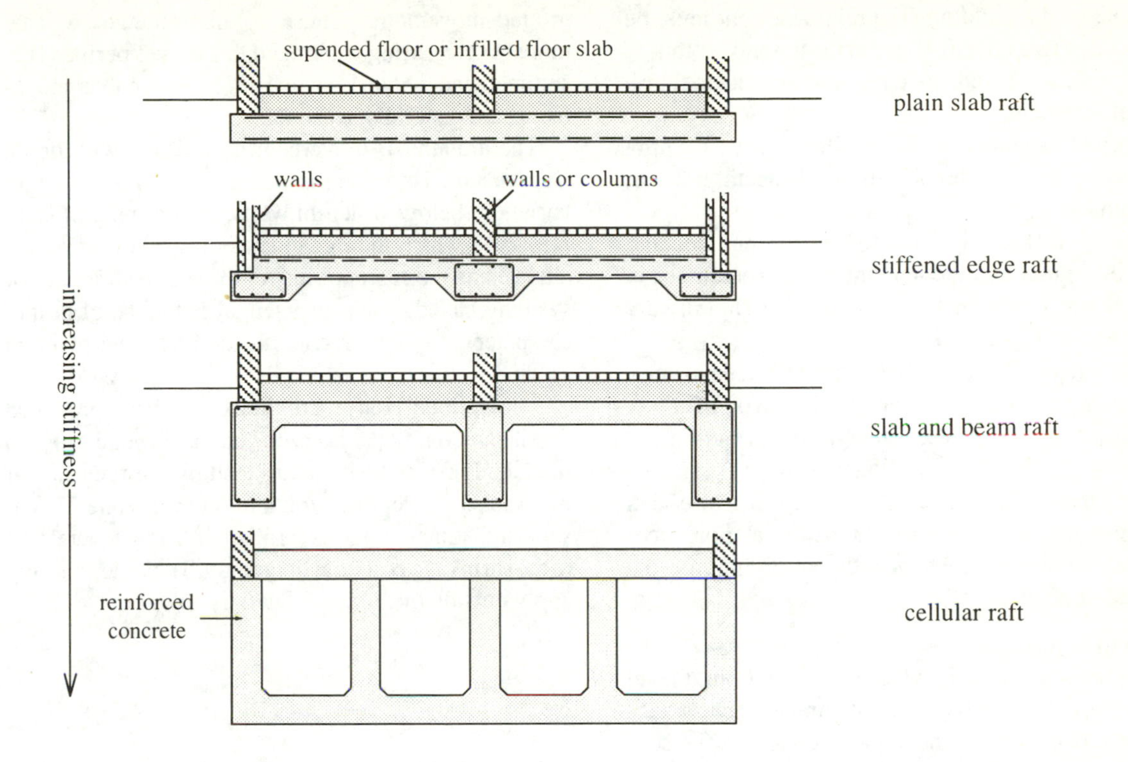

Figure 8.4 *Raft foundations*

it is quicker to construct, saves brickwork, eliminates the need for trench support and provides a good surface for the start of bricklaying. Alternatively, mass concrete piers could be constructed to support a reinforced concrete beam spanning between the piers, referred to as pier and beam foundations.

Pad foundations *(Figure 8.3)*

Unreinforced pad foundations are only used for small point loads otherwise excessive thicknesses, T, are necessary to provide the width required.

Pad foundations are often square to enable easier placing of reinforcement. However, where eccentric or inclined loading occurs, rectangular pads with the long axis in line with the eccentricity will be more effective.

Where an external column has to be placed close to a site boundary it may be preferable to support it on a combined pad foundation where the balancing force of the internal column can be used.

When pad foundations are fairly close to each other it may be easier to construct them as continuous pads (by making them rectangular), to provide simpler excavation and easier formwork. They may be individually reinforced and, therefore, independent or they may be continuously reinforced to provide a strip foundation with greater overall stiffness to minimise differential settlements. This foundation will then provide support for the infill walls.

Smaller pads which are further apart can be made to interact, to transfer and share horizontal loading, moment loading and additional loading due to differential settlements, by providing interconnecting ground beams.

Raft foundations *(Figure 8.4)*

Raft foundations which are perfectly flexible will settle into a dish-shaped profile which may cause distress (distortion, cracking etc.) to the structure. The purpose of a raft foundation must be, therefore, to keep differential settlements to within the tolerable limits of the particular type of structure, by providing sufficient stiffness, or resistance to bending deflections. This is achieved by interaction between:

1 a continuously reinforced concrete slab beneath the

whole of the building. The reinforcement must be placed in two directions and at the top and bottom of the slab, a convenient type being mesh or fabric reinforcement.

2 concrete beams with bar reinforcement, running beneath the structural walls or beneath lines of columns.

Increasing stiffness is provided by the number and depth of beams and the amount of reinforcement in them. Some common types of raft foundations are illustrated on Figure 8.4.

If the differential settlements were reduced to zero, the foundation would be described as rigid or infinitely stiff. This may seem a desirable goal but would mean designing for very large bending moments and, since it has been established that most ordinary structures can tolerate some differential settlements, a more economical solution is obtained by designing for the required stiffness.

Depth of foundations

Foundations must not be placed close to ground level (existing or proposed) because natural agencies cause ground movements in the upper horizons of the soil. Some examples of depth considerations follow.

Adequate bearing stratum *(Figure 8.5)*

Foundations should be taken below incompetent strata, such as made ground or peat, and founded in competent strata beneath. This depth should not be considered for depth effects in settlement analyses or depth factors in bearing capacity calculations, but could be used to determine net bearing pressures.

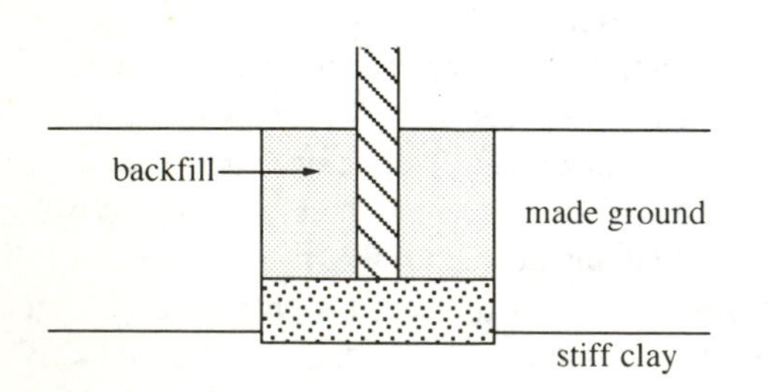

Figure 8.5 *Adequate bearing stratum*

Seasonal moisture variations *(Figure 8.6)*

In a clay soil, the moisture content of the upper horizon can reduce considerably in dry periods and increase in wet periods, producing volume changes and hence ground movements. These volume changes will be more severe during prolonged dry or wet periods (i.e. summer and winter), so the most noticeable movements are generally seasonal.

The amount of movement diminishes with depth (Figure 8.6). Thus there is a zone of seasonal moisture variation, below which movements are insignificant. Depths to which the movements occur depend on the climate and the susceptibility of the soil to shrinkage or swelling, called shrinkage potential. Estimates of shrinkage potential based on classification tests are given in Table 8.1.

The National House-Building Council has produced detailed recommendations on depths of foundations in the UK. For a clay site unaffected by the presence of tree roots, the depth of foundation (and zone of seasonal moisture variations) is given in Table 8.2, related to the shrinkage potential of the clay. These depths only apply outside the zone affected by tree roots.

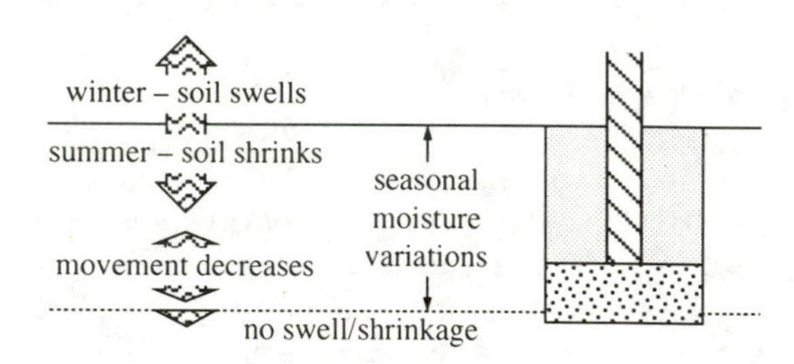

Figure 8.6 *Seasonal moisture variations*

Table 8.1 *Clay shrinkage potential*

Plasticity index %	Clay fraction %	Shrinkage potential
> 35	> 95	very high
22 - 48	60 - 95	high
12 - 32	30 - 60	medium
< 18	< 30	low

From BRE Digest 240, 1980

Table 8.2
Foundation depths outside exclusion zone (clay soils unaffected by trees)

Shrinkage potential	P. I. %	Minimum depth m
high	> 40	1.0
medium	20 - 40	0.9
low	10 - 20	0.75

From NHBC 1992

***Effects of tree roots** (Figures 8.7 and 8.8)*
In 1976, 1985 and 1990 in the UK severe droughts occurred in the summer months, forcing tree roots to extend in search of moisture beneath many brittle structures (load-bearing brickwork and plaster domestic properties). The resulting volume changes and ground movements caused varying degrees of damage (moderate to very severe) and resulted in insurance claims totalling many millions of pounds.

The radius of the zone affected by tree roots, *D*, around a tree (Figure 8.7) has been related to the mature height of the tree, *H*, and the species of the tree. The Building Research Establishment (BRE) and the NHBC have recommended typical values of this exclusion zone, some of which are given in Tables 8.3 and 8.4, respectively. Provided foundations are outside the zone affected then no special precautions will be necessary.

If shallow foundations are to be sited within the zone affected then they must be placed down to considerable depths (BRE Digest 298) or, alternatively, a specially designed bored pile foundation should be provided. The latter is probably more economical and more effective (BRE Digests 241 and 242).

The most severe mode of deformation in which a structure can be placed is the hogging mode, discussed in Chapter 9. This condition will be produced by shrinkage or swelling of a clay soil beneath part of a structure (Figure 8.8). Shrinkage will be caused by existing trees and planting new trees. Swelling or heave is caused by removing trees before construction.

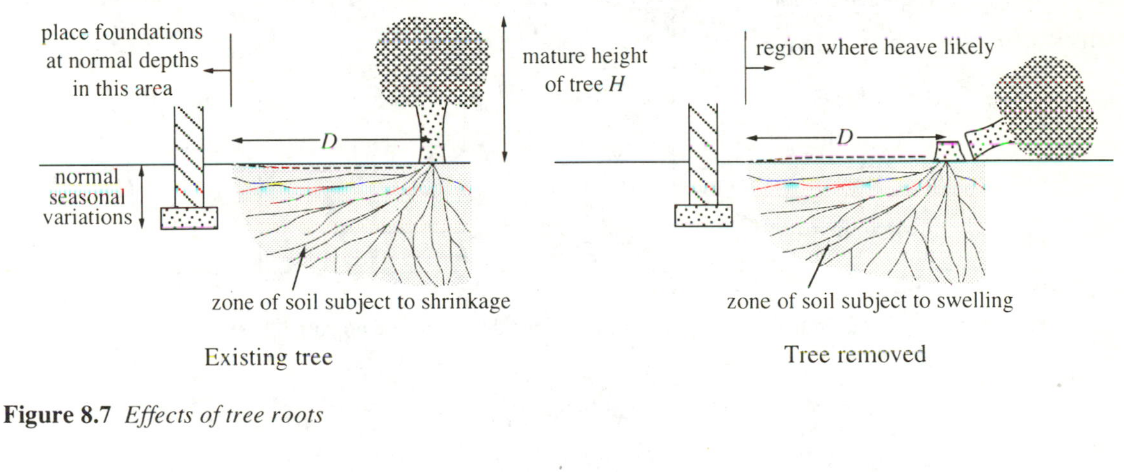

Figure 8.7 *Effects of tree roots*

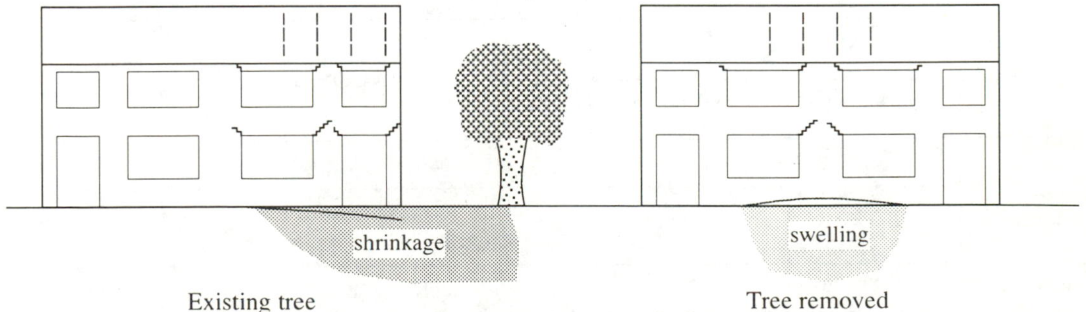

Figure 8.8 *Property movements caused by hogging*

Table 8.3 *Ranking of tree species known to have caused damage (From BRE Digest 298, 1987)*

Ranking [1]	Species	Maximum tree height H m	75 % of damage cases occurred within this distance m	Minimum recommended separation in very highly and highly shrinkable clays D m
1	Oak	16 - 23	13	1 H*
2	Poplar	24	15	1 H*
3	Lime	16 - 24	8	0.5 H
4	Common ash	23	10	0.5 H
5	Plane	25 - 30	7.5	0.5 H
6	Willow	15	11	1 H
7	Elm	20 - 25	12	0.5 H
8	Hawthorn	10	7	0.5 H
9	Maple/Sycamore	17 - 24	9	0.5 H
10	Cherry/Plum	8	6	1 H*
11	Beech	20	9	0.5 H
12	Birch	12 - 14	7	0.5 H
13	Whitebeam/Rowan	8 - 12	7	1 H*
14	Cypress	18 - 25	3.5	0.5 H

1 – in descending order of severity

* – exclusion zone could be reduced to 0.5 H for soils of low to medium shrinkage potential

Table 8.4

Zone of influence around trees - clay soils

Water demand	Maximum D/H		Typical tree species	Mature height H m
	coniferous	broad-leaved		
high	0.6	1.25	elm, oak, poplar, willow	16 - 24
moderate	0.35	0.75	ash, cherry, chestnut, lime, maple, plane, sycamore	10 - 20
low	-	0.5	beech, birch, holly	10 - 20

From NHBC 1992

A clay soil will also shrink or swell in a horizontal direction, especially at shallow depths where the horizontal movements may far exceed the vertical movements. These horizontal movements can produce severe distortion of buried structures such as trench fill foundations, basements walls and services.

Frost action *(Figure 8.9)*
Permanently frozen ground (permafrost) exists in the cold regions of northern Canada and northern Russia which are areas of sparse population, understandably. In Europe, the Gulf Stream provides a more equable climate and frost action is less severe.

Nevertheless, during prolonged cold periods when the air temperature remains below freezing, heat is conducted away from the earth's surface and a freezing front penetrates the soil. As small pockets of water freeze and crystallise, they give off latent heat. If this heat balances the heat lost by conduction to the earth's surface, the freezing front becomes stationary and ice lens formation proceeds, fed by free water or capillary water from below. Ice formation is expansive so ground heave will result and when the ice thaws a very weak morass will be formed.

Some soils are more 'frost-susceptible' than others, allowing ice lens formation more readily. In particular, soils having a permeability comparable to a silt are prone to ice lens formation e.g. silts, sands with moderate fines content and laminated or varved clays.

For highway pavements in the UK, no materials which are susceptible to frost (including the subgrade) should be present within 450 mm of ground surface, while for foundations of structures (which are less easy to repair) a depth of 600 mm is recommended. For soils less susceptible to frost action such as clean and 'dry' sands or gravels shallower depths may be permitted.

River erosion *(Figure 8.10)*
Erosion of a river bed depends on the depth of water flowing and the erodibility of the soil forming the river bed. When river levels rise, general scour of the river bed will occur, producing deeper channels, possibly in a different location to the normal deep channel. The deep flood channel could occur at different locations during subsequent floods.

When structures such as bridge piers obstruct the river flow, additional local scour will occur, which could extend below the pier foundation causing an

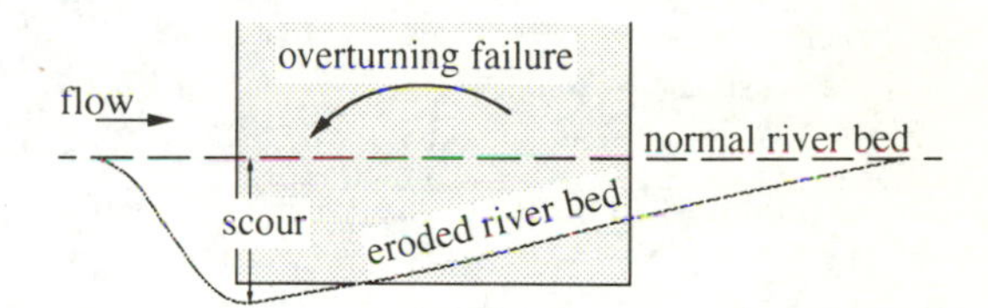

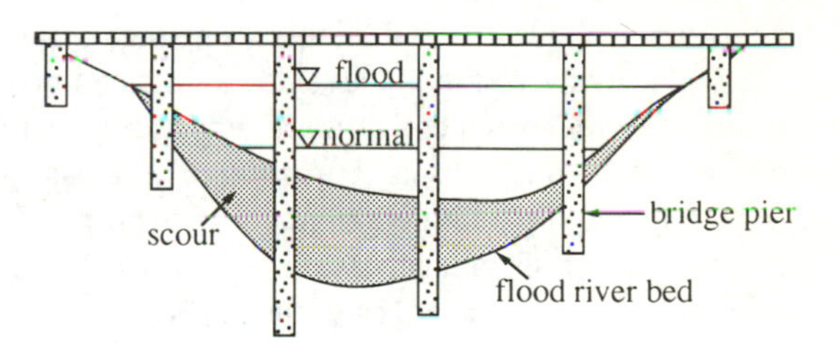

Figure 8.10 *River erosion*

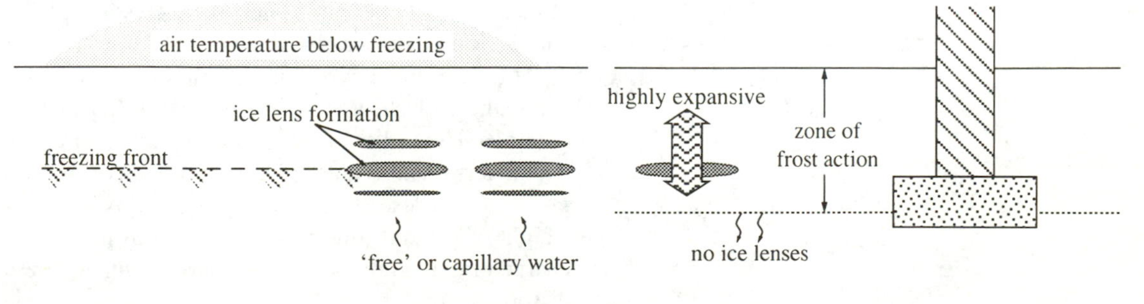

Figure 8.9 *Frost action*

overturning failure resulting in catastrophic collapse of the bridge. The foundations must be placed below the anticipated depth of scour. Additional scour protection or prevention measures such as sheet pile skirts, mattresses and rip-rap should also be considered.

Water table *(Figure 8.11)*
In granular soils, foundations should be kept as high above a water table as possible, to avoid construction problems and to minimise settlements. However, water tables can fluctuate, with higher levels during prolonged wet periods and in the winter months.

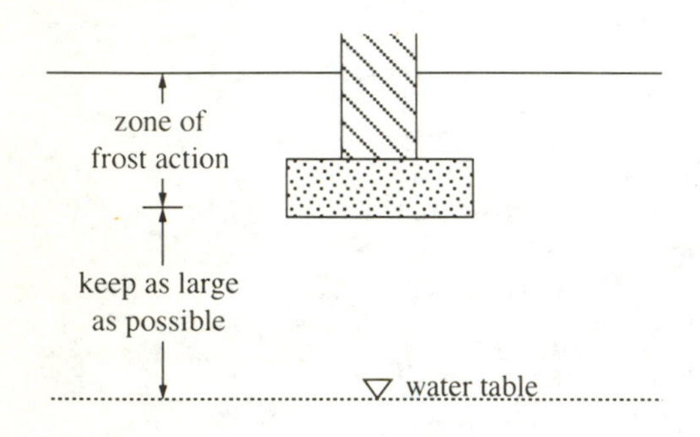

Figure 8.11 *Water table location*

Super-elevation *(Figure 8.12)*
Steel storage tanks (for oil, liquids, chemicals) are usually placed above ground level to facilitate emptying and to prevent water collecting around and beneath the tank which could cause corrosion. The existing ground may require treatment or removal, to eliminate the problems of shrinkage, swelling or frost heave, particularly beneath the edges of the tanks.

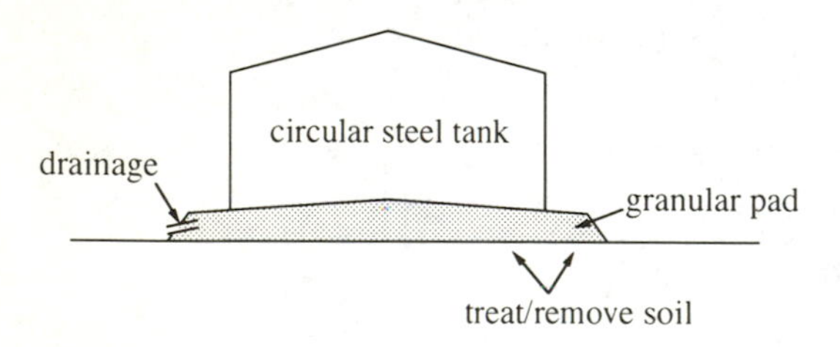

Figure 8.12 *Super-elevation*

Bearing capacity

Modes of failure *(Figure 8.13)*
Failure is defined as mobilising the full value of shear strength of the soil accompanied by large and excessive settlements. The mechanism causing failure for shallow foundations depends on the soil type, particularly its compressibility, and the type of loading.

If the soil is compressible, vertical displacements of the foundation are produced by volume reductions throughout the soil generally with limited shear distortion (punching shear).

If the soil is incompressible then volume reductions are not possible. Vertical displacements of the foundation can only be produced by mass movements with shear distortion occurring along a slip surface (general shear). The local shear mode is a transition between these two mechanisms.

When a horizontal load as well as a vertical load is applied to a foundation there is a risk of failure by overturning due to the resultant inclined load, when a reduced bearing capacity is obtained. There must also be sufficient adhesion in the case of a clay or friction in the case of a sand to prevent sliding of the foundation due to the horizontal load component.

Bearing capacity – vertical load only *(Figure 8.14)*
An expression for bearing capacity has only been derived for the general shear mode of failure (incompressible solid) and is based on superposition of three components. Nevertheless, factors can be applied to compensate for the effects of compressibility, and the superposition of the three components leads to errors which are on the safe side.

The expression is based on the failure mechanism illustrated in Figure 8.14, where the soil in zones I, II and III is in a state of plastic equilibrium. The active Rankine zone I is pushed downwards and, in turn, pushes the radial shear zones II sideways and the passive Rankine zones III sideways and upwards.

At failure, the movement of these masses is mobilising the full shear strength of the soil, which is obtained from the Mohr-Coulomb shear strength parameters c and ϕ and the total or effective stresses in the soil. These stresses are provided by the self-weight of the soil (due to gravity) and from the surcharge pressure around the foundation (due to stress distribution).

Mode Of Failure — Characteristics — Typical Soils

General Shear

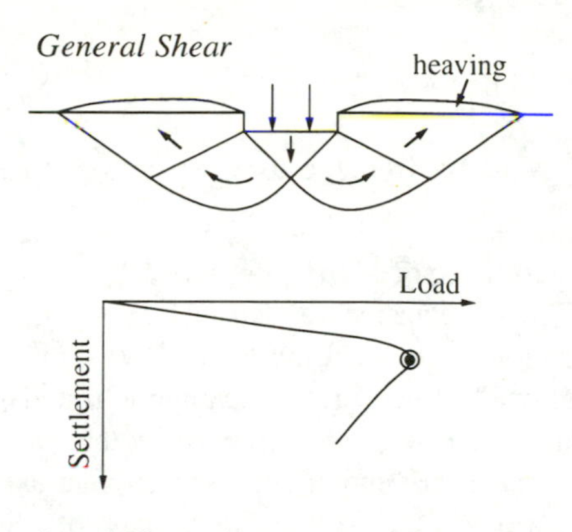

Characteristics

- Well-defined continuous slip surface up to ground level
- Heaving occurs on both sides with final collapse and tilting on one side
- Failure is sudden and catastrophic
- Ultimate value is peak value

Typical Soils

- Low compressibility soils
- Very dense sands
- Saturated clays (NC and OC) undrained shear (fast loading)

Local Shear (Transition)

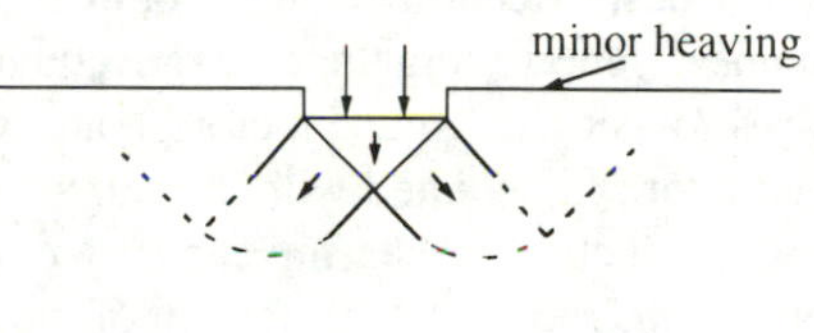

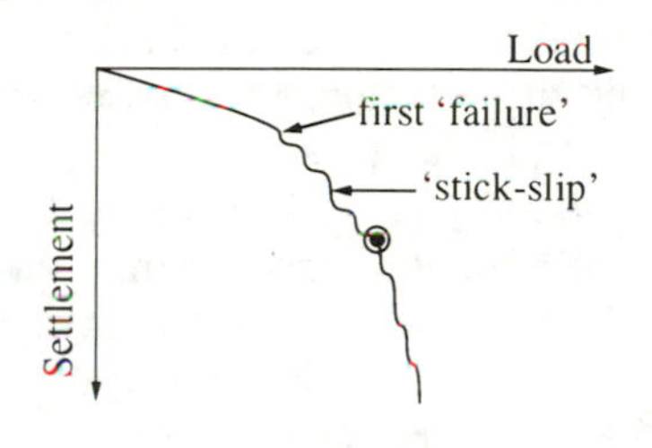

Characteristics

- Well-defined slip surfaces only below the foundation, discontinuous either side
- Large vertical displacements required before slip surfaces appear at ground level
- Some heaving occurs on both sides with no tilting and no catastrophic failure
- No peak value, ultimate value not defined

Typical Soils

- Moderate compressibility
- Medium dense sands

Punching Shear

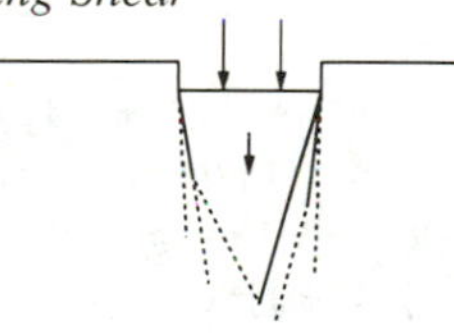

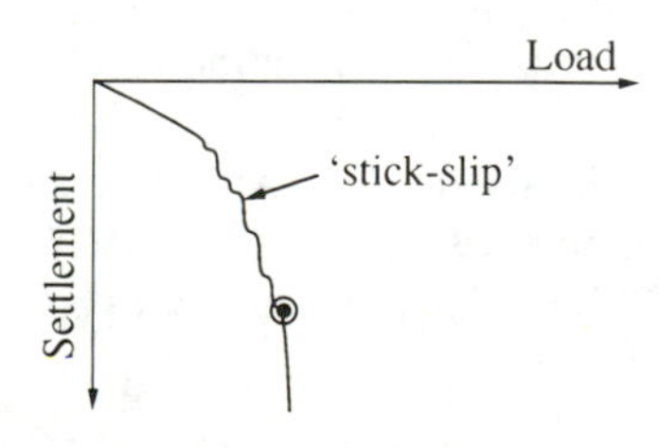

Characteristics

- Well-defined slip surfaces only below the foundation , none either side
- Large vertical displacements produced by soil compressibility
- No heaving, no tilting or catastrophic failure
- No ultimate value, increased compression densifies sand

Typical Soils

- High compressibility soils
- Very loose sands
- Partially saturated clays
- NC clay in drained shear (very slow loading)
- Peats

Figure 8.13 *Modes of failure*

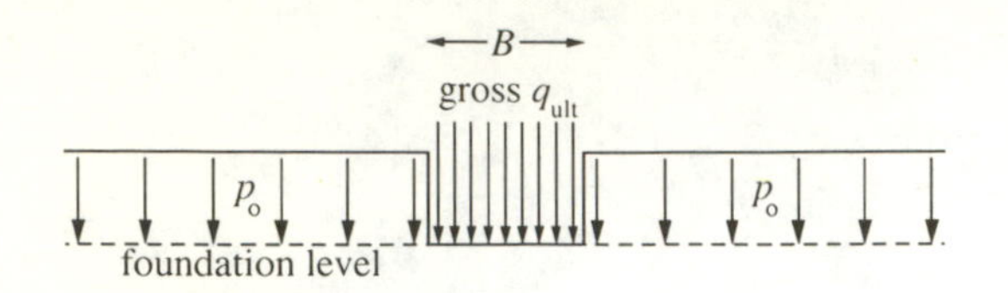

a) Foundation conditions

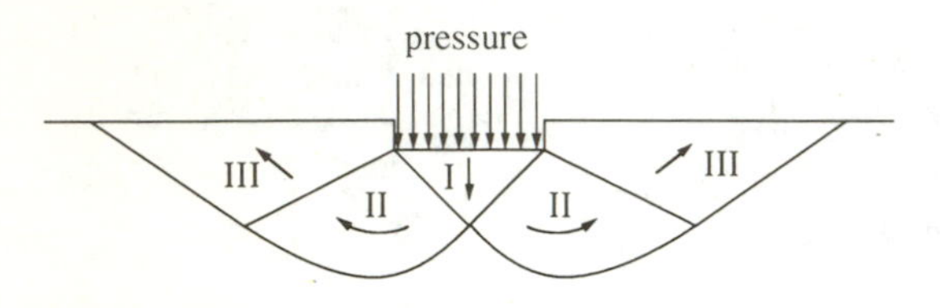

b) Failure mechanism

Figure 8.14 *Bearing capacity*

The three components of the Terzaghi (1943) bearing capacity equation are:

$$\text{gross } q_{ult} = cN_c + p_c N_q + \tfrac{1}{2}\gamma B N_\gamma \tag{8.3}$$

where:
cN_c is due to cohesion and friction in the soil
$p_o N_q$ is due to surcharge and friction in the soil
$^1/_2 \gamma B N_\gamma$ is due to self-weight and friction in the soil
p_o = total overburden pressure at foundation level around the foundation (Figure 8.14)
γ = bulk unit weight of soil
B = width of foundation
N_c, N_q and N_γ are termed bearing capacity factors and are related to the ϕ value only.

Values of N_c and N_q, attributed to Prandtl and Reissner are given by:

$$N_c = \left(N_q - 1\right)\cot\phi \tag{8.4}$$

$$N_q = \exp(\pi \tan\phi)\tan^2\left(45° + \frac{\phi}{2}\right) \tag{8.5}$$

Values of N_γ have been obtained by several authors adopting different rupture figures from the one used by Prandtl and Reissner, so superposition of the three components can only be an approximation, albeit a safe one. However, the contribution of the third term to bearing capacity is not significant for narrow foundations so the more conservative expression given by Brinch Hansen is suggested:

$$N_\gamma = 1.5\left(N_q - 1\right)\tan\phi \tag{8.6}$$

Values of N_c, N_q and N_γ from the above expressions are given in Table 8.5.

Shape and depth factors *(Tables 8.6 and 8.7)*
The original Terzaghi equation was derived for a very long (strip) foundation where shearing in only two dimensions was assumed. However, for rectangular foundations, shearing of the soil will also occur at the ends, producing an enhanced 'end effect'. For circular and square foundations, a three-dimensional mass of soil will be sheared. These effects are catered for in a semi-empirical manner by modifying the Terzaghi equation (see Equation 8.3) using the shape factors s_c and s_q.

The shape factor s_γ associated with the self-weight term provides a reduction of bearing capacity due to shape, because of the reduced confinement of the soil provided by rectangular, square or circular foundations. Values of the shape factors summarised from Vesic (1975) are given in Table 8.6.

Depth factors d_c, d_q and d_γ provide for the additional bearing capacity which could be obtained from shearing through the soil above foundation level. However, the original Terzaghi equation assumed that the soil above foundation level did not contribute to bearing capacity and there are good reasons for retaining this assumption such as:

1. The soil above foundation level is usually inferior to the supporting soil. This is the main reason for taking foundations deep below ground level e.g. made ground or shrinkable clay.
2. Mobilising shearing resistance over this section depends on the soil below foundation level being virtually incompressible.

Expressions for depth factors, d_c, d_q and d_γ have been given by Brinch Hansen (1970) and Vesic (1975) and are reproduced in Table 8.7, but these should be used with caution.

Bearing capacity – overturning *(Figure 8.15)*
Many foundations have to be designed for the effects of horizontal loading or moment loading transferred

Table 8.5

Bearing capacity factors

ϕ	N_c	N_q	N_γ
0	5.14	1.0	0
1	5.4	1.1	0
2	5.6	1.2	0
3	5.9	1.3	0
4	6.2	1.4	0
5	6.5	1.6	0.1
6	6.8	1.7	0.1
7	7.2	1.9	0.2
8	7.5	2.1	0.2
9	7.9	2.3	0.3
10	8.4	2.5	0.4
11	8.8	2.7	0.5
12	9.3	3.0	0.6
13	9.8	3.3	0.8
14	10.4	3.6	1.0
15	11.0	3.9	1.2
16	11.6	4.3	1.4
17	12.3	4.8	1.7
18	13.1	5.3	2.1
19	13.9	5.8	2.5
20	14.8	6.4	3.0
21	15.8	7.1	3.5
22	16.9	7.8	4.1
23	18.1	8.7	4.9
24	19.3	9.6	5.7
25	20.7	10.7	6.8
26	22.3	11.9	7.9
27	23.9	13.2	9.3
28	25.8	14.7	10.9
29	27.9	16.4	12.8
30	30.1	18.4	15.1
31	32.7	20.6	17.7
32	35.5	23.2	20.8
33	38.6	26.1	24.4
34	42.2	29.4	28.8
35	46.1	33.3	33.9
36	50.6	37.8	40.0
37	55.6	42.9	47.4
38	61.4	48.9	56.2
39	67.9	56.0	66.8
40	75.3	64.2	79.5
41	83.9	73.9	95.1
42	93.7	85.4	114.0
43	105.1	99.0	137.1
44	118.4	115.3	165.6
45	133.9	134.9	200.8
46	152.1	158.5	244.7
47	173.6	187.2	299.5
48	199.3	222.3	368.7
49	229.9	265.5	456.4
50	266.9	319.1	568.5

via the structure to the foundation both of which could produce overturning. In order to proceed with the analysis it is necessary to resolve the loading into a single total vertical load V_T acting at the underside of the foundation (foundation level), with the moment and horizontal load effects catered for by placing the total load at an eccentricity, e (eccentric loading).

If a horizontal load exists, it is then applied at foundation level where it no longer contributes to overturning, but it does produce a resultant inclined loading. Both eccentric and inclined loading produce a reduced bearing capacity and they are dealt with in separate ways.

Table 8.6

Shape factors (From Vesic, 1975)

Shape of foundation	s_c	s_q	s_γ
strip	1.0	1.0	1.0
rectangle	$1+\frac{B'}{L'}\frac{N_q}{N_c}$	$1+\frac{B'}{L'}\tan\phi$	$1-0.4\frac{B'}{L'}$
circle or square	$1+\frac{N_q}{N_c}$	$1+\tan\phi$	0.6

Eccentric loading *(Figure 8.16)*

The location of V_T may be eccentric along both the long and short axes, producing eccentricities e_L and e_B, respectively. A convenient and conservative approach attributed to Meyerhof (1953) is suggested for the analysis of eccentric loading.

This assumes that V_T acts centrally on an effective foundation area (ignoring the area outside of this effective area), so that the bearing capacity equation for central vertical loading can be used. From then on, only the effective width B' and effective length L' are used in the equations. Expressions for B' and L' are given on Figure 8.16 derived from:

$$\tfrac{1}{2}B=\tfrac{1}{2}B'+e_B \tag{8.7}$$

Table 8.7 *Depth factors (From Vesic, 1975)*

ϕ value	d_c		d_q		d_γ
$\phi = 0$ Clay undrained	$D/B' \leq 1$	$1 + 0.4\, D/B'$	1.0		1.0
	$D/B' > 1$	$1 + 0.4 \tan^{-1} D/B'$ D/B' in radians			
$\phi > 0$ Clay drained Sand	$d_q - \dfrac{1 - d_q}{N_c \tan\phi}$		$D/B' \leq 1$	$1 + 2 \tan\phi (1 - \sin\phi)^2 D/B'$	1.0
			$D/B' > 1$	$1 + 2 \tan\phi (1 - \sin\phi)^2 \tan^{-1} D/B'$ D/B' in radians	

Use these factors with caution – see text

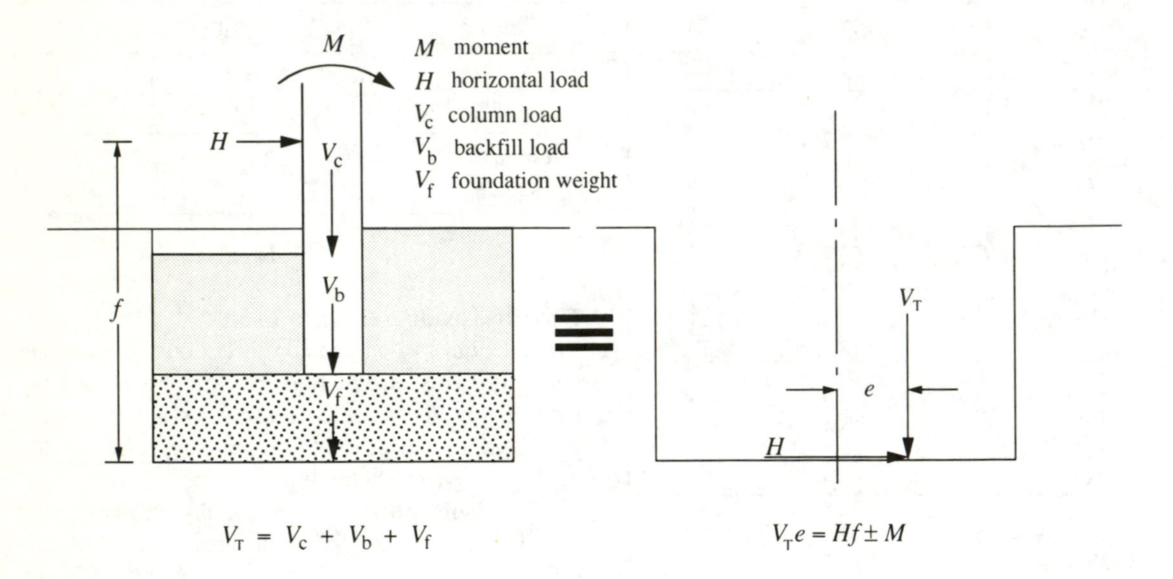

Figure 8.15 *Overturning (eccentric and inclined loading)*

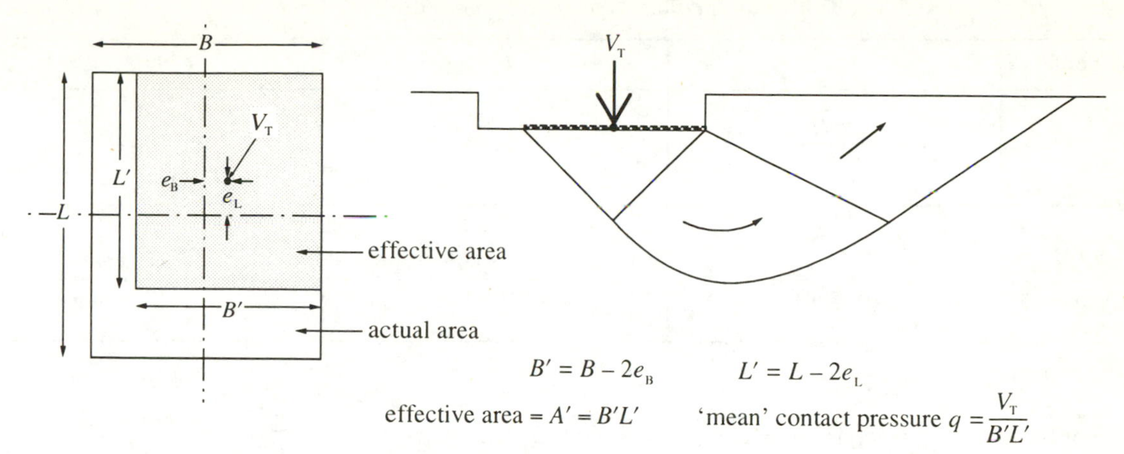

Figure 8.16 *Eccentric loading – effective area method*

Inclined loading *(Figure 8.17)*
This produces a smaller failure zone and hence reduced bearing capacity. Following the work of several authors (summarised in Vesic, 1975), the general bearing capacity expression has been modified to:

$$\text{gross } q_{\text{ult}} = cN_c s_c d_c i_c + p_o N_q s_q d_q i_q + \tfrac{1}{2}\gamma B' N_\gamma s_\gamma d_\gamma i_\gamma \quad (8.8)$$

where i_c, i_q and i_γ are inclination factors, expressions for which are given in Table 8.8.

Different soil strength cases
There are three cases usually considered in soil mechanics:
Case (a) is the *undrained* condition in clay (short-term case) and is represented by the shear strength parameters:

cohesion $c = c_u$ and $\phi = \phi_u = 0°$

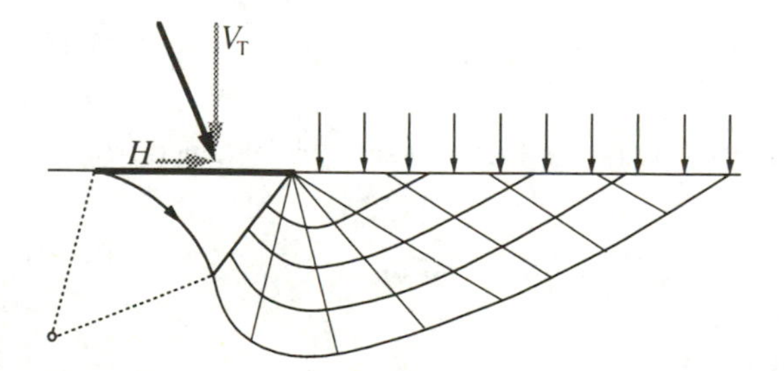

Figure 8.17 *Inclined loading*

giving the bearing capacity expression, Equation 8.7 (with the factors removed for clarity) as:

$$\text{gross } q_{\text{ult}} = c_u N_c + p_o \quad (8.9)$$

Case (b) is the *drained* condition in clay (long-term case) and is represented by the shear strength parameters:
cohesion $c = c'$ and $\phi = \phi'$ (> 0)

giving the bearing capacity expression, Equation 8.7 as:

$$\text{gross } q_{\text{ult}} = c' N_c + p_o N_q + \tfrac{1}{2}\gamma B' N_\gamma \quad (8.10)$$

Case (c) is the *drained* condition in sand (short and long-term case) and is represented by the shear strength parameters:

cohesion $c = 0$ $\phi = \phi'$ (> 0)

giving the bearing capacity expression, Equation 8.7 as:

$$\text{gross } q_{\text{ult}} = p_o N_q + \tfrac{1}{2}\gamma B' N_\gamma \quad (8.11)$$

The shape, depth and inclination factors are also dependent on the choice of c and ϕ values, so different factors will be obtained depending on the soil strength case assumed.

Table 8.8 *Inclination factors (From Vesic, 1975)*

ϕ value	i_c	i_q	i_γ
$\phi = 0$ Clay undrained	$1 - \dfrac{mH}{B'L'c_u N_c}$	1.0	1.0
$\phi > 0$ Clay drained Sand	$i_q - \dfrac{1 - i_q}{N_c \tan\phi}$	$\left[1 - \dfrac{H}{V_T + B'L'c'\cot\phi}\right]^m$	$\left[1 - \dfrac{H}{V_T + B'L'c'\cot\phi}\right]^{m+1}$
Exponent m For horizontal load acting along the short axis (side B') use $m = m_B = \dfrac{2 + B'/L'}{1 + B'/L'}$ For horizontal load acting along the long axis (side L') use $m = m_L = \dfrac{2 + L'/B'}{1 + L'/B'}$			

Effect of water table

If the water table lies B m or more below the foundation, then Equation 8.7 does not require modification.

If the water table lies at foundation level then Equation 8.7 (with the factors removed for clarity) should be modified to:

$$\text{gross } q_{ult} = cN_c + p_o N_q + \tfrac{1}{2}\gamma' B' N_\gamma \qquad (8.12)$$

where γ' is the submerged unit weight = $\gamma - \gamma_w$.

If the water table lies within B m below foundation level, then a value of γ' could be obtained by linear interpolation between γ and γ', depending on the actual depth of water table below foundation level. Since the water table is likely to fluctuate during the life of the structure it would be prudent to adopt γ' irrespective of water table depth.

If the water table lies at a height h_w above foundation level then Equation 8.7 should be modified to:

$$\text{gross } q_{ult} = cN_c + p_o' N_q + \tfrac{1}{2}\gamma' B' N_\gamma + \gamma_w h_w$$

$$= cN_c + p_o'(N_q - 1) + \tfrac{1}{2}\gamma' B' N_\gamma + p_o \qquad (8.13)$$

where p_o' is the *effective* overburden pressure at foundation level.

Net ultimate bearing capacity

In Equations 8.7 to 8.12, q_{ult} is referred to as the gross ultimate bearing capacity. The net ultimate bearing capacity is the maximum additional pressure the soil can support in excess of the stress at foundation level which existed before placing the foundation. Thus:

$$\text{net } q_{ult} = \text{gross } q_{ult} - p_o \qquad (8.14)$$

Factor of safety

This is used to obtain the net *safe* bearing capacity q_s:

$$\text{net } q_s = \frac{\text{net } q_{ult}}{F} \qquad (8.15)$$

$$\text{gross } q_s = \frac{\text{net } q_{ult}}{F} + p_o \qquad (8.16)$$

Alternatively, the factor of safety is obtained from:

$$F = \frac{\text{gross } q_{ult} - p_o}{\text{gross } q_{app} - p_o} = \frac{\text{net } q_{ult}}{\text{net } q_{app}} \qquad (8.17)$$

where:

q_{app} = applied pressure

and the other values are defined above.

A value of F is chosen to ensure that the risk of failure of the foundation and consequently the structure is minimal so a combination of structural factors and geotechnical factors must be considered. These could include:

1. *uncertainty of loading* – especially with non-routine buildings and live loading effects which are difficult to quantify, e.g. wind, water forces, moving loads, dynamic forces.
2. *likelihood of maximum design load* – for non-routine structures it is likely that unfavourable variations of loading will occur, whereas routine buildings are often designed on nominal loading, which is unlikely to occur.
3. *consequences of failure* – the public expect less risk to be taken with structures the failure of which could result in catastrophic consequences. More risk is taken with temporary works than permanent works.
4. *uncertainty of soil model* – inaccuracy of strength values, water table fluctuations, mode of failure, limitation of analytical method all provide uncertainty.
5. *extent of investigation* – sufficient depth, layering of deposits, uniformity of ground conditions, number of tests. The more extensive the site investigation, the more confidence there will be in the choice of the soil model.

Values for factor of safety taking into consideration the above points have been suggested by Vesic (1975) and are reproduced in Table 8.9.

Effect of compressibility of soil

The above expressions assume that the soil is incompressible, that only mass movements of zones I, II and III (Figure 8.14) occur and the general shear failure case applies. Clays sheared in an undrained manner could be considered incompressible and to fail in the general shear mode so no reduction of bearing capacity need be imposed.

However, clays sheared slowly enough to attain drained conditions or sands will be compressible and the full amount of bearing capacity, as given by Equation 8.7, will not be achieved. Vesic (1975) has suggested the use of compressibility factors to modify this equation, but these require knowledge of soil parameters which are not readily available and the method is somewhat tentative. The following simple approaches may be adopted.

Table 8.9
Minimum safety factors for design of shallow foundations
(After Vesic, 1975)

Category	Characteristics of the category	Extent of soil investigation		Typical structures
		thorough, complete	limited	
A	Maximum design load likely to occur often; consequences of failure disastrous	3.0	4.0	railway bridges warehouses blast furnaces hydraulic structures retaining walls silos
B	Maximum design load may occur occasionally; consequences of failure serious	2.5	3.5	highway bridges light industrial and public buildings
C	Maximum design load unlikely to occur	2.0	3.0	apartment and office buildings

1 For clays sheared in drained conditions, the approach suggested by Terzaghi (1943) of using reduced strength parameters c^* and ϕ^* in Equation 8.7 where:

$$c^* = 0.67c' \qquad (8.18)$$

$$\text{and} \quad \tan\phi^* = 0.67\tan\phi' \qquad (8.19)$$

2 For sands, Vesic (1975) proposed:

$$\tan\phi^* = \left(0.67 + D_r - 0.75D_r^{\,2}\right)\tan\phi' \qquad (8.20)$$

where D_r is the relative density of the sand, recorded as a fraction. Typical values of D_r are given in Table 1.6. This reduction factor only applies for loose and medium dense sands. For dense sands ($D_r > 0.67$) the strength parameters need not be reduced, since the general shear mode of failure is likely to apply.

Sliding
Failure by sliding will occur when the applied horizontal force H (or component) exceeds the maximum resisting force, H_{max}. The factor of safety against sliding is then given as:

$$F = \frac{H_{max}}{H} \qquad (8.21)$$

In some cases a proportion of the passive resistance of the ground, P_p, acting alongside the foundation may be included to give:

$$F = \frac{H_{max} + P_p}{H} \qquad (8.22)$$

but there may be good reasons to ignore this contribution such as shrinkage of the soil, poorly compacted backfill, possible future excavations and the large movements required to mobilise passive resistance.

For the undrained case in clay H_{max} would be given by:

$$H_{max} = A'c_a \qquad (8.23)$$

where
A' = effective area = $B'L'$
and c_a = adhesion = αc_u
$\alpha = 1$ for soft and firm clays
$\alpha = 0.5$ for stiff clays

For a sand and the drained case in clay (c' assumed to be zero) H_{max} would be given by:

$$H_{max} = V_T'\tan\delta \qquad (8.24)$$

where:
V_T' is the vertical effective load.
If the water table lies above foundation level then V_T' will be given by

$$V_T' = V_T - uA' \qquad (8.25)$$

where:
u is the pore water pressure at foundation level.
The skin friction angle δ can be taken as equal to ϕ'.

Allowable bearing pressure of sand

Settlement limit
A serviceability limit for a foundation is the allowable (or tolerable, or permissible) settlement of the supported structure. For routine buildings on sands this is commonly taken as a total settlement of 25 mm on the understanding that differential settlements will then be within tolerable limits (Terzaghi and Peck, 1967).

The foundations must be designed to ensure that this settlement is not exceeded by using the allowable bearing pressure. For fairly wide foundations on sand it would be generally found that the factor of safety at this pressure will be far greater than the normal values of 3 – 4 so the foundation will certainly be 'safe' (see Figure 8.1).

Allowable bearing pressure *(Figures 8.18 – 8.20)*
Given the difficulties of obtaining undisturbed samples of sand, the Standard Penetration Test (SPT) was developed in the United States as an empirical way of assessing the degree of compactness of a sand. The test procedure has since been standardised (BS 1377:1990 and ASTM D1586).

However, given the wide variety of equipment still used to carry out the test and the many factors which can affect the result, the repeatability (one operator obtaining the same result several times over) and reproducibility (several different operators obtaining the same result) are not good.

Terzaghi and Peck (1948) developed an empirical chart relating allowable bearing pressure q_a to the SPT 'N' value. However, this chart also related q_a to the width of foundation which complicated design because of the iterative process required. Peck, Hanson and Thornburn (1974) simplified the chart into two regions, one where the design pressure was related to bearing capacity (for small foundation widths) and one where the design pressure was limited by settlement (Figure 8.18).

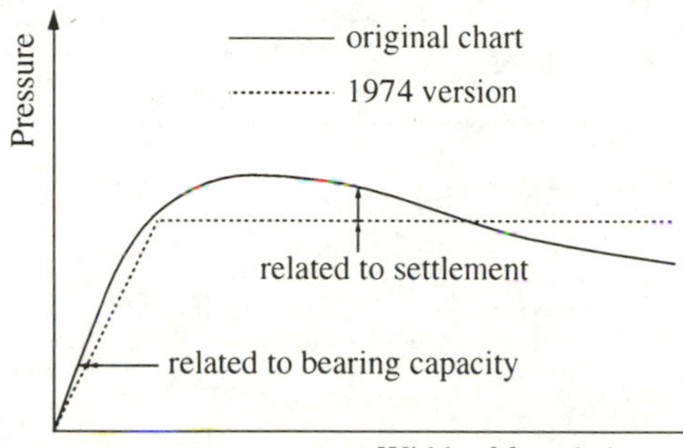

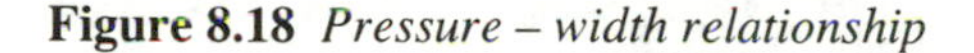

Figure 8.18 *Pressure – width relationship*

The chart is still very much empirical and any over-sophistication of approach will give an impression of accuracy which is unwarranted. A modified version of the Peck, Hanson, Thornburn chart is given in Figure 8.19 which has been simplified so that allowable bearing pressure can be reported in the form:

$$q_a = 10.5N'B \quad \text{for width } B < 1 \text{ m} \tag{8.26}$$

and

$$q_a = 10.5N' \quad \text{for width } B \geq 1 \text{ m} \tag{8.27}$$

where N' is a corrected value of the field value N given by:

$$N' = NC_N C_W \tag{8.28}$$

N is usually taken as the loosest value on the site and within a depth equal to the width of the foundation B below foundation level. The depth of foundation does not seem to have a significant effect.

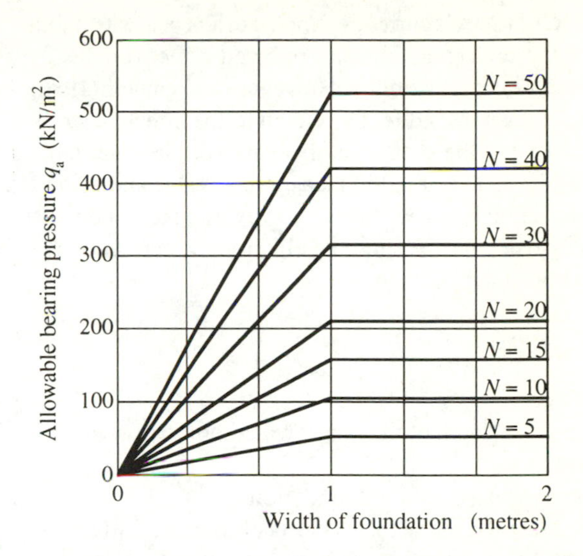

Figure 8.19 *Design chart for allowable bearing pressures (Adapted from Peck, Hanson and Thornburn, 1974)*

The chart was derived for square or rectangular foundations. Strip foundations can produce greater settlements for the same width and applied pressure so the allowable bearing pressure should be reduced by about 20% for strip foundations.

C_N is the correction for the effects of overburden pressure given by Peck *et al*, 1974:

$$C_N = 0.77 \log_{10} \frac{20}{p'} \tag{8.29}$$

where
p' = effective stress at the depth of the test in ton/sq ft (1 ton/sq ft = 95.7 kN/m^2).

This expression normalises the N values to a standard stress level of 1 ton/sq ft or about 100 kN/m^2 for which the chart was derived.

Penetration resistance is related to confining stress so at shallow depths (p' less than 100 kN/m^2) the actual N value will underestimate the compactness or density of the sand so the values must be increased to obtain an N value as though it had been carried out at a confining stress of 100 kN/m^2. C_N from the above expression is plotted in Figure 8.20 (overleaf).

C_w is Peck's correction for the effect of a water table on the understanding that reduced effective stresses increase compressibility. However, as Skempton (1986) has shown, N is directly related to the effective stress existing in the ground so the effect of the water table will be reflected in the measured N value and further correction is unnecessary. However, if the water table may rise after the site investigation the correction C_w could be applied:

$$C_w = 0.5 + \frac{0.5D_w}{D_f + B} \quad (8.30)$$

where:
D_w is the depth of the raised water table below ground level,
D_f is the depth of the foundation.

Many other factors have been found to affect the SPT N value such as the diameter of the borehole, the use of liners in the split spoon, grain size, crushability of particles, overconsolidation, ageing of the deposit, fluctuating load and time after construction. Judgement must be exercised in the choice of an allowable bearing pressure.

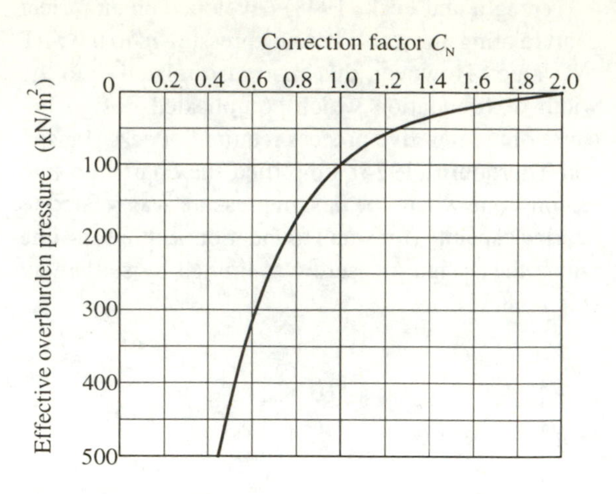

Figure 8.20 *Correction factor* C_N

Worked Example 8.1 *Vertical loading*

A rectangular foundation 2.5 m wide and 3.5 m long is to be placed at a depth of 1.7 m below ground level in a thick deposit of firm saturated clay. The water table is at 1.2 m below ground level. Determine the gross and net ultimate bearing capacities for

a) undrained and

b) drained conditions.

Investigate the contribution made by including depth factors.

The soil parameters are $c_u = 65$ kN/m² $\quad \phi_u = 0°$

$c' = 3$ kN/m² $\quad \phi' = 27°$

$\gamma_{sat} = 21.5$ kN/m³

a) Undrained condition

Without depth factors

for $\phi = 0°$ $\quad N_c = 5.14 \quad N_q = 1 \quad N_\gamma = 0$ $\quad$ From Table 8.6, $s_c = 1 + 0.2 \times 2.5/3.5 = 1.14$

$p_o = 21.5 \times 1.7 = 36.6$ kN/m²

From Equation 8.9, gross $q_{ult} = c_u N_c s_c + p_o$

$= 65 \times 5.14 \times 1.14 + 36.6 = 417.5$ kN/m²

net $q_{ult} = 417.5 - 36.6 = 380.9$ kN/m²

Assuming a factor of safety of 3 the net safe bearing capacity is $\frac{1}{3} \times 381 = 127$ kN/m²

With depth factors

$$\frac{D}{B'} = \frac{1.7}{2.5} = 0.68\ (<1)$$ From Table 8.7, $d_c = 1 + 0.4 \times 0.68 = 1.27$

gross $q_{ult} = 380.9 \times 1.27 + 37 = 520.3$ kN/m² $\quad$ net $q_{ult} = 520.3 - 36.6 = 483.7$ kN/m²

b) Drained condition

Without depth factors

Assume the soil to be incompressible ∴ no reduction of strength parameters is considered.

Using Equations 8.8 and 8.13:

From Table 8.5, for $\phi' = 27° N_c = 23.9 \quad N_q = 13.2 \quad N_\gamma = 9.3$

From Table 8.6,

$$s_c = 1 + \frac{2.5}{3.5} \times \frac{13.2}{23.9} = 1.40 \quad s_q = 1 + \frac{2.5}{3.5} \tan 27° = 1.36 \quad s_\gamma = 1 - 0.4 \times \frac{2.5}{3.5} = 0.71$$

$p_o' = 36.6 - 0.5 \times 9.8 = 31.7$ kN/m²

$\gamma' = 21.5 - 9.8 = 11.7$ kN/m³

gross $q_{ult} = 3 \times 23.9 \times 1.40 + 31.7 \times 12.2 \times 1.36 + 0.5 \times 11.7 \times 2.5 \times 9.3 \times 0.71 + 36.6$

$= 100.4 + 526.0 + 96.6 + 36.6 = 759.6$ kN/m²

net $q_{ult} = 759.6 - 36.6 = 723$ kN/m²

If the foundation supports a net applied pressure of 127 kN/m² the factor of safety in the long-term would be

$$F = \frac{723}{127} = 5.7$$

With depth factors

$$\frac{D}{B'} = 0.68\ (<1)$$ From Table 8.7,

$d_q = 1 + 2 \tan 27° (1 - \sin 27°)^2 \times 0.68 = 1.21$ $\quad$ $d_c = 1.21 - \dfrac{1 - 1.21}{23.9 \tan 27°} = 1.23$

$d_\gamma = 1.00$

gross $q_{ult} = 100.4 \times 1.23 + 526 \times 1.21 + 96.6 \times 1.00 + 36.6 = 893.2$ kN/m²

net $q_{ult} = 893.2 - 36.6 = 856.6$ kN/m²

Worked Example 8.2 *Effect of compressibility*
For the foundation in Example 8.1 assess the effect of compressibility on the bearing capacity values.

(a) Undrained condition
Assume the clay is incompressible for the undrained case so the bearing capacity is unchanged

(b) Drained condition
The soil is assumed to be compressible so reduced strength parameters are used.
From Equations 8.18 and 8.19
$c^* = 0.67 \times 3 = 2$ kN/m^2
$\phi^* = \tan^{-1}(0.67 \times \tan 27^\circ) = 18.8^\circ$
Consider the case without depth factors only.
for $\phi = 18.8^\circ$ $\quad N_c = 13.7 \quad N_q = 5.7 \quad N_\gamma = 2.4$

$$s_c = 1 + \frac{2.5}{3.5} \times \frac{5.7}{13.7} = 1.30 \quad s_q = 1 + \frac{2.5}{3.5} \tan 18.8° = 1.24 \quad s_\gamma = 0.71$$

gross $q_{ult} = 2 \times 13.7 \times 1.30 + 31.7 \times 4.7 \times 1.24 + 0.5 \times 11.7 \times 2.5 \times 2.4 \times 0.71 + 36.6$
$= 35.6 + 184.8 + 24.9 + 36.6 = 281.9$ kN/m^2
net $q_{ult} = 281.9 - 36.6 = 245.3$ kN/m^2
The factor of safety in the long-term would be F = 245/127 = 1.9
Since the clay at shallow depths is of a firm consistency it is likely to be overconsolidated so the effects of compressibility are probably overstated.

Worked Example 8.3 *Eccentric loading*
The rectangular foundation in Example 8.1 is 0.5 m thick and backfilled with soil of unit weight 19.5 kN/m^3. A column transmitting a vertical load of 850 kN is placed 0.6 m from the centre of the foundation on the long axis. Determine the bearing capacity and factor of safety against general shear failure.
Unit weight of concrete is 25 kN/m^3
Weight of backfill = $2.5 \times 3.5 \times 1.2 \times 19.5 = 204.8$ kN
Weight of foundation = $2.5 \times 3.5 \times 0.5 \times 25 = 109.4$ kN
Total vertical load = $204.8 + 109.4 + 850 = 1164$ kN

$$\text{eccentricity } e_L = \frac{850 \times 0.6}{1164} = 0.44 \text{ m} \qquad B' = B = 2.5 \text{ m} \qquad L' = 3.5 - 2 \times 0.44 = 2.62 \text{ m}$$

a) Undrained condition
Without depth factors

$$s_c = 1 + 0.2 \times \frac{2.5}{2.62} = 1.19$$

$\therefore$ gross $q_{ult} = 65 \times 5.14 \times 1.19 + 36.6 = 434.2$ kN/m^2 $\qquad$ net $q_{ult} = 434.2 - 36.6 = 397.6$ kN/m^2

$$\text{Gross applied pressure} = \frac{1164}{2.5 \times 2.62} = 1.78 \text{ kN / m}^3$$

Net applied pressure = $177.7 - 36.6 = 141$ kN/m^2

$$\text{Factor of safety} = \frac{397.6}{141.1} = 2.8$$

(b) Drained condition
Without depth factors
for $\phi' = 27^{\circ}$ $N_c = 23.9$ $N_q = 13.2$ $N_\gamma = 9.3$

$$s_c = 1 + \frac{2.5}{2.62} \times \frac{13.2}{23.9} = 1.53 \quad s_q = 1 + \frac{2.5}{2.62} \tan 27^{\circ} = 1.49 \quad s_\gamma = 1 - 0.4 \times \frac{2.5}{2.62} = 0.62$$

gross $q_{ult} = 3 \times 23.9 \times 1.53 + 31.7 \times 12.2 \times 1.49 + 0.5 \times 11.7 \times 2.5 \times 9.3 \times 0.62 + 36.6$
$= 109.7 + 576.2 + 84.3 + 36.6 = 806.8$ kN/m^2
net $q_{ult} = 806.8 - 36.6 = 770.2$ kN/m^2

Factor of safety $= \dfrac{770}{141} = 5.5$

Compared with Example 8.1 the net bearing capacity has increased slightly but the net applied pressure has also increased. If the column load had been placed at the centre of this foundation the factor of safety would be given by:

Net applied pressure $= \dfrac{1164}{2.5 \times 3.5} - 37 = 133 - 37 = 965$ kN / m^2

(a) Undrained condition

Factor of safety $= \dfrac{381}{96} = 4.0$

(b) Drained condition

Factor of safety $= \dfrac{381}{96} = 4.0$

Worked Example 8.4 *Eccentric and inclined loading*
A horizontal load of 80 kN is applied 1.5 m above ground level to the column in Example 8.3 in the direction of the short side. Determine the bearing capacity and factor of safety against general shear failure.

eccentricity $e_L = \dfrac{80 \times 3.2}{1164} = 0.22$ m

(a) undrained condition
Without depth factors

$$s_c = 1 + 0.2 \times \frac{2.06}{2.62} = 1.16$$

The inclination of the load is in the direction of the short side so

$$m_B = \frac{2 + \dfrac{2.06}{2.62}}{1 + \dfrac{2.06}{2.62}} = 1.56 \quad i_c = 1 - \frac{1.56 \times 80}{2.06 \times 2.62 \times 65 \times 5.14} = 0.93$$

gross $q_{ult} = 65 \times 5.14 \times 1.16 \times 0.93 + 36.6 = 397$ kN/m^2
net $q_{ult} = 397 - 36.6 = 360.4$ kN/m^2

Gross applied pressure $= \dfrac{1164}{2.06 \times 2.62} = 216$ kN / m^2

Net applied pressure $= 215.7 - 36.6 = 179.1$ kN/m^2

Factor of safety $= \dfrac{360}{179} = 2.0$

(b) Drained condition

$s_c = 1 + \frac{2.06}{2.62} \times \frac{13.2}{23.9} = 1.43 \quad s_q = 1 + \frac{2.06}{2.62} \tan 27° = 1.40 \quad s_\gamma = 1 - 0.4 \times \frac{2.06}{2.62} = 0.69$

$m_B = 1.56$

$i_\gamma = 0.933^{2.56} = 0.84$

$i_c = 0.90 - \frac{1 - 0.90}{23.9 \tan 27°} = 0.89$

From Equations 8.8 and 8.13

gross $q_{ult} = 3 \times 23.9 \times 1.43 \times 0.89 + 31.7 \times 12.2 \times 1.40 \times 0.90 + 0.5 \times 11.7 \times 2.06 \times 9.3 \times 0.69 \times 0.84 + 36.6$
$= 91.3 + 487.3 + 65.0 + 36.6 = 680.2$ kN/m^2

net $q_{ult} = 680.2 - 36.6 = 643.6$ kN/m^2

Factor of safety $= \frac{643}{179} = 3.6$

Factors of Safety – Summary

Load case	Short-term	Long-term
Vertical, central load	4.0	7.5
Vertical, eccentric load	2.8	5.5
Eccentric and inclined load	2.0	3.6

Worked Example 8.5 *Sliding*

For the foundation detailed in Example 8.4 determine the factor of safety against sliding for the undrained and drained condition.

(a) Undrained condition

Assume $\alpha = 1$

From Equation 8.23

$H_{max} = 2.06 \times 2.62 \times 65 \times 1 = 351$ kN

Factor of safety $= \frac{351}{80} = 4.4$

The factor of safety would still be adequate (2.2) if $\alpha = 0.5$ was used.

(b) Drained condition

Assume $c' = 3$ kN/m^2 and $\delta = \phi'$.

From Equation 8.25

$V_T' = 1164 - 0.5 \times 9.8 \times 2.06 \times 2.62 = 1138$ kN

$H_{max} = 3 \times 2.06 \times 2.62 + 1138 \times \tan 27° = 16 + 580 = 596$ kN

Factor of safety $= \frac{596}{80} = 7.5$

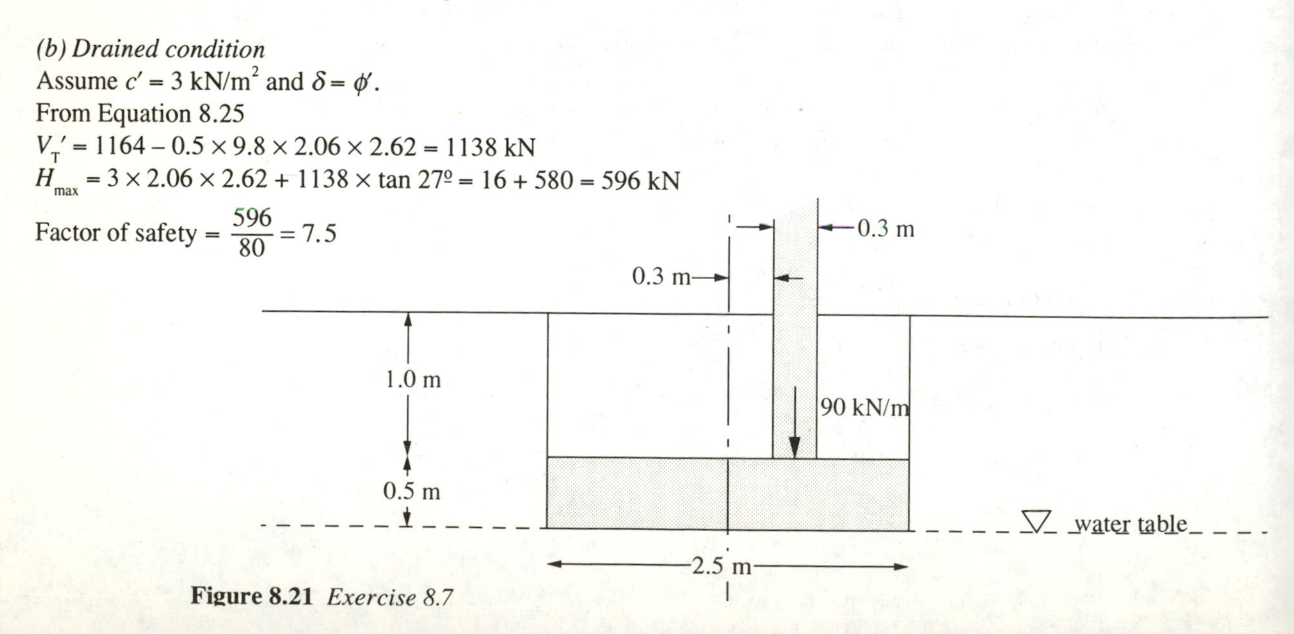

Figure 8.21 *Exercise 8.7*

Exercises

8.1 A square foundation, 3.5 m wide, is founded at 1.5 m below ground level in a stiff clay with undrained shear strength of 95 KN/m². Determine the net ultimate bearing capacity:
a) with the depth factors included
b) without the depth factors.

8.2 In Exercise 8.1, assuming the foundation and backfill have the same unit weight as the soil removed from the excavation determine the maximum central column load which can be applied. Assume a factor of safety of 3.

8.3 Determine the width required of a square foundation to support a central column load of 1450 kN placed at 1.2 m below ground level in a firm clay with undrained shear strength of 75 kN/m². Adopt a factor of safety of 3 and ignore depth factors. Assume that the foundation and backfill have the same unit weight as the soil removed.

8.4 For the foundation designed in Exercise 8.3 determine the factor of safety for long-term conditions given that the water table lies at 1.2 m below ground level.
$c' = 4$ kN/m², $\phi' = 32°$ and the unit weight of the clay is 21.5 kN/m³.

8.5 A rectangular foundation, 3.0 m wide and 4.5 m long is to be placed at 2.0 m below ground level in a dense sand with $\phi' = 38°$ and unit weight of 19.5 kN/m³. A water table exists at 0.5 m below ground level.
a) Determine the net ultimate bearing capacity for a central vertical load.
b) To keep the settlements within permissible limits the maximum column load which can be applied is estimated to be 3500 kN. Determine the factor of safety against bearing capacity failure.
Assume that the foundation and backfill have the same unit weight as the soil removed.

8.6 Repeat Exercise 8.5, assuming the water table to lie below the influence of the foundation.

8.7 A reinforced concrete retaining wall is to be founded at 1.5 m below ground level in a clay with the water table at 1.5 m below ground level as shown in Figure 8.21.
Properties of the materials are as follows:
clay: $c' = 4$ kN/m² $\phi' = 23°$ $\gamma = 22$ kN/m³ $c_u = 40$ kN/m² $\phi_u = 0$
backfill: $\gamma = 21$ kN/m³
concrete: $\gamma = 25$ kN/m³.
The stem of the wall applies a line load of 90 kN/mrun to the base of the wall. Ignoring the effects of depth factors determine the factor of safety against bearing capacity failure for:
a) short-term conditions and b) long-term conditions.

8.8 A reinforced concrete pad foundation, 3.5 m square and 0.6 m thick is to be founded at 1.8 m below ground level in a clay with the water table at 1.2 m below ground level. Properties of the materials are as follows:
clay: $c' = 2$ kN/m² $\phi' = 28°$ $\gamma = 21.5$ kN/m³ $c_u = 85$ kN/m² $\phi_u = 0$
backfill: $\gamma = 20$ kN/m³
concrete: $\gamma = 25$ kN/m³.
The foundation supports a vertical column load of 1600 KN at its centre and a horizontal load of 150 kN acts along one of the axes of the foundation at 1.5 m above ground level. Determine the factor of safety against bearing capacity failure for:
a) short-term conditions and b) long-term conditions.

9 Shallow Foundations – Settlements

Introduction

Foundations supported on sand are designed using the allowable bearing pressure as shown in Chapter 8 so it is expected that the foundation will not settle more than the allowable settlement. Semi-empirical methods to estimate the settlement of a foundation on sand are given in this chapter.

Foundations supported on clay are designed using the factored safe bearing capacity, with the factor of safety chosen for reasons other than limiting settlements, see Chapter 8. It is then necessary to estimate the settlement of a foundation on clay to determine whether it will lie within the permissible settlements of the structure.

For a proposed structure, simple methods have been established to determine settlements or vertical downward movement, and these are covered in this chapter. However, methods to predict other movements are less well developed.

Foundations may move downwards because of shrinkage, erosion, subsidence, thawing, etc. but the foundation engineer usually prefers to avoid these situations rather than quantify them and accept them as inevitable. Settlements which must be quantified because they cannot be avoided are those caused by changes in stresses in the ground as a result of engineering works.

Settlements produced by applied stresses originate from:

- immediate settlement, ρ_i
- consolidation settlement, ρ_c
- secondary compression, ρ_s

and amounts of each must be determined to give the total settlement, ρ_T as:

$$\rho_T = \rho_i + \rho_c + \rho_s \qquad (9.1)$$

Some definitions of ground and foundation movement are given later in this chapter and these are useful in the assessment of the movement and damage of an existing structure.

Clays

Immediate settlement

General method *(Figure 9.1)*

This is also termed undrained settlement, occurring with no water entering or leaving the soil (no volume change means the Poisson's ratio $\nu = 0.5$). The settlements are produced by shear strains within the soil resulting in the soil surface changing shape. These strains are assumed to be elastic, so the settlements should be recovered on removal of the load.

The foundation or loaded area is assumed to be flexible, producing a dish-shaped settlement profile with maximum settlement at the foundation centre.

A rigorous solution for immediate settlements with the normal assumptions listed in Table 9.1 was given by Ueshita and Meyerhof (1968) for the conditions illustrated in Figure 9.1 using the expression:

$$\rho_i = \frac{qB}{E_u} I \qquad (9.2)$$

where:

ρ_i = immediate settlement at the corner of the loaded area

q = uniform applied pressure

B = width of loaded area

I = influence factor, given on Figure 9.1

E_u = undrained modulus of soil.

Principle of superposition *(Figure 9.2)*

The above method gives the settlement at the corner of a loaded area. To determine settlements at other points beneath the foundation such as the maximum settlement at the centre, the principle of superposition must be used, illustrated in Figure 9.2.

Principle of layering *(Figure 9.3)*

Where there are two or more layers of soil with different modulus values the principle of layering can be used, as illustrated in Figure 9.3.

Where the settlement at a point within the soil layer is needed, such as the effect of a surface loaded area on a buried structure, e.g. a pipe, this principle is used.

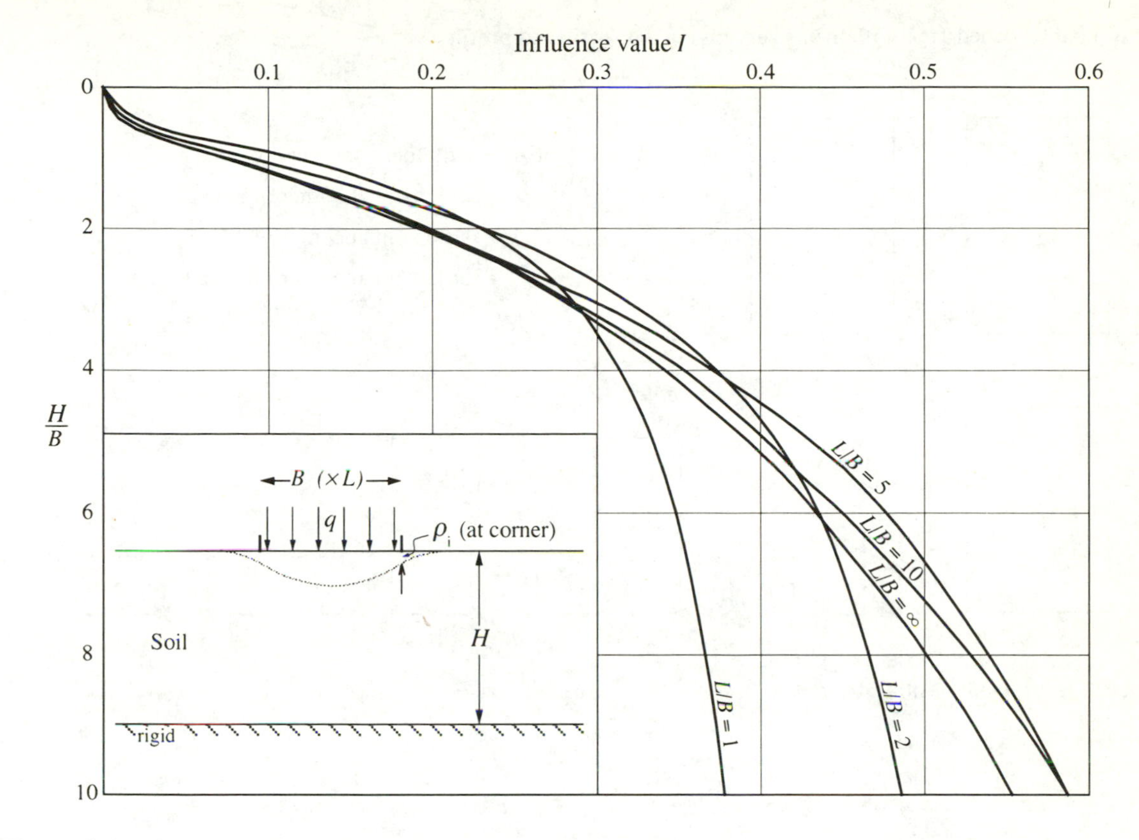

Figure 9.1 *Influence values for immediate settlement (From Ueshita and Meyerhof, 1968)*

Table 9.1
Immediate settlement – assumptions

Normal assumptions	Modifications
Classical methods give settlements:	
at the corner of	principle of superposition
a flexible loaded	average settlement (Christian and Carrier) rigidity correction
rectangular area	principle of superposition
on the surface of	depth correction factor (Christian and Carrier)
a homogeneous	principle of layering modulus increasing with depth (Butler, Meigh)
isotropic soil	
with linear stress-strain relationship	effect of plastic yield (D'Appolonia et al)

For a flexible loaded area settlements vary giving a dish-shaped profile

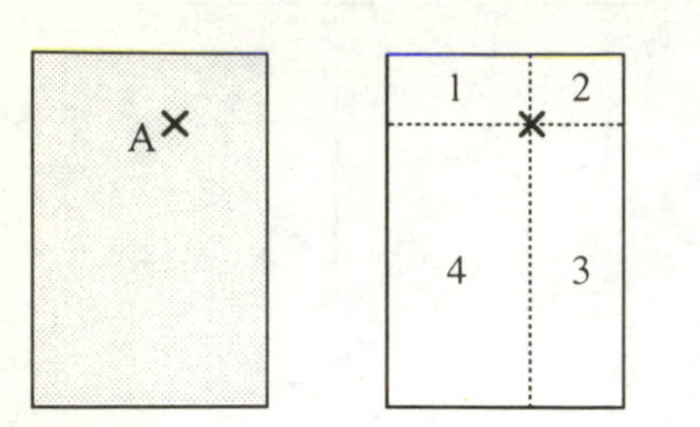

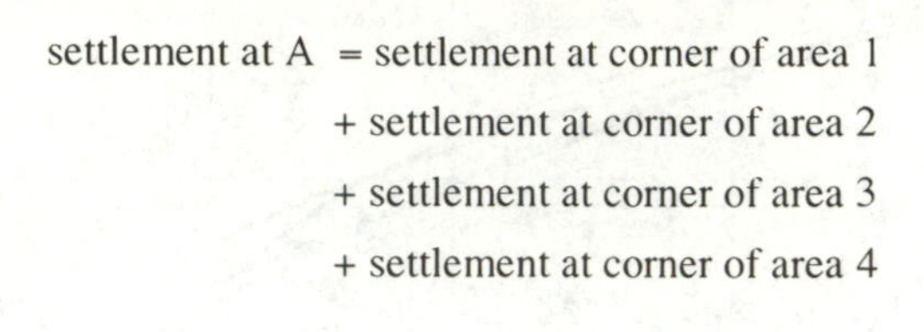

Examples

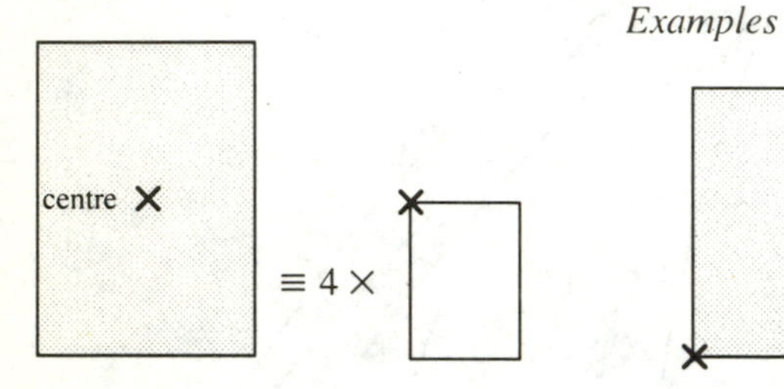

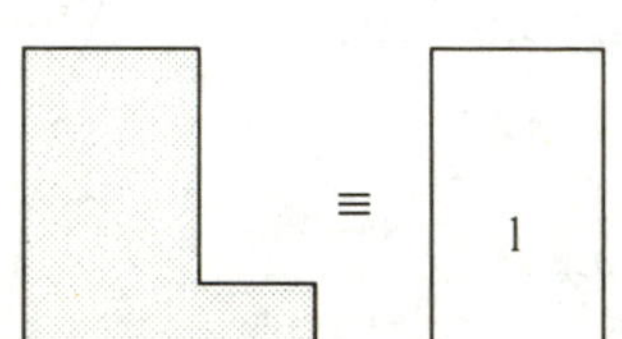

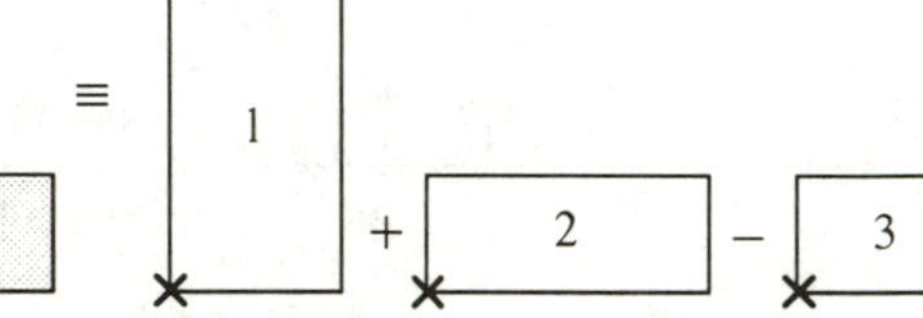

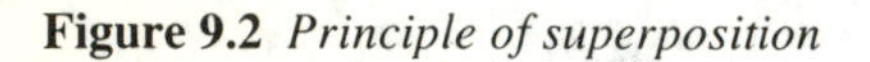

Figure 9.2 *Principle of superposition*

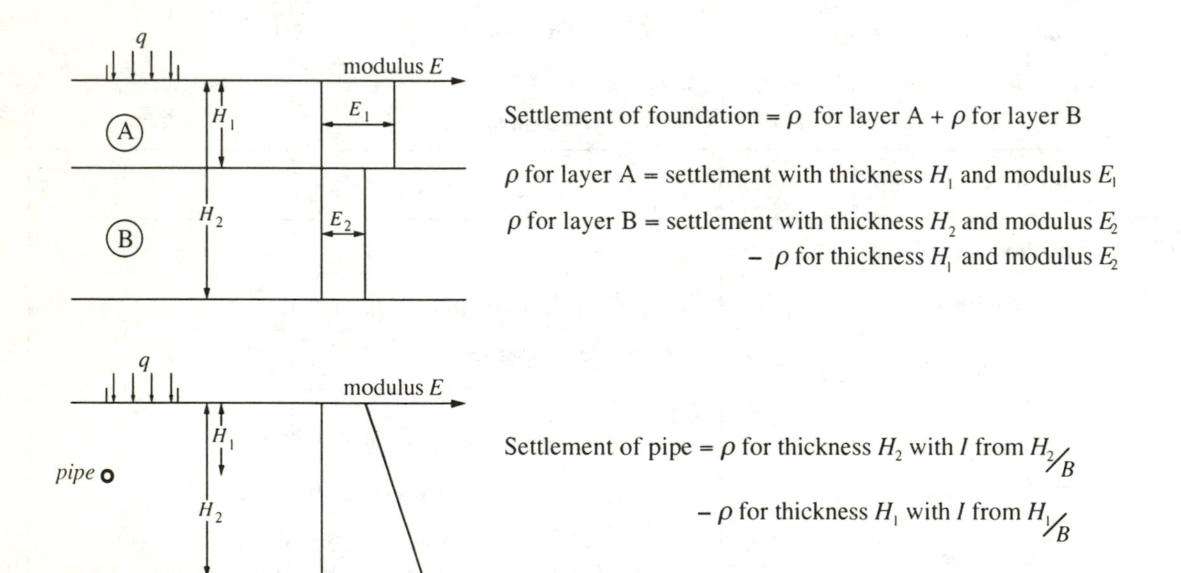

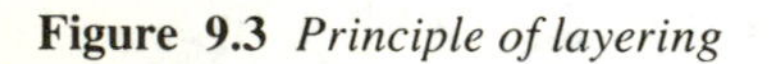

Figure 9.3 *Principle of layering*

Rigidity correction *(Table 9.2)*
The settlement beneath a rigid foundation is the same at all points. A sufficiently accurate correction to the above method for the case of a rigid foundation is often quoted as:

$$\rho_{\text{rigid}} \approx 0.8\rho_{\text{maximum flexible}} \tag{9.3}$$

The general correction for rigidity can be represented as:

$$\rho_{\text{rigid}} = \mu_r \rho_{\text{maximum flexible}} \tag{9.4}$$

where $\rho_{\text{maximum flexible}}$ is the settlement at the centre of the foundation assuming it to be flexible.

Fraser and Wardle (1976) demonstrated the effect of varying stiffness of a foundation on settlements and for a rigid foundation (infinite stiffness) the values of μ_r given in Table 9.2 have been derived from their results.

Table 9.2
Rigidity correction factor μ_r

L/B	*H/B* = 1	*H/B* = ∞
1	0.68	0.77
2	0.72	0.78
3, 4, 5	0.79	0.80

Depth correction *(Figure 9.4)*
The effect of depth of embedment on settlement can be included using the factor μ_0 (from Burland, 1970) given in Figure 9.4. This method assumes that settlement is first determined by taking the loaded area to the surface of the soil layer and then correcting for depth:

$$\rho_{\text{at depth}} = \mu_0 \rho_{\text{at surface}} \tag{9.5}$$

Average settlement *(Figure 9.4)*
The average settlement of a flexible foundation is often assumed to be similar to the settlement of a rigid foundation. The average settlement of a flexible loaded area was determined by Janbu *et al* (1956) and modified later by Christian and Carrier (1978). They gave:

$$\rho_{\text{ave}} = \mu_0 \mu_1 \frac{qB}{E_u} \tag{9.6}$$

where μ_0 and μ_1 are factors for depth of embedment and thickness of soil layer beneath the foundation, respectively, see Figure 9.4. The principle of layering could be applied with this method.

Modulus increasing with depth *(Figure 9.5)*
It has been found that for most soils the modulus increases with depth so assuming a constant modulus (the homogeneous case) will over-estimate settlements. Butler (1974) produced an approximate analysis based on Steinbrenner's influence values for a soil with modulus increasing with depth giving the immediate settlement at the corner of the loaded area as:

$$\rho_i = \frac{qB}{E_0} I \tag{9.7}$$

where *I* is an influence factor related to:
- shape (*L*/*B*)
- thickness (*H*/*B*)
- a coefficient *k* given by:

$$k = \frac{E_H - E_0}{E_0} \frac{B}{H} \tag{9.8}$$

Values of the influence factor *I* can be obtained from Figure 9.5.

The method assumes that the foundation is placed on the surface of the compressible layer. A correction for a foundation at depth, μ_0, could be applied as described above.

The principle of superposition must be used for points other than the corner of the loaded area and the principle of layering can be used if layers with different modulus variation occur.

Effect of local yielding *(Figures 9.6 and 9.7)*
The above methods assume that the stress-strain relationship is linear which is a reasonable assumption provided the applied stress levels are low enough to prevent local plastic yielding of the soil. D'Appolonia *et al* (1971) carried out finite element analyses of the problem assuming non-linear stress strain properties after a yield condition had occurred (Figure 9.6). They incorporated the effect of yielding by modifying the

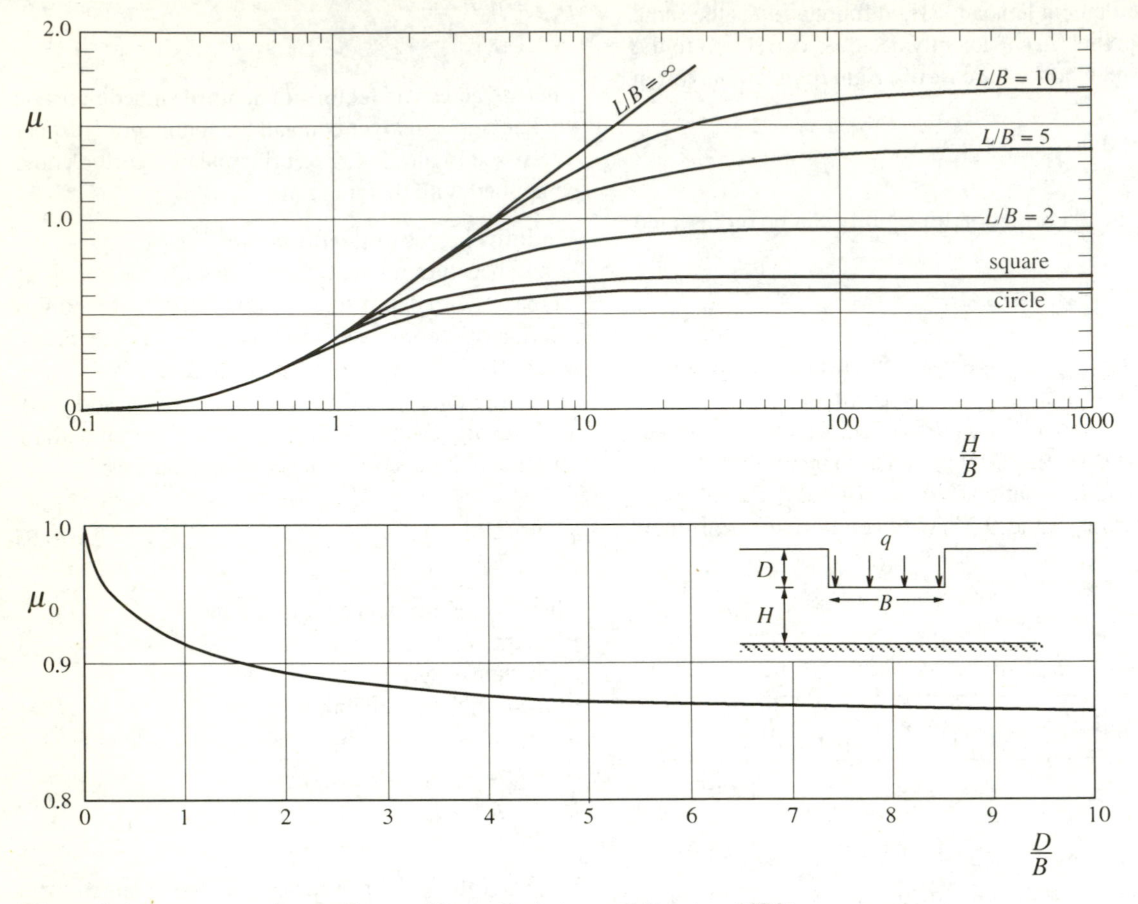

Figure 9.4 *Average settlement factors (From Christian and Carrier, 1978)*

elastic settlement ρ_i determined as above. This modification can be written in the form:

$$\rho_y = \rho_i \times f_y \tag{9.9}$$

where:
ρ_y = immediate settlement including yield
ρ_i = immediate settlement from elastic theory
f_y = yield factor, given in Figure 9.7.

Significantly, they found that yield and hence increased settlements can occur for normally consolidated clays if the factor of safety against bearing capacity failure is less than 4 to 8 whereas for heavily overconsolidated clays this does not occur unless the factor of safety is below 2.

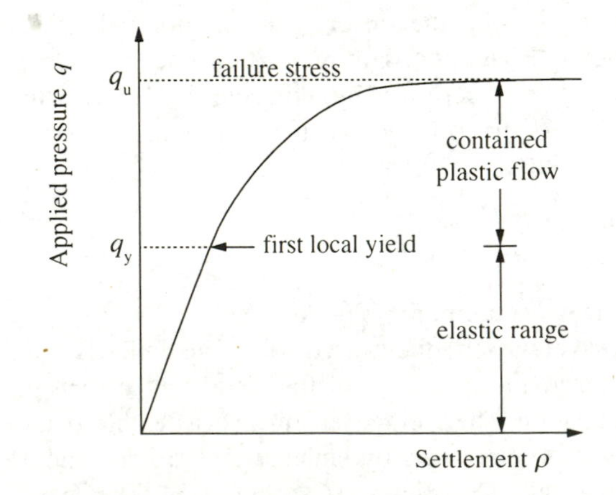

Figure 9.6 *Local yielding*

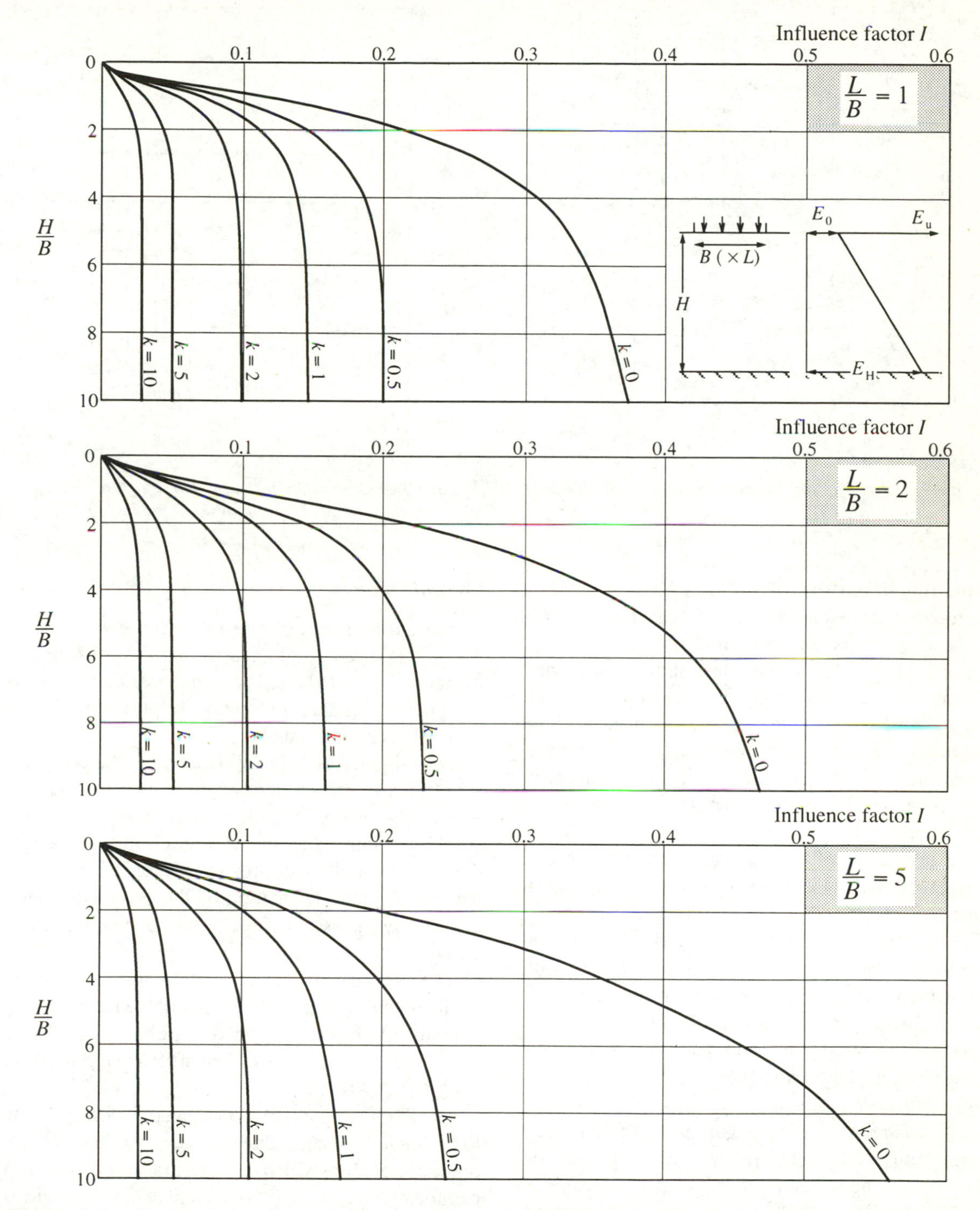

Figure 9.5 *Influence factors for modulus increasing with depth – immediate settlement (From Butler, 1974)*

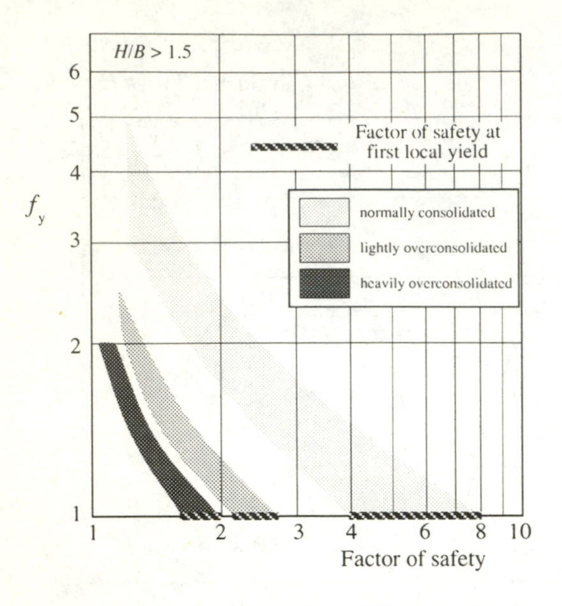

Figure 9.7 *Yield factor (Adapted from D'Appolonia et al, 1971)*

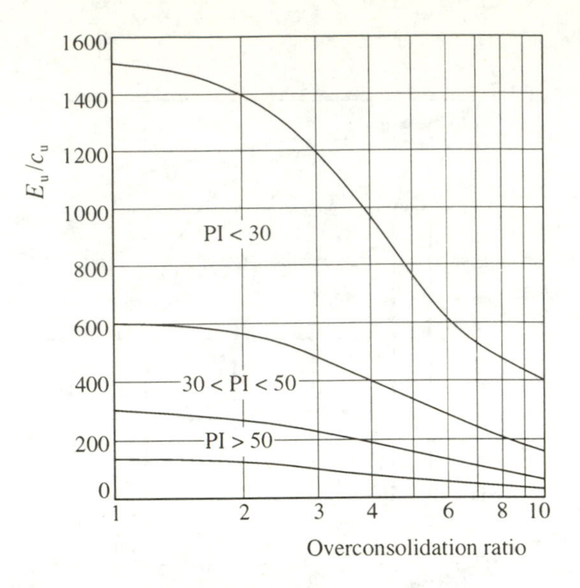

Figure 9.8 *Undrained modulus correlation (From Jamiolkowski et al, 1979)*

Estimation of undrained modulus E_u *(Figure 9.8)*
Due to sampling disturbance, the effects of stress relief and bedding errors laboratory tests are considered to give inaccurate values of the stress-strain relationship for soils. At best, they may be considered a lower bound estimate. *In situ* testing such as plate loading tests or pressuremeter tests may reduce some of these effects but these tests can still suffer from the effects of stress relief and there are uncertainties and assumptions made in their analysis.

For an estimate of settlement it is generally considered that correlations between modulus E_u and undrained shear strength c_u are the most useful approach. The E_u values are derived from back-analysis of settlement observations on actual structures on a wide variety of soils which would eliminate most of the effects of sampling disturbance, stress relief, scale effects and stress path. They have been correlated with the c_u value measured in the laboratory triaxial test.

A useful correlation presented by Jamiolkowski *et al* (1979) is reproduced in Figure 9.8. The modulus values represent the secant modulus at $0.5c_u$ (factor of safety = 2). The stress-strain relationship for soils is non-linear so higher modulus values could be expected at lower stress levels.

Consolidation settlement

Compression index C_c method *(Figure 9.9)*
This method can be adopted for normally and lightly overconsolidated clays. The compression index C_c is the gradient of the void ratio-log pressure plot for normally consolidated clay.

For a lightly overconsolidated clay, the compression index is the gradient of the *e*-log pressure plot *beyond* the preconsolidation pressure p_c'. For pressures lower than p_c' a smaller value C_s must be used for the overconsolidated portion of the plot. It can be seen that the accurate estimation of the preconsolidation pressure, p_c' is essential for lightly overconsolidated clays.

The method of determining an amount of settlement is illustrated in Figure 9.9. The soil deposit is split up into suitable layers and the void ratio change Δe for each layer is determined from which the change in thickness ΔH is obtained.

C_c values obtained from oedometer tests are likely to be underestimated due to sampling disturbance. Some correlations which relate C_c to a soil composition parameter have been published and two of these are as follows.

$$\Delta e = C_c \log_{10}\left(\frac{p'_0 + \Delta\sigma}{p'_0}\right) \qquad \therefore\ \Delta e = C_c P$$

$$e_0 = w_0 G_s \qquad \Delta H = \frac{\Delta e}{1 + e_0} H$$

Normally consolidated clay

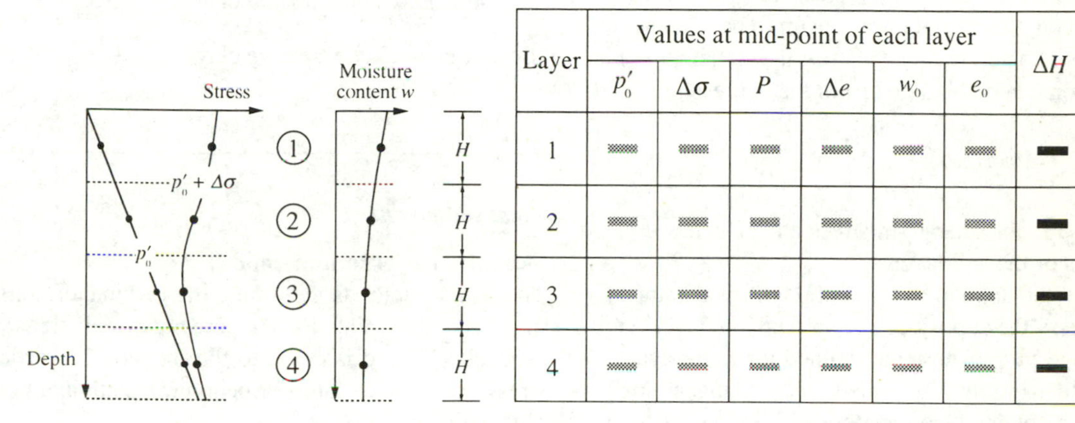

Layer	Values at mid-point of each layer						ΔH
	p'_0	$\Delta\sigma$	P	Δe	w_0	e_0	
1							
2							
3							
4							

consolidation settlement = Σ

Lightly overconsolidated clay

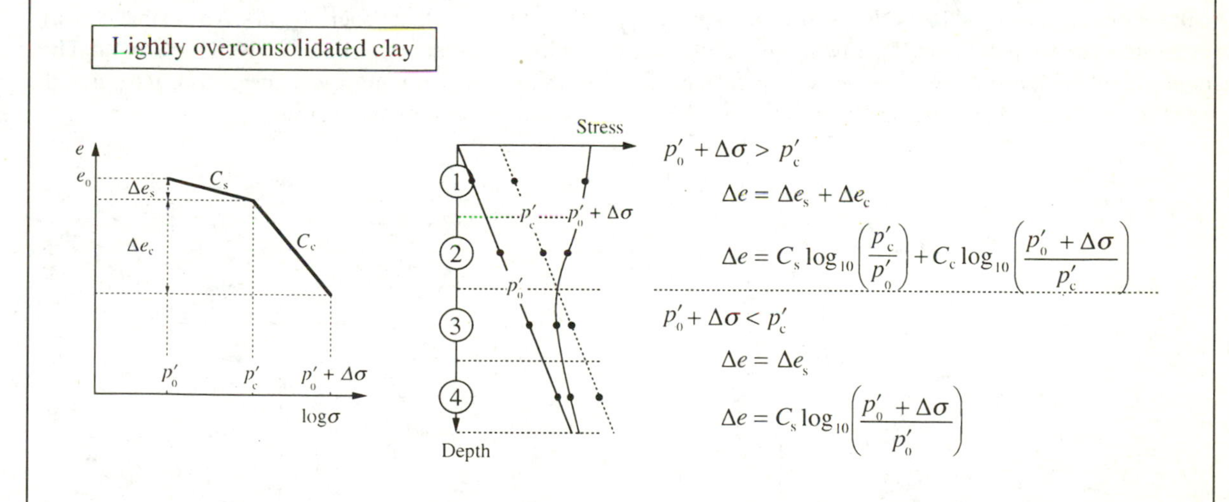

$p'_0 + \Delta\sigma > p'_c$

$$\Delta e = \Delta e_s + \Delta e_c$$

$$\Delta e = C_s \log_{10}\left(\frac{p'_c}{p'_0}\right) + C_c \log_{10}\left(\frac{p'_0 + \Delta\sigma}{p'_c}\right)$$

$p'_0 + \Delta\sigma < p'_c$

$$\Delta e = \Delta e_s$$

$$\Delta e = C_s \log_{10}\left(\frac{p'_0 + \Delta\sigma}{p'_0}\right)$$

Figure 9.9 *Compression index method*

$C_c \approx 0.009\,(\text{LL} - 10)$ (Terzaghi and Peck, 1948)

$C_c \approx 0.5\rho_s \dfrac{\text{PI}}{100}$ (Wroth, 1979)

where:
LL = liquid limit
PI = plasticity index
ρ_s = particle density

Oedometer or m_v method *(Figure 9.10)*
During an oedometer test values of m_v are determined for each pressure increment applied above the vertical effective stress or overburden pressure p_o' at the depth from which the sample was taken. The change in thickness of a soil layer and hence the settlement of a foundation can be obtained from the expression:

$$\Delta H = \rho_{oed} = m_v H \Delta\sigma \tag{9.10}$$

where $\Delta\sigma$ is the change in stress and H is the initial thickness of the soil layer.

Where the applied stress, and possibly also the m_v values, vary the deposit can be split up into layers and the change in thickness determined for each layer as illustrated in Figure 9.10. Values of the change in stress $\Delta\sigma$ can be obtained from methods given in Chapter 5.

m_v is not a true soil property, it depends on the pressure increments adopted. It is usually reported for a pressure range of $p_o' + 100$ kN/m² but where applied pressures are smaller than 100 kN/m² m_v should be determined for appropriate increments. Typical values of m_v for different clay types are given in Table 9.3.

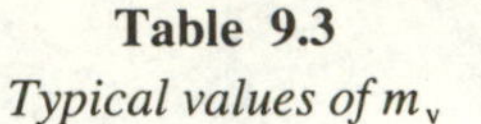
Table 9.3
Typical values of m_v

Type of clay	m_v m²/MN
Very stiff heavily overconsolidated clay	< 0.05
Stiff overconsolidated clay	0.05 – 0.1
Firm overconsolidated clay, laminated clay, weathered clay	0.1 – 0.3
Soft normally consolidated clay	0.3 – 1.0
Soft organic clay, sensitive clay	0.5 – 2.0
Peat	> 1.5

Total settlement

Skempton-Bjerrum method
The above methods determine the oedometer settlement, ρ_{oed} assuming that the pore pressure increase is produced by and is equal to the increase in vertical stress, i.e. $\Delta u = \Delta\sigma_v$, and that oedometer settlements are obtained from $m_v H \Delta\sigma_v$.

Skempton and Bjerrum (1957) realised that the soil beneath structures is not laterally confined as in the oedometer and that generally $\Delta u < \Delta\sigma_v$ so that consolidation settlements will be less, given by $m_v H \Delta u$. They introduced a semi-empirical correction, μ to give the consolidation settlement ρ_c as:

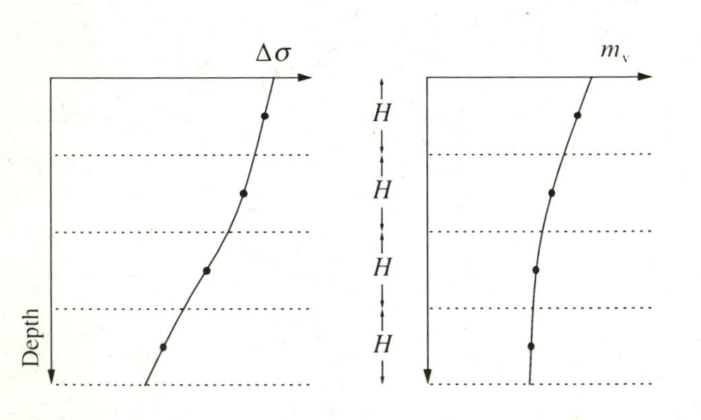

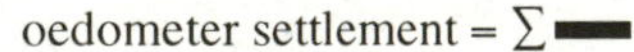

Figure 9.10 *Oedometer or m_v method*

$$\rho_c = \mu \rho_{oed} \tag{9.11}$$

with $\mu = A + \alpha(1 - A)$

A is Skempton's pore pressure parameter (Equation 4.12), which is related to the overconsolidation ratio, and α depends on the geometry of the loaded area (shape L/B and thickness H/B). However, this expression for μ relies on A and m_v being constant with depth which does not occur. Some typical values of μ, suggested by Tomlinson (1986) are given in Table 9.4.

Table 9.4

Skempton - Bjerrum correction μ
(From Tomlinson, 1986)

Type of clay	μ
Very sensitive clays (Alluvial, estuarine and marine clays)	1.0 – 1.2
Normally consolidated clays	0.7 – 1.0
Overconsolidated clays (London Clay, Weald, Kimmeridge, Oxford, Lias clays)	0.5 – 0.7
Heavily overconsolidated clays (Glacial Lodgement till, Keuper Marl)	0.2 – 0.5

The μ value for London Clay is usually taken as 0.5

The total settlement of a foundation is then given by:

$$\rho_T = \rho_i + \mu \rho_{oed} \tag{9.12}$$

Elastic drained method *(Figures 9.11– 9.13)*
For soils with behaviour that can be assumed to be linear elastic within the range of working stresses, an elastic settlement procedure to determine total settlements is quite simple and attractive. Many overconsolidated clays behave in this way, so by using drained values of modulus E' and Poisson's ratio ν' the total settlement (immediate and consolidation combined) can be obtained directly.

Using the same procedure as Butler (above), Meigh (1976) gave the expression for total settlement beneath the corner of a flexible rectangular area on a soil with modulus increasing with depth (Figure 9.11) as:

$$\rho_T = \frac{qB}{E_f'} I F_D F_B \tag{9.13}$$

where:
q = net applied pressure
E_f' = drained modulus at foundation level
I = influence factor determined from H/B, L/B and k, Figure 9.12, but see below
F_D = correction for depth of embedment, Figure 9.13
F_B = correction for roughness at underside of foundation, Figure 9.14.

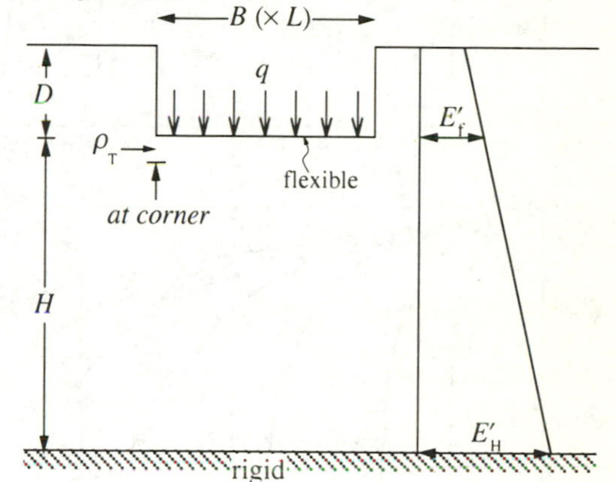

Figure 9.11 *Elastic drained method – definitions*

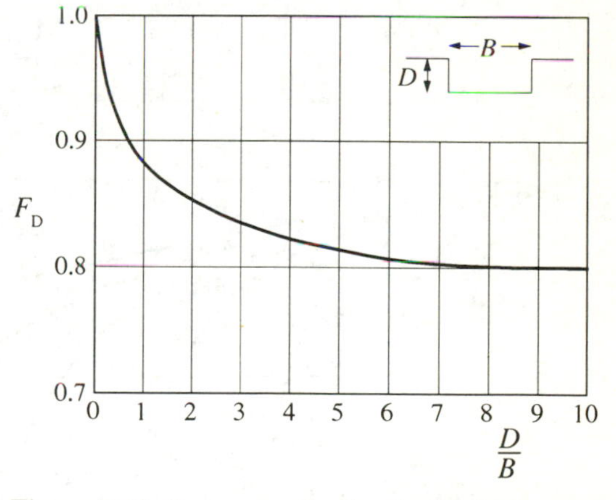

Figure 9.13 *Correction for depth of embedment*

The influence factor I is given assuming the base of the soil layer to be smooth. This tends to underestimate settlement so a simple adjustment for the effect of roughness at the base of the deposit is adopted, entering the charts at a value of $1.2H/B$ instead of H/B.

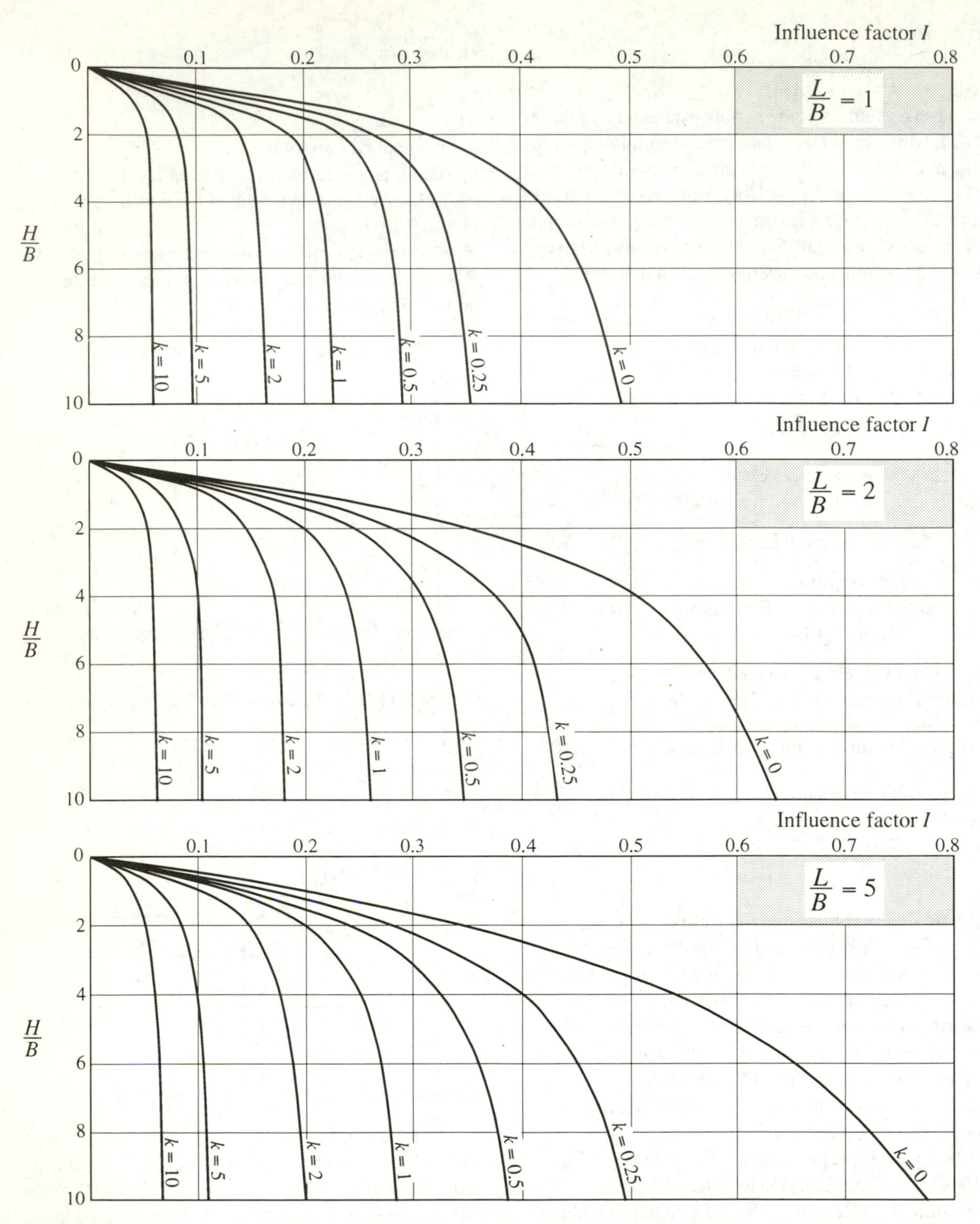

Figure 9.12 *Influence factors for modulus increasing with depth - Total settlement (From Meigh, 1976)*

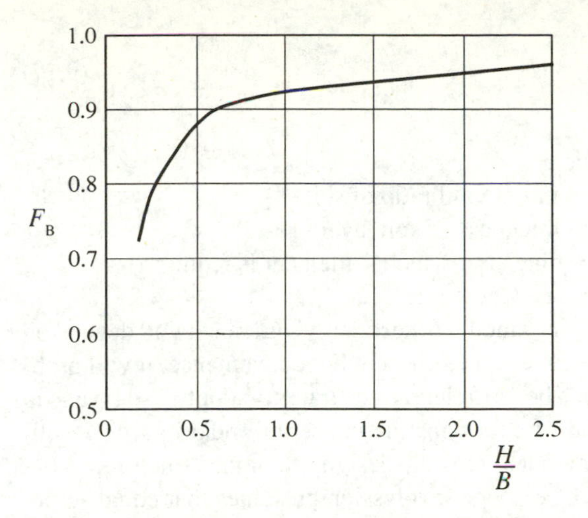

Figure 9.14 *Correction for roughness of base of foundation*

k represents the rate of change of modulus with depth and is given by:

$$k = \frac{E_H - E_f}{E_f} \frac{B}{H} \qquad (9.14)$$

ν' is assumed to be 0.2 and is incorporated into the values of I.

The principle of superposition must be used for points other than the corner of the loaded area and the principle of layering can be applied. Compared to the homogeneous, constant modulus case ($k = 0$) it can be seen that even a modest increase of modulus with depth can significantly reduce settlements.

For the case of a rigid foundation, Meigh suggested using the mean settlement of the flexible foundation as:

$$\rho_{\text{rigid}} = \tfrac{1}{3}\left(\rho_{\text{centre}} + \rho_{\text{corner}} + \rho_{\text{centre long edge}}\right)_{\text{flexible}} \qquad (9.15)$$

A simpler approach and probably no less accurate would be to determine ρ_{centre} and multiply this value by μ_r, obtained from Table 9.2.

Estimation of drained modulus E'

As for the undrained modulus reasonable correlations have been developed between the drained modulus E' derived from back-analysis of settlement observations and the undrained shear strength, c_u.

Butler (1974) derived a correlation for London Clay (of high plasticity) as $E' = 130\, c_u$ and Stroud and Butler (1975) gave correlations between, E', c_u and SPT 'N' for soils of lower plasticity. Values which relate to their 'suggested design line' which errs on the conservative side are given in Table 9.5.

Table 9.5

Drained modulus values (From Stroud et al, 1975)

Plasticity index %	E'/c_u
10 – 20	270
20 – 30	200
30 – 40	150
40 – 50	130
50 – 60	110

Secondary compression

Introduction *(Figures 9.15 and 9.16)*

With some soil types volume reductions and hence settlements have been found to continue even after primary consolidation has finished and when all pore water pressures have dissipated (Figure 9.15). These settlements are referred to as secondary compression or drained creep.

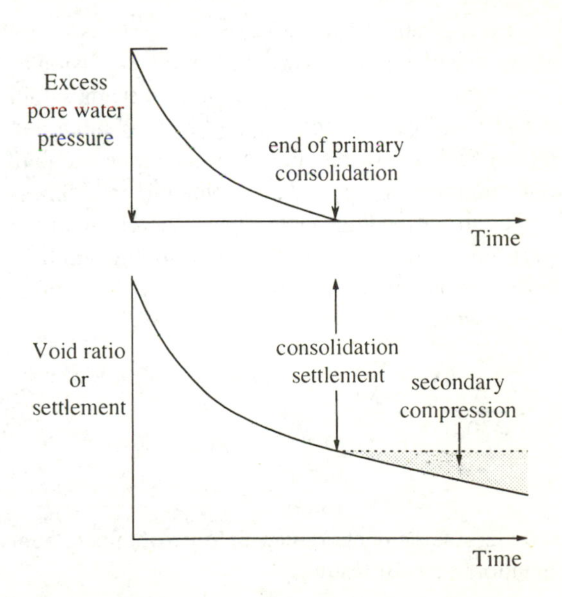

Figure 9.15 *Definition of secondary compression*

Various mechanisms and models have been suggested to explain the phenomenon but it seems most likely to be associated with a redistribution of the interactions (forces) between particles following the large structural rearrangments which occurred during the *normal* consolidation stage of the soil. This is supported by the fact that secondary compressions are insignificant for stress levels below the preconsolidation pressure (when the soil is overconsolidated) but can be large when stress levels exceed the preconsolidation pressure (Figure 9.16).

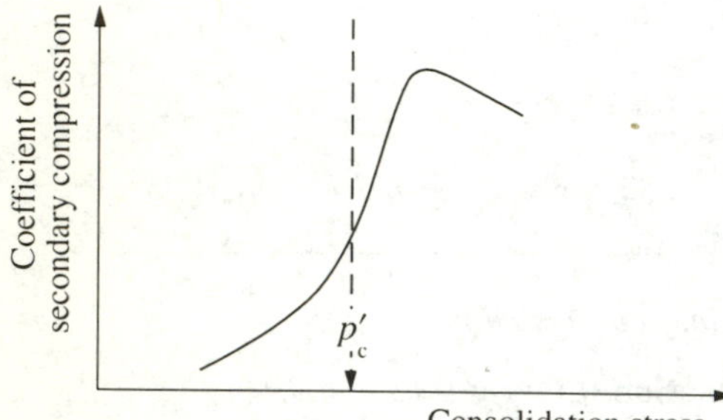

Figure 9.16 *Effect of preconsolidation pressure*

It is also dependent on the type of particles, for example secondary compressions can be large when organic material is present, see below.

General method *(Figure 9.17)*
For volume changes to occur due to the expulsion of pore water there must be pore pressure gradients within the soil but these are considered to be so small and occur at such a slow rate that they are immeasurable. Quite commonly it is found that void ratio or thickness changes in the oedometer test plot linearly with the logarithm of time (Figure 9.17) with the gradient referred to as the coefficient of secondary compression, C_α where:

$$C_\alpha = \frac{\Delta e}{\Delta \log t} = \frac{\Delta e}{\log_{10} \frac{t_2}{t_1}} \tag{9.16}$$

Secondary settlements ρ_s can then be obtained from this laboratory test result as:

$$\rho_s = H \frac{C_\alpha}{1+e_0} \log_{10} \frac{t_2}{t_1} \tag{9.17}$$

where:
e_o = initial void ratio of soil
H = thickness of soil layer
t_2 = time after which settlement is required
t_1 = reference time.

The amount of secondary settlement will depend on the time values chosen. For convenience, it will probably be sufficiently accurate to adopt t_1 = 1 year to allow for the construction period and primary consolidation and t_2 as the design life of the structure.

The above expression presumes that equal settlements occur for each log cycle of time.
e.g. settlement from 1 to 10 months (first year) =
settlement from 10 to 100 months (first decade) =
settlement from 100 to 1000 months (first century)
and so on.

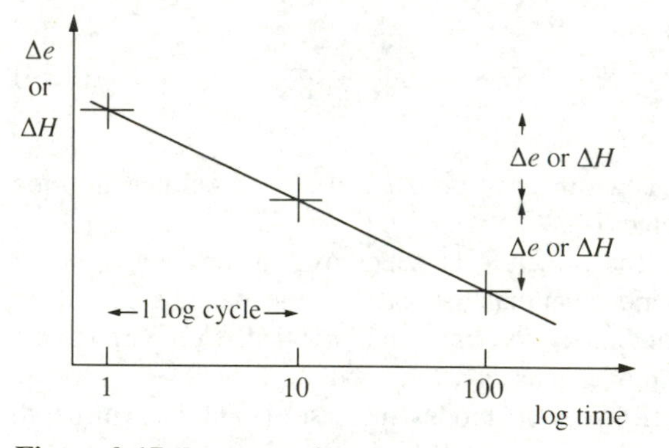

Figure 9.17 *Log time plot – secondary compression*

Estimation of C_α or ε_α values *(Figure 9.18)*
Values of C_α could be obtained from the laboratory oedometer test by continuing readings beyond the primary consolidation stage. However, as the times are plotted on a logarithmic scale to establish the secondary compression line the test will be time-consuming.

It may be sufficient to use an estimate from a correlation such as the one published by Mesri (1973), Figure 9.18, who reported the coefficient of secondary compression as ε_α:

$$\varepsilon_\alpha = \frac{C_\alpha}{1+e_0} \tag{9.18}$$

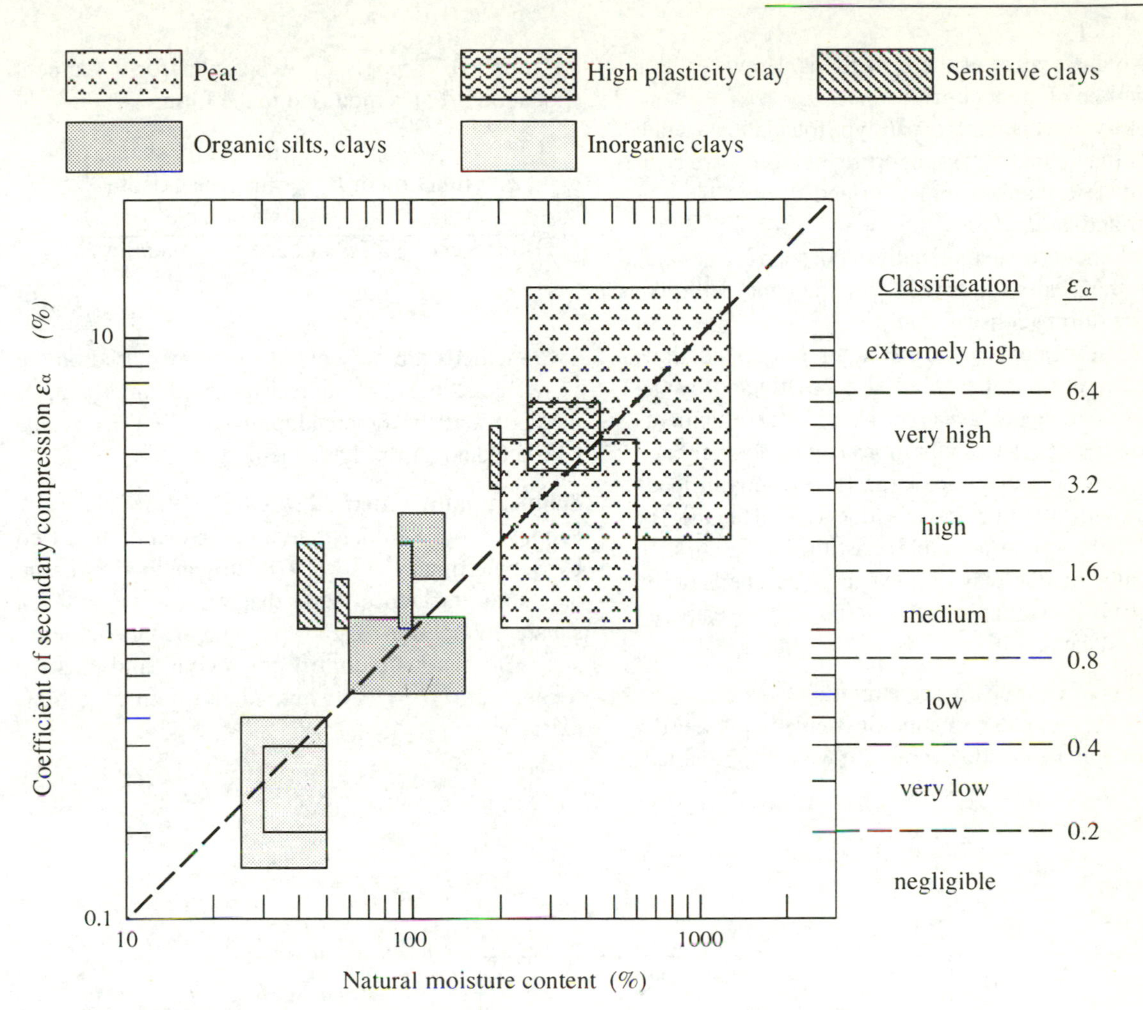

Figure 9.18 *Coefficients of secondary compression ε_α for natural deposits (After Mesri, 1973)*

This value of the coefficient of secondary compression may then be used directly in equation 9.17. The chart (Figure 9.18) shows the significant effect on secondary compression of:

- organic content, especially peats
- clay mineralogy as represented by high plasticity, especially montmorillonite
- metastable mineral grain structures as represented by sensitive clays.

Sands

Introduction

There exists a large number of methods of estimating settlements of foundations on sands developed over the last 50 years. They are all based on empirical correlations between observed settlements and other appropriate parameters but probably because of the large range of variables involved not one of these methods can predict settlements particularly accurately. It should be accepted that only an approximate estimate is the best we can expect of these methods. However, this need not be of great concern since:

1 Observed settlements of pad-type foundations (width less than about 5 m) supporting buildings are typically less than 40 mm (Burland *et al*, 1978 and Burland *et al*, 1985).
2 Settlements in sands usually occur soon after applying the load so most of the settlements will take place during construction.

Greater settlements beneath larger raft foundations will be likely but these can also be affected by the presence of deeper clay layers. Additional settlements can be expected due to vibrations from machinery and traffic or where there are large fluctuations in loads such as with silos or chimneys (due to wind loading) or where there are compressible constituents present such as organic matter, clay or mica particles, shells or the sand itself is crushable such as coral sand, weathered sands, volcanic ash.

Methods of estimating settlements

Most of the methods for estimating settlements take the form of a quasi-elastic expression with empirical factors applied. The elastic expression for settlement (Equation 9.13) is modified to the form:

$$\rho = \frac{q \times (\text{function of } B) \times \begin{pmatrix}\text{factors for shape,}\\ \text{thickness, depth,}\\ \text{water table, time}\end{pmatrix}}{\text{function of SPT } N \text{ or cone } q_c \text{ or modulus } E'} \qquad (9.19)$$

Two methods are presented below, one based on the Dutch cone penetration resistance (Schmertmann's method) and the other adopting the SPT '*N*' value (Burland and Burbridge's method).

Schmertmann's method *(Figure 9.19)*

Proposed by Schmertmann (1970) and modified by Schmertmann *et al* (1978) this method is based on a strain influence factor diagram for two cases, a square foundation (L/B = 1) where axi-symmetric stress and strain conditions occur and a strip foundation (L/B = 10) where plane strain conditions exist.

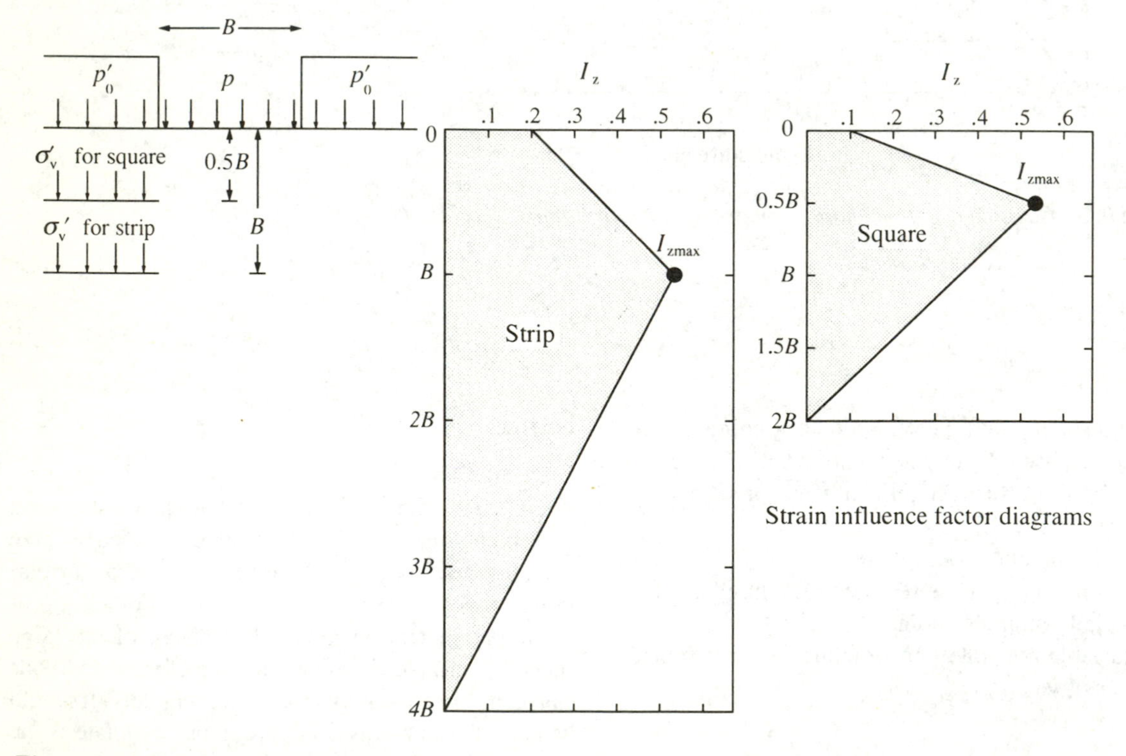

Figure 9.19 *Schmertmann's method*

The expressions for settlement are:

a) *square foundation* ($L/B = 1$)

$$\rho = \frac{C_1 C_2}{2.5} \Delta p \sum_0^{2B} \frac{I_z \Delta z}{q_c} \qquad (9.20)$$

b) *strip foundation* ($L/B = 10$)

$$\rho = \frac{C_1 C_2}{3.5} \Delta p \sum_0^{4B} \frac{I_z \Delta z}{q_c} \qquad (9.21)$$

where:
p = gross applied pressure
p_o' = effective stress at foundation level around the foundation
Δp = net applied pressure = $p - p_o'$ kN/m²
q_c = cone end resistance, kN/m², for each soil layer
Δz = thickness of each soil layer, metres
C_1 = correction for depth of foundation or confinement

$$= 1 - 0.5 \frac{p_0'}{\Delta p} \qquad (C_1 \text{ should be} \geq 0.5)$$

C_2 = correction for creep or time related settlement

$$= 1 + 0.2 \log_{10} 10t$$

where
t = time in years after construction
I_z = average strain influence factor for each soil layer.

The average strain influence factor, I_z, is obtained as the value at the mid-point of each soil layer from a diagram drawn alongside the q_c-depth graph with a depth of $2B$ for a square foundation and $4B$ for a strip foundation as shown in Figure 9.19. The maximum value of I_z is given as:

$$I_{Z\max} = 0.5 + 0.1 \left(\frac{\Delta p}{\sigma_v'} \right)^{0.5} \qquad (9.22)$$

where:
σ_v' = vertical effective stress at a depth of $B/2$ for a square foundation and B for a strip foundation, see Figure 9.19.

Values of Δz, average q_c and average I_z for each soil layer are required for the summation term and these can be conveniently presented in a table, as illustrated in the Worked Example 9.11.

Settlements for shapes intermediate between square and strip can be obtained by interpolation.

Burland and Burbridge method *(Figures 9.20 and 9.21)*
Based on a statistical analysis of a large number of settlement observations from case studies, Burland and Burbridge (1985) proposed a method for normally consolidated and overconsolidated sand.

1 *For foundations on the surface of normally consolidated sand* the average immediate settlement ρ_i at the end of construction is given by:

$$\rho_i = f_s f_l q' B^{0.7} I_c \qquad (9.23)$$

where
ρ_i = average immediate settlement, mm
f_s = shape factor

$$= \left(\frac{1.25 \frac{L}{B}}{\frac{L}{B} + 0.25} \right)^2$$

= 1.0 for square or circle ($\frac{L}{B} = 1$) and 1.56 for strip ($\frac{L}{B} = \infty$)

f_l = thickness factor

$$= \frac{H_s}{Z_I} \left(2 - \frac{H_s}{Z_I} \right)$$

H_s = thickness of sand below foundation, metres (when $H_s < Z_I$)
Z_I = depth of influence, metres, see Figure 9.20
If $H_s > Z_I$ then the thickness factor $f_l = 1$
q' = average gross effective foundation pressure, kN/m²
B = width of foundation, metres
I_c = compressibility index

$$= \frac{1.71}{\overline{N}^{1.4}}$$

$\overline{N}$ = average SPT value over depth of influence Z_I

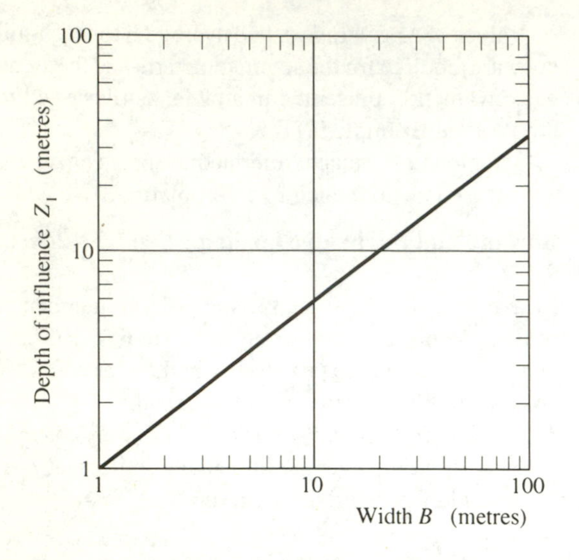

Figure 9.20 *Depth of influence*

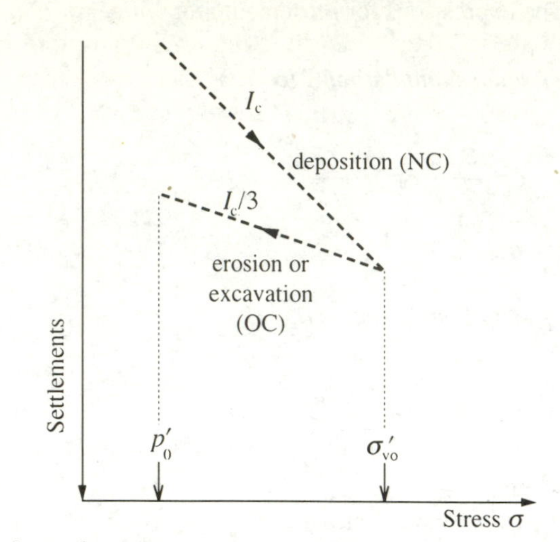

Stress history

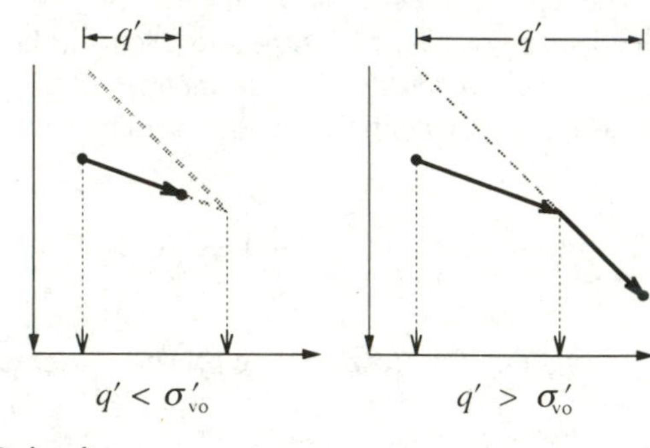

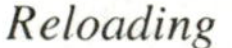

Reloading

Figure 9.21 *Settlements of overconsolidated sand*

The N values obtained from the site are not corrected for the effect of overburden pressure but they are corrected (N') for grain size effects:

$N' = 15 + 0.5\ (N - 15)$

for very fine or silty sands below the water table and

$N' = 1.25N$

for gravel or sandy gravel

2 *For foundations on the surface of an overconsolidated sand or for loading at the base of an excavation in normally consolidated sand* the average immediate settlement at the end of construction is given by:

$$\rho_i = f_s f_l q' B^{0.7} \frac{I_c}{3} \qquad (9.24)$$

when q' is less than σ_{vo}'

and by:

$$\rho_i = f_s f_l \left(q' - \tfrac{2}{3}\sigma_{vo}'\right) B^{0.7} I_c \qquad (9.25)$$

when q' is greater than σ_{vo}'.

All of the terms are as described in 1) above. σ_{vo}' is the past maximum stress for an overconsolidated sand or the effective stress at foundation level or excavation level for a normally consolidated sand. The method assumes that the compressibility of a sand is reduced by a factor of 3 when it has become overconsolidated, either by a geological removal process such as erosion or by a construction process such as excavation, Figure 9.21. There exists a past maximum stress σ_{vo}' below which compressibility is given by $I_c/3$ and above which it is given by I_c. Settlements will be small if q' is less than σ_{vo}' but much larger if q' is greater than σ_{vo}'.

Unfortunately, there is no sure way of determining whether a sand is normally consolidated or overconsolidated nor of determining σ_{vo}' for an overconsolidated sand. In the absence of sufficient evidence it would be prudent, therefore, to assume that a sand is normally consolidated and to introduce the effect of σ_{vo}' only when a foundation is placed at the base of an excavation.

Many of the cases studied displayed time-dependent settlement after the end of construction. A correction f_t should be applied to obtain the long-term settlement, thus:

$$\rho_t = \rho_i f_t \qquad (9.26)$$

where

$$f_t = 1 + R_3 + R_t \log_{10} \frac{t}{3} \qquad (9.27)$$

t = time after end of construction, years.
Values of R_3 and R_t are given in Table 9.6.

Table 9.6
Time correction factors

Loading condition	R_3	R_t
static loads	0.3	0.2
fluctuating loads	0.7	0.8

The apparent accuracy of the above expressions for settlements should be tempered by the moderate accuracy of the statistical correlation and the inherent variability of all sand deposits. It is stated that the actual settlement could lie between ± 50% of the predicted value.

Permissible settlements

Introduction

Having determined an amount of settlement for a foundation it is then necessary to know whether it will be acceptable or not. It should be less than the permissible (or tolerable, or allowable) settlement. However, there is no straightforward approach to this problem because the structure-foundation-soil interaction phenomenon is a complex and uncertain subject.

It must be remembered that structures may not move due to ground settlements alone. There may be movements and associated damage due to dimension changes in the structural materials such as due to moisture or temperature changes, creep or chemical reactions. Some examples of these are illustrated in BRE Digest 361, 1991. There may be structural movements due to ground movements other than settlements as calculated above, such as mining subsidence, shrinkage or swelling of clay soils, erosion of sandy soils, slope instability, poorly compacted backfill, vibrations etc.

Definitions of ground and foundation movement *(Figure 9.22)*

Burland and Wroth (1974) have proposed some useful definitions of ground, foundation and structural movement to enable a detailed investigation of the various modes of movement of a structure. These are illustrated on Figure 9.22 where points A to D could represent points on a raft foundation, the locations of isolated foundations or points along a loadbearing wall. The definitions are given as:

- *settlement* ρ
  Downward movement at a point. This will vary across a non-rigid structure.
- *heave* ρ_h
  Upward movement at a point. This will vary across an excavation or beneath a structure.
- *differential settlement or differential heave* – $\delta\rho$ or $\delta\rho_h$ The difference in settlement or heave between two points. Usually two adjacent points are chosen but the choice may be arbitrary, Figure 9.22a.
- *horizontal displacement u*
  Extension or contraction of a building in the horizontal direction will result in tensile or compressive strains.
- *rotation* θ
  The change in gradient of a line joining two reference points, e.g. between A and B on Figure 9.22a.
- *tilt* ω
  The rigid body rotation of the whole of a structure or a well-defined part of it. This is obtained by drawing a straight line between the two edges of a building or a well-defined part of it such as points A and D on Figure 9.22b. This is referred to as the tilt plane and will relate to both horizontal and vertical components of a building.

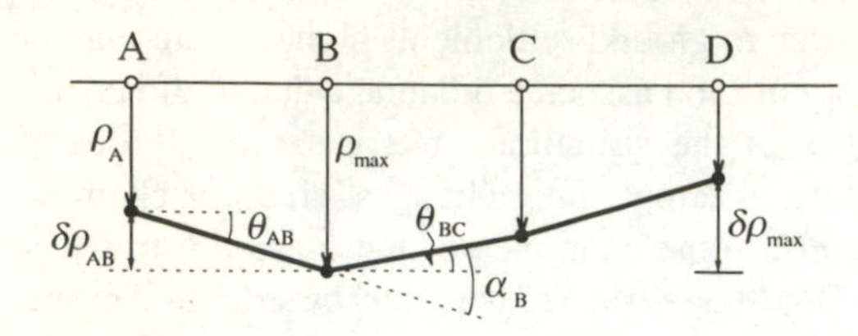

a) Definitions of settlement ρ, relative settlement or differential settlement $\delta\rho$, rotation θ and angular strain α

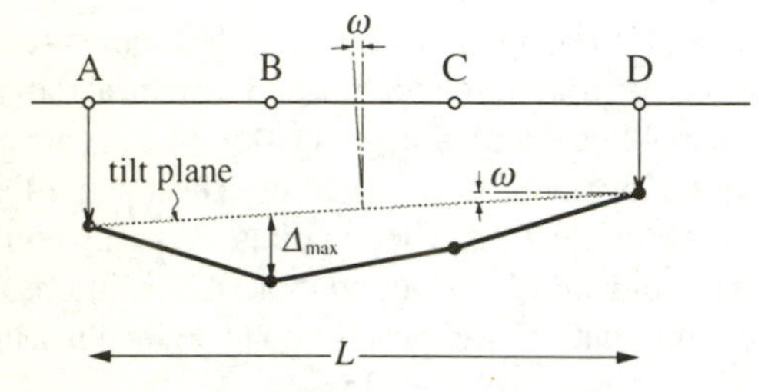

b) Definitions of tilt ω, relative deflection Δ and deflection ratio Δ/L

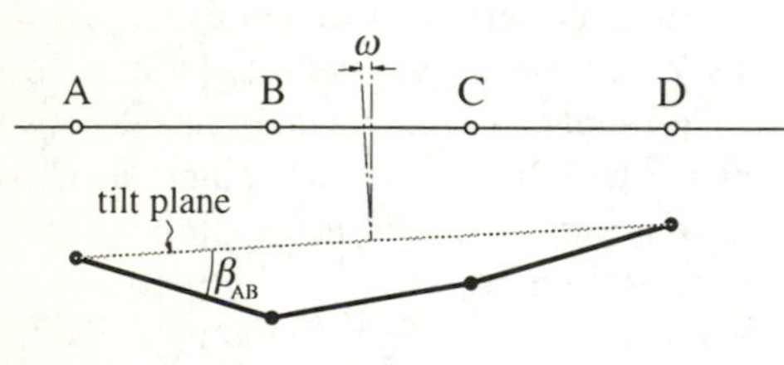

$$\beta_{AB} = \theta_{AB} + \omega$$

$$\beta_{CD} = \theta_{CD} - \omega$$

c) Definitions of relative rotation or angular distortion β

Figure 9.22 *Definitions of foundation movement (From Burland et al, 1978)*

- *relative deflection* Δ
  The displacement relative to the 'tilt' plane, Figure 9.22b. Downward displacement is described as sagging and upward movement as hogging.
- *deflection ratio* Δ/L
  This denotes the degree of curvature to which the building or a part of it has been subjected and can represent a sagging ratio or a hogging ratio. The degree of curvature a building is subjected to will determine the amount of distortion and hence the degree of damage of the structure. The deflection ratio is the parameter adopted for Figure 9.23. Deflection ratio is preferred to angular distortion β since the latter is affected by the amount of tilt, Figure 9.22c.
- *angular strain* α
  This represents the total rotation at a point of a structure. For example, at B on Figure 9.22a it is given by:

$$\alpha_B = \theta_{AB} + \theta_{BC}$$

  It is positive if it produces sag and negative if it produces hog. This parameter is particularly useful for assessing localised movements of brick walls since cracks are often concentrated where angular strains are high.
- *relative rotation (or angular distortion)* β
  The rotation of the line joining two reference points relative to the tilt line, Figure 9.22c.

Criteria for movements

From the work of Burland, Broms and de Mello (1978) there appear to be four criteria which must be satisfied:

1. *visual appearance of the structure as a whole*
   Tilting of walls, floors and the whole building could be unpleasant or even alarming for the occupants and visitors. A deviation in excess of about 1 in 250 from the vertical or horizontal would probably be noticeable.
2. *visual appearance of the architectural materials*
   Visible damage such as cracking or distortions of claddings can vary from being unsightly to alarming. A classification for the assessment of damage has been suggested by the Building Research Establishment (BRE Digest 251, 1989). Although cracking is indicative of movements the classification is based on the ease of repair of the damage.

3 *serviceability or function of the structure*
Movements can occur which affect the overall efficiency of the building such as reduced weathertightness, rain penetration, dampness, draughts, heat loss, reduced sound insulation, windows and doors sticking. Movements can also occur which affect the basic function or purpose of the structure such as with the operation of lifts or precision machinery, access ramps or steps, fracturing of service pipes.

4 *stability*
Large movements and very severe damage to the cladding and fittings will have occurred before the structure itself fails due to instability so the above criteria are used to dictate permissible settlements.

Routine settlement limits *(Figure 9.23)*
Based on a large number of observations of structures Terzaghi and Peck (1948) and Skempton and MacDonald (1956) gave values of permissible settlements for framed structures and these are summarised in Table 9.7. It is emphasised that these refer to 'routine' buildings with fairly uniform distribution of loading and uniform ground conditions.

With regards to the interaction between a structure and its supporting soil a number of important points emerge:

- permissible settlements for the same structure can be larger on clay soils than on sands. This is probably due to the longer period of time over which settlements occur on clay soils allowing the structure to gradually adjust to the settlements whereas if the structure was placed on sand it must respond immediately to settlements.
- frame buildings and their claddings can tolerate more distortion than loadbearing walls. Many of the claddings in frame buildings are less sensitive to movement and they are installed at a time when much of the settlement has already elapsed whereas loadbearing walls are more brittle and are subjected to settlements from commencement of construction.
- frame buildings without infill panels (open frames) can tolerate more settlement than infilled frames.
- frame buildings on isolated foundations may distort differently from buildings on raft foundations.
- load bearing walls undergoing sagging can tolerate more distortion (nearly twice as much) than walls undergoing hogging.
- longer structures can tolerate more relative deflection Δ.
- a stiff soil layer overlying the compressible soil causing the settlements will not prevent the total settlement occurring but will significantly reduce differential settlements.

Burland and Wroth (1974) proposed a simple criterion to relate the distortion of a structure (relative deflection Δ/L) to the onset of visible cracking of the cladding or finishes and this relationship (Burland *et al*, 1978) is reproduced in Figure 9.23 for frame buildings and loadbearing walls. If Δ/L lies above the criterion lines shown it is likely that the buildings will suffer architectural damage.

The purpose of foundation design must be to ensure that Δ/L lies below the criterion line. This can be achieved by considering the stiffness of the foundation and structure in a soil/structure interaction assessment.

Table 9.7

Routine guides to permissible settlements

Settlement	SAND		CLAY
	Reference 1	Reference 2	Reference 2
maximum differential settlement $\delta\rho$	20	25	40
maximum settlement ρ (isolated foundations)	25	40	65
maximum settlement ρ (raft foundations)	50	40 – 65	65 – 100

Reference 1 – Terzaghi and Peck,1948 Reference 2 – Skempton and MacDonald, 1956

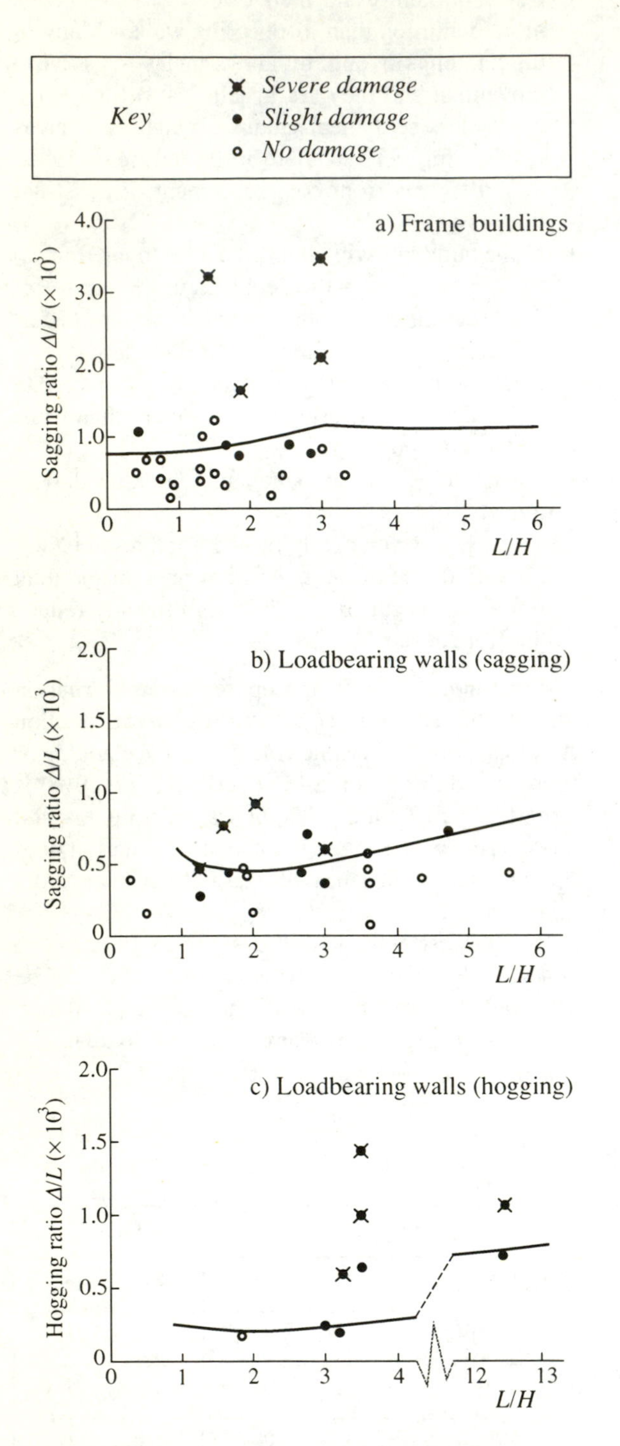

Figure 9.23 *Damage criterion for different types of structure (From Burland et al, 1978)*

Worked Example 9.1 *Immediate settlement – general method*
A flexible foundation 10 m long, 5 m wide applies a uniform pressure of 75 kN/m² on the surface of a saturated clay, 20 m thick. The undrained modulus of the clay is 8 MN/m². Determine the immediate settlement at the corner of the foundation.

Using Figure 9.1, $B = 5$ m $\quad \frac{H}{B} = \frac{20}{5} = 4 \quad \frac{L}{B} = \frac{10}{5} = 2 \quad I = 0.37$

From Equation 9.2

$$\rho_i = \frac{75 \times 5 \times 0.37 \times 1000}{8 \times 1000} = 17 \text{ mm}$$

Worked Example 9.2 *Immediate settlement – principle of superposition*
Determine the immediate settlement at the centre of the foundation in Example 9.1.

From Figure 9.2, the settlement at the corner of a quarter foundation is required.

$B = 2.5$ m $\quad \frac{H}{B} = \frac{20}{2.5} = 8 \quad \frac{L}{B} = \frac{5}{2.5} = 2$ From Figure 9.1, $I = 0.47$

$$\rho_i = \frac{75 \times 2.5 \times 0.47 \times 4 \times 1000}{8 \times 1000} = 44 \text{ mm}$$

Worked Example 9.3 *Immediate settlement – principle of layering*
The foundation in Example 9.1 is placed on two layers of clay, both 10 m thick. The modulus of the upper layer is 8 MN/m² and for the lower layer it is 16 MN/m². Determine the immediate settlement at the centre of the foundation.

From Figure 9.3
Settlement of upper layer

$B = 2.5$ m $\quad H = 10$ m $\quad \frac{H}{B} = \frac{10}{2.5} = 4 \quad \frac{L}{B} = 2 \quad$ From Figure 9.1, $I = 0.37$

$$\rho_i = \frac{75 \times 2.5 \times 0.37 \times 4 \times 1000}{8 \times 1000} = 35 \text{ mm}$$

Settlement of lower layer

$B = 2.5$ m $\quad H = 20$ m $\quad \frac{H}{B} = \frac{20}{2.5} = 8 \quad \frac{L}{B} = 2 \quad$ From Figure 9.1, $I = 0.47$

$B = 2.5$ m $\quad H = 10$ m $\quad \frac{H}{B} = \frac{10}{2.5} = 4 \quad \frac{L}{B} = 2 \quad$ From Figure 9.1, $I = 0.37$

$$\rho_i = \frac{75 \times 2.5 \times (0.47 - 0.37) \times 4 \times 1000}{16 \times 1000} = 5 \text{ mm}$$

Settlement of foundation = 35 + 5 = 40 mm

Worked Example 9.4 *Immediate settlement – correction for depth*
The foundation in Example 9.2 is placed 2 m below ground level. Determine the immediate settlement at the centre of the foundation.

From Example 9.2 the settlement of the foundation on the surface of the clay is 44 mm.
Using Figure 9.4, $D = 2$ m $\quad B = 5$ m $\quad D/B = 0.4 \quad \mu_0 = 0.94$
$\rho_{\text{at depth}} = 44 \times 0.94 = 41$ mm

Worked Example 9.5 *Average settlement*
Determine the average settlement for the foundation in Example 9.4.

From Figure 9.4, for $\frac{H}{B} = \frac{20}{5} = 4 \quad \frac{L}{B} = 2 \qquad \mu_1 = 0.77$

From before, $\frac{D}{B} = 0.4 \; \mu_0 = 0.94$

$$\rho_{\text{ave}} = \frac{0.94 \times 0.77 \times 75 \times 5 \times 1000}{8 \times 1000} = 34 \text{ mm}$$

This compares reasonably with settlement for a rigid foundation obtained by Equation 9.4 and μ_r from Table 9.2:
$\mu_r \approx 0.77$
$\rho_{\text{rigid}} = 0.77 \times 41 = 32$ mm

Worked Example 9.6 *Modulus increasing with depth*
Determine the immediate settlement at the centre of the foundation in Example 9.1 when the undrained modulus of the clay increases from 4 MN/m² at ground level to 12 MN/m² at the base of the clay layer.

$B = 2.5$ m $\quad H = 20$ m $\quad E_0 = 4$ MN/m² $\qquad E_H = 12$ MN/m²

$k = \frac{12 - 4}{4} \times \frac{2.5}{20} = 0.25 \quad \frac{H}{B} = \frac{20}{2.5} = 8$

From Figure 9.5, $\frac{L}{B} = 2 \qquad I = 0.30$ (Note: non-linear interpolation)

$$\rho_i = \frac{75 \times 2.5 \times 0.30 \times 4 \times 1000}{4 \times 1000} = 56 \text{ mm}$$

Comparing this answer with Example 9.2, it is not acceptable to take the 'average' modulus of 8 MN/m² and assume the homogeneous condition.

Worked Example 9.7 *Effect of local yielding*
The ultimate bearing capacity of the clay in the above examples is 150 kN/m². Determine the effect of yielding assuming the clay to be lightly overconsolidated.

The factor of safety is $\frac{150}{75} = 2$

From Figure 9.7 an average yield factor of about 1.15 is indicated so the above settlements should be increased by 15% to allow for the effect of yielding.

Worked Example 9.8 *Consolidation settlement – compression index method*
A flexible rectangular raft foundation, 4 m wide and 5 m long is to be constructed on the surface of a layer of soft normally consolidated clay, 8 m thick and will support a uniform pressure of 60 kN/m². The properties of the clay which are constant throughout the deposit are:
Bulk unit weight = 18.8 kN/m³
Specific gravity = 2.72
Compression index $C_c = 0.12$
Assume $\gamma_w = 9.8$ kN/m³
The water table is at ground level. The moisture content of the clay decreases linearly from 36% at ground level to 28% at the base of the clay layer. Determine the maximum consolidation settlement.
The maximum settlement will occur at the centre of the foundation where the maximum stress increases occur. The stress increases $\Delta\sigma$ have been determined using Figure 5.7 and are given in the table below.
Splitting the clay into 4 no. 2 m thick sub-layers the values of p_0', $\Delta\sigma$ and w_0 are determined at the mid-point of each sub-layer and assumed to be the average values for each sub-layer, as shown on Figure 9.9.

Layer	Depth m	p_0' kN/m²	$\Delta\sigma$ kN/m²	P	Δe	w_0 %	e_0	ΔH m
1	1	9.0	56.4	0.861	0.103	35	0.952	0.106
2	3	27.0	32.4	0.342	0.041	33	0.898	0.043
3	5	45.0	16.8	0.138	0.017	31	0.843	0.018
4	7	63.0	9.6	0.062	0.007	29	0.789	0.008
								Σ = 0.175 m

The maximum consolidation settlement is 175 mm

Worked Example 9.9 *Consolidation settlement – oedometer method*
A flexible rectangular raft foundation, 4 m wide and 5 m long is to be constructed on the surface of a layer of firm overconsolidated clay, 8 m thick and will support a uniform pressure of 60 kN/m². The water table lies at 2 m below ground level. The bulk unit weight of the clay is 19.8 kN/m³, both above and below the water table. Using the oedometer test result in Example 6.1 determine the maximum consolidation settlement.
The stress increases in the soil due to the applied pressure at the mid-point of 2 m thick sub-layers are obtained from Example 9.8. Using the oedometer test result plotted as thickness versus log pressure values of m_v have been obtained for each sub-layer using the average values of p_0' and $\Delta\sigma$. The oedometer settlement for the foundation has been determined using Equation 9.10. The oedometer test gave a value of m_v = 0.135 m²/MN for the typical pressure increment of 50–100 kN/m² so the oedometer settlement has also been calculated with this value for all sub-layers. It can be seen that a sufficiently accurate estimate is obtained.

Layer	Depth m	p_0' kN/m²	$\Delta\sigma$ kN/m²	H_0 mm	ΔH mm	m_v m²/MN	ρ_{oed} [1] mm	ρ_{oed} [2] mm
1	1	19.8	56.4	19.840	0.202	0.18	20	15
2	3	49.6	32.4	19.700	0.072	0.11	7	9
3	5	69.6	16.8	19.652	0.032	0.10	3	5
4	7	89.6	9.6	19.615	0.014	0.07	1	3
				Oedometer settlement =			Σ 31	32

[1] Using individual oedometer test m_v values
[2] Using one m_v value = 0.135 m²/MN for the pressure increment 50–100 kN/m²
Alternatively, the settlements could be determined from the pressure-void ratio plot using

$$\Delta H = \frac{\Delta e}{1+e_0} H$$

where e_0 and Δe are the initial and change in void ratio, respectively, for each change in pressure p_0' to $p_0' + \Delta\sigma$ and H is the thickness of each sub-layer = 2000 mm.

Layer	Depth m	p_0' kN/m²	$\Delta\sigma$ kN/m²	e_0 mm	Δe mm	ΔH mm
1	1	19.8	56.4	0.635	0.0166	20
2	3	49.6	32.4	0.623	0.0059	7
3	5	69.6	16.8	0.619	0.0026	3
4	7	89.6	9.6	0.616	0.0012	1
				Oedometer settlement =		Σ 31 mm

Worked Example 9.10 *Total settlements – elastic drained method*
A flexible rectangular foundation, 4 m wide and 8 m long is to be constructed 2 m below the surface of a layer of firm overconsolidated clay, 10 m thick and will support a uniform pressure of 60 kN/m². The drained modulus increases from 10 MN/m² at ground level to 20 MN/m² at the base of the clay. Determine the maximum total settlement.

The modulus at foundation level, $E_f' = 12$ MN/m²
From Figure 9.2 the settlement at the corner of a quarter foundation is obtained:

$B = 2$ m $\quad \dfrac{H}{B} = \dfrac{8}{2} = 4 \quad \dfrac{L}{B} = 2$

From Equation 9.14,

$$k = \frac{20-12}{12} \times \frac{2}{8} = 0.167$$

Enter the charts, Figure 9.12, at $\dfrac{H}{B} = 1.2 \times 4 = 4.8$ $\quad \therefore I = 0.44$ (Note: non-linear interpolation)

$\dfrac{D}{B} = \dfrac{2}{4} = 0.5$ From Figure 9.13 $F_D = 0.92$

$\dfrac{H}{B} = \dfrac{8}{4} = 2$ From Figure 9.14 $F_B = 0.95$

$$\therefore \rho_T = \frac{60 \times 2 \times 0.44 \times 0.92 \times 0.95 \times 4 \times 1000}{12 \times 1000} = 15 \text{ mm}$$

Worked Example 9.11 *Settlements on sand – Schmertmann's method*
A square foundation, 4 m wide is placed at 2 m below ground level in a layered sand and applies a uniform gross pressure of 140 kN/m². The bulk unit weight of the sand is 18 kN/m³ and the water table lies at ground level. The cone end resistance of the layers is plotted on Figure 9.24. Determine the immediate and long-term settlement of the foundation.

Due to buoyancy from the water pressure at the underside of the foundation the resultant gross pressure will be
$140 - 2 \times 9.8 = p = 120.4$ kN/m²
$p_0' = (18 - 9.8) \times 2 = 16.4$ kN/m²
$\Delta p = 120.4 - 16.4 = 104.0$ kN/m²

$$C_1 = 1 - 0.5 \times \frac{16.4}{104.0} = 0.92$$

At $0.5B = 2$ m below foundation level $\sigma_v' = (18 - 9.8) \times 4 = 32.8$ kN/m²

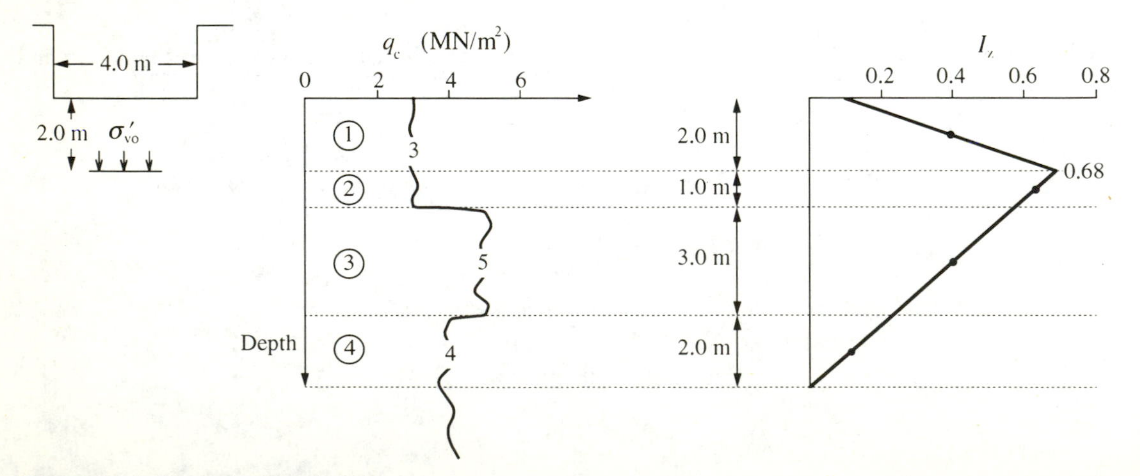

Figure 9.24 *Worked Example 9.11*

The variation of I_{zmax} has been plotted against depth for this value of I_{zmax} on Figure 9.24. Splitting the cross-section into suitable layers values of Δz and average values of q_c and I_z are obtained:

Layer	Δz	q_c	I_z	$\frac{I_z \Delta z}{q_c}$
	m	MN/m²		
1	2.0	3	0.39	0.26
2	1.0	3	0.62	0.21
3	3.0	5	0.40	0.24
4	2.0	4	0.11	0.06
				$\Sigma = 0.77$

From Equation 9.20, the immediate settlement will be

$$\rho_i = \frac{0.92 \times 104.0 \times 0.77 \times 1000}{2.5 \times 1000} = 29 \text{ mm}$$

For a design life of say, 30 years

$C_2 = 1 + 0.2 \log_{10} 10 \times 30 = 1.5$

$\therefore$ the long-term settlement will be

$\rho_t = 29 \times 1.5 = 44$ mm

Worked Example 9.12 *Settlements on sand – Burland and Burbridge method*

A square foundation 8 m wide is placed at 2 m below ground level in a sand 6 m thick and applies a uniform gross pressure of 140 kN/m². The bulk unit weight of the sand is 18 kN/m³ and the water table lies at ground level. The average SPT 'N' value of the sand is 16. Determine the immediate and long-term settlements of the foundation.

Shape factor $f_s = 1$

From Figure 9.20, depth of influence, $Z_I = 5$ m

$H_s = 4$ m and is less than Z_I

$$\therefore \; f_l = \frac{4}{5}\left(2 - \frac{4}{5}\right) = 0.96$$

Assuming the sand to be normally consolidated

$\sigma_{v0}' = (18 - 9.8) \times 2 = 16.4$ kN/m²

$$I_c = \frac{1.71}{16^{1.4}} = 0.035$$

Due to buoyancy from the water pressure at the underside of the foundation the resultant gross pressure will be $140 - 2 \times 9.8 = q' = 120.4$ kN/m²

From Equation 9.25

$$\rho_i = 0.96\left(120.4 - \frac{2}{3} \times 16.4\right) \times 8^{0.7} \times 0.035 = 16 \text{ mm}$$

Assuming fluctuating loads and a design life of 30 years, from Table 9.6 and Equation 9.27

$$f_t = 1 + 0.7 + 0.8 \log_{10} \frac{30}{3} = 2.5$$

$\therefore$ the long-term settlement will be

$\rho_t = 16 \times 2.5 = 40$ mm

Exercises

9.1 A flexible rectangular foundation, 5 m wide and 10 m long is to be placed on the surface of a layer of clay, 20 m thick with undrained modulus of 20 MN/m^2. The foundation will support a uniform pressure of 120 kN/m^2. Determine the immediate settlement at the centre and corner of the foundation.

9.2 An existing pipe is buried at 5 m below ground level and will lie on the diagonal of the foundation in Exercise 9.1. Determine the immediate settlement of the pipe:
(a) beneath the centre of the foundation
(b) beneath the corner of the foundation.

9.3 For the foundation in Exercise 9.1 determine:
(a) the immediate settlement of the foundation assuming it to be rigid
(b) the average immediate settlement of the foundation assuming it to be flexible.

9.4 A flexible rectangular foundation, 5 m wide and 10 m long, is to be placed on the surface of a layer of clay, 20 m thick with undrained modulus increasing from 10 MN/m^2 at ground level to 30 MN/m^2 at the base of the clay. The foundation will support a uniform pressure of 120 kN/m^2. Determine the immediate settlement at the centre and corner of the foundation.

9.5 A flexible rectangular foundation, 4 m wide and 8 m long, is to be placed on the surface of a layer of clay, 16 m thick with undrained modulus of 8 MN/m^2. The foundation will support a uniform pressure of 40 kN/m^2. Determine the maximum immediate settlement assuming:
(a) linear elastic behaviour
(b) local yield occurs. With a factor of safety of 3 assume the yield factor f_y is 1.5.

9.6 A foundation is to be placed on the surface of a normally consolidated clay, 8 m thick, with the water table at 4 m below ground level. The following properties of the clay are constant with depth:
Unit weight = 18.5 kN/m^3
Specific gravity = 2.70
Compression index = 0.15
The increase in vertical stress provided by the foundation and the water content of the clay vary with depth:

Depth (m)	1	3	5	7
$\Delta\sigma_v$ (kN/m^2)	70	40	25	15
w_o (%)	40	35	31	28

Using the compression index method determine the consolidated settlement of the foundation.

9.7 For the foundation described in Exercise 9.6 determine the consolidation settlement using the m_v method given that the coefficient of compressibility m_v decreases with depth in the form $m_v = 0.80 - 0.04z$ where z is the depth below ground level and the Skempton-Bjerrum μ value is 0.8. Assume 2 m thick sub-layers.

9.8 A flexible rectangular foundation, 8 m wide and 16 m long is to be placed at 4 m below ground level in a layer of clay, 24 m thick and will impose a uniform pressure of 70 kN/m^2. The drained modulus of the clay increases from 4 MN/m^2 at ground level to 28 MN/m^2 at the base of the clay. Determine the total settlement at the centre and the corner of the foundation.

9.9 Determine the amount of secondary compression in the first 5 years and after a period of 30 years for a deposit of soft organic clay, 8 m thick with a coefficient of secondary compression ε_{α} of 1.5%. Assume t_1 is 1 year.

9.10 A square foundation 10 m wide is to be placed at 2 m below ground level in a sand deposit 25 m thick and will impose a gross pressure of 90 kN/m^2. The water table exists at 4 m below ground level and the unit weight of the sand above and below the water table is 19.5 kN/m^3. The static cone resistance of the sand is constant with depth at 4 MN/m^2. Determine the immediate settlement of the foundation and the long-term settlement after a period of 30 years using the Schmertmann method.

9.11 For the foundation described in Exercise 9.10 determine the immediate and long-term settlements using the Burland and Burbridge method. Assume the sand to be normally consolidated with a mean standard penetration resistance of $N = 10$ and the loads are static.

10 Pile Foundations

Pile Foundations

Introduction

If a structure cannot be satisfactorily supported on a shallow foundation, for reasons such as incompetent ground, swelling/shrinking soil or open water, then a pile foundation can provide an economical alternative.

A single pile can be defined as a long, slender structural member, used to transmit loads applied at its top to the ground at lower levels. This is achieved via:

- a shear stress mobilised on the surface of the shaft of the pile, called skin friction in sands and adhesion in clays, and
- development of failure by bearing capacity at the base of the pile, called end bearing.

Loading conditions *(Figure 10.1)*

The loads applied by a structure are quite large and usually cannot be supported by a single pile, so a number of piles are placed together to form a pile group, with a substantial reinforced concrete pile cap placed on top to transfer and distribute the loadings from the structure to the piles. The various types of pile and loading conditions which piles may be subjected to are illustrated in Figure 10.1.

Types of pile

Piles can be classified according to:

- their method of installation, i.e. driven, bored, jacked,
- their material type, i.e. timber, steel or concrete, precast or cast in situ, full length or segmental,
- the plant used to install them such as driven, tripod, auger, under-reamed, continuous flight auger,
- their size, i.e. small diameter bored, large diameter bored, under-reamed, mini-piling,
- their effects during installation, i.e. displacement or replacement,
- the way they provide load capacity, i.e. end bearing, friction piles, uplift piles, raking piles.

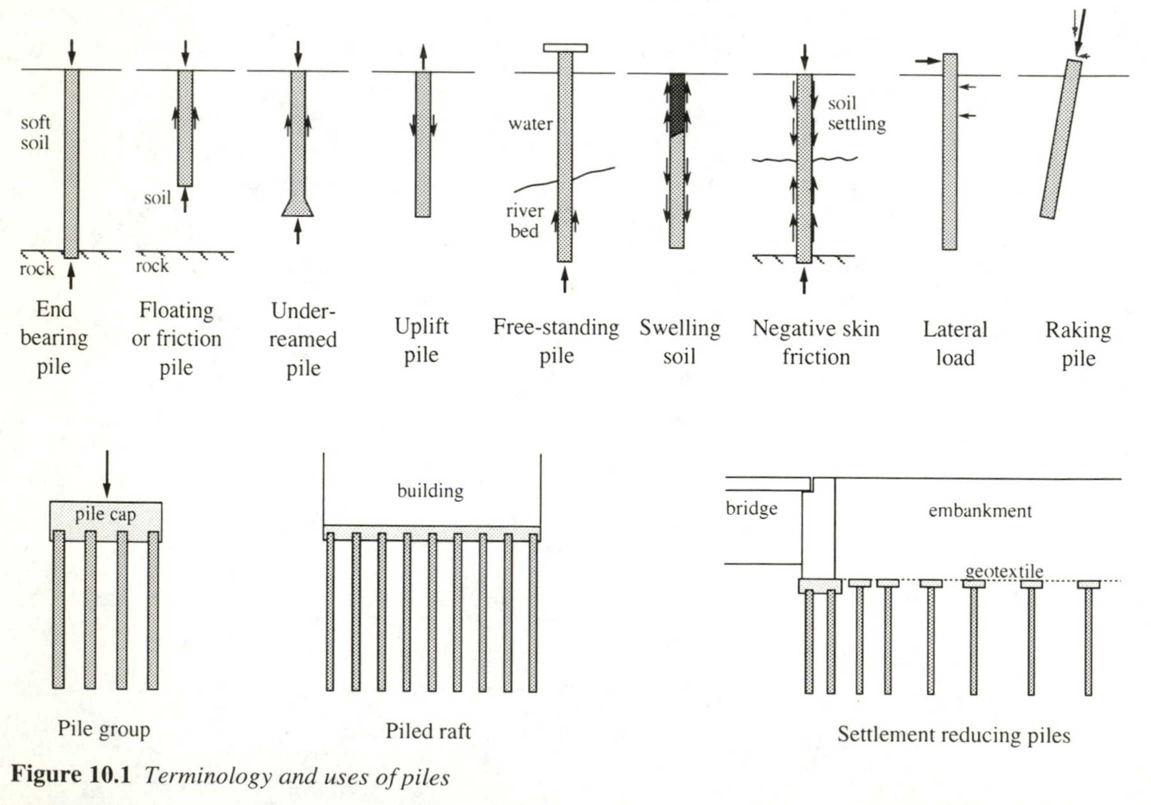

Figure 10.1 *Terminology and uses of piles*

For further reading on this subject the reader is recommended to refer to the books by Tomlinson, Poulos, Randolph *et al* or Whitaker.

Design of single piles

With so many different types of piles, different soil conditions and various effects of installation such as remoulding, softening, compaction and others, the development of design processes has evolved in a theoretical manner, but there has always been a need for empiricism to obtain correlations and 'adjustment factors' for various effects. Thus the predictions of load-carrying capacity of a pile will not be particularly accurate and should never be relied upon fully without some proof-load testing.

A particular type of pile is chosen to suit the ground and other conditions, and a preliminary estimate is made of its diameter and length to support a working load using the methods given below. Following this 'design' process, a number of these piles should be installed on the site and tested under load, to confirm their adequacy and to assess the most economical solution.

If proof-load testing is not carried out, then the design would have to be based on conservative values of soil strengths with higher factors of safety, and an economical design is less likely to be obtained.

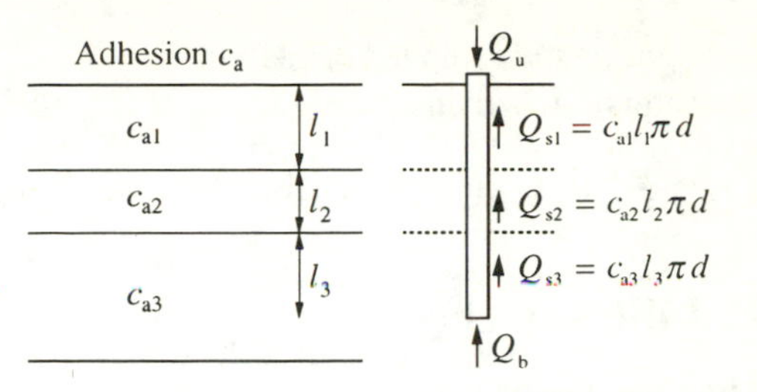

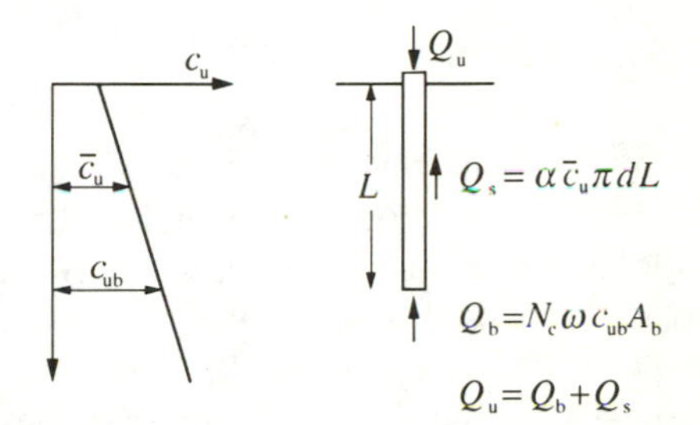

Figure 10.2 *Layered or non-homogeneous clays*

Shaft resistance is obtained as a load by multiplying the surface area of the shaft of the pile by a shear stress acting between the pile surface and the soil called skin friction in sands, f_s, and adhesion in clays, c_a.

$$Q_s = \sum_0^L c_a \pi dl \quad \text{(clays)} \qquad (10.3)$$

$$Q_s = \sum_0^L f_s \pi dl \quad \text{(sands)} \qquad (10.4)$$

Piles usually penetrate several different soil types, each providing different shaft resistances as shown in Figure 10.2, so the total shaft resistance is the sum of the individual values.

The weight of the pile, W_p, should be included in Equation 10.1 but this can be ignored if the net bearing capacity is used to obtain the base resistance since:

net bearing capacity = gross bearing capacity – q

where q, as for shallow foundations, represents the surcharge or vertical stress at pile base level, σ_{vb} and

$$W_p \approx A_b \sigma_{vb} \qquad (10.5)$$

or $W_p \approx W_s$

Load capacity of single piles *(Figure 10.2)*

There are two resistances offered by a pile to the vertical load applied:

- shaft resistance
- base resistance.

At failure the ultimate values of these loads are mobilised and give:

$$Q_u = Q_s + Q_b \qquad (10.1)$$

where:

Q_u = ultimate applied load
Q_b = ultimate base load
Q_s = ultimate shaft load.

Base resistance is obtained as a load by multiplying the cross-sectional area of the base of the pile by a bearing capacity stress.

$$Q_b = q_b A_b \qquad (10.2)$$

where:

q_b = end bearing resistance
A_b = base cross-sectional area

where

W_s = weight of soil removed/displaced.

The full equation is then:

$$Q_u = Q_s + Q_b - W_p + W_s = Q_s + Q_b \qquad (10.1)$$

Bored piles in clay

End bearing resistance q_b

This can be obtained from:

$$q_b = N_c \omega c_{ub} \quad (\text{kN/m}^2) \qquad (10.6)$$

The quantity N_c is a bearing capacity factor which for the undrained condition for a deep foundation can be shown to be theoretically between 8.0 and 9.8 (Whitaker and Cooke, 1966) and a value of 9 suggested by Skempton (1959) is widely adopted.

The quantity c_{ub} is the undrained shear strength at the base of the pile taken from the mean line of the shear strength/depth plot. These strengths used to be obtained from triaxial compression tests on 38 mm diameter specimens prepared in the conventional manner by jacking tubes into a U100 sample. If fissures exist within the clay these tests will be affected to varying degrees by fissuring within the specimens tested, as illustrated in Figure 7.22 and the mean results when plotted will not reflect the en masse fissure strength which would be operative beneath a pile. The factor ω is applied to convert these laboratory strengths into fissure strength and values suggested by Skempton (1966) are:

non-fissured clay $\omega = 1.0$
fissured clay, $d < 0.9$ m $\quad \omega = 0.8$
fissured clay, $d > 0.9$ m $\quad \omega = 0.75$

However, if the undrained shear strength is obtained from triaxial tests on 100 mm diameter samples which is more common nowadays then these tests will more likely represent the fissure strength and the ω factor need not be applied, especially if the lower bound of the data is used.

Before assuming the full value of end bearing, it should be confirmed that the pile bore has been thoroughly cleaned out before concreting. If debris, slurry or any bentonite may remain at the base of the pile then a reduced end bearing resistance should be adopted.

Adhesion c_a

The shear stress developed between the pile surface and the soil will occur within a narrow zone adjacent to the pile. The adhesion c_a will always be less than the initial undrained shear strength, c_u, due to:

- remoulding caused by the action of the drilling tools
- softening caused by stress relief with moisture migrating from the nearby soil and from the wet concrete towards the annulus of soil around the pile. Any water present in the borehole will aggravate this problem.

The value of adhesion is likely, therefore, to be dependent on the drilling techniques adopted and, in particular, any delay before concrete is inserted.

Adhesion is often obtained from a total stress approach using an empirical modification of the undrained shear strength:

$$c_a = \alpha c_u \qquad (10.7)$$

where α is an adhesion factor.

For soft and firm clays, a value of $\alpha = 1.0$ can be used (Vesic, 1977). For London Clay, Skempton (1959) found values of α between 0.3 and 0.6, from back-analyses of a number of pile loading tests, and recommended a value of 0.45 for piles extending beneath the highly fissured and weathered upper horizon of the London Clay. For short piles existing mostly within this upper zone and where delay before concreting is likely, such as for large-diameter under-reamed bored piles, a value of $\alpha = 0.3$ is recommended.

When bentonite fluid or drilling mud is used to support the sides of the borehole then, provided it is completely displaced by the concrete during the tremie process, it should have no detrimental effects on the adhesion value (Fleming and Sliwinski, 1977). Tomlinson (1987) recommends that the adhesion values should be reduced by 20% to allow for drilling mud effects, since it cannot be guaranteed that the mud will be removed entirely.

A maximum value of adhesion of 100 kN/m^2 is recommended for many soils (Skempton, 1959, Vesic, 1977) although for piles in glacial clays a maximum value of 70 kN/m^2 is recommended (Weltman and Healy, 1978).

Driven piles in clay

End bearing resistance q_b

Because of the limiting diameter of conventional driven piles of about 450–600 mm, and the small cross-sectional area, the base load obtainable tends to be a small amount in relation to the shaft load. Nevertheless, it could be calculated from:

$$q_b = N_c c_{ub} \tag{10.8}$$

where N_c can be taken as 9 and c_{ub} is the undisturbed undrained shear strength at the base of the pile.

Adhesion c_a *(Figures 10.3 – 10.6)*

Driving a pile into clay requires considerable displacement and causes major changes in the clay. The effects of installation are different for soft clays and stiff clays.

Driving a pile into soft clay increases the total stresses, which are transferred to a large rise in pore water pressure in the annulus of soil around the pile. This increase in pore pressure is larger for piles with a greater volumetric displacement such as solid piles compared to H-section piles and for soils with a tendency for their mineral grain structures to collapse such as sensitive clays.

The time taken for this pore pressure to dissipate will depend on the initial excess pore pressure, the permeability of the soil, the permeability of the pile material and the number of piles and the spacing between them.

As the consolidation process occurs the effective stresses around the pile increase and the pile load capacity increases. Thus, the initial load carrying capacity of a pile may be quite small but will increase with time. However, from measurements which have been carried out, several weeks or months may elapse before the full load capacity is achieved.

Driving piles into stiff, overconsolidated clays can produce three significant effects:

- expansion of the soil surrounding the pile with associated radial cracking and opening of macro-fabric features such as fissures. Any positive pore pressures set up during driving will rapidly dissipate into this open structure and expansion of the soil is more likely to produce negative pore pressures at least in the upper levels. Relatively short piles, therefore, may provide an initially high load carrying capacity but this could diminish with time. Longer piles are more likely to produce positive pore pressures in their lower regions.

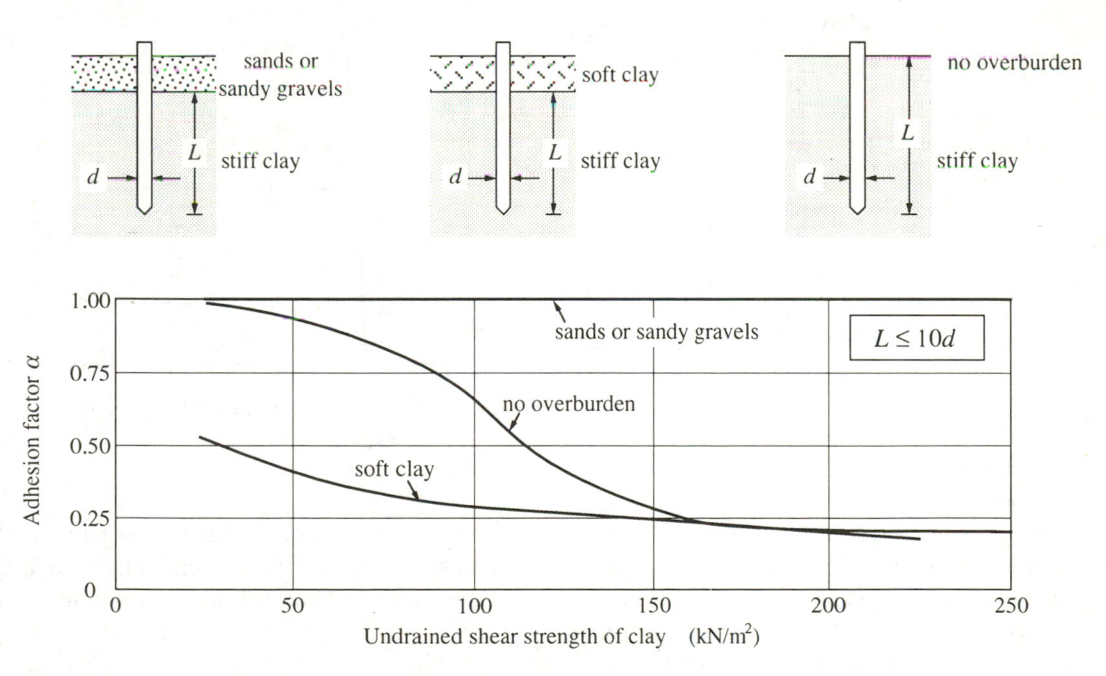

Figure 10.3 *Adhesion factors for piles driven into stiff clay – short penetration L ≤ 10d (From Tomlinson, 1987)*

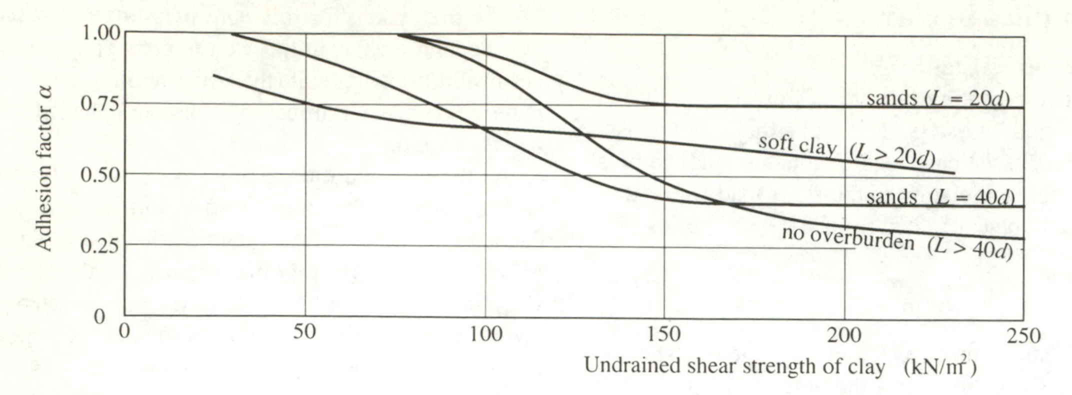

Figure 10.4 *Adhesion factors for piles driven into stiff clay – length > 20 to 40d (From Tomlinson, 1987)*

- ground heave comprising upward and outward displacement of the soil around a pile being driven. This effect can occur up to ten pile diameters away from a pile (Cole, 1972), so driving piles in groups can magnify the effect and cause damage to existing buried structures and previously driven piles, due to separation or fracture. Heave is particularly detrimental when the piles are intended to provide most of their load in end bearing.
- 'whippiness' or lateral vibrations set up in the pile once it has been partly driven into the clay. This produces a gap like a 'post-hole' effect between the clay and the pile, so no adhesion can exist over this length. Tomlinson (1970, 1971) also observed that any soil above the stiff clay was dragged down into this gap so soft clay overburden would produce a lower apparent adhesion but sand would produce a higher adhesion. The penetration of the pile into the stiff clay and the type of overburden, is, therefore, very important. This is illustrated in Figure 10.3 which gives values of adhesion factor α for short penetration piles taken from Tomlinson (1987). For longer piles, ($L > 20$ or 40 diameters) the effect of the gap diminishes, as illustrated on Figure 10.4. It should be noted that the scatter of data points used to obtain these curves was considerable.

The average adhesion on the shaft of a pile is then calculated from Equation 10.7 using the adhesion factors from Figures 10.3 and 10.4.

For very long piles driven into stiff clays it has also been found that the shaft capacity depends on the length of the pile but for probably different reasons.

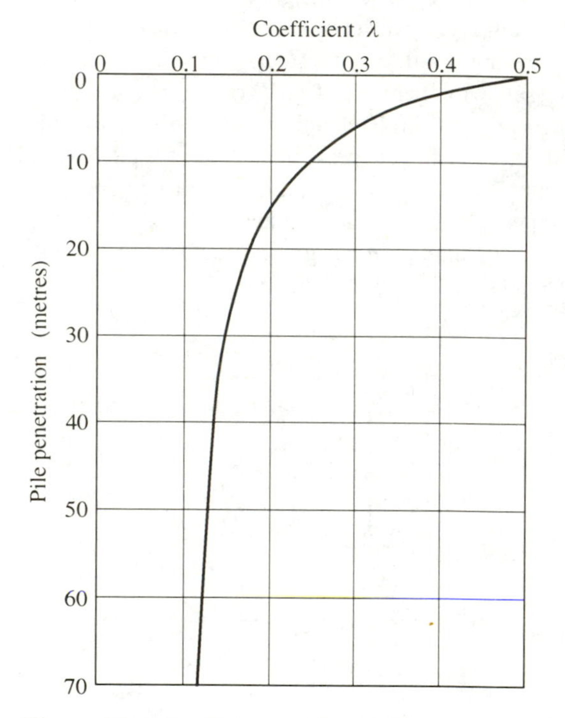

Figure 10.5 *Coefficient λ (From Vijayvergiya and Focht, 1972)*

Vijayvergiya and Focht (1972) suggested a quasi-effective stress approach for the determination of average adhesion along long steel-pipe piles in the form:

$$\bar{c}_a = \lambda\left(\bar{\sigma}_m' + 2\bar{c}_m\right) \quad (10.9)$$

$$Q_s = \bar{c}_a A_s \quad (10.10)$$

where $\bar{c}_a$ is the average adhesion along the pile, $\bar{\sigma}_m{}'$ is the mean effective vertical stress between ground level and the base of the pile, $\bar{c}_m$ is the mean undrained shear strength along the pile, and λ is an empirical factor, plotted in Figure 10.5.

Randolph and Wroth (1982) have shown that the average adhesion for long piles depends on the K_o value which tends to decrease with depth for an overconsolidated clay, see Chapter 4. They expressed this effect in terms of the $c_u/\sigma_v{}'$ ratio which is related to overconsolidation ratio *OCR*.

Semple and Rigden (1986) proposed the relationships on Figure 10.6 giving the shaft load as

$$Q_s = F\alpha_p \bar{c}_u \pi dL \qquad (10.11)$$

where $\bar{c}_u$ is the average undrained shear strength over the length of the pile and $\bar{\sigma}_v$ is the average vertical effective stress.

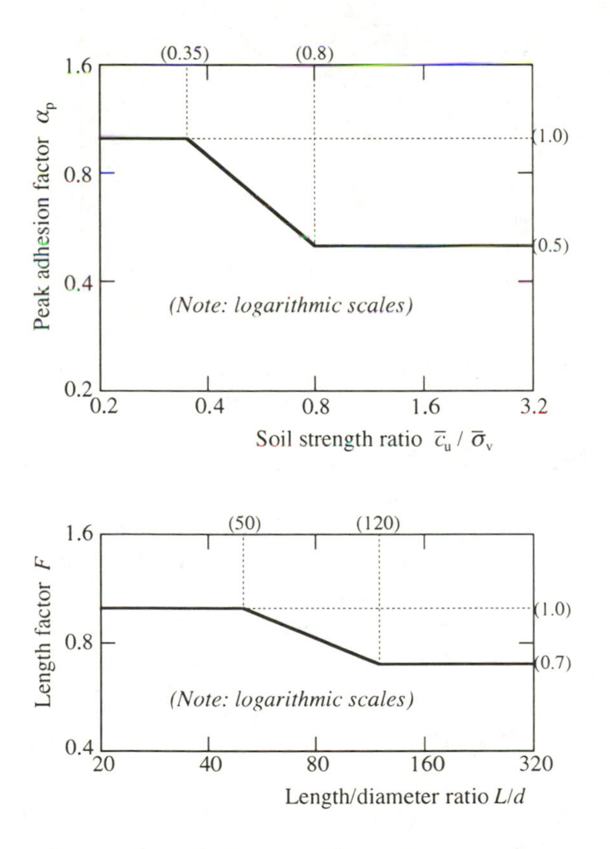

Figure 10.6 *Peak adhesion factor α_p and length factor F (From Semple et al, 1986)*

Note that α_p is the peak adhesion factor for a rigid pile. The value F is a length factor to account for flexibility and compressibility of the pile since:

- Lateral flexibility produces the whippiness or flutter effect as the pile is driven.
- The axial compressibility of the pile will permit greater displacements in the upper part of the pile than in the lower part taking the skin friction beyond any peak value and towards a lower residual value.

Effective stress approach for adhesion

On the basis that pile capacity increases with time after driving, due to dissipation of pore pressures during consolidation and, hence, increases in effective stress, it is generally accepted that adhesion is governed by the effective stress in the soil. Since the shear stress mobilised on the shaft surface is frictional it is referred to as skin friction, and is given by a general expression:

$$f_s = K_s p_o{}' \tan\phi' \qquad (10.12)$$

$p_o{}'$ is the vertical effective stress at any depth z and K_s is a coefficient of horizontal earth pressure so $K_s p_o{}'$ represents the normal stress (horizontal stress) acting on the surface of the pile. K_s has a similar function to the value of K_o, the coefficient of earth pressure at rest. For normally consolidated clay it is given by $K_s = K_o = 1 - \sin\phi'$ and can be expected to remain fairly constant with depth. Burland (1973) suggested the simple form:

$$f_s = \beta \bar{p}_o{}' \qquad (10.13)$$

where $\bar{p}_o{}'$ is the average effective vertical stress down the length of the pile and

$$\beta = K_s \tan\phi'$$
$$= (1 - \sin\phi') \tan\phi' \qquad (10.14)$$

For a typical range of ϕ' values (15° to 30°) β varies between 0.2 and 0.29 and for bored piles in soft normally consolidated clay a value of about 0.3 is suggested. Meyerhof (1976) found that β varied from about 0.2 – 0.4 for short piles driven into soft clays (less than about 15 m long) to about 0.1 – 0.25 for very long piles which may be due to some overconsolidation in the upper horizons and pile compressibility.

Meyerhof also found higher values of β for tapered piles, reflecting the higher horizontal stresses produced.

For piles bored or driven into stiff overconsolidated clays the K_o and hence K_s value can be expected to vary with depth. Meyerhof (1976) stated that, for bored piles, K_s varies from about 0.7 K_o to 1.2 K_o but for driven piles it varies from about K_o to more than $2K_o$. Randolph and Wroth (1982) suggest a relationship between β and the ratio $\bar{c}_u / \bar{\sigma}_v'$. However, although it is clear that shaft resistance is governed by effective stresses, the empirical correlations required to determine values of β make this approach no better than the traditional total stress or α method, at the present time.

Driven piles in sand

Effects of installation

Driving piles into loose sands compacts them, increasing their density and angle of internal friction and increasing the horizontal stresses around the pile.

Driving piles into dense sands may not compact them. Instead, dilatancy and negative pore pressures may temporarily increase the pile load capacity, make driving difficult and possibly result in overstressing and damage to the pile. Dissipation of this negative pore pressure after driving will cause the pile load capacity to decrease so a false impression of load capacity can be derived from the driving records. This is often referred to as relaxation.

The extent to which driving may increase the density of the sand could be up to 4–6 diameters away from the pile and 3–5 diameters below the pile (Broms, 1966). This zone of influence is larger for loose sands than dense sands and will obviously affect the driving of piles in groups where piles are typically 2–3 diameters apart.

It is also presumed that the sands are hard, clean quartz grains which will not deteriorate under driving stresses. Softer crushable grains will produce lower angles of friction after driving and be more compressible.

The design of a pile must consider installation effects and the final state of the sand. It can only be considered as approximate and should be checked by *in situ* pile loading tests.

End bearing resistance q_b *(Figure 10.7)*

The conventional approach to end bearing resistance is to use the surcharge term of the bearing capacity equation (Equation 8.11) as $c' = 0$ and the width of a pile is small compared to its length:

$$q_b = N_q \sigma_v' \quad (10.15)$$

where σ_v' is the vertical effective stress at the base of the pile and N_q is a bearing capacity factor. The values of N_q provided by Berezantsev *et al* (1961) are commonly used, Figure 10.7.

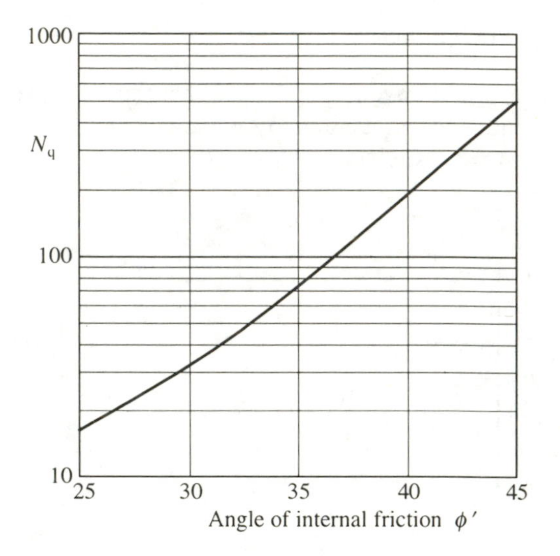

Figure 10.7 *Bearing capacity factor N_q for piles in sand (From Berezantsev et al, 1961)*

Critical depth *(Figures 10.8–10.10)*

Equation 10.15 suggests that as the pile penetrates more deeply into the sand the end bearing resistance will increase with depth. However, field tests have shown that end bearing resistance does not increase continually with depth. It seems more logical that end bearing resistance depends on the mean effective stress at pile base level rather than just the vertical stress:

$$\sigma_m' = \tfrac{1}{3}(\sigma_v' + 2\sigma_H') \quad (10.16)$$

Since $\sigma_H' = K_o \sigma_v'$ end bearing resistance will then be affected by the K_o value which for overconsolidated soils decreases with depth.

The Mohr-Coulomb criterion for soils at higher stress levels often shows some curvature rather than the straight line assumed. Thus as the stresses at pile base level increase the ϕ' value and, hence, the bearing capacity factor N_q decreases. Arching is also considered to be a contributory factor. The combined effect is to obtain decreasing end bearing resistance with depth.

The simplest way of incorporating this effect into pile design is to adopt the concept of a critical depth z_c as shown in Figure 10.8. Even though the vertical effective stress σ_v' increases with depth the end bearing resistance q_b and the skin friction f_s are considered as constant below the critical depth, having the value at the depth z_c. The critical depth has been found to be shallow for loose sands and deeper for dense sands.

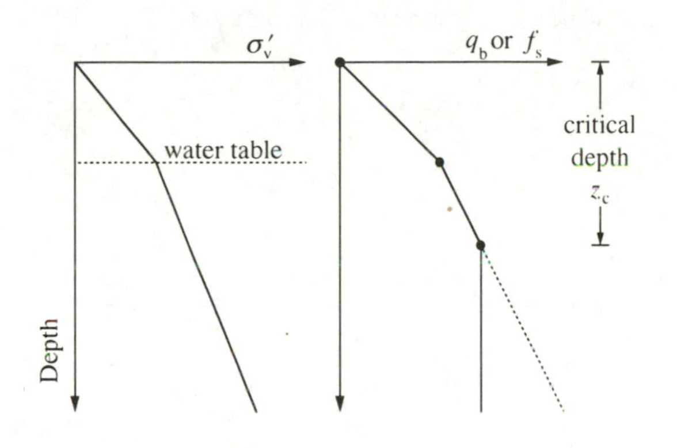

Figure 10.8 *Critical depth in sands*

At the present time, values of this critical depth are somewhat tentative, the values suggested by Vesic (1967) and Meyerhof (1976) are given in Figure 10.9.

For the determination of N_q and z_c on Figures 10.7 and 10.9, respectively, the angle of internal friction ϕ' should relate to the state of the sand after pile installation so the values given in Table 10.1 are suggested (Poulos, 1980).

The initial angle of internal friction ϕ_1', before installation of the pile, is not an easy parameter to determine, since sampling disturbance will largely destroy the initial mineral grain structure, making laboratory tests meaningless. The ϕ_1' value is usually obtained from correlations between the SPT '*N*' value or the cone penetrometer q_c as illustrated on Figure 10.10.

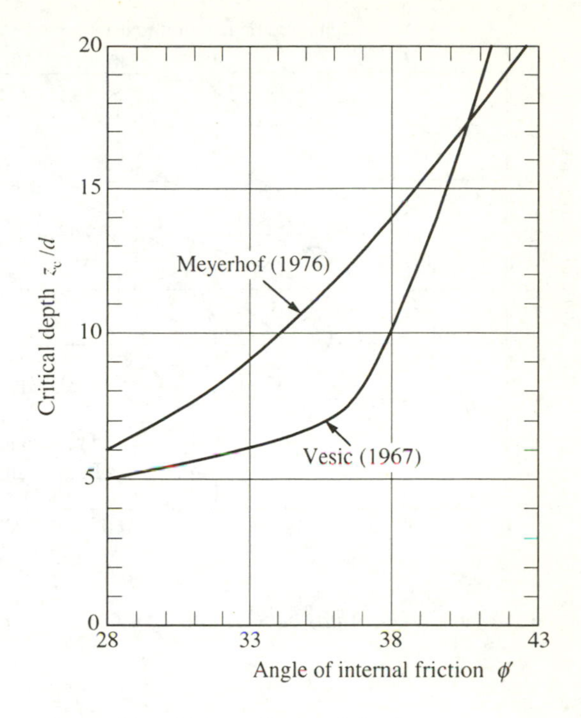

Figure 10.9 *Values of critical depth*

Table 10.1
Values of ϕ' after installation (From Poulos, 1980)

Requirement	Values of ϕ after installation	
	Bored piles	Driven piles
N_q	$\phi_1' - 3$	$\frac{\phi_1' + 40}{2}$
z_c/d	$\phi_1' - 3$	$3/4\phi_1' + 10$
$K_s \tan\delta$	ϕ_1'	$3/4\phi_1' + 10$

ϕ_1' is the ϕ value before installation

Skin friction f_s *(Figures 10.11 and 10.12)*
Assuming effective stresses acting on the pile/soil surface the unit skin friction f_s at a depth z below the top of a pile is given by:

$$f_s = K_s \sigma_v' \tan\delta \tag{10.16}$$

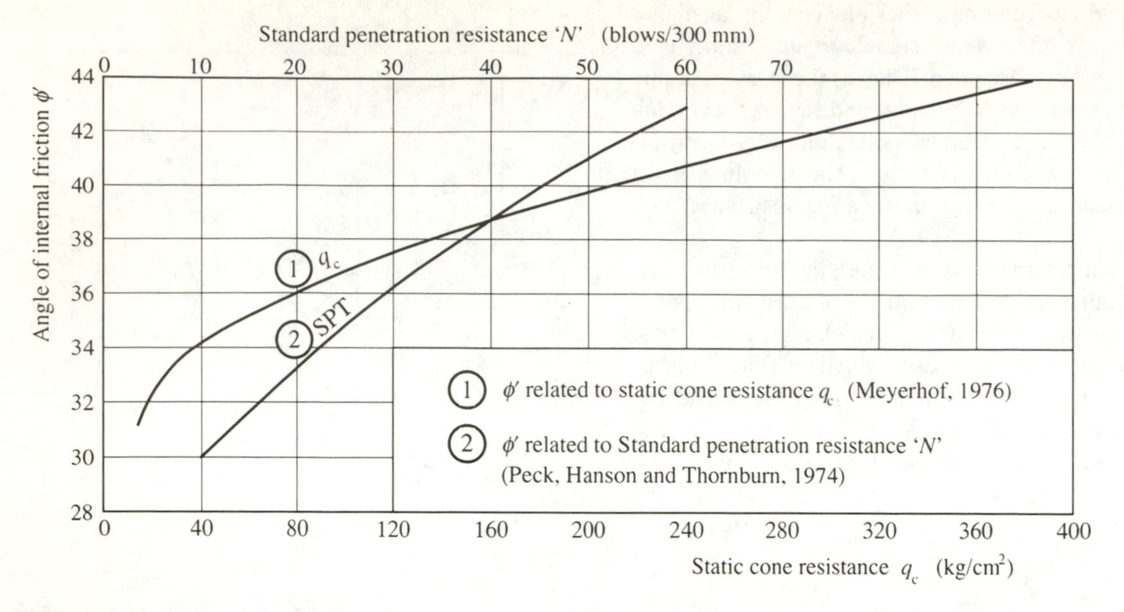

Figure 10.10 *Relationships between angle of internal friction and in situ tests*

where:
K_s = coefficient of horizontal stress
δ = angle of friction between the pile surface and the soil.

Since both K_s and $\tan\delta$ will be governed by the method of installation, and values of these factors may be difficult to assess separately, a simple approach is to consider values of the lumped parameter $K_s\tan\delta$ (cf. β for effective stress approach to piles in clay, Equation 10.14). Values of this parameter are given in Figure 10.11 related to the initial angle of internal friction. These are based on K_s values given by Meyerhof (1976), the ϕ values given in Table 10.1 and assuming $\delta = 0.75\phi'$, for a normally consolidated sand. Higher values may be possible for overconsolidated sands.

It has been found that skin friction values also decrease with depth in a similar fashion to end bearing resistance so a critical depth z_c approach should be adopted. Values of z_c can be obtained from Table 10.1 and Figure 10.9.

The total shaft load must then be summated from the shaft loads Q_{s1}, Q_{s2}, etc, as illustrated in Figure 10.12.

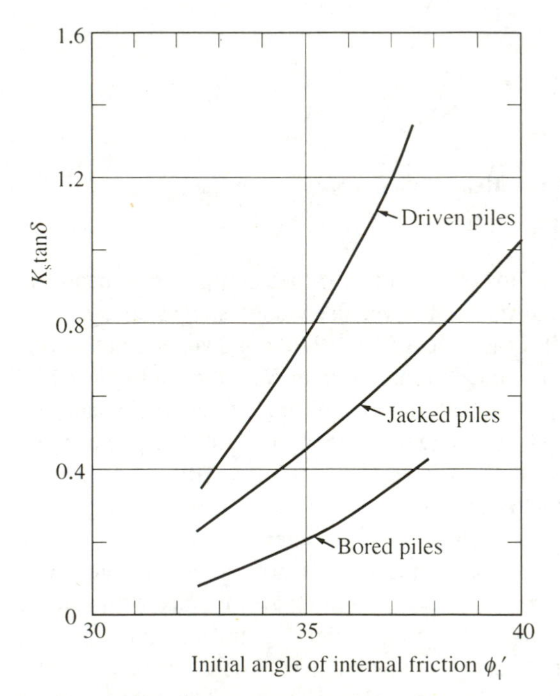

Figure 10.11 *Skin friction parameter $K_s\tan\delta$ (From Poulos, 1980)*

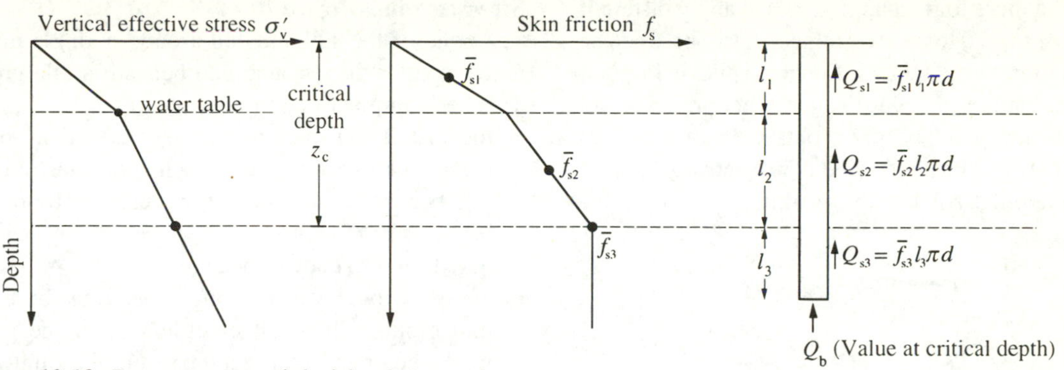

Figure 10.12 *Determination of shaft loads in sands*

Bored piles in sand

Boring holes in sands loosens an annulus of soil around the hole and reduces horizontal stresses, so bored piles constructed in initially dense sands can be expected to have low load capacity. If jetting techniques are used then the loosening can be even more severe. Casting concrete ***in situ*** will produce a rough surface but this effect is diminished by the loosening of the sand.

Poulos (1980) suggests using the methods given for driven piles but with reduced values of the final angle of internal friction as given in Table 10.1. Meyerhof (1976) suggests that for preliminary estimates the base resistance of a bored pile could be taken as one-third of the value determined for a driven pile with about one-half for the shaft resistance.

Factor of safety

A factor of safety is applied to safeguard against the uncertainties in the ground conditions and installation effects, and to limit settlement to a permissible value.

Although piles are often designed by applying a factor of safety to the ultimate load to obtain a working load, the over-riding performance criterion for a pile is that it must not settle more than a permissible amount. Tomlinson (1987) stated that, from his experience for piles up to 600 mm diameter, if an overall factor of safety of 2.5 is adopted to give:

$$\text{working load} = \frac{\text{ultimate load}}{\text{factor of safety}} \tag{10.17}$$

the settlement of the pile under the working load is unlikely to exceed 10 mm.

For piles larger than 600 mm diameter, it has been found that, in clays the two components, shaft resistance and base resistance are mobilised at different amounts of settlement. Approximately, the full shaft resistance is mobilised at a pile head settlement of about 1– 2% of the pile diameter, whereas to mobilise the full base resistance the pile must be pushed down about 10–20% of the diameter. This is illustrated in Figure 10.13.

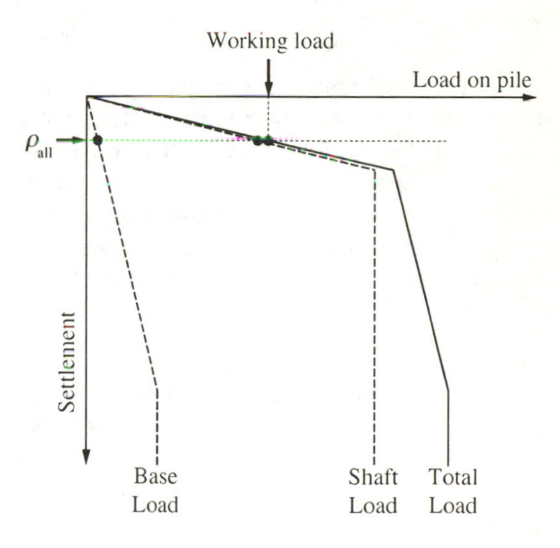

Figure 10.13 *Mobilisation of base and shaft loads*

For a typical pile diameter and a permissible settlement of ρ_{all}, it can be seen that when the working load is applied to obtain this settlement a large proportion of the shaft resistance is mobilised with only a small proportion of the base resistance acting.

Thus, a more logical approach is to apply different factors (of safety) to the two components. These can be referred to as partial factors. For bored piles in London Clay, Burland *et al* (1966) suggest that providing an overall factor of safety of 2 is obtained partial factors on the shaft and base of 1 and 3, respectively, should also be applied so that the working load Q_a is the smaller of:

$$Q_a = \frac{Q_{ult}}{2} = \frac{Q_s + Q_b}{2} \tag{10.18}$$

or

$$Q_a = \frac{Q_s}{1} + \frac{Q_b}{3} \tag{10.19}$$

Burland *et al* stated that the latter expression generally governs the design for large under-reamed piles and the former commonly governs straight-shafted piles.

When there is less certainty about the ground conditions, applied loadings and installation effects, higher factors of safety should be applied such as 2.5 overall, and 1.5 and 3.5 for the shaft and base load, respectively.

Pile groups

It is not common for a single pile to support a structural load on its own because of overturning effects, lateral loading and difficulty of transferring the load axially down the pile. Large-diameter piles would be the exception. Piles are grouped together usually on a square grid pattern with the structural load transferred to and shared between the piles by a thick reinforced concrete pile cap.

Pile spacing

The centre to centre spacing between piles, *s*, is typically three diameters for friction piles and may be two diameters for end bearing piles. For under-reamed piles the spacing should be no less than two pile base diameters.

If the piles are too far apart the bending stresses in the pile cap are large necessitating thicker concrete and more reinforcement to distribute the applied loads to the piles. If the piles are too close then the soil between can become excessively disrupted and disturbed leading to high pore pressure increases, ground heave and poor efficiency of loading. Interference due to misalignment may also be a problem.

Stressed zone *(Figure 10.14)*

The zone of the soil stressed around a single pile is much smaller than around and beneath a pile group. This has a number of consequences:

- the installation method has less effect on group behaviour than on single pile behaviour since around and beneath a group the zone affected by disturbance is relatively small and the stresses will be transferred to undisturbed soil.
- compressible layers existing beneath the base of a pile group will produce settlement of the group while they may not affect the result of a single pile load test.
- because of the above and other factors extrapolation from the performance of a single pile load test to the behaviour of a group must be treated with caution.

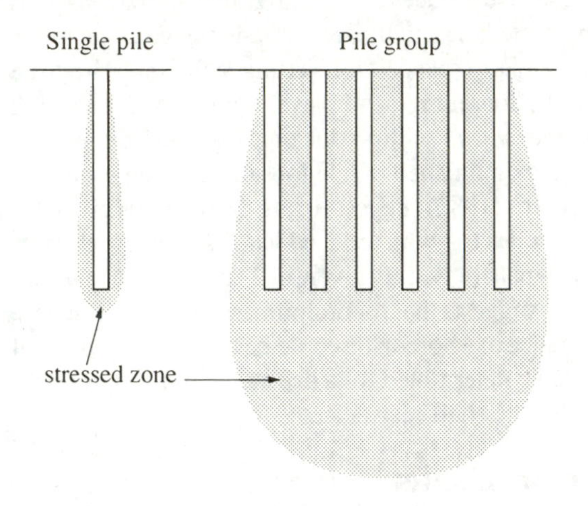

Figure 10.14 *Stressed zone around piles*

Load variation *(Figure 10.15)*

It is a fallacy to presume that each pile in a pile group carries the same load. The loads in the piles differ and this is well illustrated by reference to Figure 5.1.

If a flexible pile cap is provided and a uniform load is applied, then the contact pressures will be fairly uniform for both clays and sands and piles placed beneath the flexible pile cap will carry fairly similar loads. This could arise for piles placed beneath an embankment to reduce the settlement of the embankment. It is to be noted, however, that a dish-shaped profile of settlement is obtained which many structures cannot tolerate.

The purpose of a rigid pile cap is to even out the settlement profile and to produce similar settlements. From Figure 5.1 it can be seen that if the piles are embedded in a clay then the rigid cap will increase the loads on the outer piles and decrease loads on the centre piles. This was confirmed for model piles in a clay tested by Whitaker (1957) who showed that for typical pile spacings of less than four diameters the corner piles carry the largest load and the centre piles carry least load (Figure 10.15).

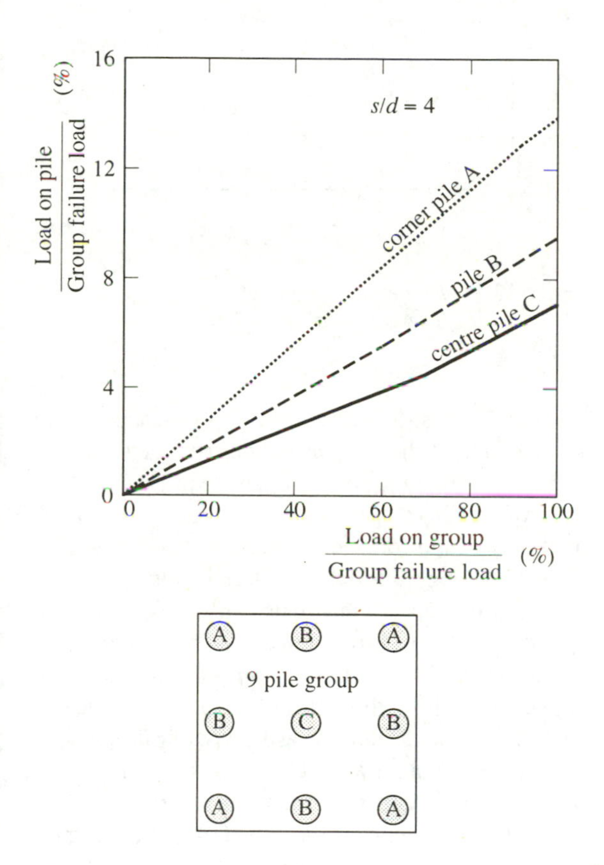

Figure 10.15 *Load distribution in a 9 pile group (From Whitaker, 1970)*

From Figure 5.1 for pile groups in sands with rigid pile caps the distribution of load can be expected to be reversed with the centre piles carrying the largest loads and the corner piles the least load and this has been confirmed by Vesic (1969).

Efficiency *(Figure 10.16)*
It is usually incorrect to assume that the group failure load equals the sum of the pile failure loads with the piles acting individually. The difference is represented by an efficiency factor η:

$$\eta = \frac{\text{average load per pile at failure of group}}{\text{failure load of single isolated pile}} \tag{10.20}$$

For piles driven into loose or medium dense sands the effect of compaction will lead to an efficiency η greater than 1 with higher efficiencies for closer pile spacing and in looser sands. For piles driven into dense sands the efficiency is unlikely to exceed 1 and could be less if driving causes disturbance. For bored pile groups in sand the overlapping disturbance zones between piles is likely to reduce efficiency.

The model tests of Whitaker (1957) on piles in clay (Figure 10.16) showed the significant effect of the spacing of the piles on efficiency. For a square group of piles the efficiency at a spacing of eight diameters or more even with a rigid pile cap was close to unity giving equal pile loads. As the spacing decreases the efficiency decreases but not too significantly.

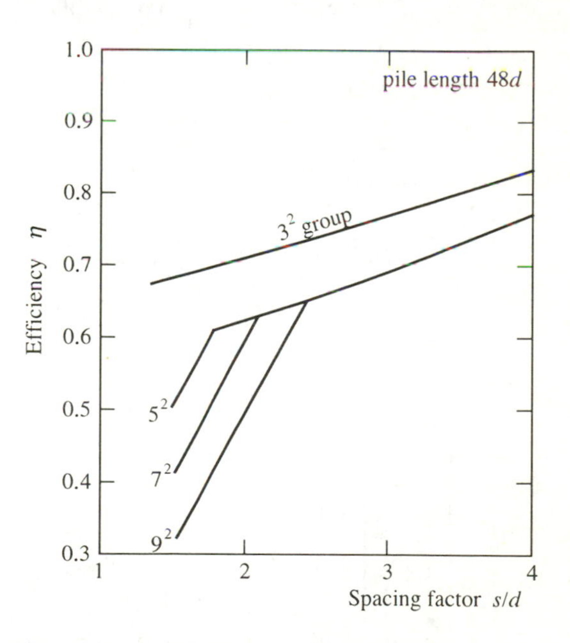

Figure 10.16 *Efficiency of freestanding pile groups (From Whitaker, 1970)*

With wider spacing the pile group reaches a state of failure referred to as individual pile penetration (Figure 10.17) with the soil remaining static and the piles alone moving down. This, of course, assumes a 'free-standing' group. However, at a critical pile spacing the soil between the piles moves down with the piles causing 'block failure' when the efficiency diminishes dramatically. The efficiency was also found to decrease for longer piles and for larger pile groups.

Ultimate capacity *(Figures 10.17 and 10.18)*
The ultimate capacity of a free-standing pile group in clays should be taken as the lesser of:

a) the sum of the failure loads of the individual piles in the group, or
b) the bearing capacity of a block of soil bounded by the perimeter of the pile group, i.e.

$$P_{ult} = c_u N_c s_c d_c B_g L_g + 2\left(B_g + L_g\right) L\bar{c} \qquad (10.21)$$

where:
$c_u N_c s_c d_c$ is the bearing capacity of the clay beneath the pile group
B_g and L_g are the plan width and length of the pile group
L is the length of piles
$\bar{c}$ is the average c_u value over the length of the piles.

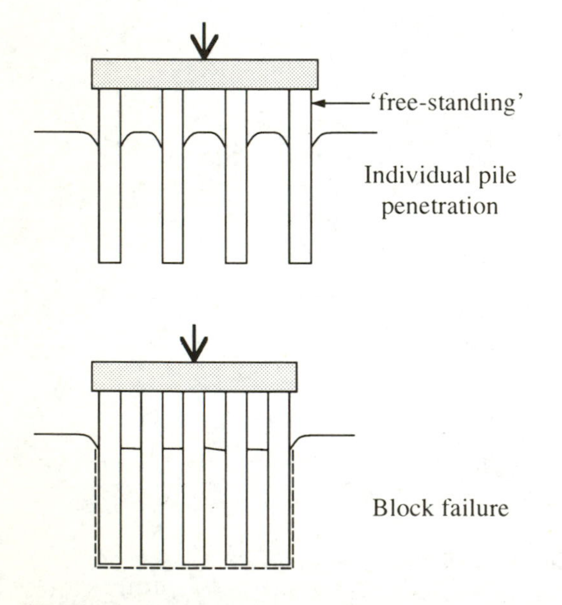

Figure 10.17 *Modes of failure for free-standing pile groups*

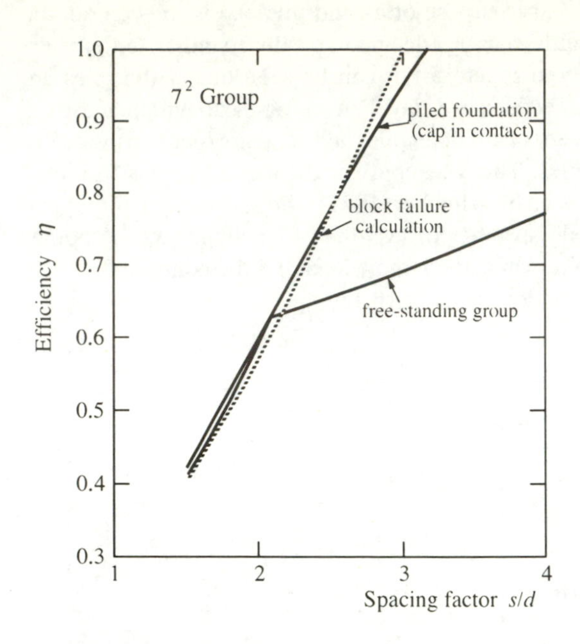

Figure 10.18 *Efficiency of pile groups (From Whitaker, 1970)*

The above refers to 'free-standing' pile groups where the underside of the pile cap is not in contact with the supporting soil or the soil between the pile cap and the supporting soil is compressible. This latter situation is commonly found for pile foundations where the piles are taken through inferior superficial soils.

If the pile cap is constructed in contact with the supporting soil, this is referred to as a 'pile foundation' and the cap itself can support a proportion of the load applied. At failure, the effect of a contacting pile cap is to induce block failure, irrespective of pile spacing, as shown in Figure 10.18. The ultimate capacity of a pile foundation with its pile cap in contact with the supporting soil can then be obtained from the lesser of:

a) the block failure mode as given by Equation 10.21 above

or

b) the sum of the failure loads of the individual piles plus the failure capacity of the remainder of the pile cap contact surface

$$P_{ult} = n\left(Q_s + Q_b\right) + c_{uc} N_c s_c d_c \left(B_c L_c - nA_p\right) \qquad (10.22)$$

where:
n = number of piles
Q_s = shaft load of a single pile obtained from the average shear strength along the pile
Q_b = base load of a single pile obtained from the shear strength at the base of the pile
c_{uc} = undrained shear strength at pile cap level
$N_c s_c d_c$ are factors obtained from bearing capacity theory, Chapter 8.
B_c and L_c are the plan width and length of the pile cap
A_p = cross-sectional area of a pile.

Settlement ratio *(Figures 10.19 – 10.21)*
As a pile is pushed downwards shear stresses and strains will develop in the soil around the pile. Cooke (1974) assumed that these shear stresses are transferred radially between successive annular soil elements and diminish with distance from the shaft of the pile, as shown in Figure 10.19. The point where shear strains were insignificant was found theoretically to be at n = 22, assuming a continuous medium, and by experiment, in a discontinuous fissured clay using sensitive instruments, to be at n = 10.

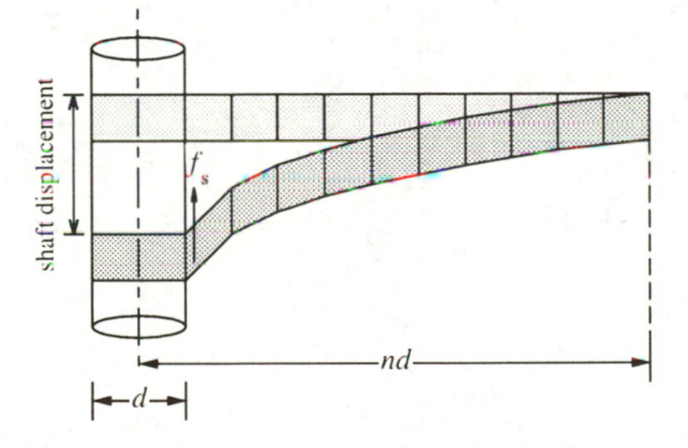

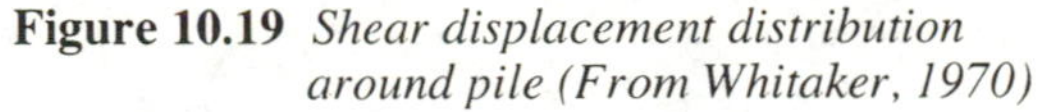

Figure 10.19 *Shear displacement distribution around pile (From Whitaker, 1970)*

Normal pile spacings are much less than these values so the piles within a pile group will effect a complex pile to pile interaction when they are loaded producing additional settlements on surrounding piles and vice versa. This additional settlement has been computed by Poulos (1968) using an interaction factor α_F given by

$$\alpha_f = \frac{\text{additional settlement caused by adjacent pile}}{\text{settlement of pile under its own load}} \qquad (10.23)$$

Values of α_F are given in Figure 10.20 for rigid piles, infinite soil thickness and homogeneous undrained soil (ν=0.5). They are smaller for flexible piles, finite layer depths and for soil modulus increasing with depth but slightly higher for under-reamed piles and drained soil ($\nu' < 0.5$). Adopting a super-position principle the interaction factors can be used to determine the cumulative settlement of a group of piles and give a settlement ratio R_s:

$$R_s = \frac{\text{settlement of group}}{\text{settlement of single isolated pile}} \qquad (10.24)$$

when the loads in the group and the single pile are at the same proportion of their failure load (and same factor of safety).

Settlement ratios have been computed by Cooke *et al* (1980) and are illustrated in Figure 10.21. They also stated for pile spacings greater than two diameters settlement ratios are unlikely to exceed six for rectangular pile groups with any number of piles, in soil increasing in stiffness with depth.

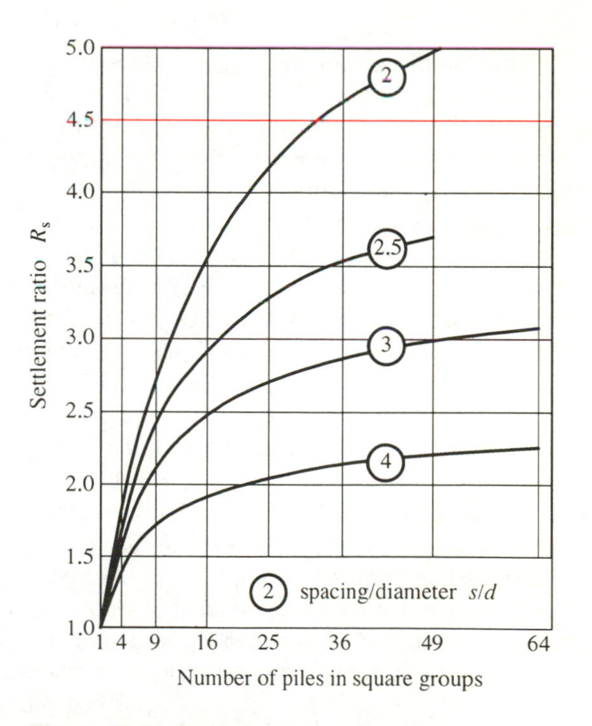

Figure 10.21 *Settlement ratios for pile groups (From Cooke et al, 1980)*

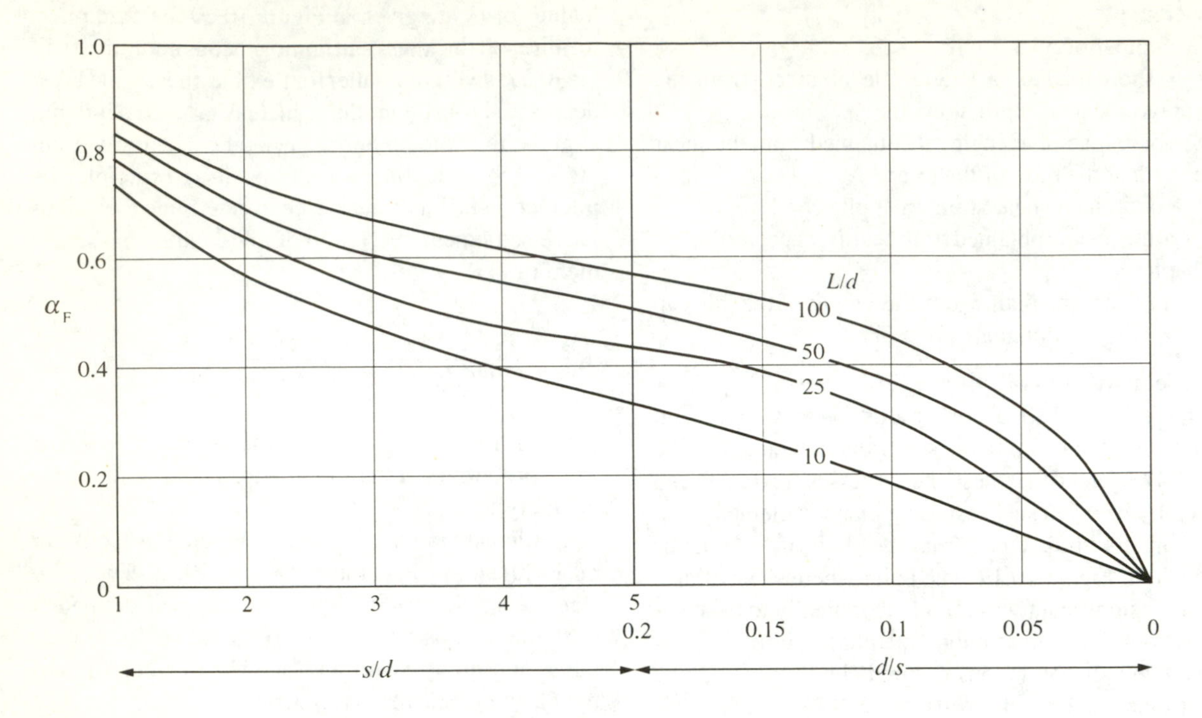

Figure 10.20 *Interaction factors for floating piles (From Poulos, 1980)*

Settlement of pile groups *(Figures 10.22 and 10.23)*
A convenient method of estimating settlement of a pile group is to assume that the group is represented by an 'equivalent raft', at some depth below ground level, and then to determine settlements using the conventional methods given in Chapter 9.

Uniform loading over this raft is assumed. This is obtained by dividing the total load (net) by the area of the equivalent raft. The size and depth of the equivalent raft are determined as shown in Figure 10.22 for different soil conditions.

The assumed raft is embedded within the compressible deposit so the depth correction factor, μ, given by Fox (1948) may be used, Figure 10.23. However, this would entail moving the equivalent raft to the surface of the deposit to determine the settlements for a surface loaded foundation. This may be appropriate for immediate settlement but not for consolidation settlement. The consolidation settlement should be determined from the stresses applied by the equivalent raft where it exists in the ground and the thickness of soil beneath this location.

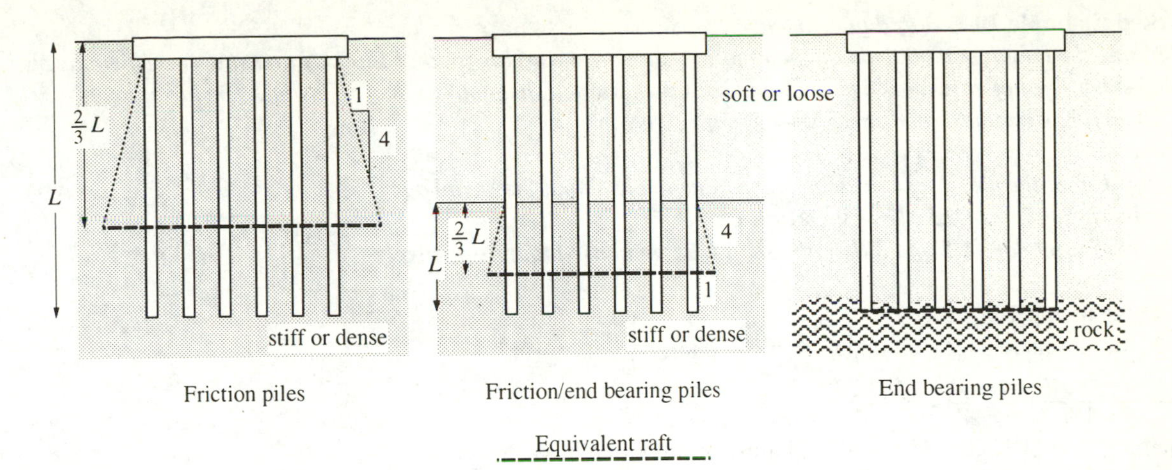

Figure 10.22 *Equivalent raft foundation*

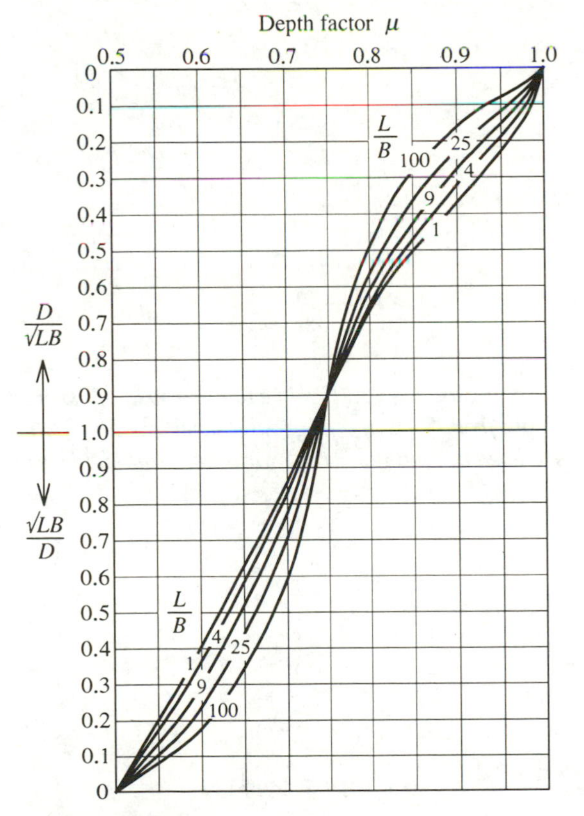

Figure 10.23 *Depth correction factor μ (From Fox, 1948)*

Worked Example 10.1 *Bored piles in clay*

Determine the length of a 600 mm diameter pile to support a working load of 1200 kN at a site where two layers of clay exist. The upper layer is 8 m thick and has an undrained shear strength of 80 kN/m². The lower layer is 20 m thick and has an undrained shear strength of 120 kN/m².

Assume:

- *the top 1 m of the pile does not support load due to clay/concrete shrinkage*
- *an adhesion factor of 0.5* $N_c = 9$ $\omega = 1.0$
- *factors of safety are 1.5 and 3.0 on the shaft load and base load, respectively*

Shaft load in upper layer Q_{s1}

Length of pile in contact with upper layer is 7.0 m

$$\frac{0.5 \times 80 \times \pi \times 0.6 \times 7.0}{1.5} = 352 \text{ kN}$$

remaining load of 1200 – 352 = 848 kN must be supported in the lower layer

Base load in lower layer Q_b

$$\frac{9 \times 1.0 \times 120 \times \pi \times 0.6^2}{3 \times 4} = 102 \text{ kN}$$

remaining load of 848 – 102 = 746 kN must be supported by the shaft in the lower layer.

Assuming the length of pile in the lower layer to be L

$$\frac{0.5 \times 120 \times \pi \times 0.6 \times L}{1.5} = 746 \text{ kN}$$

$L = 9.9$ m

Length of pile required is 8.0 + 9.9 = 17.9 m, say 18 m

If end bearing is ignored or cannot be relied on then

$$\frac{0.5 \times 120 \times \pi \times 0.6 \times L}{1.5} = 848 \text{ kN}$$

$L = 11.25$ m

Length of pile is 8.0 + 11.25 = 19.25 m

Worked Example 10.2 *Bored piles in fissured clay*

Determine the length of a pile, 1200 mm diameter, to support a working load of 4500 kN in a thick deposit of fissured clay with undrained shear strength increasing linearly with depth from 55 kN/m² at ground level and at 5 kN/m² per metre depth. The same assumptions as in Example 10.1 are used but with $\omega = 0.75$ *for a large diameter pile.*

At 1 m below ground level $c_u = 55 + 5 = 60$ kN/m²

Let L be the length of pile below 1 m

At the base of the pile $c_{ub} = 60 + 5L$ kN/m²

Mean shear strength over effective length of shaft $L = 60 + \frac{5L}{2} = 60 + 2.5L$

$$4500 = \frac{9 \times 0.75 \times (60 + 5L) \times \pi \times 1.2^2}{3.0 \times 4} + \frac{0.5 \times (60 + 2.5L) \times L \times \pi \times 1.2}{1.5}$$

giving $L^2 + 28.05L - 1383.6 = 0$ and $L = 25.73$ m

$$\text{Check: } \frac{9 \times 0.75 \times 188.7 \times \pi \times 1.2^2}{3.0 \times 4} + \frac{0.5 \times 124.3 \times 25.73 \times \pi \times 1.2}{1.5} = 480 + 4019 = 4499 \text{ kN}$$

∴ length of pile is 1.0 + 25.7 = 26.7 m, say 27 m

Worked Example 10.3 *Short driven pile in clay*
A 250 mm square concrete pile is driven 2.5 m into a very stiff clay with undrained shear strength of 150 kN/m².
Determine the working load of the pile assuming an overall factor of safety of 2.0 for:
a) soft clay overlying
b) sand overlying.
For both cases the base resistance can be calculated from
$Q_b = 9 \times 150 \times 0.25^2 = 84.4$ kN

a) Soft clay overlying
From Figure 10.3, $\alpha = 0.25$
$Q_s = 0.25 \times 150 \times 4 \times 0.25 \times 2.5 = 93.8$ kN

$$\text{Working load} = \frac{84.4 + 93.8}{2} = 89 \text{ kN}$$

b) Sand overlying
From Figure 10.3, $\alpha = 1.0$
adhesion = $1.0 \times 150 = 150$ kN/m²
Assuming a maximum adhesion of 100 kN/m²
$Q_s = 100 \times 4 \times 0.25 \times 2.5 = 250$ kN

$$\text{Working load} = \frac{84.4 + 250}{2} = 167 \text{ kN}$$

Worked Example 10.4 *Long driven pile in clay*
A closed end steel tubular pile, 0.9 m diameter, is driven into a stiff clay with a penetration of 50 m. The undrained shear strength of the clay is 120 kN/m² and the submerged unit weight is 11 kN/m³. Determine the working load assuming an overall factor of safety of 2.0.
Base load is $9 \times 120 \times \pi \times 0.9^2/4 = 687$ kN
a) Using the λ method
Assuming a water table at ground level or submerged conditions
mean effective stress = $25 \times 11 = 275$ kN/m²
From Figure 10.5, $\lambda = 0.13$
$c_a = 0.13\,(275 + 2 \times 120) = 67$ kN/m²
Shaft load = $67 \times \pi \times 0.9 \times 50 = 9472$ kN

$$\text{Working load} = \frac{687 + 9472}{2.0} = 5080 \text{ kN}$$

b) Using the $\frac{c_u}{\sigma_v'}$ ratio method:

$$\frac{c_u}{\sigma_v'} = \frac{120}{275} = 0.44 \quad \frac{L}{d} = \frac{50}{0.9} = 55.6$$

From Figure 10.6 $\alpha_p = 0.82$ and $F = 0.95$
Shaft load = $0.95 \times 0.82 \times 120 \times \pi \times 0.9 \times 50 = 13215$ kN

$$\text{Working load} = \frac{687 + 13\,215}{2.0} = 6951 \text{ kN}$$

Worked Example 10.5 *Driven pile in sand*

A concrete pile, 350 mm square, is to be driven into a thick deposit of medium dense sand with a SPT 'N' value of 20 and a bulk unit weight of 19 kN/m³. The water table lies at 2.0 m below ground level. Estimate the length of pile required to support a working load of 450 kN assuming an overall factor of safety of 2.5.

From Figure 10.10 an estimate of the initial angle of friction is 33°

From Table 10.1 the angle of friction after driving can be obtained

$$\phi' = \frac{33+40}{2} = 36.5° \quad \text{or} \quad \phi' = \frac{3}{4} \times 33 = 34.75°$$

From Figure 10.9, using Meyerhof's relationship a critical depth at $\frac{z_c}{d} = 10.5$ is indicated

$z_c = 10.5 \times 0.35 = 3.7$ m

depth, m	σ_v' kN/m²
0	0
2.0	38.0
3.7	53.6

From Figure 10.7, $N_q = 95$

$q_b = 95 \times 53.6 = 5092$ kN/m²

$$\text{Base load (factored)} = \frac{5092 \times 0.35^2}{2.5} = 249.5 \text{ kN}$$

∴ pile must be longer than the critical depth

From Figure 10.11, $K_s \tan\delta = 0.43$

Shaft load from ground level to 2.0 m = Q_{s1}

$$Q_{s1} = \frac{38.0 \times 0.43 \times 2.0 \times 0.5 \times 0.35 \times 4}{2.5} = 9.2 \text{ kN}$$

Shaft load from 2.0 to 3.7 m = Q_{s2}

$$Q_{s2} = \frac{(38.0 + 53.6) \times 0.5 \times 0.43 \times 1.7 \times 0.35 \times 4}{2.5} = 18.7 \text{ kN}$$

Shaft load required below 3.7 m = Q_{s3}

$Q_{s3} = 450 - (249.5 + 9.2 + 18.7) = 172.6$ kN

Length of pile below 3.7 m = L

$$L = \frac{172.6 \times 2.5}{53.6 \times 0.43 \times 0.35 \times 4} = 13.4 \text{ m}$$

Pile length = 3.7 + 13.4 = 17.1 m, say 17.5 m

Exercises

10.1 A bored pile, 750 mm diameter and 12 m long is to be installed on a site where two layers of clay exist. The upper firm clay layer is 8 m thick and has an undrained shear strength of 50 kN/m². The lower stiff clay layer is 12 m thick and has an undrained shear strength of 120 kN/m². Determine the working load the pile could support assuming the following:
a) $\alpha = 1.0$ for firm clay and 0.5 for stiff clay $N_c = 9$ $\omega = 1.0$.
b) Factors of safety of 1.5 and 3.0 are applied to the shaft load and base load, respectively.
c) The top 1 m of the firm clay is ignored due to shrinkage.

10.2 For the ground conditions and assumptions described in Exercise 10.1 determine the length of pile required to support a working load of 1500 kN.

10.3 A bored pile, 900 mm diameter and 15 m long, is to be installed in a fissured clay whose strength increases with depth from 75 kN/m² at ground level and at 4 kN/m² per metre of depth. Determine the working load the pile could support assuming the following:
a) $\alpha = 0.45$ and $\omega = 0.8$
b) $F = 1.5$ for the shaft load and $F = 3$ for the base load
c) the top 1 m is ignored.

10.4 A large diameter under-reamed bored pile is to be installed in stiff clay with undrained shear strength of 125 kN/m². The main shaft of the pile is 1.5 m diameter and the base of the under-ream is 4.5 m diameter with a height of 3.0 m and the total length of the pile from ground level to the base of the under-ream is 27 m. Determine the working load the pile could support assuming the following:
a) $\alpha = 0.3$ $N_c = 9$ $\omega = 1.0$.
b) Adhesion should be ignored over the height of the under-ream, over the main shaft of the pile up to 2 shaft diameters above the top of the under-ream and over the top 1 m of the pile.
c) A factor of safety of 3.0 should be applied to the base load but full mobilisation of shaft adhesion can be assumed.

10.5 A closed end pipe pile, 1.2 m diameter is to be installed offshore with a penetration of 40 m into the stiff clay below the water. The undrained shear strength of the clay is 140 kN/m² and its submerged unit weight is 11.5 kN/m³. Determine the working load the pile could support with an overall factor of safety of 2.5 using:
a) the λ method,
b) the c_u/σ_v' method.

10.6 A jacked pile, 0.35 m diameter and 9 m long is to be installed into a dense 'dry' sand with an initial angle of friction of 40º and bulk unit weight of 20.5 kN/m³. The water table lies well below the base of the pile. Assuming the critical depth as given by Meyerhof's curve in Figure 10.9, estimate the working load the pile could support with an overall factor of safety of 2.5.

10.7 A cylindrical pile, 450 mm diameter and 15 m long is to be driven into a medium dense sand with a water table at 1.5 m below ground level. The initial angle of internal friction of the sand is 34º and its bulk unit weight is 19.5 kN/m³. Assuming the critical depth as given by Meyerhof's curve in Figure 10.9, estimate the working load the pile could support with an overall factor of safety of 2.5.

11 Lateral Earth Pressure and Design of Retaining Structures

Lateral Earth Pressure

Introduction

The pressure in the pore water in a fully saturated soil is hydrostatic, i.e. $u_H = u_V = u$ so there is only one value. The pressure within the mineral grain structure (effective stress) is not the same in all directions. The vertical effective stress in a soil can be obtained from simple considerations of depth multiplied by the bulk or submerged unit weight and is treated as a principal stress, $\sigma_V{}'$. For the design of vertical walls the horizontal stress acting $\sigma_H{}'$ is required and a coefficient of earth pressure, K is used to relate the two stresses:

$$K = \frac{\sigma_H{}'}{\sigma_V{}'} \tag{11.1}$$

Various factors affect the horizontal stress acting on a wall, but initially the assumption of a smooth wall is considered. The amount and type of movement of the wall has a major effect on the horizontal stresses developed, as described below.

Effect of horizontal movement

1 ***None*** *– 'at rest' condition*

Consider an element of soil in the ground which is at equilibrium with no movement of any kind. There will be a vertical effective stress $\sigma_V{}'$ and a different horizontal effective stress $\sigma_H{}'$ which are both principal stresses and can, therefore, be represented on the Mohr circle diagram (Figure 4.10). The soil is obviously not in a state of failure and the ratio of the stresses is given by the coefficient of earth pressure 'at rest' K_o (equation 4.2). The values and variation of K_o are discussed in Chapter 4.

2 ***Horizontal expansion*** *– active pressure (Rankine Theory) (Figures 11.1 and 11.2)*

This theory considers the ratio of the two principal stresses when the soil is brought to a state of shear failure throughout its mass (plastic equilibrium).

The vertical effective stress $\sigma_V{}'$ in the ground will remain constant and since it will be the larger value it will be the major principal stress. As horizontal expansion of the soil located behind the wall occurs when the wall moves away from the soil the horizontal stress on the wall decreases and more of the strength of the soil is mobilised. When the failure strength of the soil is mobilised the minimum horizontal stress is termed the active pressure, p_a, and will represent the minor principal stress.

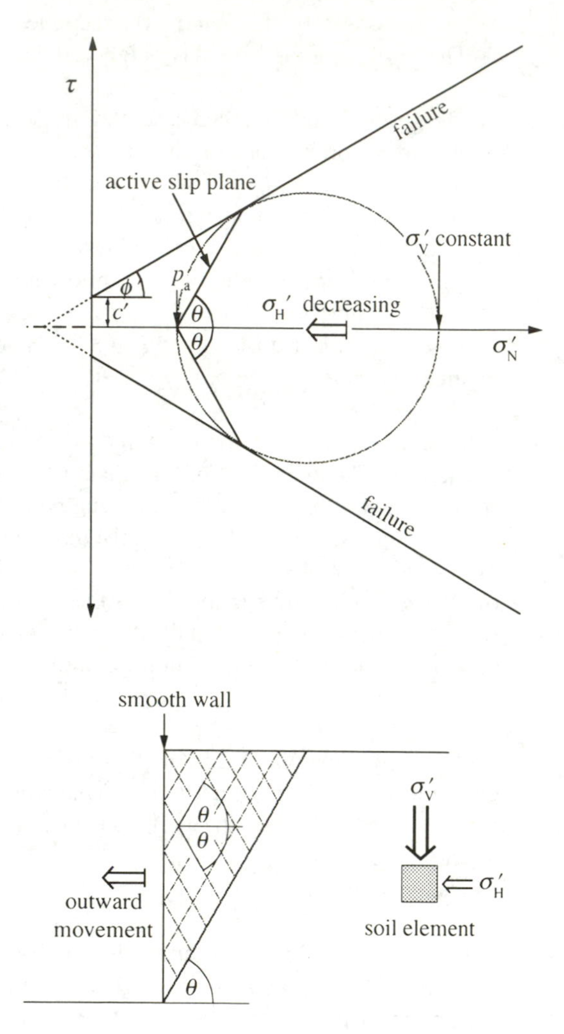

Figure 11.1 *Active Rankine state*

This state can be represented by a Mohr circle (Figure 11.1) which touches the failure envelope. Shear failure will occur at angles θ to the major principal plane so that a network of shear planes will form at angles θ to the horizontal behind the wall where

$$\theta = 45° + \frac{\phi'}{2} \tag{11.2}$$

The horizontal stress (active pressure p_a) can be obtained in terms of the vertical stress from the geometry of the Mohr-Coulomb failure envelope (Figure 11.2) given by:

$$\text{minimum } \sigma_H{}' = p_a = \sigma_V{}' K_a - 2c'\sqrt{K_a} \tag{11.3}$$

where

$$K_a = \frac{1-\sin\phi'}{1+\sin\phi'} = \tan^2\left(45° - \frac{\phi'}{2}\right) \tag{11.4}$$

This 'local' Rankine state of stress will only occur within a wedge defined by q to the horizontal. The soil outside this wedge is considered to be unstrained.

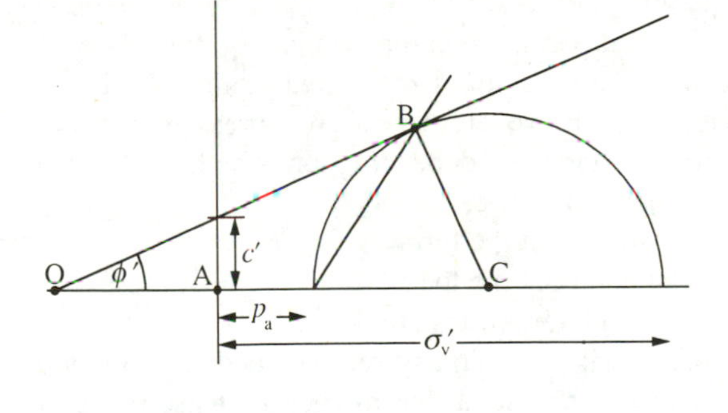

$$\sigma_v' + OA = OC + BC = OC(1 + \sin\phi')$$

$$p_a + OA = OC - BC = OC(1 - \sin\phi')$$

$$\therefore \quad \frac{\sigma_v' + OA}{p_a + OA} = \frac{1 + \sin\phi'}{1 - \sin\phi'}$$

$$OA = \frac{c'}{\tan\phi'}$$

giving

$$p_a = \sigma_v'\left(\frac{1 - \sin\phi'}{1 + \sin\phi'}\right) - 2c'\sqrt{\frac{1 - \sin\phi'}{1 + \sin\phi'}}$$

Figure 11.2 *Active Rankine pressure*

3 *Horizontal compression* – *passive pressure (Rankine Theory) (Figure 11.3)*

Consider a wall moving (or being pushed) towards the soil behind. The vertical effective stress $\sigma_V{}'$ in the ground will remain constant but the horizontal stress $\sigma_H{}'$ must increase until the soil behind the wall is brought to the state of plastic equilibrium. $\sigma_H{}'$ will be greater than $\sigma_V{}'$ so $\sigma_V{}'$ will be the minor principal stress. The maximum horizontal stress required to produce failure of the soil is termed the passive pressure p_p and will represent the major principal stress.

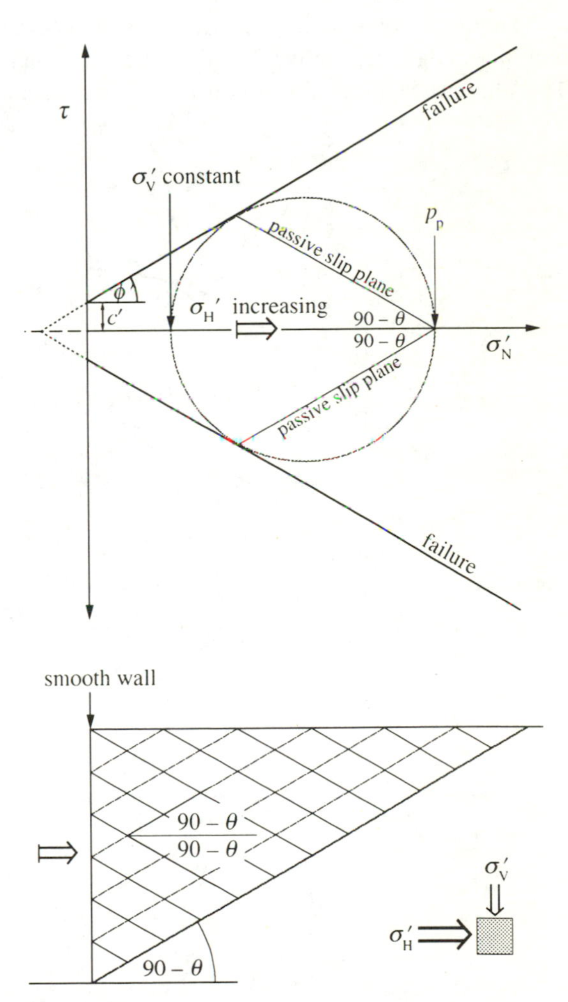

Figure 11.3 *Passive Rankine state*

This state can be represented by another Mohr circle (Figure 11.3) which touches the failure envelope. Shear failure will occur at angles θ to the major principal plane so that a network of shear planes will form at angles θ to the vertical.

The horizontal stress (passive pressure p_p) can be obtained in terms of the vertical stress from the geometry of the Mohr-Coulomb failure envelope given by:

$$\text{maximum } \sigma_H' = p_p = \sigma_V' K_p - 2c'\sqrt{K_p} \qquad (11.5)$$

where

$$K_p = \frac{1+\sin\phi'}{1-\sin\phi'} = \tan^2\left(45° + \frac{\phi'}{2}\right) \qquad (11.6)$$

This 'local' Rankine state of stress will only occur within a wedge defined by 90° – θ to the horizontal. The soil outside this wedge is considered unstrained.

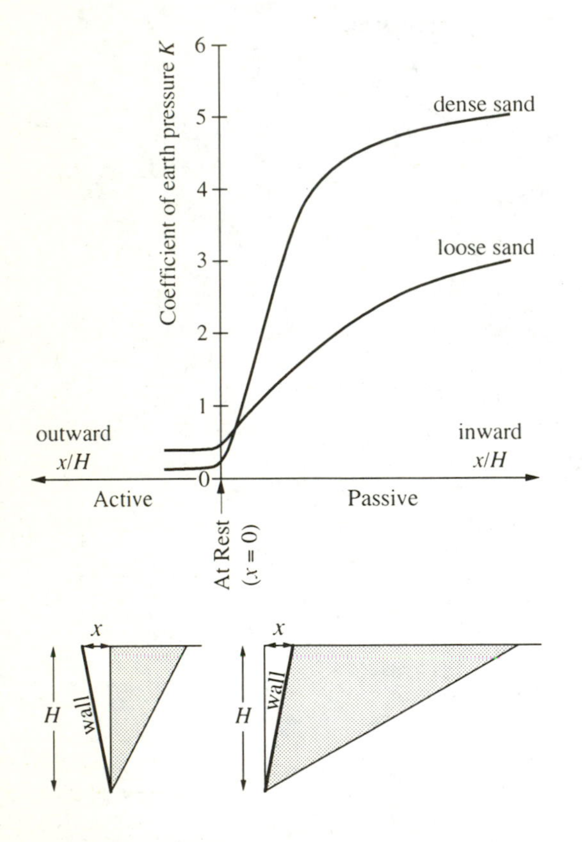

Figure 11.4 *Movements required to mobilise earth pressure*

4 ***Amount of movement required*** *(Figure 11.4)*
Generally, much greater movement of the wall is required to mobilise the full value of the passive pressure compared to the small movements required to mobilise the full value of the active pressure. A typical relationship for sands is illustrated in Figure 11.4 where x represents the inward or outward movement of the wall. It is observed that loose sands provide greater active pressures (overturning) and lower passive pressures (restraint) than dense sands.

Gravity walls, cantilever walls, sheet-pile walls and timbered walls could be considered to yield sufficiently so that the full active pressure is mobilised. The strain required to mobilise the active pressure behind a wall will mobilise only a portion of the passive pressure in front of the wall so the full amount of passive pressure should never be relied on.

With some structures where yielding is restricted, such as bridge abutments, propped or anchored basement walls and rectangular culverts the horizontal pressure acting could be greater than the active pressure and nearer the 'at rest' condition.

5 ***Type of movement*** *(Figure 11.5)*
Equations 11.3 and 11.5 suggest that the active pressure p_a and passive pressure p_p increase linearly with depth as vertical stress σ_V' increases linearly. It has been found, however, that different pressure distributions are obtained depending on whether the wall movement comprises:

- rotation about the top
- rotation about the toe
- uniform lateral translation

Typical variations of pressure developed behind a rigid wall in a dense sand due to each of these types of movements are illustrated in Figure 11.5. These are based on the distributions given in Padfield and Mair (1984) and Anon (1989).

It has also been shown (Anon, 1989) that the full passive and active thrusts are mobilised at small movements for rotation about the top and translation with much larger rotations required (about 2–3 times as much) to fully mobilise these thrusts when rotation about the toe occurs.

Effect of wall flexibility and propping *(Figure 11.6)*
Steel sheet pile walls are more flexible than reinforced concrete cantilever walls, embedded diaphragm or contiguous bored pile walls. If a wall deflects due to

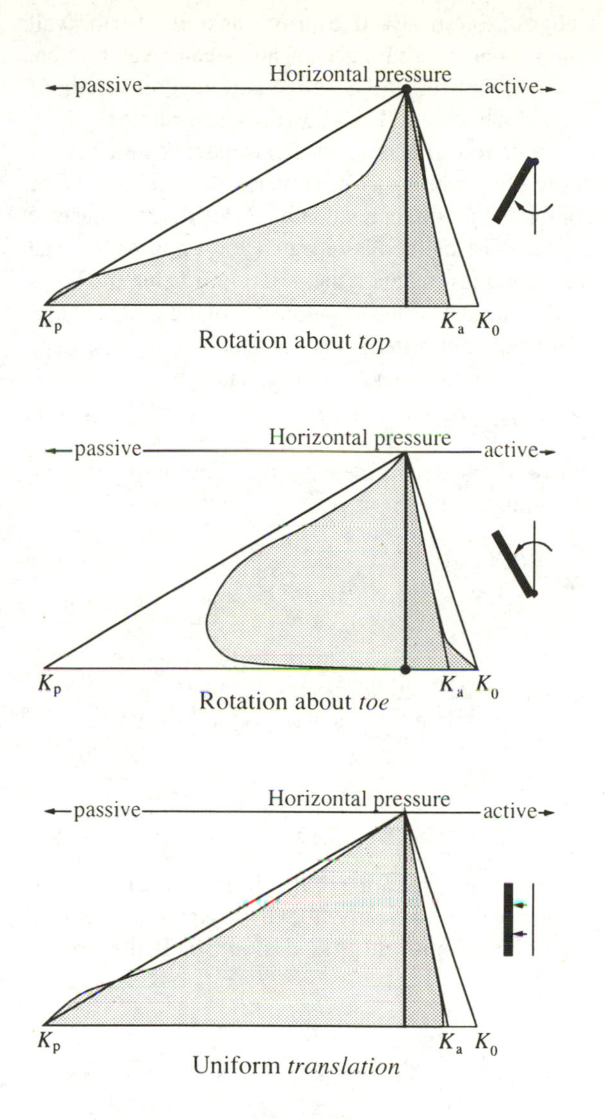

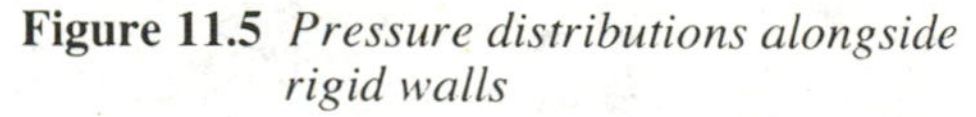
Figure 11.5 *Pressure distributions alongside rigid walls*

lateral stress there is a redistribution of stresses due to stress transfer and arching. This redistribution of stress is greater as the wall deflects more.

If the top of the wall is restrained by a prop, strut or anchor then load is attracted to this area with an increased pressure behind the wall which may reach the passive pressure. As the depth of embedment of the wall increases the bottom of the wall becomes more fixed and rotation is restricted. This fixity is provided by the passive resistance behind the wall at its toe.

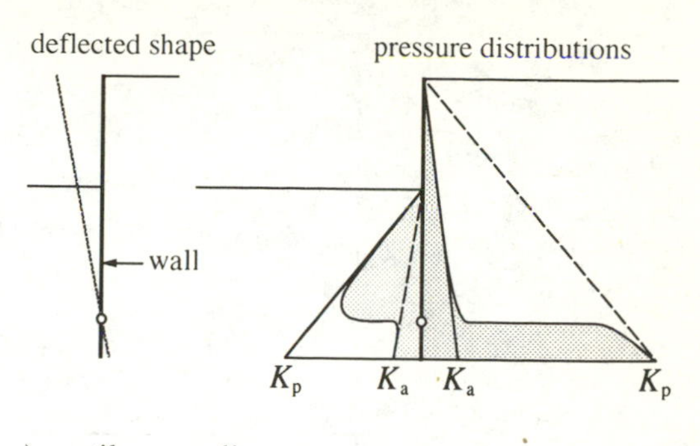

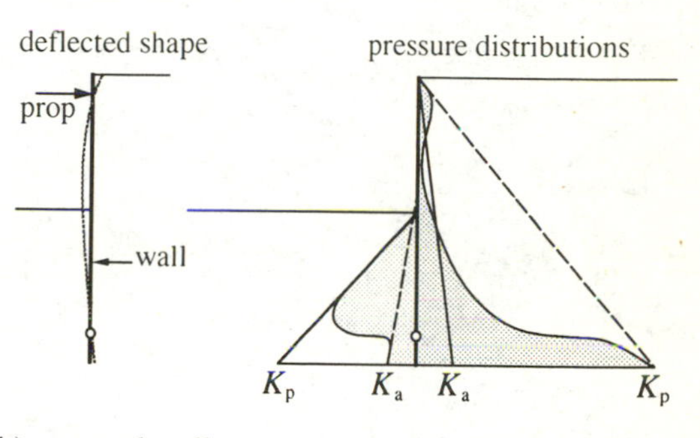

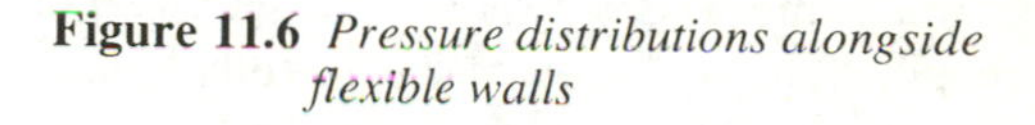
Figure 11.6 *Pressure distributions alongside flexible walls*

Effect of wall friction *(Figure 11.7)*

The Rankine theory assumes that the surface of a wall is smooth but in practice it is rough. If the soil moves downwards or upwards against the wall a shear stress is transmitted producing wall friction $f_s = \sigma_H' \tan\delta$ if it is frictional and adhesion $c_a = c_w$ if it is cohesive.

If the settlement of the wall is negligible but it rotates or moves sideways the active wedge will settle relative to the wall and the passive wedge will rise relative to the wall. The forces applied are then forces P_{an} and P_{pn} normal to the wall and $P_{an}\tan\delta$ acting downwards on the active side and $P_{pn}\tan\delta$ acting upwards on the passive side.

Coulomb theory – active thrust *(Figure 11.8)*

To a certain extent the effects of wall friction, sloping wall and sloping ground surface can be included using the method proposed by Coulomb (1776). A straight trial surface bounding a wedge of soil of weight W is considered as shown in Figure 11.8.

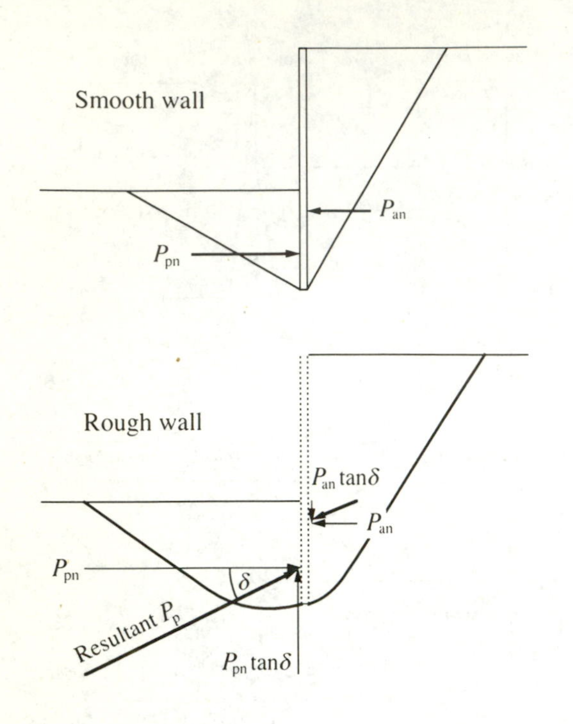

Figure 11.7 *Effect of wall friction*

As the wedge moves downwards due to gravity the shear strength of the soil is assumed to be fully mobilised on the presumed failure plane and wall friction or adhesion is mobilised on the back of the wall. The shear strength and the wall friction act in support of the wedge of soil so the active thrust transmitted to the wall will be smaller for stronger soil and greater wall friction.

From the area of the wedge and the unit weight of the soil the weight W is known. The directions of the resultant forces acting on the wedge, R and P_a, are known so assuming $c' = 0$ the triangle of forces can be completed to obtain a value of P_a for the trial surface chosen. The method is repeated for a number of trial failure planes to obtain the maximum value of P_a.

By considering the trigonometry of the wedge values of P_a and W can be determined as functions of α, β, θ and δ. The maximum value of the resultant P_a is then given by:

$$P_a = \tfrac{1}{2} K_a \gamma H^2 \tag{11.7}$$

where K_a is determined by the Coulomb equation assuming:

$$\frac{\partial P}{\partial \theta} = 0$$

$$K_a = \left(\frac{\sin(\alpha - \phi) / \sin\alpha}{\sqrt{\sin(\alpha + \delta)} + \sqrt{\sin(\phi + \delta)\sin(\phi - \beta) / \sin(\alpha - \beta)}} \right)^2 \tag{11.8}$$

The point of application of the thrust P_a (or P_{an}) can be taken as $^1/_3H$ vertically above the base of the wall assuming a uniform ground slope β. If the ground

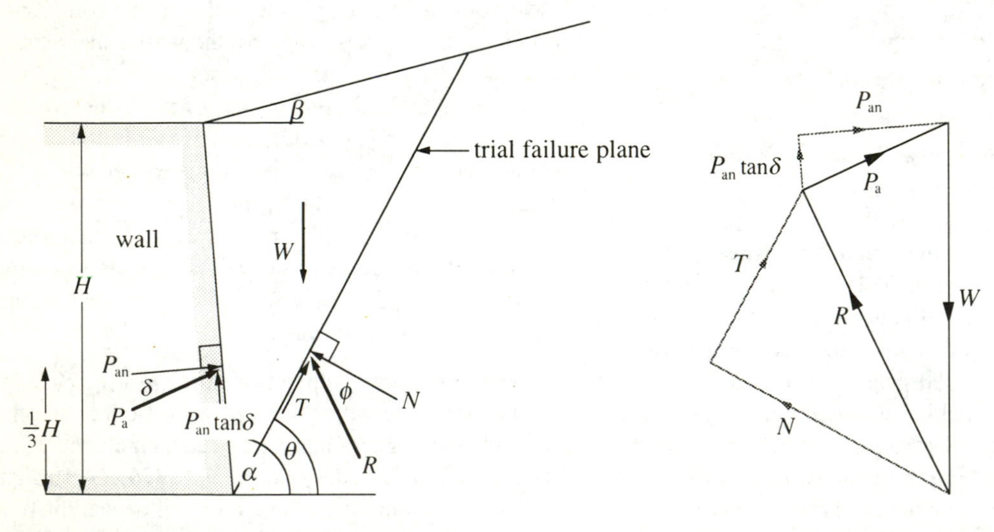

Figure 11.8 *Coulomb theory – active thrust*

surface is irregular then the centre of gravity of the critical failure wedge (giving the maximum thrust) must be determined. The point of application of the active thrust is then given as the point where a line drawn through the centre of gravity of the wedge parallel to the failure plane cuts the back of the wall.

For the case of a smooth, vertical wall ($\delta = 0$, $\alpha = 90°$) and a horizontal soil surface ($\beta = 0°$), Equation 11.8 reduces to the Rankine condition, given by Equation 11.4.

If a water table exists behind the wall then it is likely that seepage will be occurring to a vertical wall drain or a toe drain. To assess the effects of this a flow net should be constructed and the pore pressure distribution determined to obtain the variation of effective stresses. Alternatively, a simple approach not requiring a flow net is to assume that the difference in total head either side of the structure is distributed evenly around the structure. The pore pressures are then obtained from the expression, pressure head = total head – elevation head, as described in Chapter 3.

Coulomb theory – passive thrust *(Figure 11.9)*
Passive thrust is produced on the back of the wall as it is pushed towards the wedge of soil of weight W. Assuming a plane trial surface, the shear strength of the soil on this 'failure' plane is fully mobilised as the wedge is forced upwards and wall friction or adhesion is mobilised on the back of the wall. The shear strength and the wall friction resist upward movement of the wedge so the passive thrust transmitted to the wall will be larger for stronger soil and greater wall friction.

The directions of the resultant forces acting on the wedge, R and the passive thrust P_p are known so assuming $c' = 0$ the triangle of forces can be completed to obtain a value of P_p for the trial surface chosen. The method is repeated for a number of trial failure planes to obtain the minimum value of P_p.

By considering the trigonometry of the wedge values of P_p and W can be determined as functions of α, β, θ and δ. The minimum value of the resultant P_p is then given by:

$$P_p = \tfrac{1}{2} K_p \gamma H^2 \tag{11.9}$$

where K_p is

$$K_p = \left(\frac{\sin(\alpha + \phi) / \sin \alpha}{\sqrt{\sin(\alpha - \delta)} - \sqrt{\sin(\phi + \delta)\sin(\phi + \beta) / \sin(\alpha - \beta)}} \right)^2 \tag{11.10}$$

Limitations of the Coulomb theory *(Figure 11.10)*
The trial failure surfaces are assumed to be planes for both the active and passive cases whereas in practice the actual failure surfaces have curved lower portions due to wall friction. For the active case the error in assuming a plane surface is small and K_a is underestimated slightly.

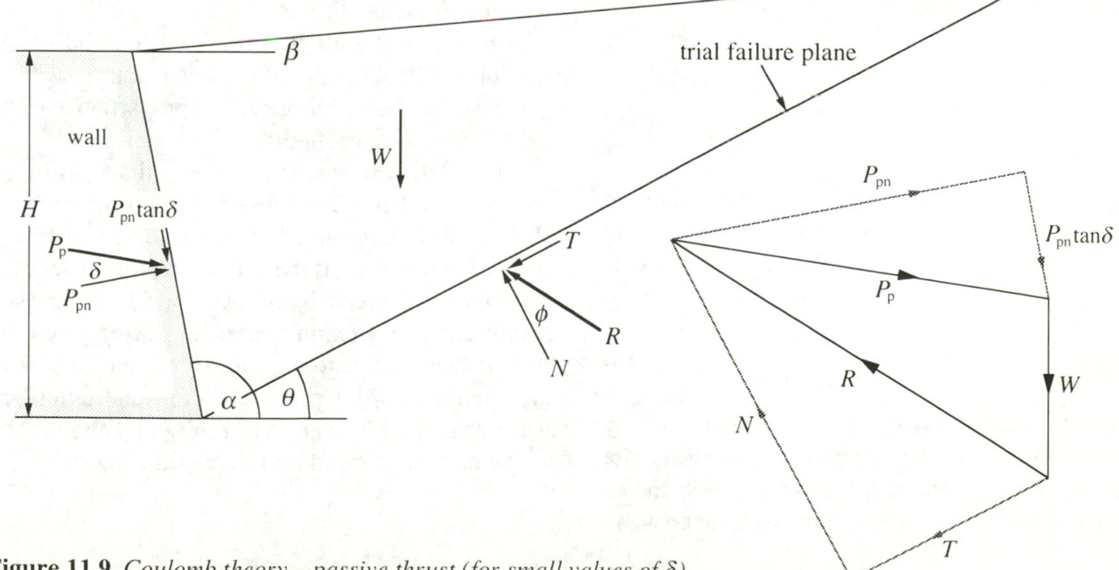

Figure 11.9 *Coulomb theory – passive thrust (for small values of δ)*

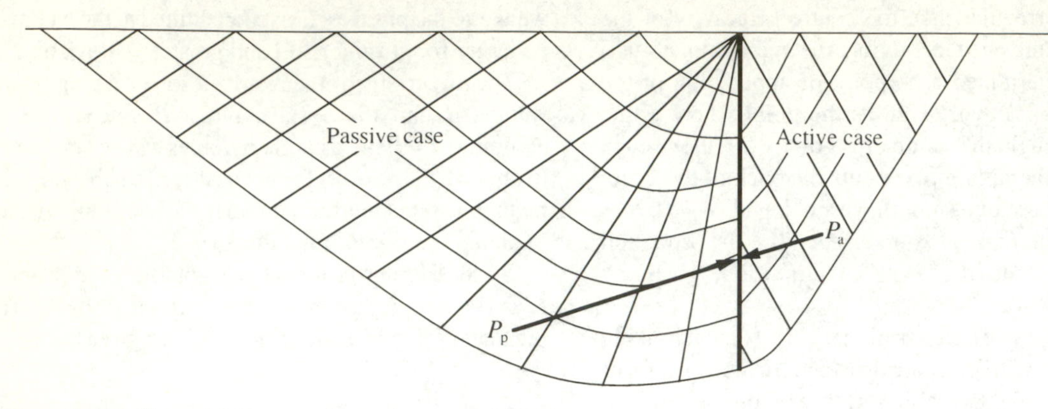

Figure 11.10 *Curved failure surfaces due to wall friction*

For the passive case the error is also small providing wall friction is low, but for values of $\delta > \phi'/3$ the error becomes large with K_p significantly over-estimated. Because of this the approach usually adopted is to use earth pressure coefficients, see below.

Earth pressure coefficients *(Figure 11.11)*
To take account of the effects of wall friction Equations 11.3 and 11.5 have been generalised to:

$$p_{an}' = K_a \sigma_V' - K_{ac} c' \qquad (11.12)$$

$$p_{pn}' = K_p \sigma_V' + K_{pc} c' \qquad (11.13)$$

The coefficients of earth pressure K_a and K_p are given for the horizontal component of pressure p_{an} or force P_{an} so that for active conditions:

$$\text{pressure} \quad p_{an} = p_a \cos\delta \qquad (11.14)$$

$$\text{force} \quad P_{an} = P_a \cos\delta \qquad (11.15)$$

where p_a and P_a are the resultant values of pressure and force acting at an angle δ to the horizontal (for a vertical wall). The shear force acting on the back of the wall is given by

$$P_{an} \tan\delta \qquad (11.16)$$

The coefficients K_a and K_p have been determined by Caquot and Kerisel (1948) assuming the curved failure surface to be a logarithmic spiral. Values of K_a and K_p for a horizontal backfill and vertical wall are given on Figure 11.11. They give the horizontal components of active and passive pressures. Wall friction forces $P_{an} \tan\delta$ or $P_{pn} \tan\delta$ then may occur on the back of the wall.

Values of K_{ac} and K_{pc} can be obtained with sufficient accuracy from the expressions

$$K_{ac} = 2\sqrt{K_a\left(1 - \frac{c_w'}{c'}\right)} \qquad (11.17)$$

$$K_{pc} = 2\sqrt{K_p\left(1 + \frac{c_w'}{c'}\right)} \qquad (11.18)$$

The above expressions will be appropriate for granular soils and overconsolidated clays where the critical condition will be the drained case and effective stress parameters are applicable.

The angle of wall friction δ will depend on the frictional characteristics of the soil and the roughness of the wall and is usually given as a proportion of ϕ', the value δ never exceeding ϕ'.

The relative movement of the wall and soil must also be considered. For active conditions wall friction should only be considered if the soil moves downwards relative to the wall. If the wall also has a tendency to settle then it is safer to ignore wall friction. For passive conditions wall friction can be considered where the wall settles relative to the soil such as a load bearing basement wall. Maximum values of wall friction and wall adhesion c_w' are given in Tables 11.1 and 11.2 for active and passive conditions, respectively.

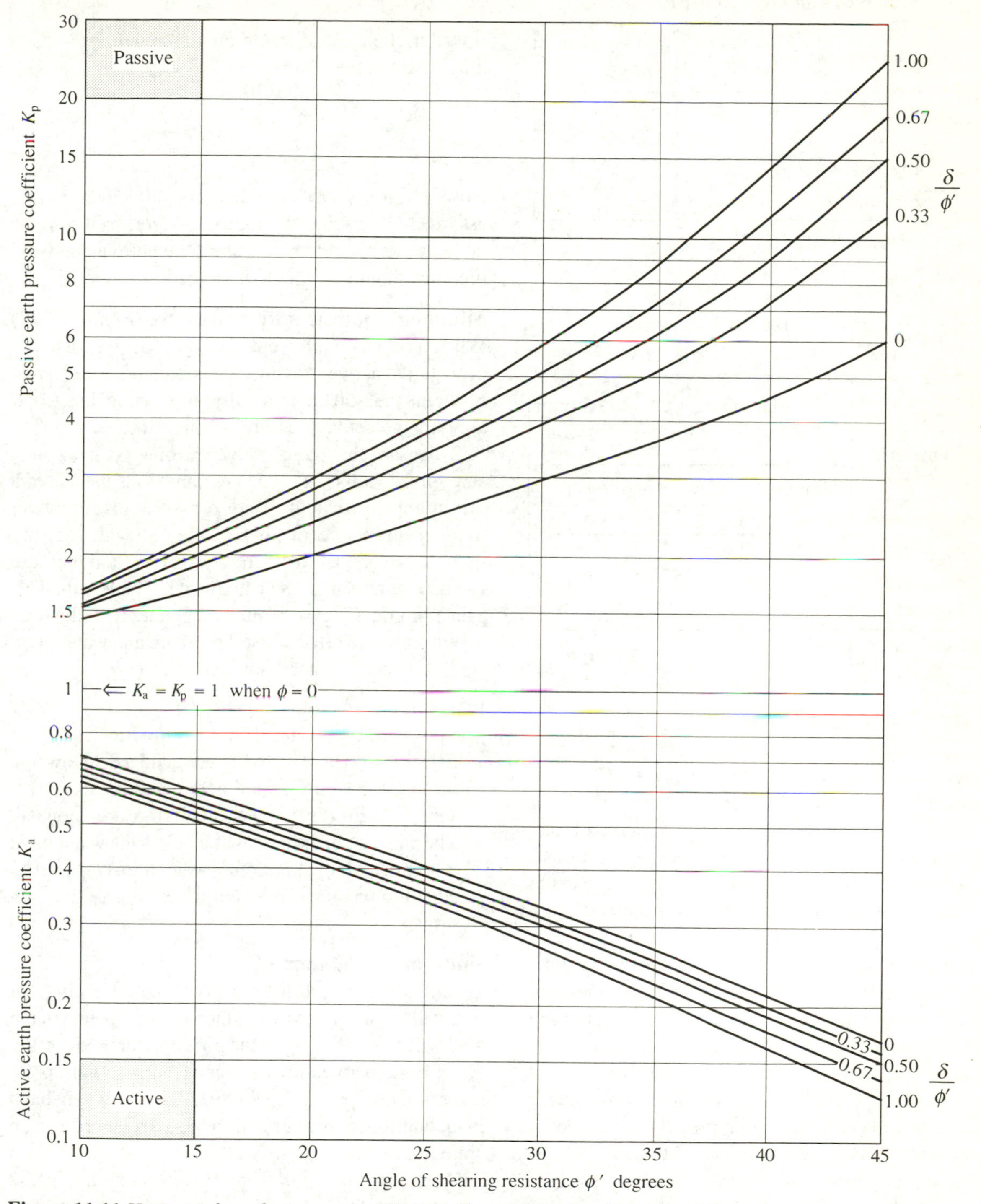

Figure 11.11 *Horizontal earth pressure coefficients K_a and K_p (After Caquot and Kerisel, 1948)*

Table 11.1

Values of δ and c_w' – Active case

Soil type	Maximum δ/ϕ	Maximum c_w'/c'
Granular	2/3	0
Overconsolidated clay	1/2 – 2/3	1/2 – 1

From Draft B.S. 8002 (1987)

Table 11.2

Values of δ and c_w' – Passive case

Wall material	Maximum δ/ϕ		Maximum c_w'/c'
	Granular soil	Overconsolidated clay	
Timber, steel, precast concrete	1/2	1/2	0.5
Cast in situ concrete	2/3	2/3	0.7

From Draft B. S. 8002 (1987)

Effect of cohesion intercept c' *(Figure 11.12)*
Although the value of c' is typically small for overconsolidated clays, it may have a marked effect on the pressures produced with lower active thrusts and greater passive thrusts. On the active side, a depth of theoretical negative pressure is obtained from Equation 11.12. This pressure cannot act in support of the wall so it is presumed to be zero over this depth.

On the passive side, the amount of movement required to fully mobilise passive thrust will be large and the shear strength may have dropped to the critical state value so the use of $c' = 0$ will give the safest, albeit conservative approach. For a normally consolidated uncemented clay and for a compacted clay the cohesion intercept could also be expected to be zero.

The effect of a cohesion intercept is shown in Figure 11.12. For the active case there is a depth of soil z_o over which the active pressure is theoretically negative. This depth is given when $p_{an} = 0$ in Equation 11.12:

$$z_0 = \frac{K_{ac}c'}{\gamma K_a} \tag{11.19}$$

If wall friction and adhesion are ignored this reverts to the Rankine case when:

$$z_0 = \frac{2c'}{\gamma\sqrt{K_a}} \tag{11.20}$$

However, it is considered that the wall should not be assumed to be subjected to no pressure, so if a water table is not present the minimum equivalent fluid pressure should be adopted, see below.

Minimum equivalent fluid pressure *(Figure 11.12)*
When a cohesion intercept c' or c_u is assumed a depth of negative active pressure z_0 is obtained, which may result in the soil theoretically supporting itself and applying no active pressure on the wall.

To ensure that there is always some positive pressure on the wall CP2 (1951) recommends the use of a minimum equivalent fluid pressure given by an 'equivalent fluid' acting behind the wall with a density of 5 kN/m^3 (or 30 lb/ft^3). The equivalent fluid pressure at a depth z m behind the wall must not be less than $5z$. This minimum pressure must be greater than the total active pressure (effective soil pressure and water pressure), otherwise it need not be used.

Effect of water table *(Figure 11.13)*
The presence of a water table has two effects:

1. Effective vertical stresses are reduced below the water table so horizontal active and passive pressures (which are effective stresses) are also reduced.
2. The pressure in the pore water below the water table is hydrostatic so a horizontal water thrust P_w must be added to the horizontal soil thrust to give the total thrust.

Undrained conditions
When a low permeability clay exists behind or in front of a wall the shear stresses induced by movement of the wall will cause changes in the pore pressures within the clay. If the permeability is very low these pore pressures will dissipate only slowly so the clay will behave in an undrained manner and total stress theory can be applied for design of the wall.

This condition could apply for temporary works design where the soil requires support for a short period of time. In the event, however, this time may be very short due to:

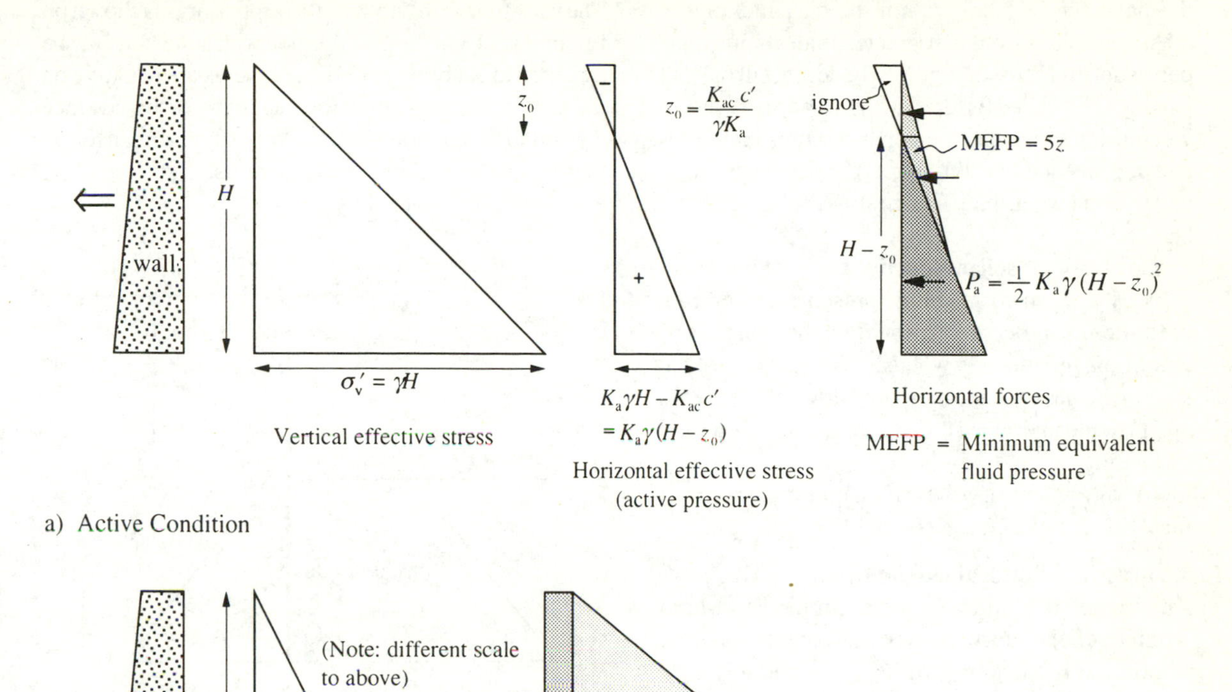

Figure 11.12 *Effect of cohesion intercept* c'

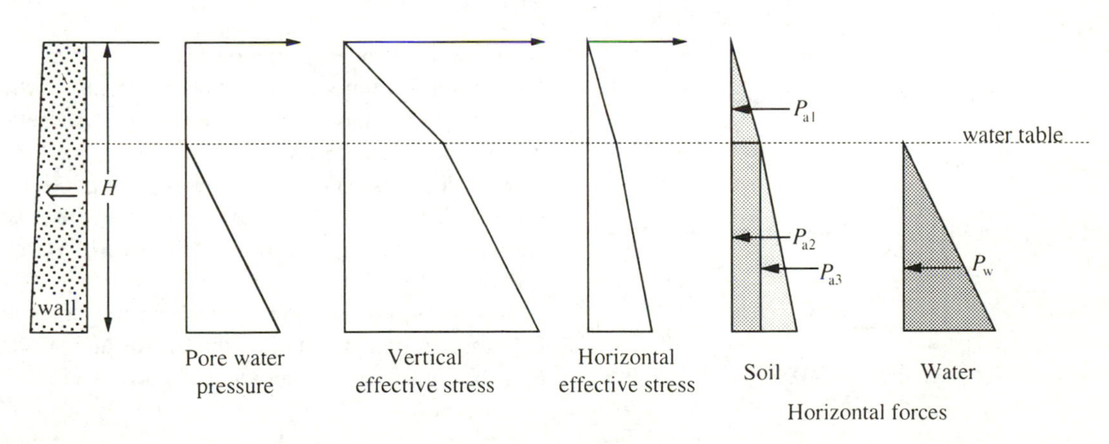

Figure 11.13 *Effect of water table – active case*

- the presence of fabric within the clay making its mass permeability much greater than its intrinsic permeability. If fissures, joints, bedding, silt or sand partings, silt-filled fissures, a higher porosity due to weathering exist then the pore pressures can dissipate rapidly and the 'long-term' condition will soon be obtained when the effective stress approach must be used.
- expansion of the soil in the active state behind a wall is likely to open up any fabric present, accelerating the softening process and providing the 'long-term' condition very quickly whereas compression of the soil on the passive side may slow down this process and provide undrained conditions during the period of loading.
- the development of vertical tension cracks which may fill with water, see below.

Earth pressures–undrained condition *(Figure 11.14)*
The undrained condition will occur in the short-term for a homogeneous intact clay, so this condition is only appropriate for temporary works. It is generally considered that the long-term condition will soon apply so it is safest to assume this latter condition.

Nevertheless, if the appropriate soil parameters, c_u (> 0) and $\phi_u = 0$, are inserted in Equations 11.4 and 11.6 then

$$K_a = K_p = 1 \tag{11.21}$$

and in Equations 11.17 and 11.18:

$$K_{ac} = K_{pc} = 2\sqrt{\left(1+\frac{c_w}{c_u}\right)} \tag{11.22}$$

The vertical stress becomes the total stress

$$\sigma_V = \gamma z$$

so from Equation 11.12 the active normal pressure is:

$$p_{an} = \sigma_V - K_{ac}c_u \tag{11.23}$$

and from Equation 11.13 the passive normal pressure is:

$$p_{pn} = \sigma_V - K_{pc}c_u \tag{11.24}$$

The variation with depth of these pressures is shown on Figure 11.14. On the passive side wall adhesion may be assumed to act but where the pressures are negative on the active side wall adhesion cannot be assumed since the soil in this region is effectively supporting itself in tension as the wall deflects outwards.

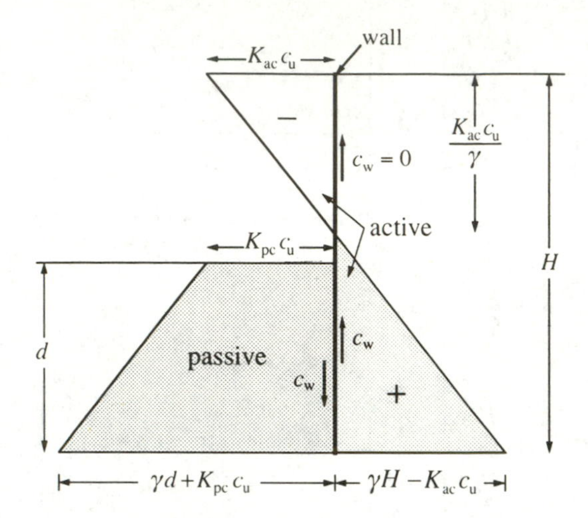

Figure 11.14 *Earth pressures – undrained condition (ϕ_u = 0º)*

Tension cracks *(Figure 11.15)*
On the active side the theoretical pressure is negative down to a depth where $p_{an} = 0$. From Equation 11.23 this depth is given as:

$$z_c = \frac{K_{ac}c_u}{\gamma} \tag{11.25}$$

where K_{ac} is given by Equation 11.22.

As the soil cannot readily support tension, vertical tension cracks may occur down to this theoretical depth. The actual depth to which tension cracks develop is likely to be affected by the support provided to the wall. They are unlikely to extend below the excavation level and may be limited to the level of a strut or anchor.

It is commonly assumed that the tension crack will fill completely with water so the hydrostatic pressure must be considered, as shown in Figure 11.15.

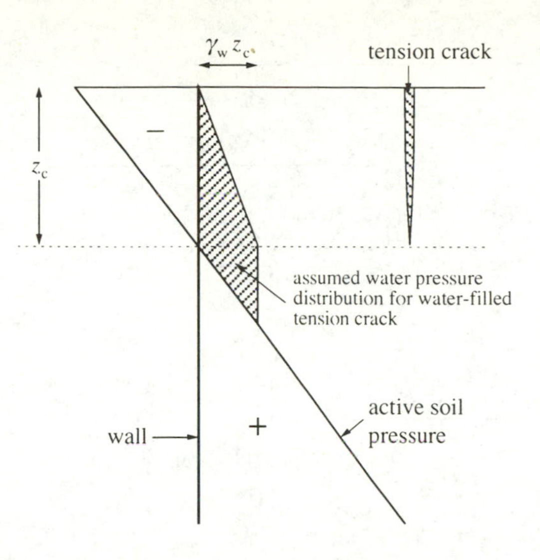

Figure 11.15 *Water-filled tension crack*

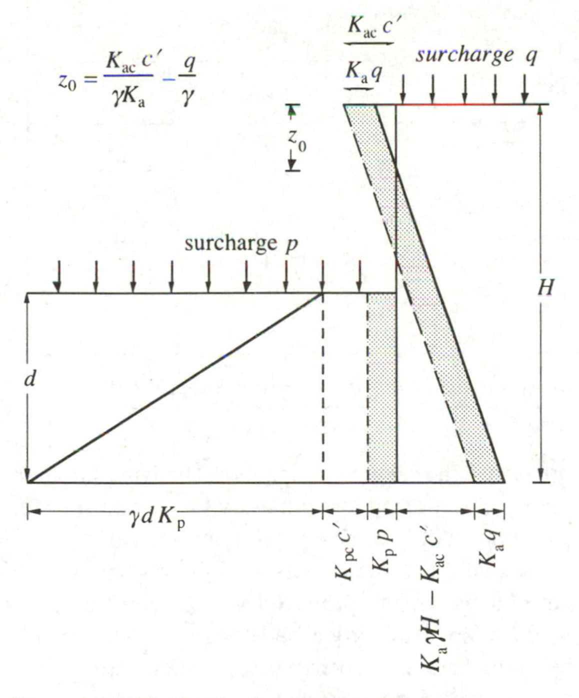

Figure 11.16 *Uniform surcharge*

Loads applied on soil surface

a) uniform surcharge (Figure 11.16)

If a surcharge is applied uniformly over the soil surface on the active or passive side the vertical stresses in Equations 11.12, 11.13 and 11.23 and 11.24 are increased by the surcharge pressure. Equation 11.12 becomes:

$$p_{an}' = K_a(\gamma z + q) - K_{ac}c' \quad (11.26)$$

and Equation 11.23 becomes:

$$p_{an} = (\gamma z + q) - K_{ac}c_u \quad (11.27)$$

The depth of the theoretical negative pressure is then altered, as shown on Figure 11.16.

b) line loads and point loads (Figure 11.17)

These are not usually considered on the passive side. On the active side they will produce an increase in the horizontal pressure acting on the back of the wall. The Boussinesq theory has provided a method for obtaining the horizontal pressure distribution on the back of a wall assuming the soil to be elastic and incompressible. Unfortunately it is neither of these.

The modifications suggested by Terzaghi (1954) have been adopted in the NAVFAC Design Manual (1982) and these are reproduced on Figure 11.17. It is likely that these horizontal pressures are underestimated (Padfield *et al*, 1984) so a conservative approach is suggested.

Retaining structures

Introduction *(Figure 11.18)*

There is a wide variety of structures used to retain soil and/or water for both temporary works and permanent works. Some of the more common types of retaining structures for different purposes are illustrated in Figure 11.18.

Mass concrete or masonry walls rely largely on their massiveness for stability against overturning and sliding. They are unreinforced so their height must be limited to ensure internal stability of the wall in bending and shear when subjected to the lateral stresses.

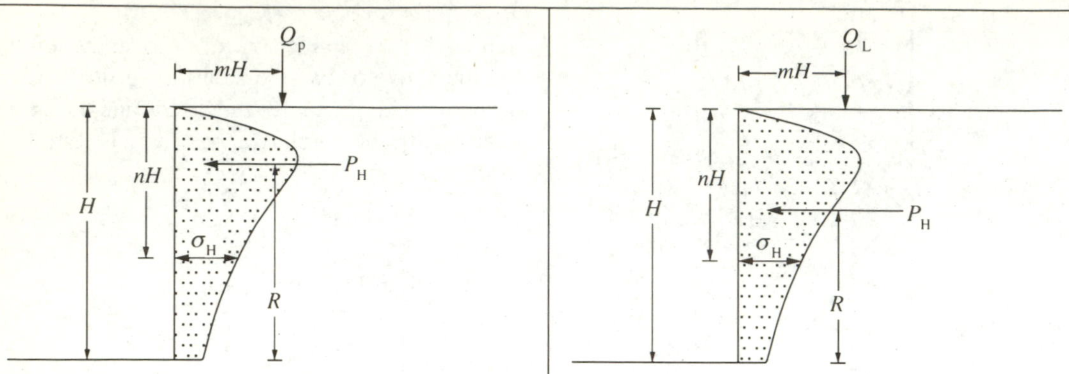

Q_p – point load (kN)

Pressure $\sigma_H = \dfrac{Q_p}{H^2} I_\sigma$

Thrust $P_H = \dfrac{Q_p}{H} I_p$

Q_L – line load (kN/mrun)

Pressure $\sigma_H = \dfrac{Q_L}{H} I_\sigma$

Thrust $P_H = Q_L I_p$

Point Load				Line Load			
m	I_p	R	I_σ	m	R	I_p	I_σ
0.2	0.78	0.59H	For $m \leq 0.4$: $\dfrac{0.28n^2}{(0.16+n^2)^3}$	0.1	0.60H	For $m \leq 0.4$: 0.55	$\dfrac{0.20n}{(0.16+n^2)^2}$
0.4	0.78	0.59H		0.3	0.60H		
0.6	0.45	0.48H	For $m > 0.4$: $\dfrac{1.77m^2n^2}{(m^2+n^2)^3}$	0.5	0.56H	For $m > 0.4$: $\dfrac{0.64}{(m^2+1)}$	$\dfrac{1.28m^2n}{(m^2+n^2)^2}$
				0.7	0.48H		

Figure 11.17 *Horizontal pressures and thrusts on rigid walls due to surface loads (From Navfac, 1982)*

They are typically no more than about 3 m high. Providing a minimum slope of 1:50 (horizontal:vertical) on the front face avoids the illusion of a vertical wall tilting forwards.

Reinforced concrete walls are more economical in concrete, with the reinforcement enabling the stem and base sections to be designed as cantilevered structural elements. Overall stability is provided by the weight of backfill resting on the base slab behind the stem.

Basement walls

Unlike the above wall types which are free-standing, basement walls are restrained by embedment in the ground, a base slab, suspended basement floors and possibly external ground anchors. The latter are more commonly adopted for temporary support during construction, with permanent propping provided by the subsequent basement slab and floor construction. Ground movements produced around deep basements by the removal of vertical and horizontal stresses must be minimised, particularly if there are existing structures nearby. The top-down method of construction (Anon, 1975) has been developed to ensure minimal ground movements.

There are two basic approaches to the construction of a basement:

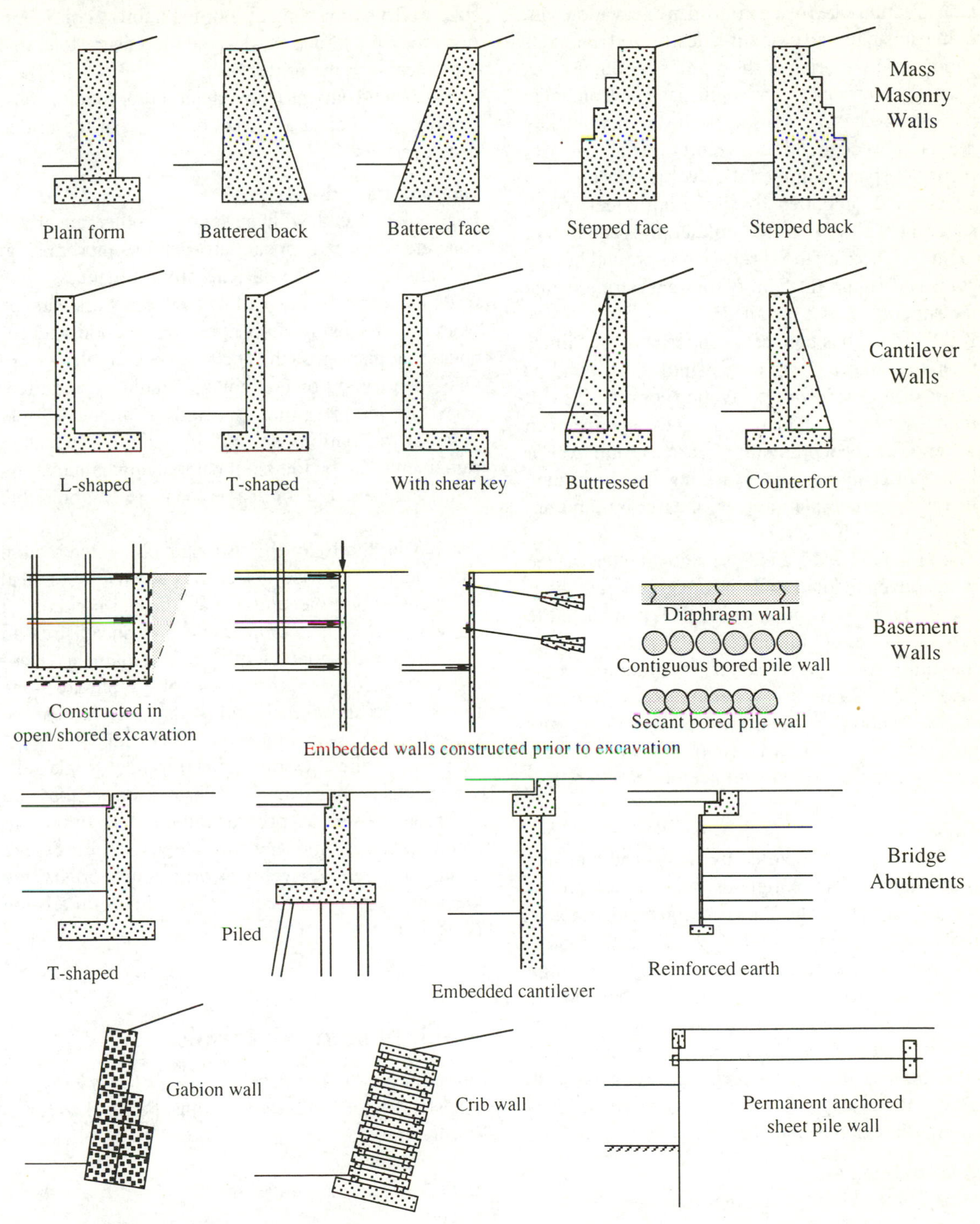

Figure 11.18 *Typical retaining structures*

1 *Backfilled basements*

Construction takes place in an open excavation with either unsupported sloping sides or vertical sides supported by shoring or sheet piling. Sloping sides occupy a large space around the basement and may require dewatering to ensure their stability but they are the most economical method for shallow basements. If space is limited then vertical faces could be cut and supported by timbering, steel trench sheeting or H-section steel soldier piles and timber lagging. These methods rely on the ground having some self-supporting ability for a short time so that the supports can be installed.

If the ground has poor self-supporting capabilities then steel sheet piling driven into the ground as vertical support *before* excavation commences will retain and exclude both soil, groundwater and open water. The sheet piling then acts as a cantilever or is supported by a system of walings and horizontal or raking struts internally or stressed ground anchors externally.

The basement walls and base slab are constructed with conventional in situ reinforced concrete. This should be of good quality and well-compacted to provide a dense, impermeable structure and to maximise resistance to water penetration into the basement. To ensure water-tightness this form of construction can be surrounded by an impermeable membrane such as a layer of asphalt tanking or cardboard panels filled with bentonite.

2 *Embedded walls*

Excavations are supported by reinforced concrete diaphragm walls, contiguous bored pile walls or secant bored pile walls. These are constructed around the basement perimeter *before* excavation commences, occupying minimal space but providing support to the soil and groundwater both in the temporary condition during excavation and construction of the basement and in the permanent condition as the final structural basement wall. They may also provide support to vertical loads such as the external columns and walls of a building.

Bridge abutments

The many types of bridge abutment are well illustrated in Hambly (1979), the more common forms are shown in Figure 11.18. These walls provide support to the retained soil and act as foundations for the bridge deck so apart from providing the normal stability considerations they must also be designed to ensure tolerable settlements for the bridge deck.

Horizontal outward movement and/or rotation must be minimal to ensure correct operation of the bridge deck bearings.

Gabions and cribwork

Even when faced with masonry or other materials, concrete walls can appear hard and uncompromising. Gabions can blend with the environment as they resemble open stone walling and cribwork can be 'softened' by using timber for construction and encouraging plant growth. They are both highly permeable so no additional drainage should be required. They are very flexible, especially gabions, so only nominal foundations are usually required and large settlements can be tolerated without apparent distress so they are suitable for use on the more compressible soils.

They both rely for their strength on the interaction from the tensile properties of the gabion wire or steel mesh cages, and the stretcher and header bond of the cribwork with the compressive and shear strength properties of the contained stone. The main disadvantages are that the wire mesh cages of gabions are prone to corrosion and abrasion although their life can be extended by galvanising and PVC coating.

They require the soil retained to have some self-supporting abilities during construction so they are commonly used to provide additional support to steep cuttings and natural slopes. Gabions are commonly used for river bank protection works where they are easy to construct and provide useful erosion protection.

Stability of gravity walls

Introduction *(Figure 11.19)*

The stability of a gravity wall must be checked for:

- rotational failure
- overturning
- bearing pressure under the toe
- sliding
- internal stability.

Rotational failure
The factor of safety against overall failure along a deep-seated slip surface extending beneath the wall can be obtained using the methods of analysis given in Chapter 12 – Slope Stability. If the wall is associated with loading applied to the ground, such as a wall at the toe of an embankment, then the short-term conditions (for clays the undrained case) will be the more critical. If the wall is constructed within an excavation then the long-term drained condition will be the more critical case.

Adequate drainage measures (permeable blankets, pipes etc.) behind the wall and within the backfill can provide a lower equilibrium phreatic surface. However, the long-term effectiveness of this drainage must not be in doubt.

Considerations of the value of factor of safety to adopt are given in Chapter 12. However, the consequences of failure of a retaining wall are likely to be much more serious than a slope. In Hambly (1979) factors of safety are given on the basis of confidence in the accuracy of soil strength values, i.e.:

$F \geq 1.25$ – for soil strengths based on back analysis of failure of the same type of soil
$F \geq 1.5$ – for soil strengths based on laboratory or *in situ* tests.

Overturning
The factor of safety against overturning about the toe can be obtained from:

$$F = \frac{\sum \text{resisting moments}}{\sum \text{overturning moments}}$$

It is recommended (CP2 : 1951) that a minimum factor of safety of 2 be obtained. Passive resistance in front of the wall is usually ignored because considerable rotation is required before it is fully mobilised and this mode of movement may not achieve the maximum value expected, see Figure 11.5. If a wall is supported at a higher level by a prop, tie or anchor then the reaction force provided at this level may be added to the restraining moments.

Bearing pressure *(Figure 11.20)*
If it is assumed that soil can sustain a linear stress distribution and that it remains elastic, without plastic

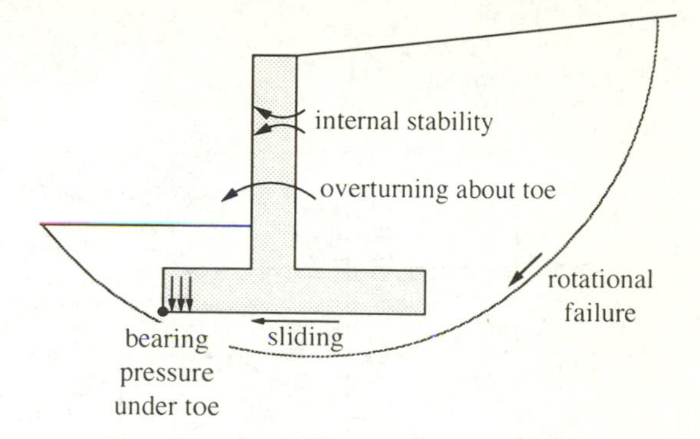

Figure 11.19 *Stability of gravity walls*

yielding, a trapezoidal distribution of pressure can be analysed as in Figure 11.20. The maximum pressure would lie beneath the toe of the wall so two options (Draft BS 8002, 1987) could be considered to satisfy the bearing pressure requirement.

1 Design the wall with a factor of safety against overturning of 2 or more. The maximum bearing pressure, q_{max} should not exceed the allowable bearing pressure of the soil. To ensure this condition or to ensure that 'uplift' or tension at the heel is prevented factors of safety against overturning greater than 2 may be required.

2 Design the wall so that the resultant vertical thrust V lies within the middle third of the base of the wall. In this case, q_{max} will be no more than twice q_{ave}. Hambly (1979) summarises the recommendations of Huntington (1957) that overturning stability should be controlled by keeping the vertical thrust:
a) within the middle third for walls on firm soils
b) within the middle half for walls on rock
c) at or behind the centre of the base for walls on very compressible soils to avoid forward tilting.

The above approaches only consider the effects of eccentric loading and ignore inclined loading. The horizontal load combined with the vertical load produces an inclined resultant applied to the soil.

From Chapter 8 (Shallow Foundations – Stability) a more rational approach is to adopt the effective area (Meyerhof) method to account for eccentric loading and to modify the bearing capacity equation with the not insignificant inclination factors.

a) V inside the middle third

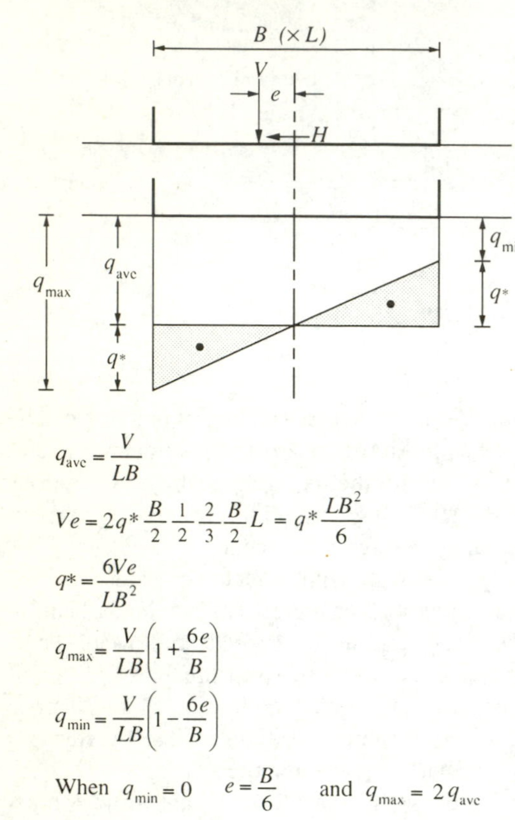

∴ to ensure no 'tension' ($q_{min} < 0$) V must lie within the middle third of the foundation

b) V outside the middle third

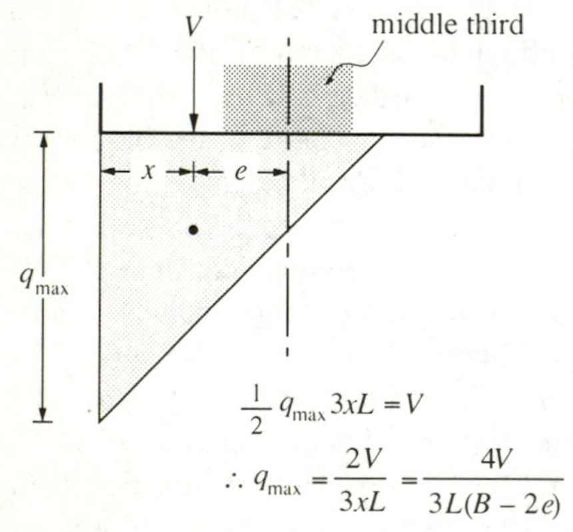

Figure 11.20 *Middle third rule*

Sliding

Excessive horizontal movement of a gravity wall could occur if there is an insufficient factor of safety against sliding. General expressions for the factor of safety are given in Chapter 8, Equations 8.19 to 8.22, and reproduced below.

1 *Granular soils*

$$F = \frac{V \tan \delta}{H_0} \quad \text{(no passive resistance)} \qquad (11.28)$$

$$F = \frac{V \tan \delta + P_p}{H_0} \quad \text{(with passive resistance)} \qquad (11.29)$$

where H_0 is the horizontal load acting at foundation level.

The Draft BS 8002 (1987) suggests values of δ, the angle of 'base' friction and recommended minimum factors of safety which are given in Table 11.3. The ϕ' value is the triaxial peak angle of shearing resistance. The ϕ' value appropriate beneath a wall would be the plane strain value which is somewhat higher than the triaxial value (see Chapter 7). However, using a lower value will compensate for the mobilisation of shear strengths beyond and lower than the peak value as the wall moves forwards. Disturbance of the soil formation level is also likely to reduce the ϕ' values.

There is also some doubt concerning the full mobilisation of any passive resistance from soil in front of the wall as large movements are required. This resistance cannot be relied on if the soil shrinks or is excavated at some time after construction.

Table 11.3

Values of δ and F for base sliding in granular soils

Type of construction	Maximum δ/ϕ'
precast concrete units	2/3
cast in situ concrete	1
Type of soil	**Minimum F**
loose	2.0 to 3.0
dense	1.5 to 2.0

From Draft B. S. 8002 (1987)

The lower recommended factor of safety in Table 11.3 would be appropriate where disturbance is minimised, no passive resistance is assumed and some horizontal movement is tolerable. The higher recommended factor of safety would then apply when disturbance cannot be avoided, passive resistance is assumed and where it is desirable to limit movement.

2 *Cohesive soils*

$$F = \frac{c_b B}{H_0} \quad \text{(no passive resistance)} \qquad (11.30)$$

$$F = \frac{c_b B + P_p}{H_0} \quad \text{(with passive resistance)} \qquad (11.31)$$

where B is the width of the wall base.

The Draft BS 8002 (1987) suggests values of the adhesion c_b at the base of the wall and recommends the minimum factors of safety given in Table 11.4.

Table 11.4

Values of c_b and F for base sliding in clays

Shear strength	Maximum c_b/c_u
$c_u < 40$ kN/m^2	1.0
$c_u > 40$ kN/m^2	0.7
Type of clay	**Minimum F**
Normally and lightly overconsolidated clay	2.0 to 4.0
Overconsolidated clay	1.5 to 2.5
Fissured clays prone to softening	3.0

From Draft B.S. 8002 (1987)

If there is a likelihood that the overturning produces small minimum bearing pressures on the underside of the wall near the heel then it is suggested that the effective width B' obtained from the effective area approach (Figure 8.16) be used in the above equations for F, instead of B.

Internal stability

This is concerned with the structural integrity of the wall itself. Brickwork or masonry walls should be proportioned so that they are not in tension at any point, otherwise, buckling or bursting failures could occur. Mass concrete walls should be proportioned so that the permissible compressive, tensile and shear stresses are not exceeded. Reinforced concrete walls are designed as cantilevered structural elements.

Sheet pile walls

Introduction

These walls may be distinguished from gravity walls in that they are constructed *in situ* prior to excavation, so they support *in situ* soils whereas gravity walls are constructed first and then support backfill. Sheet pile walls are slender structures which means:

1. Their own self-weight is ignored and they do not interact vertically with any soil compared with cantilever gravity walls.
2. They do not require a check for sliding or bearing capacity failure, overturning is the main overall stability consideration.
3. They rely on mobilisation of passive resistance in front of the wall for support below excavation level.
4. Therefore, they must be expected to deflect at least below excavation level.
5. They are commonly propped or anchored over the excavation depth so the pressures which may develop behind a wall will depend on the flexibility of the wall, the amount of support provided and the stage at which it is applied. Actual pressure distributions are, therefore, complex and to some extent dependent on the method of construction.

They should be designed to prevent:

- overall deep-seated rotational failures
- structural failure due to the maximum bending moment or shear force
- excessive deformation
- overturning instability/moment equilibrium.

Only the latter is considered in this chapter.

Cantilever sheet pile wall *(Figure 11.21)*
This type of construction is more commonly used for temporary works for support to the vertical sides of excavations during construction. Following completion of the structure and backfilling they are usually removed. They should be limited to a maximum height of 3–5 m depending on the soil type supported and the presence of water. Deflections and outward movement at the top of the wall may be significant.

For construction in sands an effective stress design approach with full pore pressure conditions must be used. For clays, even though the period of construction may be small, the effective stress condition should be assumed since the equilibration of pore pressures can be quickly achieved because of the expansion of the soil occurring on the active side (tension cracks in the extreme) swelling of the soil on the passive side due to unloading and the presence of macro-fabric such as fissures and laminations which are present in most clays.

For permanent works the 'long-term' effective stress condition is assumed. Notwithstanding these considerations a greater risk is usually taken with temporary works so lower factors of safety are adopted.

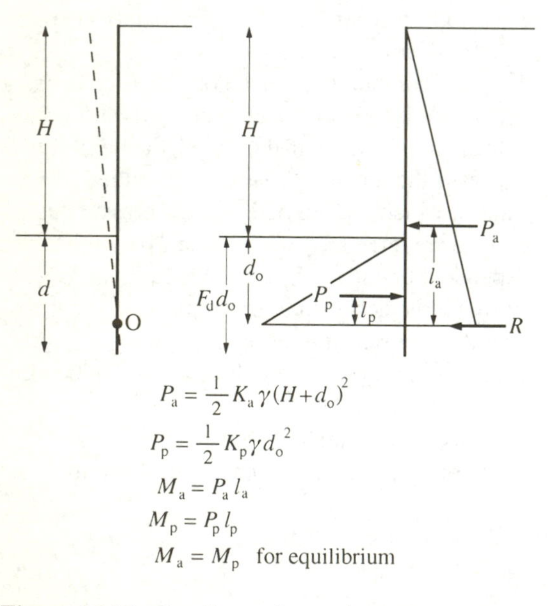

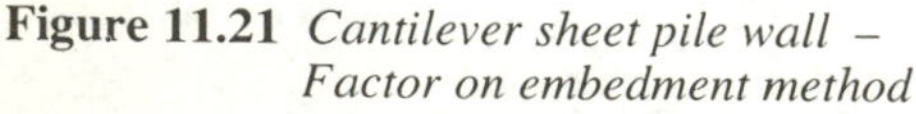
Figure 11.21 *Cantilever sheet pile wall – Factor on embedment method*

The stability of a cantilever wall is derived from the fixity obtained from the embedded portion below excavation level (see Figure 11.6). If the wall rotates about the point O on Figure 11.21 then passive resistance is mobilised in the soil above O on the excavation side and below O on the retained side. Because of the restraint below O this is referred to as the fixed earth condition. To determine the depth of embedment d the passive resistance below O is assumed to be a force R acting at O and moments about O are taken for the active and passive thrusts P_a and P_p.

There are a number of alternative ways of modifying the active and passive thrusts to ensure stability (Padfield and Mair, 1984) but the two most common methods adopted in practice are given below.

Factor on embedment method *(Figure 11.21)*
This method is described in the US Steel Design Manual and the British Steel Corporation Piling Handbook. It assumes that active and passive pressures are fully mobilised above point O (failure condition) and the depth to point O, d_o is obtained by equating moments about O for the full values of active and passive thrusts, P_a and P_p.

The depth of embedment d is then obtained from

$$d = d_o \times F_d \tag{11.32}$$

F_d is not a factor of safety but an empirically determined enlargement factor. Values of F_d are given in Table 11.5 assuming that the values of soil parameters used in the design have been chosen conservatively. Because of the empirical nature of the factor F_d it is recommended in the CIRIA Report 104 that the design should be checked using one of the other methods.

Gross pressure method *(Figure 11.22)*
This method was adopted in CP2:1951 and is recommended in the Draft Revision of CP2 (BS 8002). In this method the depth of embedment, d, is obtained by equating moments about the point O of the full value of the active thrust, P_a, but balanced by a reduced value of P_p given by:

$$\frac{P_p}{F_p} \tag{11.33}$$

Recommended values of F_p are given in Table 11.5. Water pressures and their resultant thrusts are not factored.

Table 11.5

Cantilever sheet piling - Values of F factors (From Padfield and Mair, 1984)

Method	Factor	ϕ'	Temporary works	Permanent works
Factor on embedment	F_d	All values	1.1 to 1.2 (usually 1.2)	1.2 to 1.6 (usually 1.5)
CP2 or Gross pressure	F_p	≤ 20	1.2	1.5
		20 to 30	1.2 - 1.5	1.5 to 2.0
		≥ 30	1.5	2.0

A cubic equation in d is obtained which is solved by substituting trial values of d. This value of d is then increased by 20% to give the full depth of embedment and to ensure that the passive resistance below O represented by the reaction R will be obtained. This increase is not an additional factor of safety.

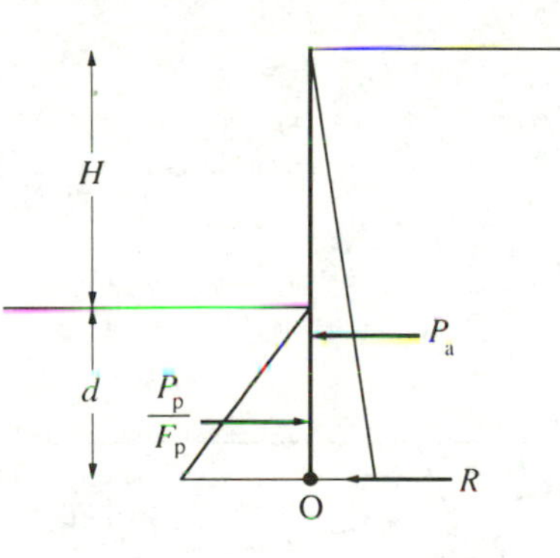

Figure 11.22 *Cantilever sheet pile wall – Gross pressure method*

Single anchor or propped sheet pile wall

A single anchor or prop near the top of the wall prevents outward deflection at this location and modifies the pressures mobilised behind the wall. The flexibility of the slender wall also allows it to deflect further modifying the pressures due to an arching action within the retained soil. The pressure distributions behind such a wall are likely to be complex.

For design purposes a simplified distribution is assumed and the overall stability of the wall is considered by taking moments about the prop level Q (Figure 11.23) of the active and passive thrusts P_a and P_p assuming a free-earth support condition. In this condition the depth of embedment is sufficient to prevent rotation, as in the fixed-earth condition. The 'factor on embedment' method and the 'gross pressure' method are described below, although alternative methods are available (Padfield and Mair, 1984). Because of the uncertainties of relative vertical movements between the wall and the soil, caution should be exercised when assuming values of wall friction δ and wall adhesion c_w.

Factor on embedment method *(Figure 11.23)*

The method assumes that active and passive pressures are fully mobilised above the point O (failure condition) and the depth to point O, d_o is obtained by equating moments about the anchor or prop level Q for the full values of active and passive thrusts, P_a and P_p. The depth of embedment d is then obtained from Equation 11.32. F_d is not a factor of safety, but an empirically determined enlargement factor, values of which are given in Table 11.5. The anchor or prop force, T, is given by:

$$T = (P_a - P_p)s \tag{11.34}$$

where $P_a - P_p$ is a force per metre length of wall (kN/m run) and s is the horizontal spacing between anchors or props.

Gross pressure method *(Figure 11.24)*

In this method the depth of embedment d is obtained by equating moments about Q of the full value of the active thrust P_a and the passive thrust P_p reduced by a

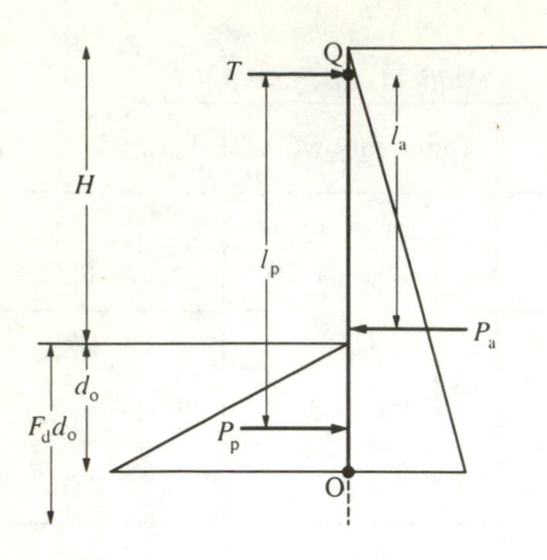

Figure 11.23 *Anchored or propped sheet pile wall – Factor on embedment method*

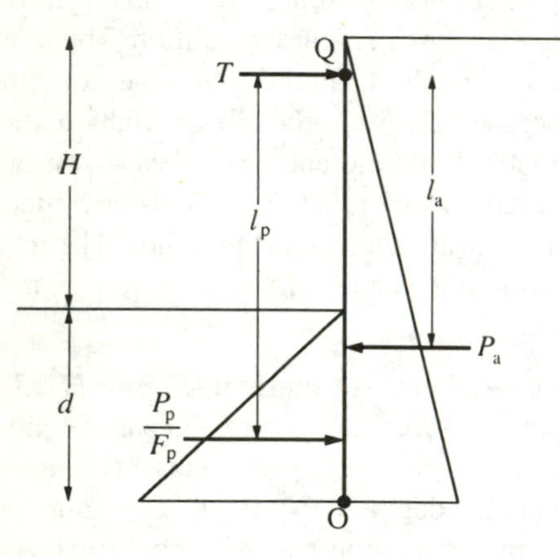

Figure 11.24 *Anchored or propped sheet pile wall – Gross pressure method*

factor of safety F_p. Recommended values of F_p are given in Table 11.5. Water pressures and their resultant thrusts are not factored.

A cubic equations in d is obtained which is solved by substituting trial values of d.

As the free-earth support condition is assumed it is not necessary to increase the depth of embedment by the 20% value adopted for cantilever sheet piling and the fixed-earth condition, see above. However, anchored sheet piling is often used for walls in harbours, river banks and canals where erosion or excessive dredging could reduce the depth of soil on the passive side of the wall so the designed depth of embedment is often increased by say 20% to allow for this.

The anchor or prop force T is given by

$$T = \left(P_a - \frac{P_p}{F_p} \right) s \tag{11.35}$$

where s is the horizontal spacing between the supports.

Anchorages for sheet piling *(Figure 11.25)*
Anchorages are essential in water-front structures. Props or struts inside an excavation provide severe restrictions to the safe and efficient construction operations while anchorages permit an unrestricted excavation. Anchorages, however, affect and occupy the ground around the excavation and behind the waterfront structure so adjacent buildings, services and other works and the rights of adjoining owners must be considered.

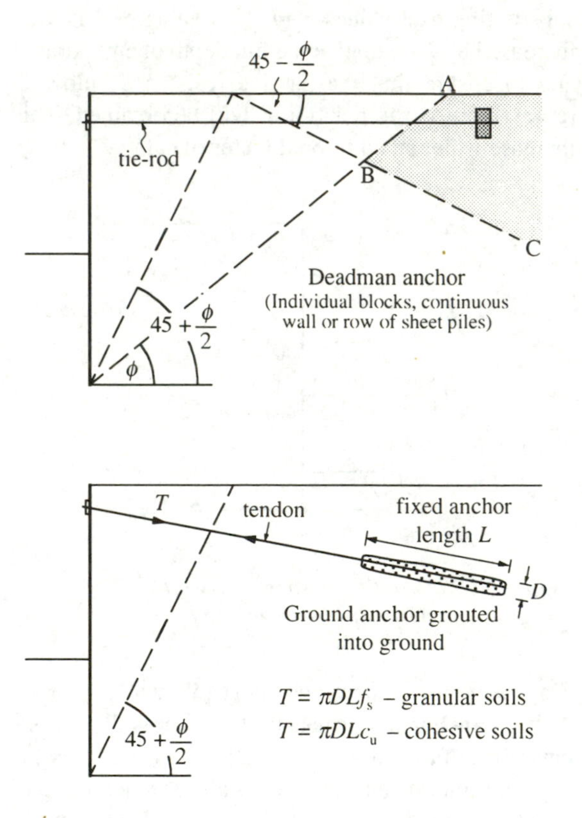

Figure 11.25 *Anchorages for sheet piling*

The most common forms of anchorages for sheet piling are shown in Figure 11.25. Deadman anchors rely on the mobilisation of passive resistance in front of them, so adequate compaction and prevention of disturbance to the backfill in front of the anchorages is essential. They must be placed beyond the lines AB and BC so that the passive restraint mobilised in front of them as they compress the soil is not affected by expansion of the soil in the active wedge behind the wall. If smaller individual anchor blocks are used for each tie rod then increased passive restraint can be expected due to the three-dimensional shear zone in front of the anchor block and shearing resistance on the sides of the block. Deflections at the top of the wall are to be expected before sufficient passive restraint in front of the anchor block can be mobilised.

Ground anchors consisting of corrosion-protected tendons of tie-bars or wire strands are inserted in boreholes drilled from the front face of the wall and bonded into the ground by various grouting techniques depending on the soil type to form a fixed anchor length. The design, construction and testing of ground anchors is described in BS 8081:1989. Their main advantage could be in restricting deflection of the wall and hence minimising both horizontal and vertical ground movements around the excavation. This is achieved by excavating a small depth to the level of the anchor position, installing the anchor, stressing the tendon and locking this force against the wall before continuing further excavation.

Strutted excavations

Introduction *(Figure 11.26)*
For excavations up to 6 m deep, the earth pressures acting on the supports can be affected by many factors, such as shrinkage and swelling of both the soil and the supports, temperature changes, the procedures adopted, materials used and quality of workmanship employed. The design of the supports, therefore, cannot be based on any reliable theory and are empirically chosen based on experience. Useful guidance is given in Tomlinson (1986) and CIRIA Report on Trenching Practice No 97 (1983).

A strutted excavation is constructed by first driving two rows of sheet piling to the full depth required. A small amount of excavation is then carried out and the first frame of walings and struts are fixed, level 1 on Figure 11.26. As excavation proceeds the level 1 strut restricts inward yielding at this level. Struts are progressively fixed as the excavation deepens so the mode of deformation is similar to a wall rotating about its top. For comparison, the likely pressure variation for this condition for a rigid wall is also shown.

Strut loads *(Figure 11.27)*
The design of strutted walls is based on a semi-empirical procedure proposed by Terzaghi and Peck (1967). They determined 'apparent pressure diagrams' which were back-analysed from the strut load measurements taken from various sites. A typical apparent pressure diagram is shown on Figure 11.26. They suggested that trapezoidal pressure envelopes could be

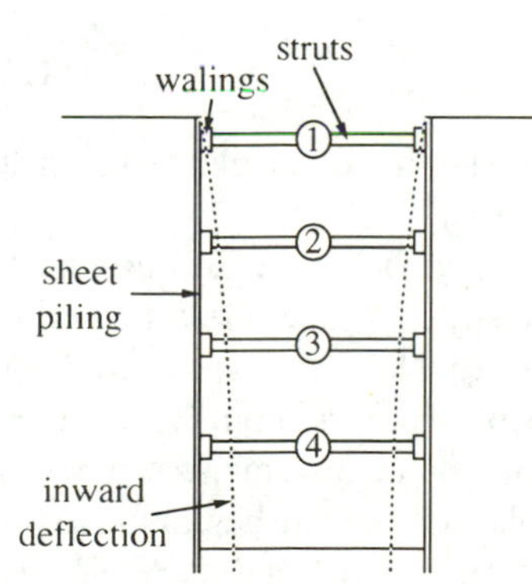

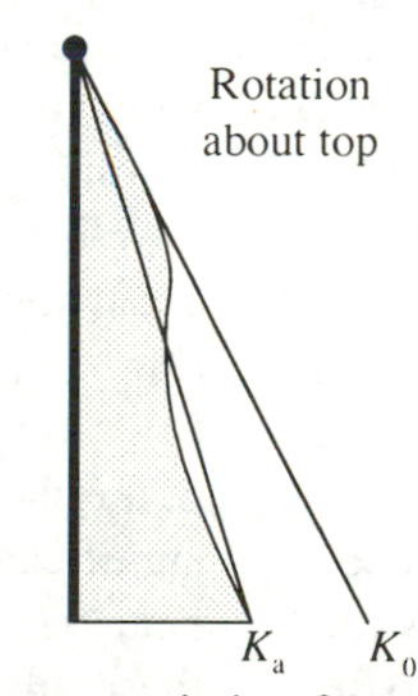

Pressure variation for stiff wall (From Figure 11.5)

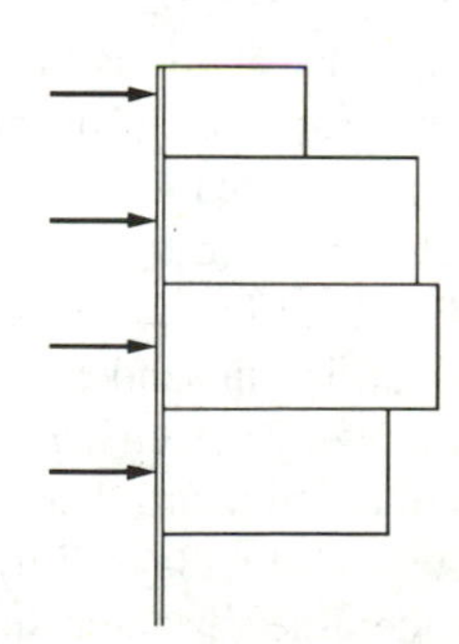

Typical apparent pressure diagram from measured strut loads

Figure 11.26 *Strutted excavations*

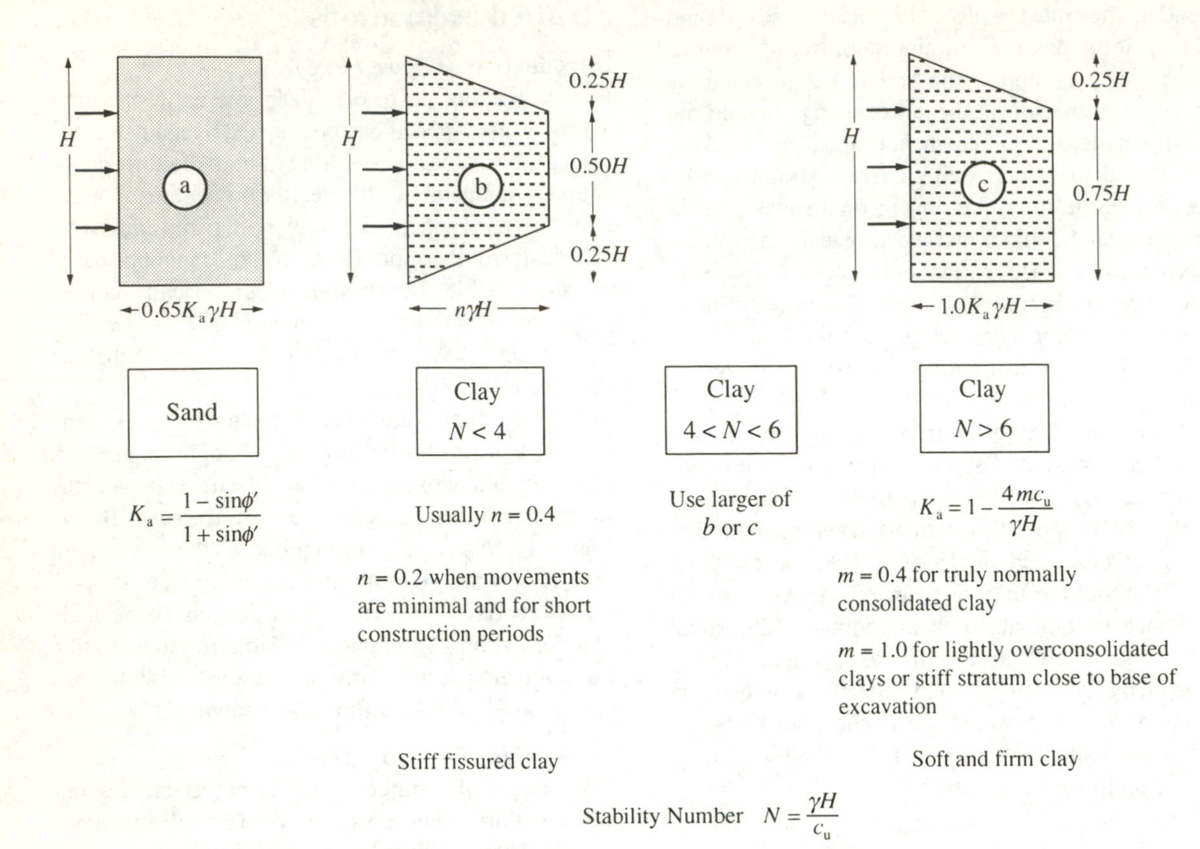

Figure 11.27 *Apparent pressure diagrams for strutted excavations (From Terzaghi and Peck, 1967)*

used to determine the strut loads. These envelopes embraced all of the distributions obtained from the field measurements to ensure that the maximum likely loads in the struts are catered for and that progressive failure should not occur due to one strut failing and shedding excess load onto other struts.

For deep excavations in sands, up to about 12 m deep the pressure envelope given in Figure 11.27, case (a), will give the maximum strut loads.

For deep excavations in clays, Terzaghi and Peck showed that considerable variations in strut loads can be obtained, up to ± 60% from the average load, so the methods proposed should be used with caution. The behaviour of a strutted excavation in clay was found to be dependent on a stability number N:

$$N = \frac{\gamma H}{c_u} \tag{11.36}$$

which is related to the stability of the clay beneath and around the excavation.

When N is less than about 4 the soil around the excavation is still mostly in a state of elastic equilibrium and the pressure envelope in Figure 11.27, case (b), can be used. When N exceeds about 6, movements of the sheet piling and ground movements can become significant because plastic zones are beginning to form near the base of the excavation, and as N increases these plastic zones and the associated movements increase. In this case higher pressures on the sheeting will occur and the pressure envelope given on Figure 11.27, case (c), should be used.

Terzaghi and Peck found that the reduction coefficient, *m*, appears to be 1.0 for most clays, provided a lowest average shear strength from the site investigation results is used. However, they showed that for a truly normally consolidated clay or a soft sensitive clay the value of *m* can be as low as 0.4.

When *N* exceeds about 7 for a long excavation, or about 8 for a circular or square excavation, then complete shear failure, base heaving and extensive collapse of the excavation will be imminent.

Reinforced earth

Introduction

Earth structures on their own are quite weak in tension, relying instead on their compression and shear strength properties for their stability. Inserting tensile reinforcement into soil, in the direction of tensile strains which are usually in the horizontal direction in earth structures, will enable vertical faced masses of soil to remain stable.

The inclusion of reinforcements to improve the stability of soil structures has been practised in simple forms for centuries such as by incorporating fibrous plant and wood materials. More recently metallic or plastic strips, bars and sheets have been used to good effect.

Reinforced earth provides a relatively cheap form of construction for retaining walls, bridge abutments, marine structures, reinforced slopes and embankments because of the speed and simplicity of construction and they can provide an aesthetic appearance. The space required for its construction is minimal so it is useful where land-take is a problem. It is a flexible form of construction in that it can follow curved lines and it can tolerate some settlements so it can be placed on poorer ground.

A reinforced earth wall is constructed using layers of compacted frictional backfill with the reinforcement placed horizontally at suitable vertical intervals and tied to interlocking precast reinforced concrete facing units with a joint filler between the units. The fill material must be frictional and free-draining, so a maximum fines content (less than 63 μm) of 10% is required with a minimum angle of internal friction of 25°. The integrity of the structure is dependent on the long-term durability of the reinforcement so the fill must not have an aggressive nature. Fills with high resistivity, high redox potential, low water content and neutral pH are preferable.

The reinforcement is placed on a compacted soil surface. A variety of reinforcements have been used, the most common consisting of galvanised mild steel strips, plain or ribbed 50 to 100 mm wide and up to 6 mm thick, and polymer geotextiles. As well as sufficient strength and bond the reinforcement must have sufficient tensile stiffness. If large extensions were required in the reinforcement before sufficient tensile force could be mobilised, then the allowable deformations of the soil structure could be exceeded.

The facing units are only intended to provide local support for the backfill to prevent spillage or erosion of the front face. In the original Vidal method, a half-round aluminium, steel or galvanised steel skin section was tied to the reinforcements.

Effects of reinforcement *(Figure 11.28)*

The reinforcement acts within the soil to improve stability by reducing the forces causing failure and increasing the overall shear force resisting failure, as illustrated on Figure 11.28. Behind the vertical face of the wall it is assumed that there is a zone of soil deforming horizontally producing strains sufficient to develop the full active condition in the soil. Internal stability within the structure is then achieved by transferring the horizontal forces in the soil to the reinforcement in the form of a surface friction or bond between the reinforcement and the fill.

Design of reinforced earth walls in the UK is carried out using the Department of Transport Technical Memorandum BE3/78 and a British Standard Code of Practice for the use of strengthened/reinforced soil is currently in preparation. As with reinforced concrete walls both the internal and external stability must be considered.

Internal stability *(Figures 11.29 and 11.30)*

Initially the overall length of reinforcing strips is assumed to be $0.8H$ where H is the full height of reinforced earth.

Assuming no pore pressures are developed as the soil is sheared (drained conditions) no friction on the back of the facing units and a granular backfill the horizontal earth pressure at any depth z is given by:

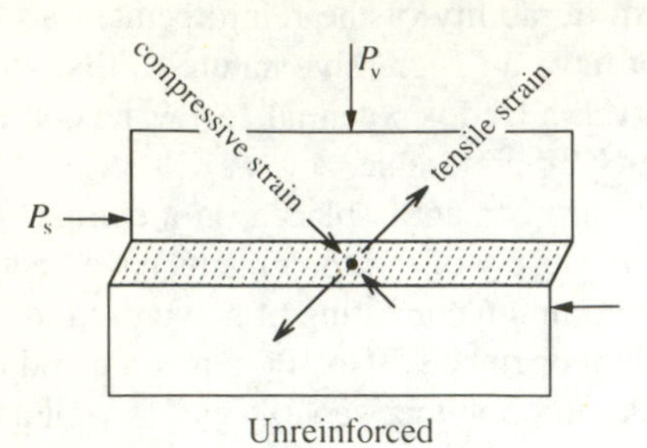

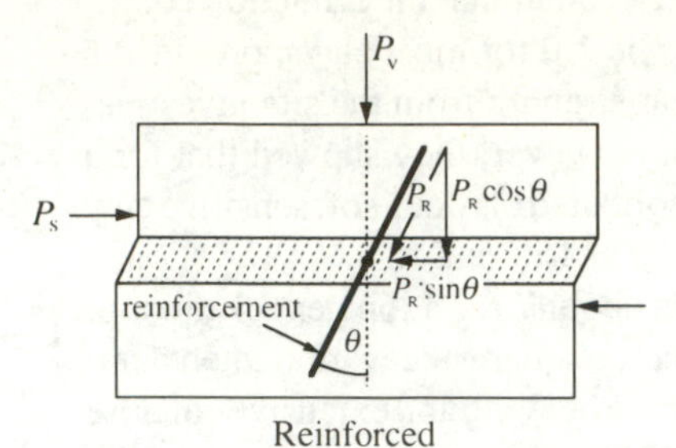

Within the shearing soil :
Force causing failure P_s
Force resisting failure $P_v \tan\phi'$

Within the shearing soil :
Force causing failure P_s
Force resisting failure $(P_v + P_R \cos\theta)\tan\phi' + P_R \sin\theta$

Figure 11.28 *Influence of reinforcement on a shear plane (From Jewell and Wroth, 1987)*

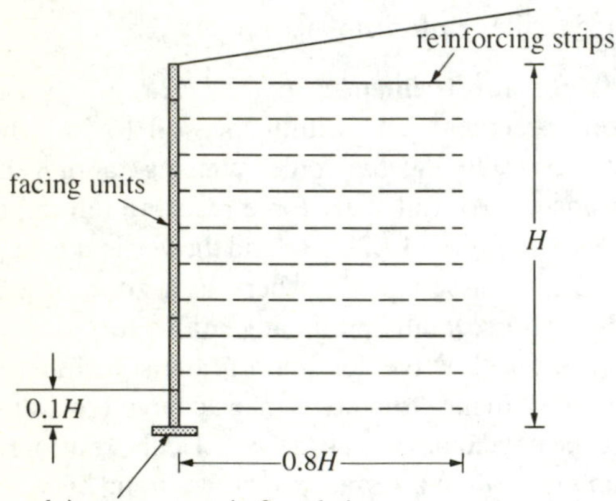

Figure 11.29 *Reinforced earth wall*

$$\sigma_z = \gamma z K_a \quad \text{where } K_a = \frac{1-\sin\phi'}{1+\sin\phi'} \tag{11.37}$$

If reinforcing strips are placed within the soil at vertical spacings of s_v and horizontal spacings s_H then the tensile force in a strip at a depth z is given by:

$$T = K_a \gamma z s_v s_H \left(1 + K_a \frac{z^2}{L^2}\right) \tag{11.38}$$

The expression in the brackets is an additional factor which is necessary because the maximum vertical stress within the reinforced section of the embankment will be greater than the overburden pressure γz due to the overturning effect produced by the active thrust on the reinforced section. This factor assumes a trapezoidal distribution of vertical stress as shown on Figure 11.20.

For strips of width b, thickness t and permissible tensile strength f_t the factor of safety against tensile failure of the strip is given by:

$$F_t = \frac{btf_t}{T} \tag{11.39}$$

For economy, thinner strips could be used at higher levels where the force T is less, although in the upper layers of fill K_o conditions may be found to apply and a minimum thickness for sacrificial corrosion must be maintained.

The maximum tensile force in the reinforcing strips has been found to occur some distance behind the face as a result of an active zone of soil attempting to move outwards resisted by an effective length L_e of reinforcement anchored within a stationary mass of soil by surface friction. The factor of safety against a pull-out failure of the strips is given by:

$$F_p = \frac{R}{T} \tag{11.40}$$

where the pull-out resistance R is provided by surface friction on both sides of the strip for a strip at depth z:

$$R = 2bL_e \gamma z \tan\delta \tag{11.41}$$

Values of the angle of friction δ between granular fill and galvanised steel strip lie between 20° and 25°. The effective length L_e can be found at various depths from:

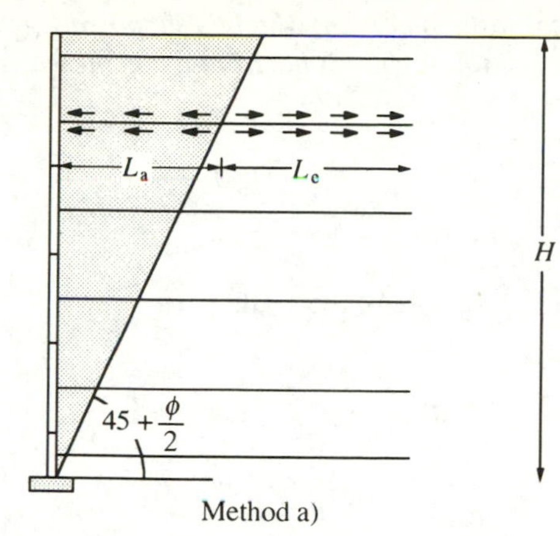

Method a)

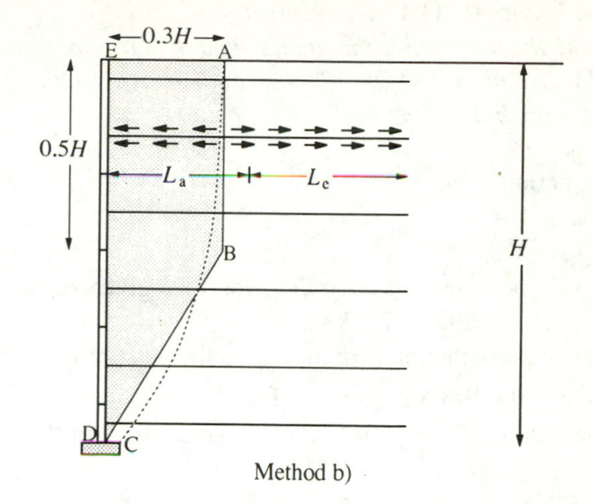

Method b)

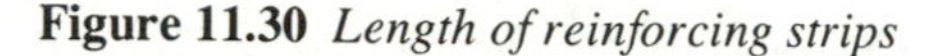

Figure 11.30 *Length of reinforcing strips*

$$L_e = \frac{F_p T}{2b\gamma z \tan\delta} \tag{11.42}$$

with a factor of safety F_p of 2.

The total length of reinforcement L_T is then obtained from the length within the active zone L_a and the effective length L_e:

$$L_T = L_a + L_e \tag{11.43}$$

There are two approaches to the determination of L_a, illustrated on Figure 11.30. Method (a) assumes the active wedge given by the Rankine theory and is somewhat more conservative than method (b). Method (b) is based on experimental work which showed that the maximum tensile stress occurred along the curve AC and L_a is given by the trapezoidal zone ABDE. If the angle of internal friction is less than about 28° then method (a) is generally more conservative than method (b) but for ϕ' values greater than 28° method (a) gives smaller values of L_a in the lower half of the wall. The Draft BS 8007 recommends method (a).

External stability

As with concrete walls the external stability of a reinforced earth wall must be checked. It is assumed that the reinforced section acts as a rigid structure for this purpose with an active thrust from Rankine theory acting on the back of this section so that rotational failure, overturning, bearing pressure and sliding can be assessed in the same manner as for Gravity walls, see above.

Minimum factors of safety of 1.5 for rotational failure and 2 for the other conditions should be obtained, otherwise, it may be necessary to increase the length of the reinforcement to provide a wider reinforced section and reduce the effect of active thrust.

Worked Example 11.1 *Active thrust*
Determine the total active thrust on the back of a smooth vertical wall, 6 m high, which is supporting granular backfill with bulk unit weight 19 kN/m³ and angle of friction 33º. The soil surface is horizontal and the water table lies below the base of the wall.

From Equation 11.4,

$$K_a = \frac{1 - \sin 33°}{1 + \sin 33°} = 0.295$$

The backfill is cohesionless, so Equation 11.3 gives the pressure at the base of the wall.

$p_a = 6.0 \times 19 \times 0.295 = 33.6$ kN/m²

The pressure distribution is triangular so the active thrust is

$\frac{1}{2} \times 33.6 \times 6.0 = 100.8$ kN/m run of wall

This thrust acts at $1/3 \times 6.0 = 2.0$ m above the base of the wall.

Worked Example 11.2 *Active thrust – with water table present*
For the same conditions as in Example 11.1 but with a water table at 1.8 m below ground level determine the total thrust on the back of the wall. Assume the saturated unit weight of the backfill is 20 kN/m³ below the water table.

The vertical effective stress and active pressure are plotted in Figure 11.31.

At 1.8 m below ground level

$\sigma_v' = 1.80 \times 19 = 34.2$ kN/m² $\quad \sigma_H' = 34.2 \times 0.295 = 10.1$ kN/m²

At 6.0 m below ground level

$\sigma_v' = 34.2 + 4.20 \times (20 - 9.8) = 77.0$ kN/m² $\quad \sigma_H' = 77.0 \times 0.295 = 22.7$ kN/m²

The active thrusts are then given by the areas of the active pressure diagram.

a) $\frac{1}{2} \times 10.1 \times 1.8 = 9.1$ kN/m run acting at 4.80 m above the base of the wall

b) $10.1 \times 4.2 = 42.4$ kN/m run at 2.10 m

c) $(22.7 - 10.1) \times \frac{1}{2} \times 4.2 = 26.5$ kN/m run at 1.40 m

The force from the water is:

$\frac{1}{2} \times 4.2 \times 9.8 \times 4.2 = 86.4$ kN/m run at 1.40 m above the base of the wall

The total force acting on the back of the wall is 164.4 kN/m run.

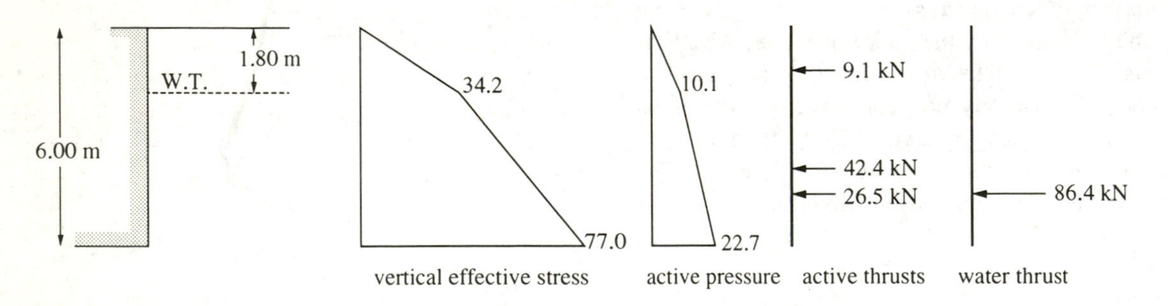

Figure 11.31 *Worked Example 11.2*

Worked Example 11.3 *Active thrust – two layers and effect of c′*
For the same conditions as in Example 11.1 but with a clay deposit 3.0 m below the top of the wall, determine the total thrust on the back of the wall.

The vertical effective stress and active pressures are plotted on Figure 11.32.
In the sand, the active pressures are
a) at 1.8 m $34.2 \times 0.295 = 10.1$ kN/m run
b) at 3.0 m $46.4 \times 0.295 = 13.7$ kN/m run in the granular soil

In the clay

$$K_a = \frac{1 - \sin 25°}{1 + \sin 25°} = 0.406$$

c) at the top of the clay, from Equation 11.3 $p_a = 46.4 \times 0.406 - 2 \times 10 \times \sqrt{0.406} = 6.1$ kN/m²
d) at the bottom of the clay $p_a = 80.0 \times 0.406 - 2 \times 10 \times \sqrt{0.406} = 19.7$ kN/m²

The active thrusts are then:

a) $\frac{1}{2} \times 10.1 \times 1.8 = 9.1$ kN/m run at 4.80 m above the base of the wall
b) $10.1 \times 1.2 = 12.1$ kN/m run at 3.60 m
c) $\frac{1}{2} \times 1.2 \times 3.6 = 2.2$ kN/m run at 3.40 m
d) $6.1 \times 3.0 = 18.3$ kN/m run at 1.50 m
e) $\frac{1}{2} \times 13.6 \times 3.0 = 20.4$ kN/m run at 1.00 m
The water force is 86.4 kN/m run at 1.40 m.
The total force acting on the back of the wall is 148.5 kN/m run.

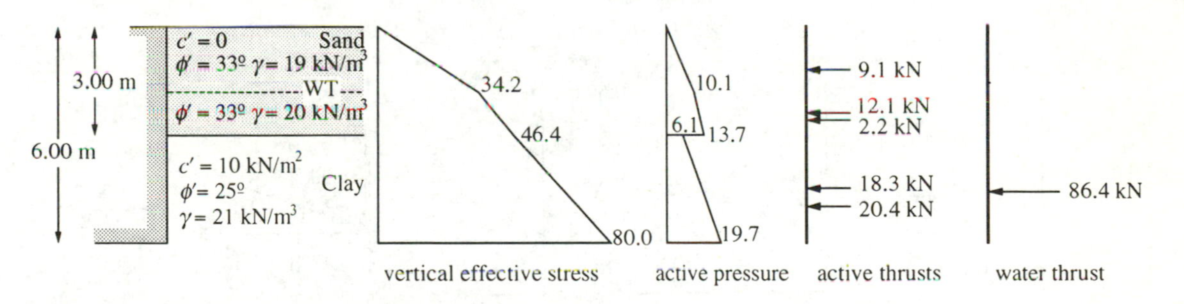

Figure 11.32 *Worked Example 11.3*

Worked Example 11.4 *Active thrust – wall friction*
For the same conditions as in Example 11.1 but with a rough wall with $\delta = 0.75\ \phi'$ determine the total thrust on the back of the wall.

From Figure 11.11, $K_a = 0.24$
From Equation 11.12 the horizontal (normal) component of the active thrust P_{an} is

$P_{an} = \frac{1}{2} \times 0.24 \times 19 \times 6.0^2 = 82.1$ kN/m run
From Equation 11.16 the shear force acting downwards on the back of the wall is
$82.1 \times \tan(0.75 \times 33) = 37.8$ kN/m run

From Equation 11.15 the resultant force acting at $\frac{3}{4} \times 33º = 24.75º$ to the horizontal will be

$$P_a = \frac{82.1}{\cos(0.75 \times 33°)} = 90.4 \text{ kN/m run}$$

Worked Example 11.5 *Active thrust – surcharge*
Determine the total active thrust on the back of a rough vertical wall, 6 m high, which is supporting a saturated clay with unit weight 21 kN/m³, c′ = 10 kN/m² and φ′ = 25º and with a uniform surcharge of 10 kN/m² acting on the soil surface. The water table lies below the base of the wall.
Assume $c_w'/c' = 0.5$ *and* $\delta = 3/4\,\phi'$

The vertical effective stresses and active pressures are plotted in Figure 11.33.

From Figure 11.11, for $\frac{\delta}{\phi'} = 0.75$ and $\phi' = 25º$ $K_a = 0.34$

From Equation 11.17 $\quad K_{ac} = 2\sqrt{(0.34 \times 0.50)} = 0.83$
From Equation 11.26 the depth to zero active pressure is given by
$0 = 0.34\,(21 \times z_o + 10) - 0.83 \times 10$
giving $z_o = 0.69$ m
Active pressure at the base of the wall is
$p_{an}' = 0.34\,(21 \times 6.0 + 10) - 0.83 \times 10 = 37.9$ kN/m²

Active thrust $P_{an} = 37.9 \times (6.0 - 0.69) \times \frac{1}{2} = 100.6$ kN/m run
acting at $\frac{1}{3}(6.0 - 0.69) = 1.77$ m above the base of the wall
From Equation 11.16 the shear force acting on the back of the wall will be
$100.6 \times \tan(\frac{3}{4} \times 25) = 34.2$ kN/m run
From Equation 11.15 the resultant force P_a acting at $\frac{3}{4} \times 25º = 18.75º$ to the horizontal will be
$P_a = 100.6 / \cos(\frac{3}{4} \times 25) = 106.2$ kN/m run
The adhesion on the back of the wall is $0.5 \times 10 \times (6.0 - 0.69) = 26.6$ kN/m run.

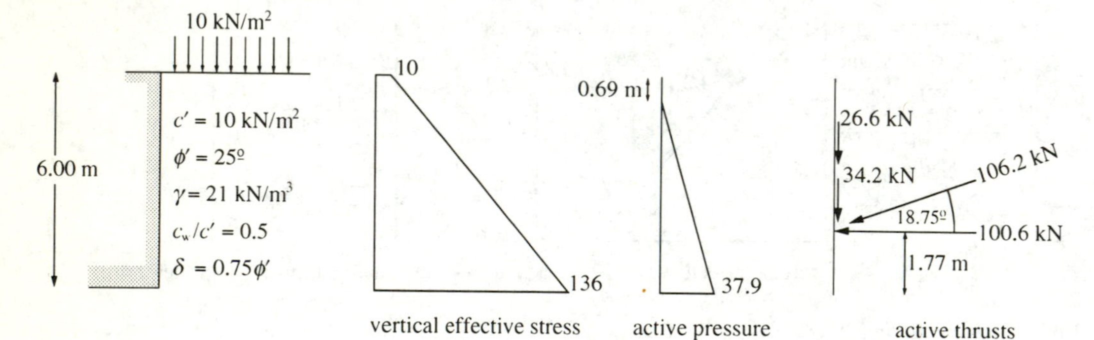

Figure 11.33 *Worked Example 11.5*

Worked Example 11.6 *Passive thrust*

For the sheet pile wall shown in Figure 11.34 determine the total passive thrust acting on the left hand side of the wall. Assume the water table lies below the base of the wall.

Assume $\delta = 0.67\ \phi'$ $\ c_w = 0.5\ c'$

From Figure 11.11, $K_p = 3.3$

From Equation 11.18, $K_{pc} = 2\sqrt{(3.3 \times 1.5)} = 4.45$

At the top of the clay, from Equation 11.13

$p_{pn}' = 3.3 \times 0 \times 21 + 4.45 \times 10 = 44.5$ kN/m²

At the bottom of the clay

$p_{pn}' = 3.3 \times 6.0 \times 21 + 4.45 \times 10 = 460.3$ kN/m²

The passive thrusts are then

a) $44.5 \times 6.0 = 267.0$ kN/m run acting 3.0 m above the base of the wall

b) $(460.3 - 44.5) \times 6.0 \times \frac{1}{2} = 1247.4$ kN/m run acting at 2.0 m

These thrusts act normal to the wall.

A shear force acts upwards of

$(267.0 + 1247.4) \tan(0.67 \times 23^\circ) = 417.4$ kN/m run

Adhesion also acts upwards.

$0.5 \times 10 \times 6.0 = 30$ kN/m run

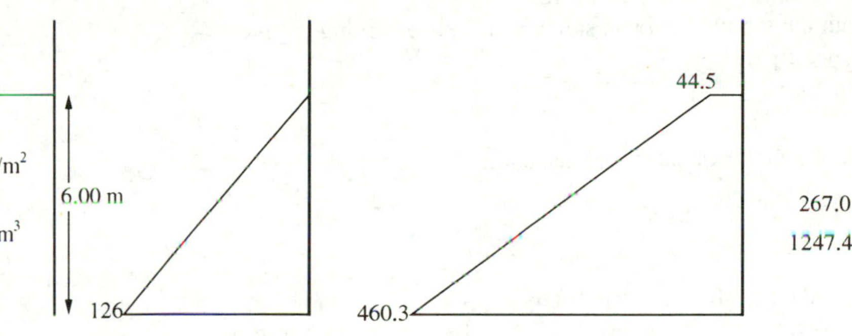

Figure 11.34 *Worked Example 11.6*

Worked Example 11.7 *Gravity wall*

The gravity wall shown in Figure 11.35 is to support backfill with unit weight 19.5 kN/m³ and shear strength parameters c′ = 0 and φ′ = 36º. The unit weight of the wall material is 24 kN/m³ and the angle of friction δ between the wall and the backfill is 27º and at the base of the wall it is 25º. Determine the factors of safety against overturning and sliding and the maximum and minimum bearing pressures.

The weight of the wall is
1.62 × 5.0 × 24 = 194.4 kN/mrun acting at 0.81 m from the toe
0.88 × 5.0 × 0.5 × 24 = 52.8 kN/mrun acting at 1.91 m from the toe.
From Equation 11.8
$\alpha = 100^\circ$ $\beta = 15^\circ$ $\delta = 27^\circ$ $\phi = 36^\circ$

$$K_a = \left(\frac{\sin 64^\circ / \sin 100^\circ}{\sqrt{\sin 127^\circ} + \sqrt{\sin 63^\circ \sin 21^\circ / \sin 85^\circ}} \right)^2 = 0.39$$

From Equation 11.7
$P_a = \frac{1}{2} \times 0.39 \times 19.5 \times 5.0^2 = 95.1$ kN/mrun
The point of application of this thrust is taken as $\frac{1}{3}H$ vertically, i.e. 1.67 m. This thrust acts at δ (=27º) to the normal to the back of the wall, i.e. 37º to the horizontal. Resolving this thrust vertically and horizontally gives
horizontal component = 95.1 cos 37º = 76.0 kN/mrun
vertical component = 95.1 sin37º = 57.2 kN/m run.
Taking moments about the toe the factor of safety against overturning is given.

$$F = \frac{194.4 \times 0.81 + 52.8 \times 1.91 + 57.2 \times 2.21}{76.0 \times 1.67} = 3.0$$

From Equation 11.28, the factor of safety against sliding is

$$F = \frac{(194.4 + 52.8 + 57.2) \tan 25^\circ}{76.0} = 1.87$$

Taking moments about the heel for the eccentricity e
$\Sigma M = 194.4 \times 1.69 + 52.8 \times 0.59 + 57.2 \times 0.29 + 76.0 \times 1.67 = 503.2$ kN/mrun.
The total vertical load = 194.4 + 52.8 + 57.2 = 304.4 kN/mrun

and acts at $\frac{503.2}{304.4} = 1.65$ m from the heel
The eccentricity e is 1.65 – 1.25 = 0.40 m on the left hand side of the wall centre-line.
2.5/6 = 0.417 m so this is just within the middle third.
From Figure 11.20:

$$q_{max} = \frac{304.4}{2.5}\left(1 + \frac{6.0 \times 0.40}{2.5}\right) = 238.7 \text{ kN/m}^2$$

$$q_{min} = \frac{304.4}{2.5}\left(1 - \frac{6.0 \times 0.40}{2.5}\right) = 4.9 \text{ kN/m}^2$$

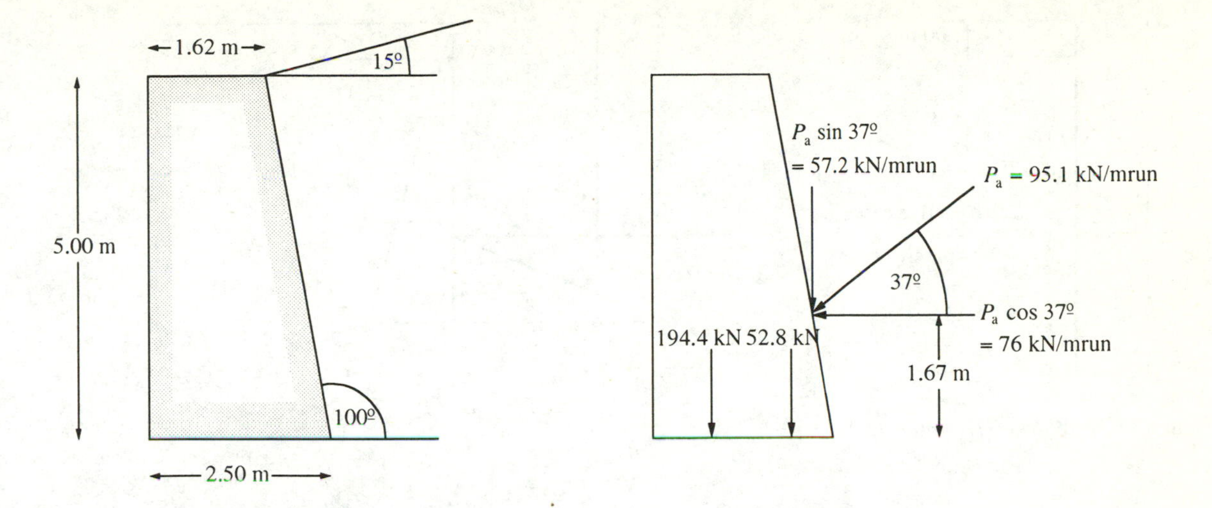

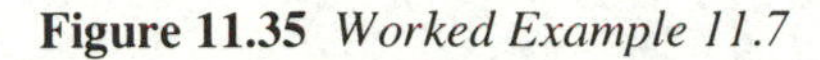
Figure 11.35 *Worked Example 11.7*

Worked Example 11.8 *Anchored sheet pile wall – Factor on embedment method*

For the anchored sheet pile wall shown in Figure 11.36 determine the depth of embedment required and the anchor force using the Factor on embedment method. The relevant soil parameters are given on the Figure and the anchors are placed at 2.5 m centres.

From Figure 11.11, $K_a = 0.32$ $K_p = 3.75$
On the active side
$P_a = 0.5 \times 0.32 \times (6.0 + d_o)^2 \times 20 = 3.2\,(6.0 + d_o)^2$
acting at $\frac{2}{3}\,(6.0 + d_o) - 1.0 = 3.0 + \frac{2}{3}d_o$ from O
On the passive side
$P_p = 0.5 \times 3.75 \times d_o^{\,2} \times 20 = 37.5\,d_o^{\,2}$
acting at $5.0 + \frac{2}{3}d_o$ from O

Equating moments gives a cubic equation
$d_o^{\,3} + 6.66d_o^{\,2} - 8.40d_o - 15.11 = 0$
By substitution $d_o = 1.90$ m
From Table 11.5, $F_d = 1.5$ From Equation 11.32 $d = 1.90 \times 1.5 = 2.85$ m
$P_a = 0.5 \times 0.32 \times 7.9^2 \times 20 = 199.7$ kN/mrun
$P_p = 0.5 \times 3.75 \times 1.9^2 \times 20 = 135.4$ kN/mrun
From Equation 11.34
Anchor force $T = (199.7 - 135.4)\,2.5 = 160.8$ kN

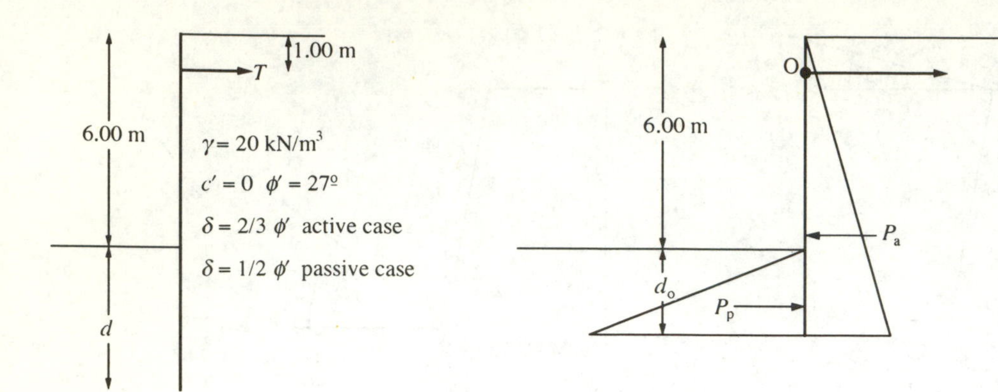

Figure 11.36 *Worked Example 11.8*

Worked Example 11.9 *Anchored sheet pile wall – Gross pressure method*
Using the same data given for Example 11.8 determine the depth of embedment required and the anchor force using the gross pressure method.

The expressions for active and passive thrust are as given in Example 11.8. With this method the moment of the passive thrust is reduced by a factor of safety F_p.
From Table 11.5, $F_p = 1.5$
From Example 11.8

$$\left(3.0+\tfrac{2}{3}d\right)\times 0.5\times 0.32\times 20\times (60+d)^2 = \frac{37.5d^2}{1.5}\times\left(5.0+\tfrac{2}{3}d\right)$$

giving $d^3 + 6.18d^2 - 13.21d - 23.78 = 0$ By substitution $d = 2.55$ m
To allow for excessive excavation etc this value is increased by 20% to give a depth of embedment of 3.06 m
$P_a = 0.5 \times 0.32 \times 8.55^2 \times 20 = 233.9$ kN
$P_p = 0.5 \times 3.75 \times 2.55^2 \times 20 = 243.8$ kN
From Equation 11.35 the anchor force is

$$T = \left(239 - \frac{243.8}{1.5}\right)\times 2.5 = 178.4 \text{ kN}$$

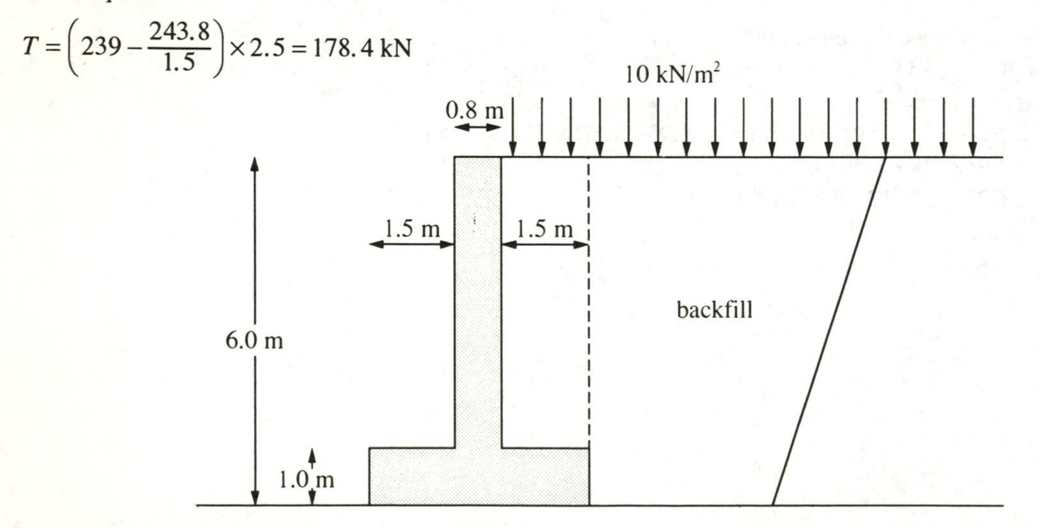

Figure 11.37 *Exercise 11.5*

Exercises

11.1 A retaining wall, 5 m high supports backfill with a horizontal surface and a water table at 2 m below ground level. The unit weight of the backfill soil is 19 kN/m^3 above the water table and 20 kN/m^3 below and the angle of internal friction of the backfill is $\phi' = 36°$ with $c' = 0$. Assuming the wall to be smooth determine the total thrust (from the soil and the water) and its point of application as the wall moves away from the soil.

11.2 For the wall and soil conditions described in Exercise 11.1 determine the total passive thrust and its point of application if the wall moves towards the soil.

11.3 For the wall and soil conditions in Exercise 11.1 but with a layer of clay below 3 m below ground level determine the total active thrust on the back of the wall and its point of application assuming drained conditions in the clay. Properties of the clay are:
$c' = 12$ kN/m^2 $\quad\phi' = 26°\quad$ $\gamma = 21.5$ kN/m^3

11.4 For the wall and soil conditions in Exercise 11.1 but with wall friction acting on the back of the wall with $\delta = 0.67\,\phi'$ determine:
a) the total horizontal thrust (active and hydrostatic) on the back of the wall and its point of application
b) the shear force acting on the back of the wall

11.5 The cantilever retaining wall shown in Figure 11.37 supports a free-draining backfill with the following properties:
$\phi' = 37°\quad c' = 0\quad \gamma = 21$ kN/m^3 $\quad$ concrete $\gamma = 25$ kN/m^3
Assuming that earth pressures are calculated on a vertical line above the heel of the wall and that soil friction acts along this line ($\delta = \phi'$) determine:
a) the factor of safety against overturning
b) the factor of safety against sliding, assuming $\delta = 25°$
c) the maximum and minimum bearing pressures beneath the base of the wall.

11.6 The anchored sheet pile wall shown in Figure 11.36 and described in Worked Example 11.8 supports the bank of a canal with a water level normally inside the canal at the same level as the water table in the ground. This water table and water level is initially at 2 m below the upper ground level. However, the water level in the canal has been lowered rapidly to its base level while the water table behind the wall has remained unchanged. Assuming hydrostatic conditions on both sides of the wall determine:
a) the depth of embedment required for stability and
b) the anchor force
using the factor on embedment method.

11.7 A reinforced earth wall 7 m high is reinforced with strip elements 8.0 m long and 60 mm wide with horizontal spacings of 1.0 m and vertical spacings of 0.6 m. The angle of friction between the elements and the fill is 25° and the ultimate tensile strength of the strips is 300 N/mm^2. The unit weight of the fill is 19 kN/m^3 and its angle of friction is 37°. The highest strip lies at 0.7 m below the top of the wall and the lowest strip is at 6.7 m below the top of the wall. Determine:
a) the minimum thickness required for the strip elements assuming a factor of safety against tensile failure of 2
b) the minimum factor of safety against bond failure.

12 Slope Stability

Slope stability

Introduction

Sloping ground can become unstable if the gravity forces acting on a mass of soil exceed the shear strength available at the base of the mass and within it. Movement of the mass of soil down the slope will then occur.

This can have catastrophic consequences to life and property if buildings exist on, above or below the slope. However, in remote, unpopulated areas mass movements may have minimal effect, merely being part of the natural degradation of the land surface. On some coastal cliffs instability involving the destruction of property is often accepted since the costs of resisting natural erosion processes with cliff stabilisation measures can be prohibitive.

Types of mass movement *(Figure 12.1)*

Mass movements, generally referred to as landslides, can take many forms. Skempton and Hutchinson (1969) have classified types of mass movement, as illustrated in Figure 12.1.

Flows are distinguished from slides, in that slides comprise the movement of large, continuous masses of soil on one or more slip surfaces, whereas flows consist of slow movement of softened or weathered debris such as from the base of a broken-up slide mass (earth flows) or somewhat faster movement of clay debris, softened and lubricated by water (mudflows), with slip surfaces less evident.

Natural slopes

These slopes have evolved as a result of natural processes, largely erosion, over a long period of time. If the soils comprising these slopes have sufficient strength to support the gravity forces, then the slope will remain stable without any mass movement.

Where they have been affected by instability in the past, the previous mass movements will have produced a 'first-time slide' resulting in considerable strains usually along a single slip surface. If the soil is prone to reduction in strength following large strains, from the peak to the residual value, this slope will be left with a slip surface along which the lowest possible shear strength exists. This slope will then remain in a state of incipient failure (only just stable) until it is affected by:

- a change in the pore pressure condition within the slope. The most common causes are extreme rainfall and drainage pattern changes. Placing soil over the toe of a slope could permit a build-up of pore water pressures within the slope, as will curtailment of pumping from aquifers close to the slope.
- a change in the geometry of the slope. This can be caused by undercutting by erosion at the toe of the slope, or accumulation of soil at the top of the slope.
- engineering works. The construction of a cutting at the toe of a slope or an embankment at the top could precipitate failure.

Artificial slopes or earthworks

Soil has been used for many construction purposes ever since human activity commenced. Major earthworks can be classified as:

- excavations
- cuttings
- embankments
- earth dams
- spoil heaps, tailings dams.

Excavations are usually made for temporary works to enable construction of a structure below ground level or for extraction and quarrying purposes. Since the ground is only open for a short period of time, a risk is usually taken on the sides of the excavation remaining stable and, for economic reasons, as steep a slope as possible is cut. This stability relies on the soil remaining in the undrained condition for the period of exposure so it is advisable to cover the slope surfaces to prevent infiltration of water. Once the structure is complete the excavation is backfilled and the stability is restored. Quarry slopes are usually left to degrade with time. The most common forms of failure would be falls, rotational slides and translational slides.

Cuttings are formed from sloping sided excavations below ground level. The soil and groundwater conditions comprising the slopes will be determined by the natural geology at the site so a slope will have to be designed for these conditions. They must be stable in the short and long-term. The long-term state is the most critical, when pore pressures have risen to equilibrium

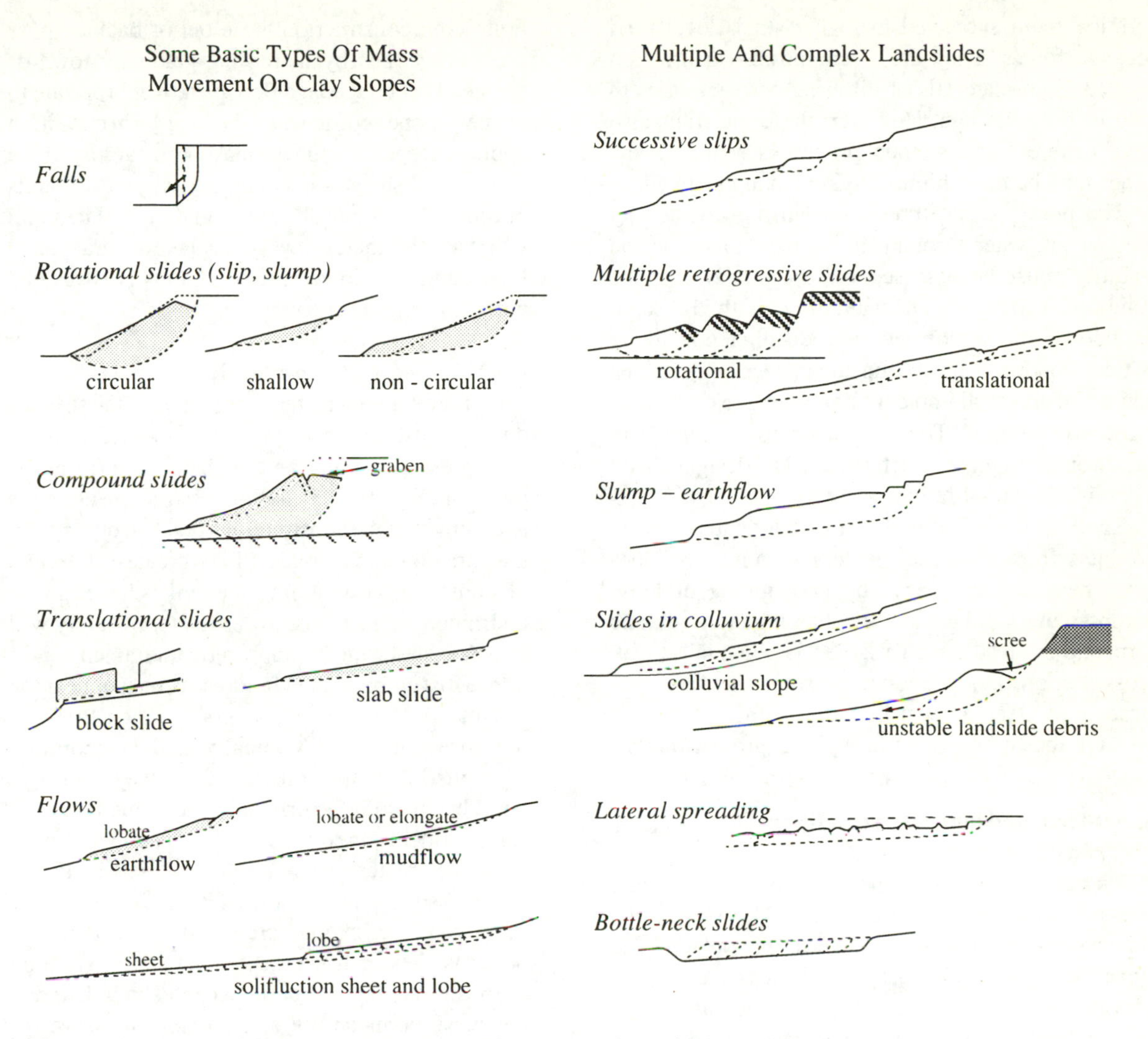

Figure 12.1 *Types of mass movement (From Skempton and Hutchinson, 1969)*

and the drained condition applies. Thus most cutting slopes are stable in the short-term but become less stable with time and may fail, sometimes many years later. Modes of failure could include falls, rotational, composite and slab slides, and slumping.

Embankments are formed by placing a mound of fill material above ground level with sloping sides. The strength and groundwater conditions within an embankment can be controlled to a certain extent by specifying good quality fill and drainage measures so the stability of the embankment slopes themselves will be more certain. However, the soil and groundwater conditions beneath the embankment will be determined by the natural geology at the site so the possibility of foundation failure must be considered.

Modes of failure would usually be rotational or composite often with circular slip surfaces, but sometimes with non-circular slip surfaces depending on the stratigraphy. Foundation failures beneath the embankment would be more likely during construction with slope failures within the embankment more likely in the long-term.

Earth dams require much greater care in their design. They are higher structures with the materials forming them subjected to higher stress levels and seepage forces. They are of composite construction formed from materials of differing permeabilities to restrict flow through. However, these materials also have different stress-strain properties so that some zones may be more highly stressed than imagined.

The pore pressure condition will be affected by seepage of water through the various zones, so the stability must be assessed for the steady seepage condition on the downstream face, with the rapid drawdown of reservoir levels producing a potentially critical condition on the upstream face. Due to the zoning of materials potential slip surfaces will tend to be non-circular. The consequences of failure are catastrophic so these structures must be designed with a negligible risk of failure.

Spoil heaps and tailings dams are mounds of waste products from industrial processes, mining residues etc. They may be placed by end tipping or loose dumping in a solid form or by pumping in a hydraulic form and can result in a variable and potentially weak cross-section. For economic reasons, they are constructed as steep as possible and, if little concern is given to the control of surface water, groundwater or process water, they can be a disaster waiting to happen.

Short-term and long-term conditions *(Figures 12.2, 12.3 and 12.4)*

When a cutting or embankment is constructed the total stresses in the ground are changed. This results in a change in pore water pressure (see Equation 4.12) and since the factor of safety of a slope decreases as the pore pressure increases the most critical condition will occur when pore pressures are greatest.

- ***Cuttings*** *(Figure 12.2)*

  In Figure 12.2 the total stress and pore pressure variations are shown for a point P within a cutting, for different types of soil.

  The reduction in total stresses during construction will lead to a decrease in the pore pressure as the soil structure attempts to expand. If the excavation is carried out quickly, then there will be no time available for redistribution of pore pressures. The pore pressure reduction will be greatest at the end of construction and its amount will depend on the type of soil present, particularly the pore pressure parameter value A.

  Following construction these out-of balance pore pressures adjust by increasing gradually towards the steady state seepage flow pattern appropriate to the new slope profile when the long-term condition applies. Depending on the mass permeability of the soil this redistribution will require varying amounts of time. With sands the redistribution will be rapid whereas with intact clays it may take several years. For a cutting the lowest factor of safety is associated with this long-term condition.

- ***Embankments*** *(Figure 12.3)*

  For an embankment the increase in total stresses during construction will lead to an increase in the pore pressure within the foundation soil (as at the point P in Figure 12.3), as the soil structure attempts to contract. If the construction is carried out rapidly the pore pressure increase will be greatest at the end of construction assuming the soil behaves in an undrained manner. The lowest factor of safety will be associated with this short-term undrained condition. Most embankment failures occur at or near the end of construction. After construction the pore pressures dissipate vertically and horizontally through the foundation soil, until they return to equilibrium with the original water table level.

  Embankments on soft, normally consolidated clays are at the greatest risk of failure associated with the largest rise in pore pressures. Their stability can be improved by allowing pore pressure dissipation to occur as construction proceeds but this will require a slower rate of construction, a good knowledge of the mass permeability and possibly measures to accelerate dissipation such as vertical drains.

- ***Earth dams*** *(Figure 12.4)*

  For an earth dam the shear stresses and the pore pressures will vary both during and after construction depending on the state of the dam and the level of the water impounded in the reservoir, see Figure 12.4. The following conditions should be checked:

  1. during and shortly after construction – both up stream and downstream faces.
  2. with the reservoir full (steady seepage) – down stream face.
  3. following rapid drawdown of the impounded water – upstream face.

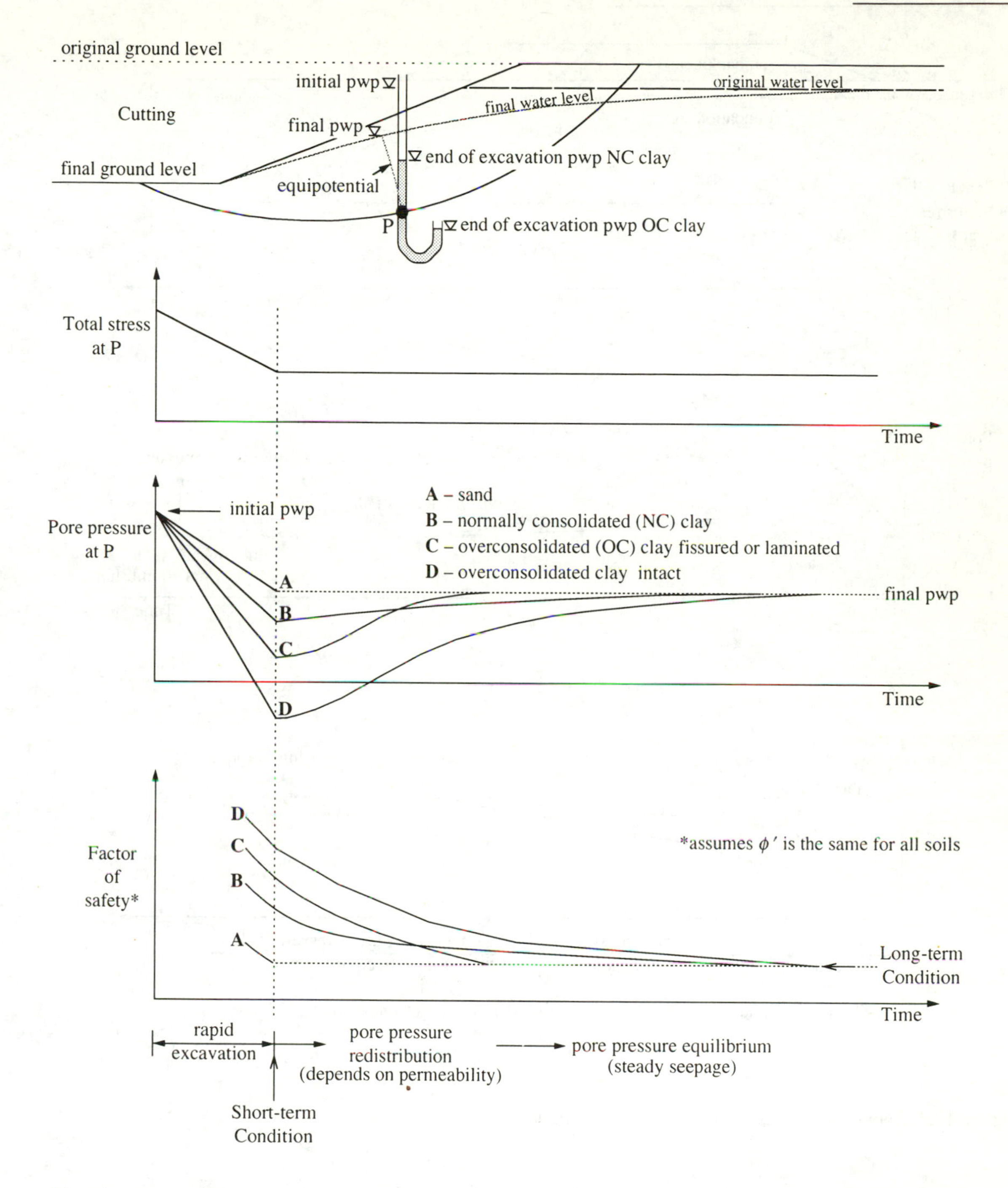

Figure 12.2 *Cuttings – short and long-term conditions*

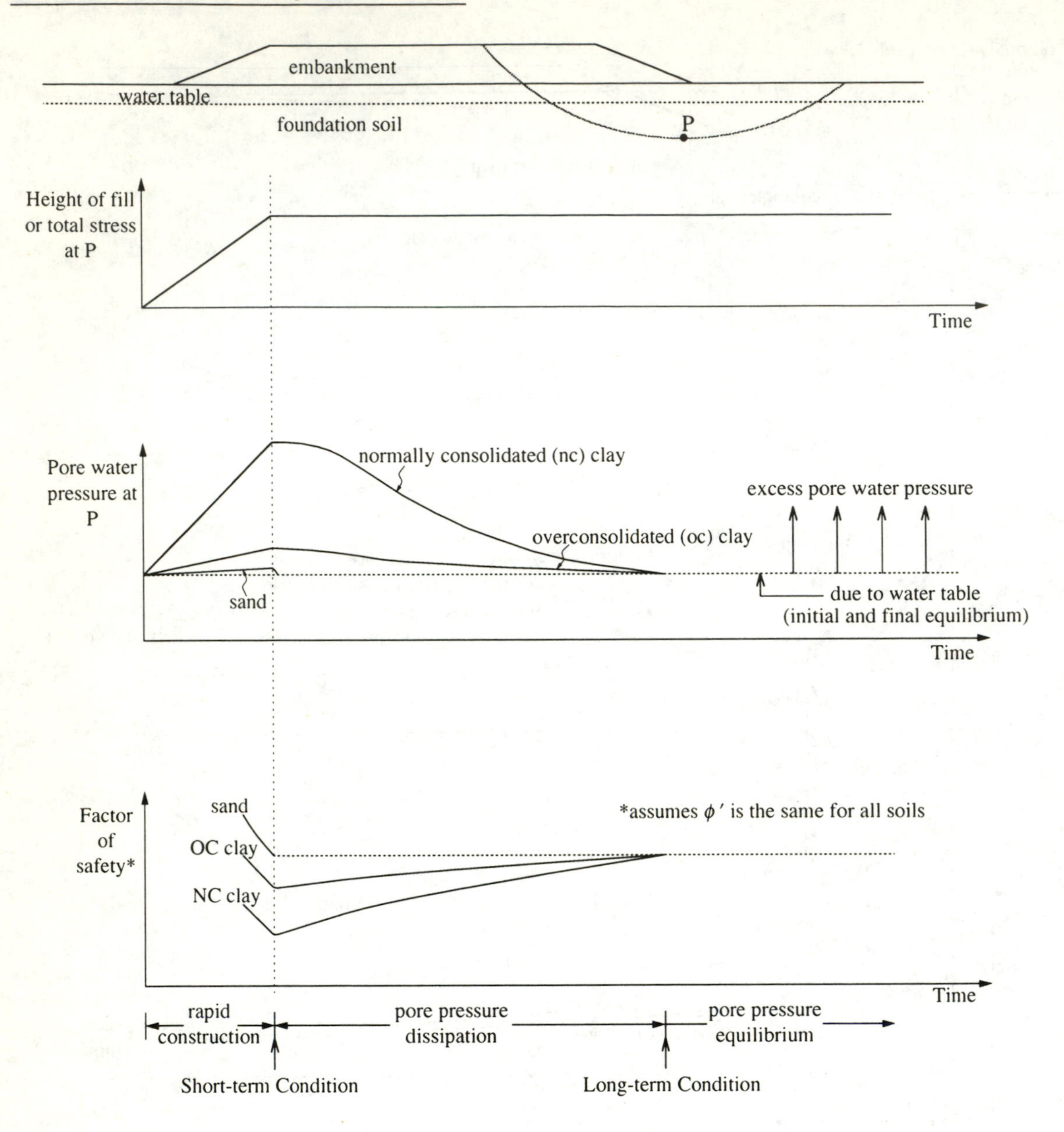

Figure 12.3 *Embankments – short and long-term conditions*

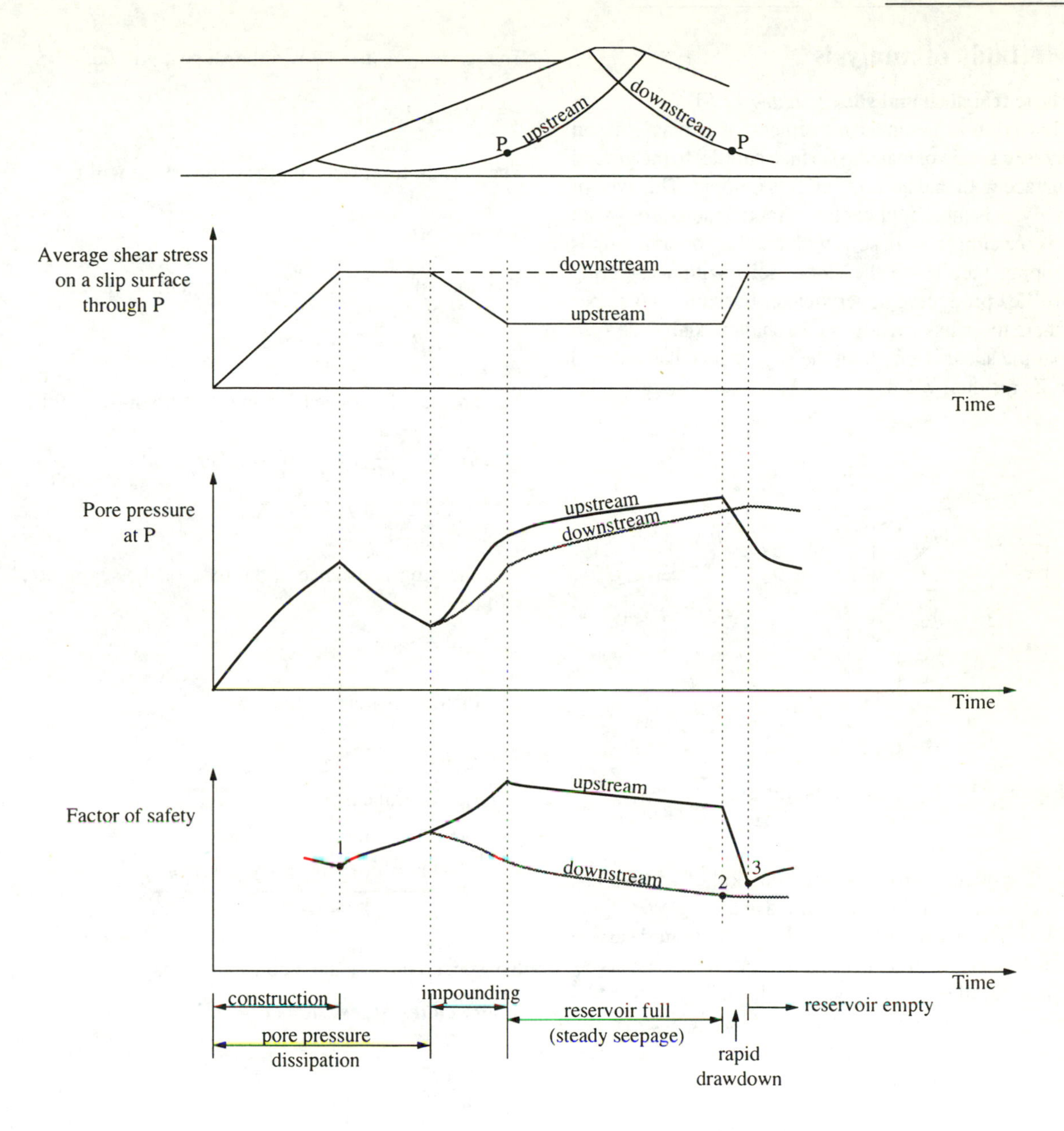

Figure 12.4 *Earth dams – stability conditions (From Lambe and Whitman, 1979)*

Methods of analysis

Plane translational slide *(Figure 12.5)*
This method assumes movement of a mass of soil above a single planar slip surface parallel to the ground surface with end and side effects ignored. This type of analysis is most applicable to granular soils, soils with no cohesion ($c' = 0$), soils with bedding or laminations dipping parallel to the slope, soils with weathering profiles producing upper weaker horizons and slopes where there has already been a shallow slab slide such that the shear strength on the slip surface has reduced to its residual value.

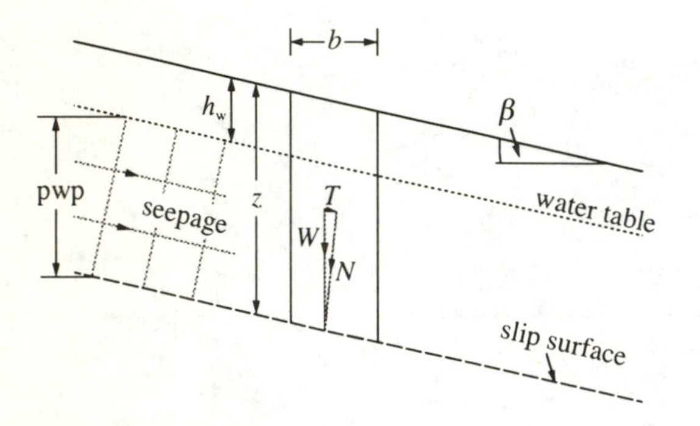

Figure 12.5 *Plane translational slide*

Consider a vertical segment of soil of width b and unit thickness in a slope inclined at an angle β (Figure 12.5). The water table is parallel to the ground surface at a depth h_w with steady seepage assumed to be taking place parallel to the ground surface. The pore water pressure u on the slip surface will be given by:

$$u = (z - h_w)\gamma_w \cos^2\beta$$

Assuming the unit weights of the soil above and below the water table (bulk and saturated unit weights) to be the same, γ, the weight of the segment W will be:

$$W = \gamma z b$$

The tangential force down the slope T will be:

$$T = W \sin\beta = \gamma z b \sin\beta$$

The tangential stress down the slope τ will be:

$$\tau = \gamma z \sin\beta \cos\beta$$

The normal total force on the slip surface will be:

$$N = W \cos\beta$$

The normal total stress on the segment σ will be:

$$\sigma = W \cos\beta \cos\beta / b = \gamma z \cos^2\beta$$

The normal effective stress in the segment σ' will be:

$$\sigma' = \sigma - u = \gamma z \cos^2\beta - (z - h_w)\gamma_w \cos^2\beta$$
$$= \cos^2\beta(\gamma z - \gamma_w z + \gamma_w h_w)$$

The shearing resistance at the base of the segment τ_f will be:

$$\tau_f = c' + \sigma' \tan\phi'$$

The factor of safety is given by:

$$F = \frac{\tau_f}{\tau}$$

For the general case:

$$F = \frac{c' + \tan\phi' \cos^2\beta(\gamma z - \gamma_w z + \gamma_w h_w)}{\gamma z \sin\beta \cos\beta} \qquad (12.1)$$

Six special cases can be considered, below.

1 ***Dry cohesionless slope*** $c' = 0, h_w = z$

$$F = \frac{\tan\phi'}{\tan\beta} \qquad (12.2)$$

so for the critical case (when $F = 1$)

$$\beta = \phi'$$

2 ***Wet cohesionless slope*** $c' = 0 \;\; h_w = 0$

$$F = \frac{\gamma_{sub} \tan\phi'}{\gamma \tan\beta} \qquad (12.3)$$

3 *Cohesionless slope with pore pressure ratio r_u*

$$r_u = \frac{u}{\gamma z}$$

$$F = \left(1 - r_u \sec^2 \beta\right)\frac{\tan \phi'}{\tan \beta} \qquad (12.4)$$

4 *Cohesionless slope with water table below slip surface producing suction*
The capillary zone above the water table in a fine sand or silt can be significant (see Chapter 4), and will produce negative pore pressures within this zone given by:

$$u_s = -\gamma_w h_s$$

where h_s is the vertical distance between the water table and the slip surface.

The factor of safety can then be obtained as:

$$F = \frac{\tan \phi'\left(\gamma z \cos^2 \beta - \gamma_w h_s\right)}{\gamma z \sin \beta \cos \beta} \qquad (12.5)$$

A simpler expression can be obtained if seepage through the suction zone parallel to the surface is assumed when:

$$u_s = -\gamma_w h_s \cos^2\beta$$

and

$$F = \left(1 + \frac{\gamma_w h_s}{\gamma z}\right)\frac{\tan \phi'}{\tan \beta} \qquad (12.6)$$

Caution must be exercised when assuming suctions to act in support of a slope.

5 Layered soil profile *(Figure 12.6)*
Apart from the completely dry or completely wet cases 1 and 2, this method requires some information or an assumption concerning the depth to the presumed slip surface, *z*. Worked Example 12.1 shows that, for a homogeneous soil, the factor of safety decreases as the presumed slip surface moves deeper, so the method has obvious limitations for a homogeneous slope. With a layered soil profile the tangential shear stress down the slope could be plotted with depth as:

$$\tau = \gamma z \cos\beta \sin\beta$$

with the available shearing resistance τ_f superimposed:

$$\tau_f = c' + \tan \phi' \cos^2 \beta\left(\gamma z - \gamma_w z + \gamma_w h_w\right)$$

as illustrated in Figure 12.6.

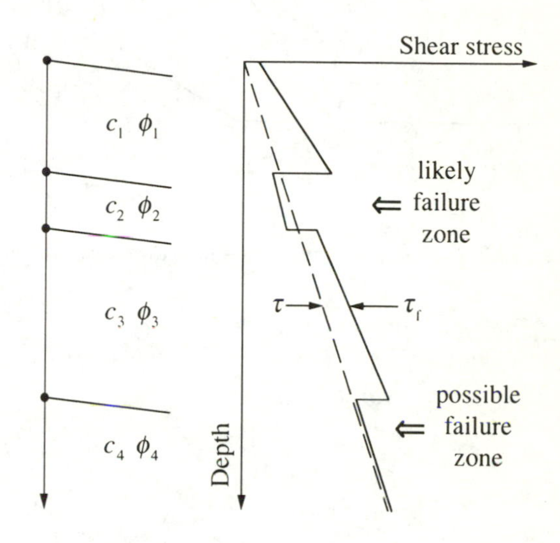

Figure 12.6 *Translational slide – layered soil profile*

6 Undrained conditions
All of the above assume the long-term condition to be critical using effective stress soil parameters. If the slope behaves in an undrained manner (with $\phi_u = 0°$) the factor of safety is given by:

$$F = \frac{c_u}{\gamma z \sin \beta \cos \beta} \qquad (12.7)$$

Circular arc analysis – undrained condition or ϕ_u = 0° analysis *(Figures 12.7 and 12.8)*
This is an analysis in terms of total stress and applies to the short-term condition for a cutting or embankment assuming the soil profile to comprise fully saturated clay.

A segment of the slope of unit thickness bounded by a circular arc (a presumed slip surface) is considered (Figure 12.7). The tendency for gravity forces to rotate the circular segment about its circle centre is resisted

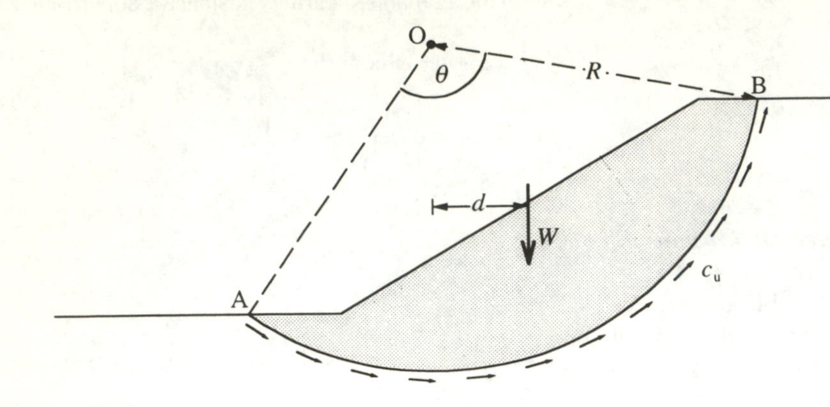

Figure 12.7 *Circular arc analysis – undrained condition*

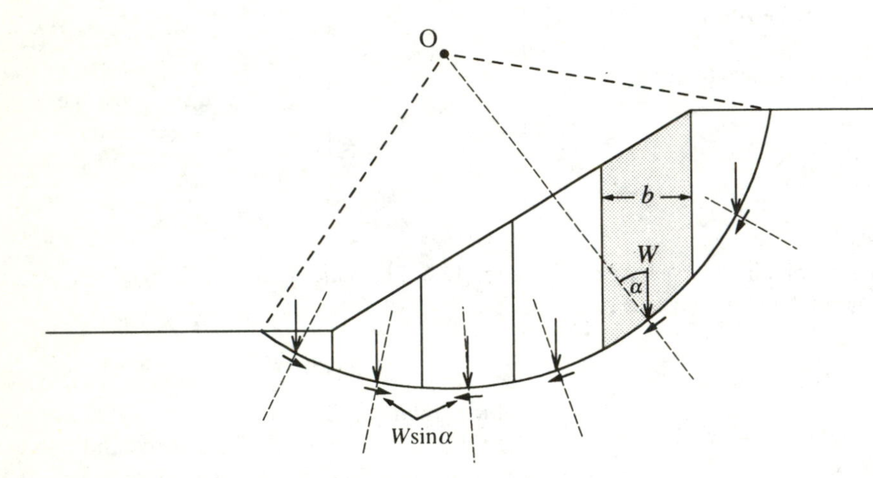

Figure 12.8 *Circular arc analysis – undrained condition using slices*

by the shear strength mobilised along the slip surface so moment equilibrium is applied. The factor of safety against instability is given by:

$$F = \frac{\text{shear resistance moment}}{\text{overturning moment}}$$

The shear resistance moment is equal to the force along the length L_{AB} of the circular arc multiplied by the radius of the circle, R:

$$\text{shear resistance moment} = c_u L_{AB} R = c_u R^2 \theta$$

where θ is in radians.

The overturning moment is Wd. W is the total weight of the segment, determined from γA where γ is the bulk unit weight of the soil and A is the area of the segment. d is the horizontal distance from the circle centre to the centroid of the segment. The factor of safety is then:

$$F = \frac{c_u R^2 \theta}{Wd} \qquad (12.8)$$

The determination of W and d can entail lengthy calculations so it will be more convenient to split the segment into a number of slices as shown on Figure 12.8 where the factor of safety is then given by:

$$F = \frac{\sum c_u b \sec\alpha}{\sum W \sin\alpha} \quad (12.9)$$

The area of each slice can be obtained from the mid-height multiplied by the width of the slice, and the angle α can be scaled off or calculated. Note how the sign of the overturning forces changes to the left and right of the circle centre. Inter-slice forces are ignored.

It is necessary to find the circular arc which gives the lowest factor of safety, the critical circle. A number of trial circles must then be analysed in the same way but with different circle centres and different points where the circle cuts the slope. This process can be time-consuming so the use of charts for homogeneous slopes and computer programs for heterogeneous slopes is recommended.

The slices approach can also be used where the soil profile consists of different layers and where there are other forces to consider such as a foundation at the top of the slope or a water-filled tension crack.

Tension crack *(Figure 12.9)*

As the condition of limiting equilibrium develops with the factor of safety close to 1, a tension crack may form near the top of the slope through which no shear strength can be developed, and if it fills with water a horizontal hydrostatic force P_w will increase the disturbing moment by $P_w y_c$. The factor of safety will be further reduced because of the shorter length of circular arc along which shearing resistance can be mobilised.

The depth of a tension crack can be taken as:

$$z_c = \frac{2c_u}{\gamma} \quad (12.10)$$

and the hydrostatic force P_w is:

$$P_w = \tfrac{1}{2}\gamma_w z_c^{\,2}$$

The expression for the factor of safety then becomes:

$$F = \frac{R\sum c_u b \sec\alpha}{\sum RW \sin\alpha + P_w y_c} \quad (12.11)$$

Note that the width of the slices b need not be the same for all slices.

Undrained analysis stability charts – Taylor's Method *(Figure 12.10)*

Taylor (1937) proposed the use of a stability number, N_s given by:

$$N_s = \frac{c_u}{F\gamma H} \quad (12.10)$$

where F is the lowest factor of safety obtained from a circular arc analysis of a homogeneous slope with undrained shear strength c_u, bulk unit weight γ and height H.

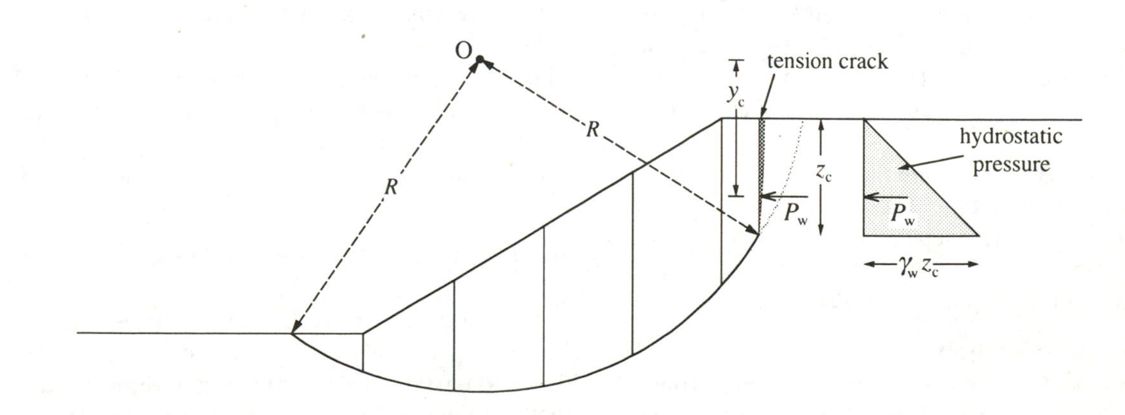

Figure 12.9 *Circular arc analysis – undrained condition with tension crack present*

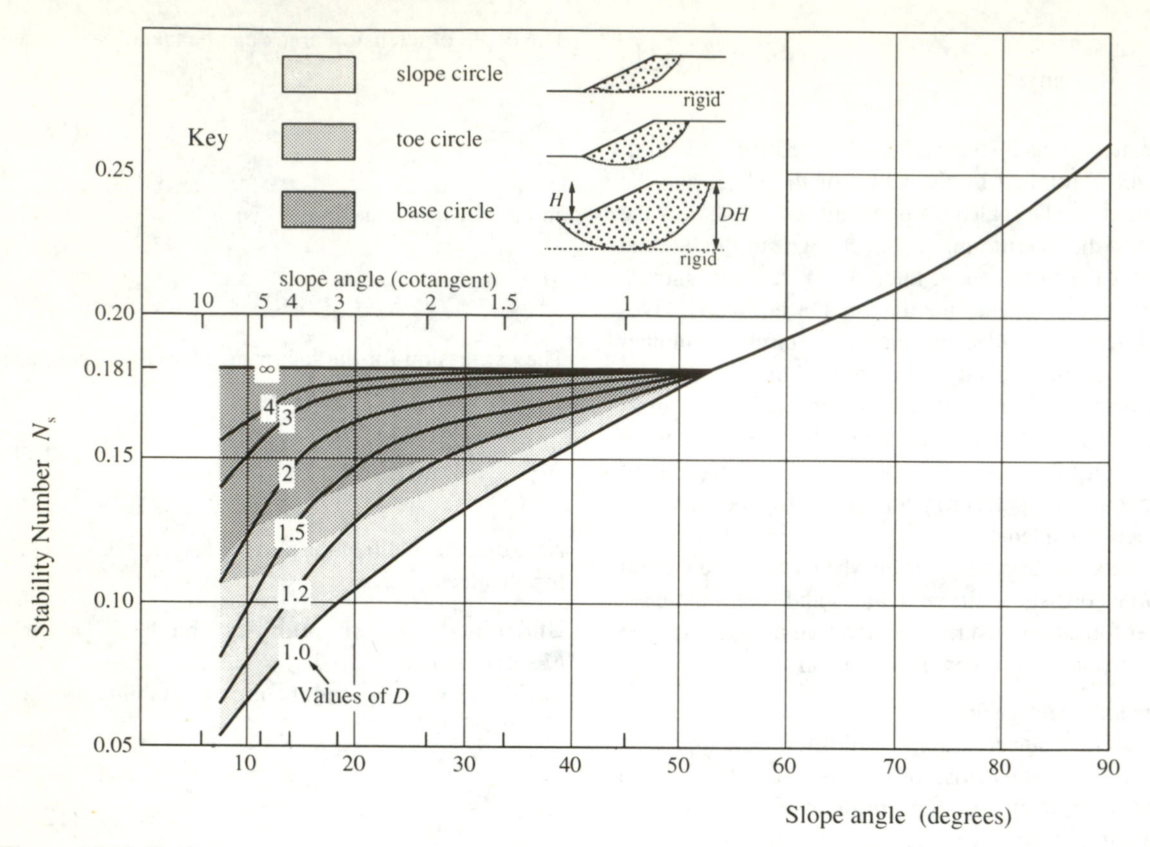

Figure 12.10 *Taylor's curves*

The stability number depends on the slope angle β, and the depth factor D where DH is the depth to a rigid stratum. Values of N_s can be obtained from Figure 12.10.

A feature of the analysis of a homogeneous slope is that when the slope angle is less than 53º the factor of safety decreases with deeper circles so more critical (lower) values are obtained for deeper-seated 'base' circles.

Shallower slip circles are likely to occur for non-homogeneous conditions such as when the shear strength increases with depth. When the slope angle is greater than 53°, all critical circles are toe circles irrespective of the depth to the rigid stratum.

Effective stress analysis

Stability analyses in terms of effective stress can only be carried out when there is a reasonably accurate assessment of pore water pressures within the slope.

For a natural slope and a cutting the more critical long-term condition can be analysed using the effective stress analysis. The pore pressures can be represented by the steady seepage state when a flow net can be drawn or when a known equilibrium water table level exists.

For an embankment the more critical short-term condition can also be analysed using an effective stress analysis but this will be limited by the accuracy with which the excess pore pressures can be estimated.

Effective stress analysis – method of slices *(Figure 12.11)*

A circular arc slip surface is presumed with radius R. The soil segment above this circular arc is of unit thickness and is divided into a number of vertical slices usually of equal width b and varying height h as shown in Figure 12.11. With a sufficiently large number of

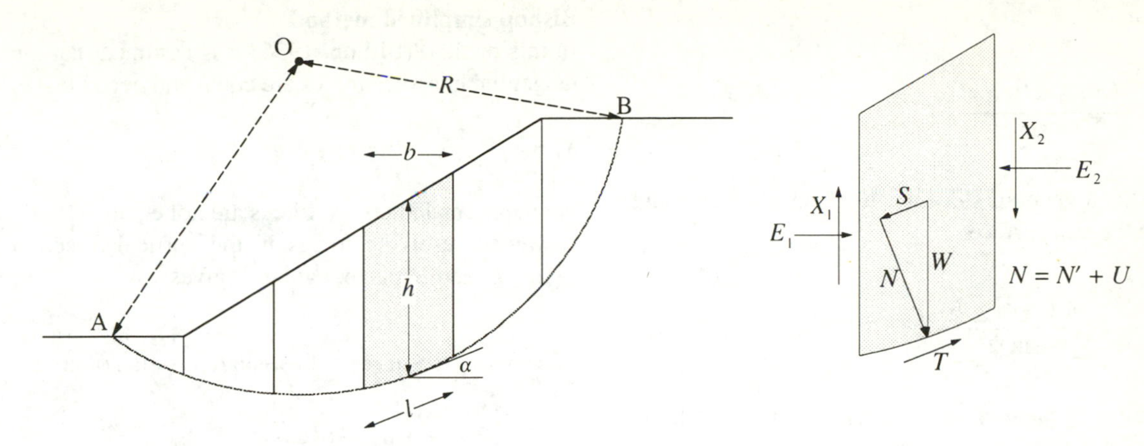

Figure 12.11 *Method of slices*

slices the base of each slice can be assumed to be a straight line inclined at an angle α to the horizontal and with a length $l = b \sec\alpha$.

The segment is separated into slices only for convenience of analysis. It is assumed that the segment (all of the slices) rotates around the circle centre O as a whole body so that the factor of safety is the same for each slice, i.e. there is one factor of safety for the trial surface chosen. This implies that forces must act between the slices, known as inter-slice forces and these are usually taken as normal and tangential to the sides of the slices.

The forces acting on a slice are:

1. The total weight of the slice, $W = \gamma bh$ where γ is the bulk unit weight of the soil.
2. The weight of each slice will induce a shear force parallel to its base $S = W \sin\alpha$
3. The total normal force on the base, $N = \sigma l$.
4. The total normal force is obtained from the total normal stress which has two components, the effective normal force $N' = \sigma' l$ and the water force $U = ul$ where u is the pore water pressure at the centre of the base of the slice.
5. The shearing resistance of the soil will provide a shear force $T = \tau_m l$.
6. The interslice forces can be represented as total normal forces E_1 and E_2 and tangential shear forces X_1 and X_2.

Other forces acting on the segment such as a foundation or surcharge at the toe or crest of the slope, or a water-filled tension crack may be included in the analysis.

For moment equilibrium the overturning moment produced by the forces S must be balanced by the resisting moment of the mobilised shear strength forces T:

$$\Sigma SR = \Sigma TR$$

$$\sum W \sin\alpha R = \sum \tau_m lR = \sum \frac{\tau_f}{F} lR$$

Since F is the same for all slices:

$$F = \frac{\sum \tau_f l}{\sum W \sin\alpha}$$

The shear strength of the soil in terms of effective stress is:

$$\tau_f = c' + \sigma' \tan\phi'$$

and $\tau_f l = c' l + N' \tan\phi'$

so that

$$F = \frac{\sum(c'l + N'\tan\phi')}{\sum W \sin\alpha}$$

For a homogeneous slope where c' and ϕ' are constant along the slip surface:

$$F = \frac{c'L_{AB} + \tan\phi'\sum N'}{\sum W \sin\alpha} \qquad (12.13)$$

where L_{AB} is the total arc length between A and B.

The solution of this expression requires assumptions to be made about the interslice forces which affects the values of the forces N'. Two methods are given below.

Fellenius method

Also known as the Swedish method, in this analysis it is assumed that the interslice forces are equal and opposite so their resultants are zero:

$E_1 = E_2$ and $X_1 = X_2$.

Resolving the forces acting normal to the base of each slice:

$$N' = W\cos\alpha - ul$$

where $u = \gamma_w h_w$ and h_w is the height to the water table above the base of the slice or related to the nearest equipotential if a flow net is drawn. The factor of safety is then:

$$F = \frac{c'L_{AB} + \tan\phi'\sum W\cos\alpha - ul}{\sum W \sin\alpha} \qquad (12.14)$$

Values of W, α and u can be determined for each slice and presented in a table, the summations obtained by addition.

This solution underestimates the factor of safety compared with more accurate methods of analysis with an error up to 20%, and will, therefore, be conservative.

Bishop simplified method

In this method (Bishop, 1955) it is assumed that the tangential interslice forces are equal and opposite, i.e.

$$X_1 = X_2$$

but the normal interslice forces are not equal, $E_1 \neq E_2$. However, resolving forces in the vertical direction (when E_1 and E_2 can be ignored) gives:

$$W = N'\cos\alpha + ul\cos\alpha + \frac{c'l}{F}\sin\alpha + \frac{N'}{F}\tan\phi'\sin\alpha$$

$$\therefore N' = \frac{W - \dfrac{c'l}{F}\sin\alpha - ul\cos\alpha}{\cos\alpha + \dfrac{\tan\phi'\sin\alpha}{F}}$$

Inserting this expression into Equation 12.13 and rearranging gives:

$$F = \frac{1}{\sum W\sin\alpha}\sum\frac{[c'b + (W - ub)\tan\phi']\sec\alpha}{1 + \dfrac{\tan\alpha\tan\phi'}{F}} \qquad (12.15)$$

Since the value of F occurs on both sides of the expression a trial value for F must be chosen on the right hand side to obtain a value of F on the left hand side. By successive iteration convergence on the true value of F is obtained.

The method is obviously better solved by using a computer program which can obtain the factor of safety of a trial circle in a matter of seconds compared to many minutes when done manually.

This speed of calculation is also beneficial in searching for the circle giving the lowest factor of safety where the computer programs will analyse circles over a grid of circle centres and with the circles passing through chosen points on the slope, over a range of radius values or tangential to particular levels. The programs also permit more complex soil profiles, groundwater conditions, external loading and seismic effects.

Pore pressure ratio r_u

The pore pressure ratio, r_u can be used to represent overall or local pore pressure conditions in a slope. It is given by the ratio of the pore water pressure to the total stress:

$$r_u = \frac{u}{\gamma h} \tag{12.16}$$

Equation 12.15 can then be written as:

$$F = \frac{1}{\sum W \sin\alpha} \sum \frac{\left[c'b + W\left(1 - r_u\right)\tan\phi'\right]\sec\alpha}{1 + \dfrac{\tan\alpha\tan\phi'}{F}} \tag{12.17}$$

Effective stress analysis – stability coefficients *(Figures 12.12–12.14)*

Bishop and Morgensten (1960) found a good relationship between the factor of safety of a slope and the pore pressure ratio r_u using the Bishop simplified method of analysis with effective stress conditions as:

$$F = m - nr_u \tag{12.18}$$

where m and n are termed stability coefficients. These are related to a number of variables, the slope angle, the soil properties ϕ' and a combined value $c'/\gamma H$ and a depth factor D, similar to Taylor's approach. Values of m and n are plotted and tabulated in their paper and have been extended by O'Connor and Mitchell (1977) and Chandler and Peiris (1989).

This method has been used for many years as a rapid means of assessing the stability of a slope. However, the method of estimating an average value of r_u for a slope can be laborious and inaccurate and representing the pore pressure condition beneath the whole of a slope by a single value can be inappropriate.

This author (Barnes, 1992) has published a method which gives the critical (minimum) factor of safety for long-term effective stress stability of a homogeneous slope in the form:

$$F = a + b\tan\phi' \tag{12.19}$$

Pore pressures are represented as a steady state condition by a water table at toe level beyond the toe and inclined at various angles within the slope given by the depth below crest level, h_w, see Figure 12.12. This allows the groundwater conditions to be represented directly by an appropriate water table level and enables the effect of water table fluctuations on the factor of safety to be readily determined.

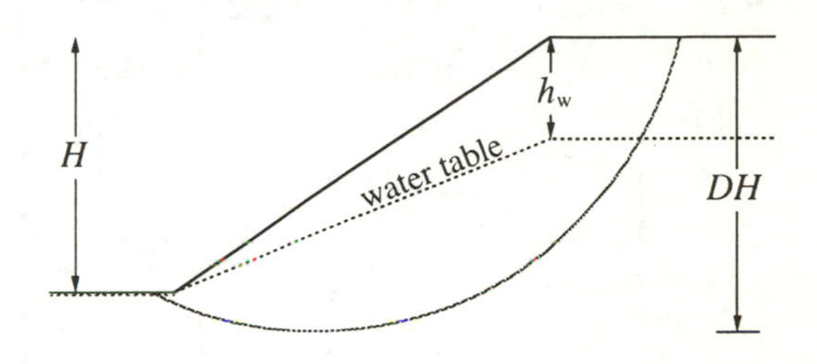

Figure 12.12 *Notation for stability coefficients (Barnes, 1992)*

The stability coefficients a and b have been found to be related to the slope angle, the cohesion soil parameter, $c'/\gamma H$, and the water table parameter, h_w/H and are given in Tables 12.1 to 12.4 for slope inclinations of 1:1, 2:1, 3:1 and 4:1, respectively. Values of the coefficients for intermediate values of all of the above parameters can be obtained from the tables with sufficient accuracy by linear interpolation.

When the critical circle passes below the water table the slope is described as 'wet' and when the critical circle lies entirely above the water table the slope can be considered as a 'dry' slope.

The water table parameter, h_w/H, at which the 'dry' slope condition is obtained has been determined and is plotted in Figures 12.13 and 12.14. If the actual water table lies below this level then the slope can be considered 'dry' with no effect from the water table. The appropriate values of a and b are then obtained from the shaded areas in the Tables.

For steeper slopes, higher values of h_w/H and higher values of ϕ' a lower factor of safety may be obtained for the 'dry' condition. These values of ϕ', above which the slope should be considered as 'dry', are given in brackets in Tables 12.1 and 12.2.

Table 12.1 *Stability coefficients a and b for slope 1:1 (Barnes, 1992)*

h_w/H	$c'/\gamma H = 0.005$		$c'/\gamma H = 0.025$		$c'/\gamma H = 0.050$		$c'/\gamma H = 0.100$		$c'/\gamma H = 0.150$	
	a	*b*	*a*	*b*	*a*	*b*	*a*	*b*	*a*	*b*
0		**0.16**		**0.27**		**0.36**		**0.45**		**0.52**
0.10		0.51		0.54		0.60		0.67		0.74
0.20		0.71		0.73		0.78		0.85		0.91
0.25	**0.06**	**0.79**	**0.22**	**0.82**	**0.38**	**0.87**	**0.68**	**0.93**	**0.97**	**1.00**
0.30		0.87		0.90		0.95		1.01		1.08
0.40		1.01		1.06		1.11		1.17		1.24
0.50		**1.15(35)***		**1.21**		**1.27**		**1.33**		**1.40**
0.60				1.29(35)*		1.36		1.42		1.50
0.70						1.41(40)*		1.50		1.60
0.75						**1.43(35)***		**1.54(45)***		**1.63(45)***
0.80								1.56(40)*		1.65(40)*
0.90										1.70(30)*
1.00										
DRY	**0.08**	**1.12**	**0.27**	**1.21**	**0.44**	**1.32**	**0.75**	**1.46**	**1.04**	**1.56**

*When ϕ' is greater than the value shown in brackets treat the slope as 'dry'. The shaded area represents the 'dry' condition.

Table 12.2 *Stability coefficients a and b for slope 2:1 (Barnes, 1992)*

h_w/H	$c'/\gamma H = 0.005$		$c'/\gamma H = 0.025$		$c'/\gamma H = 0.050$		$c'/\gamma H = 0.100$		$c'/\gamma H = 0.150$	
	a	*b*	*a*	*b*	*a*	*b*	*a*	*b*	*a*	*b*
0		**0.88**		**1.01**		**1.11**		**1.27**		**1.37**
0.10		1.24		1.31		1.38		1.50		1.60
0.20		1.46		1.51		1.57		1.69		1.78
0.25	**0.06**	**1.56**	**0.24**	**1.60**	**0.42**	**1.66**	**0.75**	**1.78**	**1.07**	**1.87**
0.30		1.64		1.69		1.75		1.86		1.95
0.40		1.81		1.85		1.91		2.02		2.11
0.50		**1.96**		**2.00**		**2.06**		**2.17**		**2.26**
0.60		2.07		2.13		2.22		2.33		2.41
0.70		2.17(45)*		2.22		2.31		2.43		2.52
0.75		**2.20(30)***		**2.27**		**2.35**		**2.48**		**2.57**
0.80		2.22(25)*		2.30(45)*		2.38		2.52		2.62
0.90				2.38(30)*		2.44		2.60		2.70
1.00				**2.39(25)***		**2.50(40)***		**2.67**		**2.79**
DRY	**0.10**	**2.08**	**0.30**	**2.24**	**0.50**	**2.37**	**0.86**	**2.56**	**1.20**	**2.69**

*When ϕ' is greater than the value shown in brackets treat the slope as 'dry'. The shaded area represents the 'dry' condition.

Table 12.3 *Stability coefficients a and b for slope 3:1 (Barnes, 1992)*

h_w/H	$c'/\gamma H = 0.005$		$c'/\gamma H = 0.025$		$c'/\gamma H = 0.050$		$c'/\gamma H = 0.100$		$c'/\gamma H = 0.150$	
	a	*b*	*a*	*b*	*a*	*b*	*a*	*b*	*a*	*b*
0		**1.48**		**1.65**		**1.77**		**1.95**		**2.09**
0.10		1.87		1.95		2.03		2.18		2.30
0.20		2.11		2.16		2.24		2.37		2.48
0.25		**2.21**		**2.26**		**2.33**		**2.45**		**2.56**
0.30		2.31		2.35		2.42		2.54		2.65
0.40		2.49		2.53		2.59		2.71		2.81
0.50	**0.07**	**2.66**	**0.25**	**2.69**	**0.44**	**2.75**	**0.79**	**2.87**	**1.12**	**2.97**
0.60		2.79		2.85		2.92		3.02		3.13
0.70		2.91		2.96		3.03		3.14		3.25
0.75		**2.97**		**3.01**		**3.09**		**3.20**		**3.31**
0.80		3.02		3.06		3.14		3.25		3.36
0.90		3.11		3.16		3.23		3.36		3.47
1.00				**3.25**		**3.32**		**3.45**		**3.57**
DRY	**0.11**	**3.10**	**0.33**	**3.27**	**0.56**	**3.41**	**0.95**	**3.63**	**1.31**	**3.80**

Table 12.4 *Stability coefficients a and b for slope 4:1 (Barnes, 1992)*

h_w/H	$c'/\gamma H = 0.005$		$c'/\gamma H = 0.025$		$c'/\gamma H = 0.050$		$c'/\gamma H = 0.100$		$c'/\gamma H = 0.150$	
	a	*b*	*a*	*b*	*a*	*b*	*a*	*b*	*a*	*b*
0		**2.05**		**2.22**		**2.35**		**2.53**		**2.67**
0.10		2.46		2.54		2.63		2.78		2.90
0.20		2.72		2.77		2.84		2.98		3.09
0.25		**2.83**		**2.88**		**2.94**		**3.07**		**3.18**
0.30		2.93		2.98		3.04		3.16		3.27
0.40		3.13		3.17		3.22		3.34		3.44
0.50	**0.07**	**3.32**	**0.26**	**3.35**	**0.47**	**3.40**	**0.84**	**3.51**	**1.18**	**3.61**
0.60		3.49		3.52		3.57		3.67		3.77
0.70		3.62		3.66		3.70		3.81		3.90
0.75		**3.69**		**3.72**		**3.76**		**3.87**		**3.97**
0.80		3.74		3.78		3.82		3.93		4.03
0.90		3.86		3.89		3.94		4.05		4.15
1.00		**3.97**		**4.00**		**4.05**		**4.16**		**4.26**
DRY	**0.12**	**4.11**	**0.37**	**4.29**	**0.60**	**4.45**	**1.01**	**4.70**	**1.38**	**4.90**

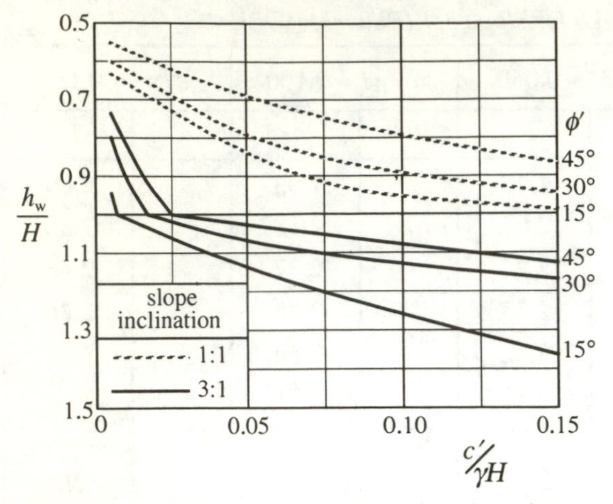

Figure 12.13 *Water table location for 'dry' slopes*

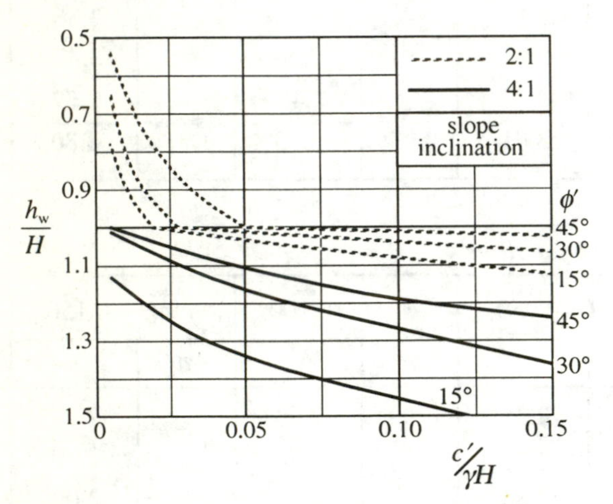

Figure 12.14 *Water table location for 'dry' slopes*

Submerged slopes (Figure 12.15)

When a slope is submerged fully or partially the weight of each slice should be calculated using the bulk unit weight, γ_b, above the external water level, A–B and the submerged or buoyant unit weight γ_{sub} below the line A–B, the shaded area in Figure 12.15. This is because the water in the slope (in the pores) below A–B and its moment about O is balanced by the water outside the slope. Thus the overturning moment is reduced so the factor of safety increases. As the external water level falls (draw-down) the factor of safety will decrease.

The shearing resistance along the slip surface is determined from effective stresses which are obtained partly by assuming submerged unit weights for the slice weights W below the line A–B. If the water table within the slope lies above the line A–B then the remaining pore pressure is determined from the difference between the piezometric head in each slice and the external water level A–B, the hatched area in Figure 12.15.

Rapid drawdown *(Figure 12.16)*

Figure 12.16 shows the upstream face of an earth dam where the steady seepage condition has become established with flow towards the downstream face. The phreatic surface and an equipotential passing through the point P on a trial slip surface are also shown. The initial pore water pressure u_o at P in this condition is:

$$u_o = \gamma_w(h + h_w - h') \tag{12.20}$$

When the reservoir level is lowered seepage will then commence towards the upstream surface, but if the permeability of the soil is low this re-adjustment will not be given sufficient time to establish, and a high water level condition and hence high pore pressures

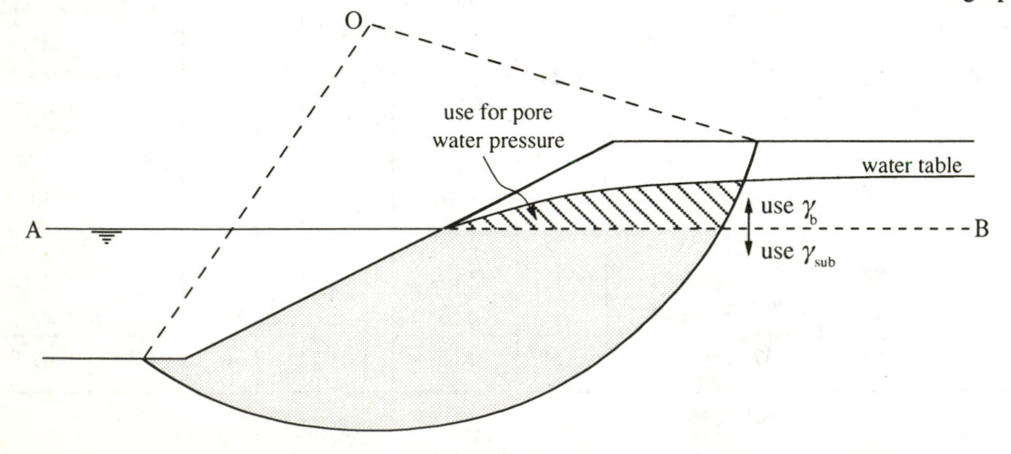

Figure 12.15 *Effect of submergence*

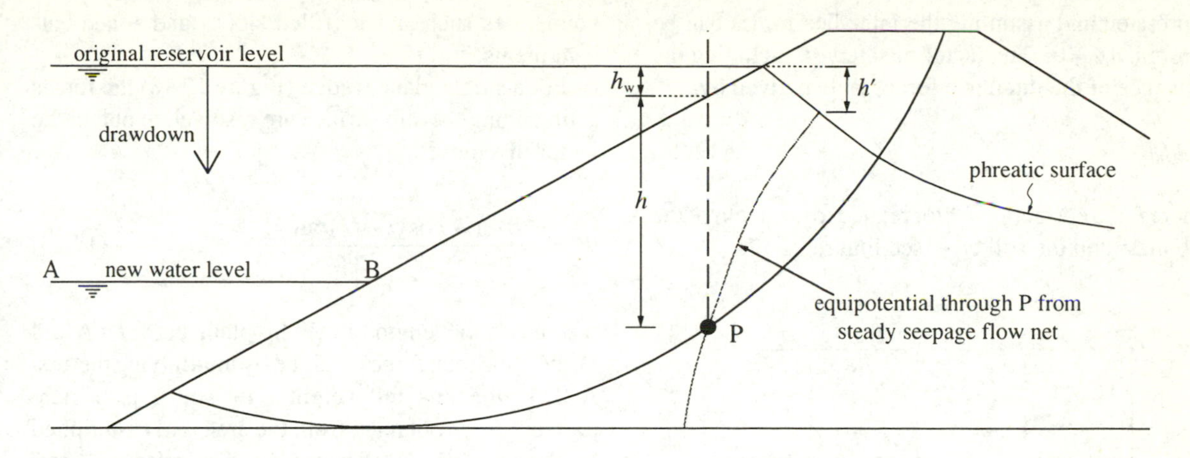

Figure 12.16 *Rapid drawdown*

will remain in the earth dam when support from the external water has been removed. It is found that because of the very slow adjustment a drawdown period of several weeks can still be considered 'rapid' with the soil behaving in an undrained manner.

Bishop and Bjerrum (1960) suggested that the change in pore water pressure u under undrained conditions could be represented by:

$$\Delta u = \bar{B}\Delta\sigma_1 \tag{12.21}$$

where $\Delta\sigma_1$ is the change in total stress and $\bar{B}$ represents a combined pore pressure parameter for a partially saturated compacted fill. In this case $\Delta\sigma_1 = -\gamma_w h_w$ so:

$$\Delta u = -\bar{B}\gamma_w h_w$$

Therefore, the pore water pressure immediately after drawdown is:

$$u = u_o + \Delta u$$

$$u = \gamma_w\left[h + h_w\left(1-\bar{B}\right) - h'\right]$$

A conservative approach is obtained by assuming $\bar{B}$ = 1 and h' can be neglected since it is generally small, in which case pore pressures on the slip surface are given by:

$$u = \gamma_w h$$

The stability can then be analysed in the manner described above assuming a partially submerged slope with an external water level A–B, as in Figure 12.15.

Non-circular slip surfaces – Janbu method

Where non-homogeneous soil profiles exist, such as with layered strata, a non-circular slip surface may be more appropriate. Using the method of slices (see Figure 12.11) and considering overall horizontal equilibrium as the stability criterion Janbu (1973) obtained the following expression for the average factor of safety along the slip surface:

$$F = \frac{\sum\left[c'b + (W + dX - ub)\tan\phi'\right]m_\alpha}{\sum(W + dX)\tan\alpha} \tag{12.22}$$

where

$$m_\alpha = \frac{\sec^2\alpha}{1 + \dfrac{\tan\phi'\tan\alpha}{F}} \tag{12.23}$$

and $dX = X_1 - X_2$, the resultant vertical interslice force.

A similar procedure for the solution of the Bishop simplified method is required using an iterative procedure to converge towards the value of F and an assumption for the position of the interslice forces on each slice (a line of thrust) and their magnitudes must be included.

A simplified procedure has been suggested by Janbu where the factor of safety F_0 is obtained using the above expression and assuming the interslice forces can be ignored, $dX = 0$. The factor of safety, F including the influence of the interslice forces is then given by:

$$F = f_0 F_0 \tag{12.24}$$

where f_0 is a correction factor related to the depth of the slip mass and the soil type, see Figure 12.17.

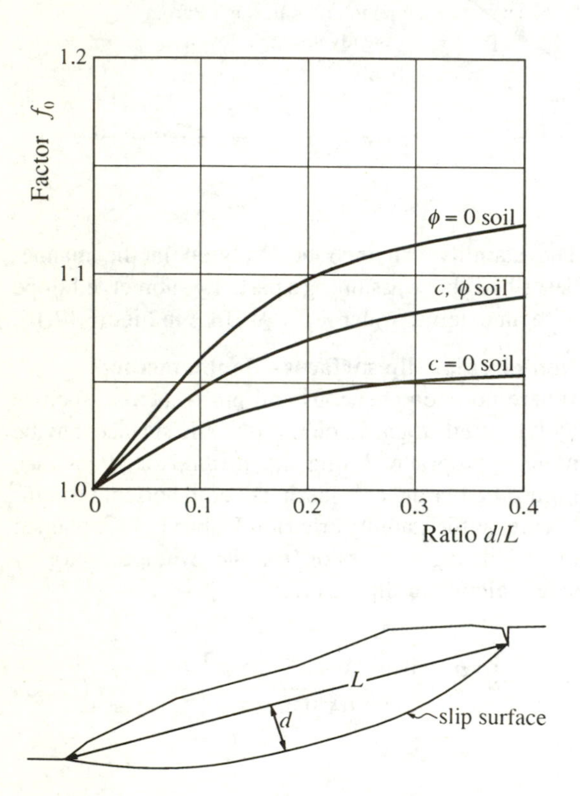

Figure 12.17 *Correction factor f_0*

Morgenstern and Price (1965) published a rigorous method of analysis for slip surfaces of any shape by considering force and moment equilibrium for each slice and assuming a relationship between the normal and tangential interslice forces, usually in the form of $X = \lambda E$. The complex iterative calculations required to obtain values of F and λ for a slip surface mean that use of a computer program is essential.

Wedge method – single plane (Figure 12.18)
Wedge analysis is a useful technique when assessing the stability of a cross-section with distinct planar boundaries such as backfilled slopes and zoned embankments.

For a single plane wedge (Figure 12.18) the forces acting along the slip surface are resolved to obtain the factor of safety:

$$F = \frac{c'L + (W\cos\alpha - U)\tan\phi'}{\sum W \sin\alpha} \tag{12.25}$$

where L is the length of the slip plane between A and D. W is the total force obtained by multiplying the area ABD by the bulk unit weight of the soil. U is the pore pressure force obtained from the area ACD multiplied by the unit weight of water, γ_w. The effective force $N' = W\cos\alpha - U$.

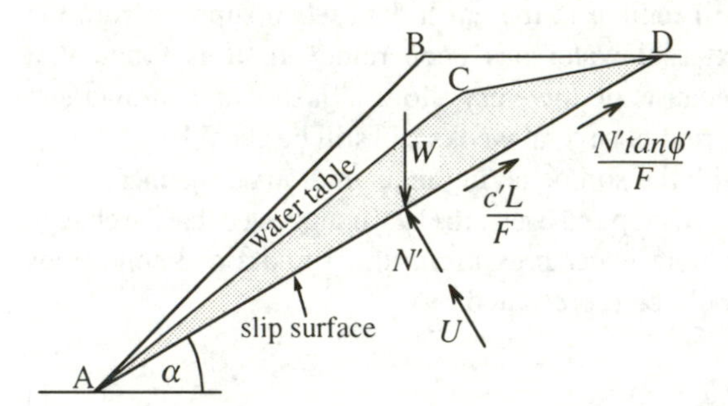

Figure 12.18 *Wedge method – single plane*

Wedge method – multi-plane *(Figure 12.19)*
When there is more than one plane surface the sliding mass can be separated into wedges with vertical interfaces with an inter-wedge force P. The magnitude of P is not known but can be found from a polygon of forces providing an assumption about its inclination θ is made. The value of the factor of safety obtained can be sensitive to the value of θ chosen. A reasonable assumption is that:

$$\theta = \phi_m \quad \text{or} \quad \tan\theta = \frac{\tan\phi'}{F} \tag{12.26}$$

where ϕ_m represents the mobilised angle of shearing resistance. Since F is not known initially a trial approach is required by adjusting the value for F until convergence is achieved.

From Figure 12.19 each wedge is considered separately. The magnitude of the total force W_1, is obtained from the area of the wedge multiplied by the unit weight of the soil. A polygon of forces can now be drawn for wedge 1. The magnitude and direction of $c'L_1/F$ (the cohesion part of T_1) are known. The direction of N_1' is known. The resultant R_1 of the frictional component is assumed to act at the angle ϕ_m' from the direction of N_1'. The direction of P_1 is assumed so the polygon of forces can be closed and a value of P_1 found.

The same procedure is adopted for wedge 2 to obtain a value of P_2. If $P_2 = P_1$ the correct value of F has been chosen. Otherwise the procedure must be repeated until $P_2 = P_1$. Alternatively, the value of $P_2 - P_1$ could be plotted versus the factor of safety as shown on Figure 12.19. The correct value of F will occur where $P_2 - P_1 = 0$. This approach will also give some indication of the adjustment to the value of F.

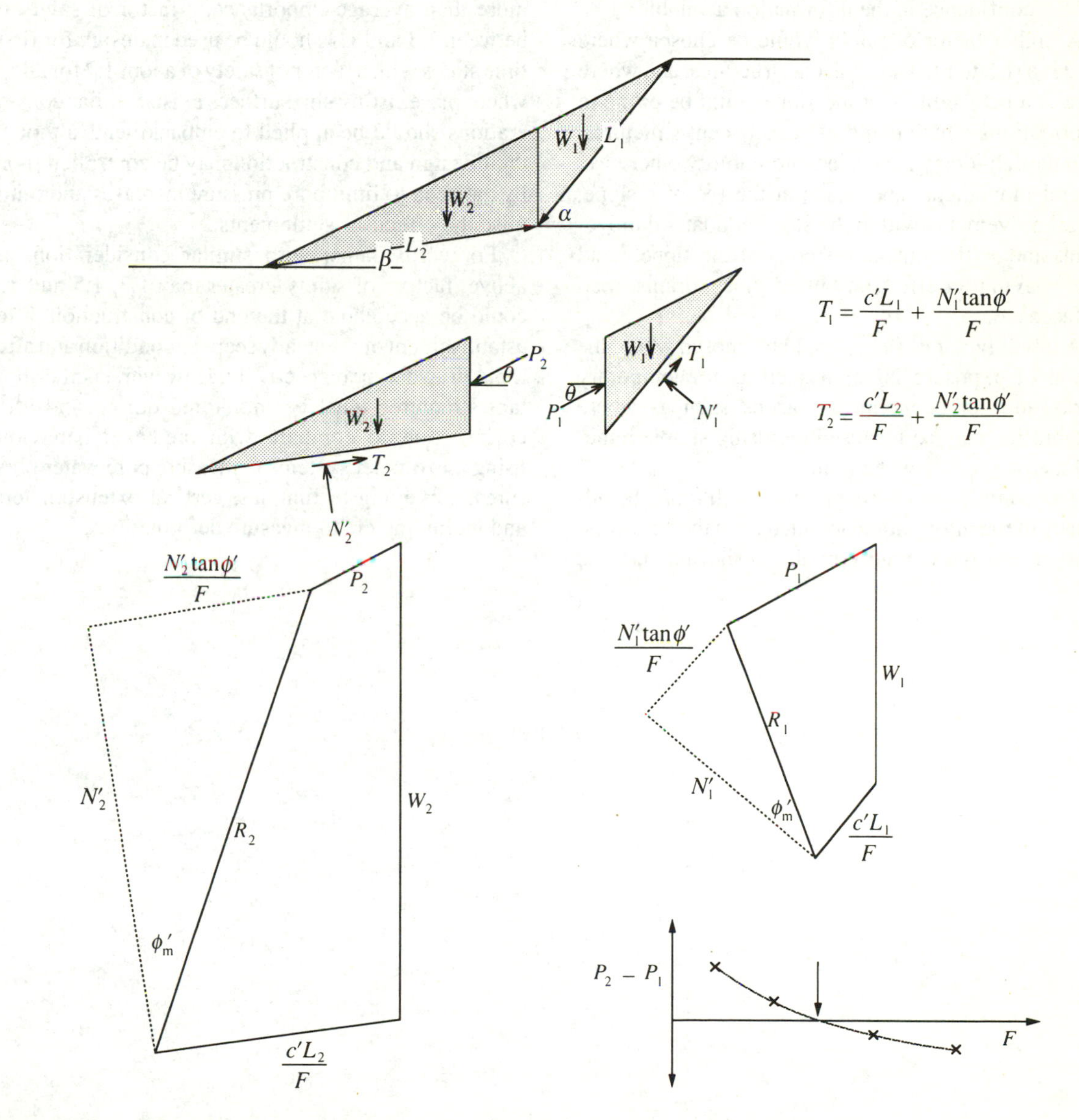

Figure 12.19 *Wedge method – two planes*

Factor of safety

The choice of an acceptable factor of safety requires sound engineering judgement due to the multitude of factors which must be considered. It must also be remembered that a factor of safety can only be determined when there is an appropriate method of analysis; some modes of failure such as flow slides cannot be readily analysed.

The factors to be considered generally fall under two headings: the consequences of a failure occurring and the confidence in the information available.

A higher factor of safety would be chosen where there is a risk to life and adjacent structures, and where there is a possibility that the slope could be prone to deformations which could affect adjacent structures, even though it may have adequate stability otherwise. Lateral movements and heave at the toe of a slope, lateral movements within the slope and lateral movements and settlements at the crest of the slope could occur having an effect on buried pipes, drains, road surfaces etc.

A lower factor of safety could be chosen where the period of exposure is small such as for temporary works and for economical reasons such as where instabilities may be localised requiring simple remedial measures only when required.

The complexity of the ground conditions, the adequacy of the information obtained from the site investigation and the certainty of the design parameters, such as shear strength and pore pressures, all affect the confidence in the chosen factor of safety. Previous local experience can be invaluable, especially if there is any knowledge about the presence of pre-existing slip surfaces. Future changes such as to water table levels, surface profiles must also be considered.

The Code of Practice for Earthworks (BS 6031:1981) suggests that for cuttings and natural slopes, provided a good standard of site investigation is obtained, and considering the factors mentioned above with no more than average importance, a factor of safety of between 1.3 and 1.4 should be used in design for first-time slides with a factor of safety of about 1.2 for slides where pre-existing slip surfaces exist. Similar considerations should be applied to embankments although their design and construction may be controlled more by the need to limit pore pressure increases and minimise the effects of settlements.

For earth dams, with similar considerations as above, factors of safety greater than 1.3, 1.5 and 1.2 could be acceptable at the end of construction, after establishment of the steady seepage condition and after rapid drawdown, respectively. However, such important structures must be monitored during and after construction to compare with predicted behaviour using piezometer systems to measure pore water pressures, surveying techniques, vertical extensometers, and inclinometers to measure deformations.

Worked Example 12.1 *Translational slide*
A long slope comprising fine sand exists at an inclination of 4:1 (horizontal:vertical) with a water table parallel to the slope. The unit weight of the sand is 20 kN/m³ and the angle of internal friction is 26.6º. Assume the unit weight of water to be 10 kN/m³. Determine the factor of safety of the slope assuming:
a) the water table exists at ground level (worst case)
b) the slope is dry (best case)
c) the water table exists at 2.0 m below ground level

$$\tan\beta = \frac{1}{4} = 0.25$$

a) waterlogged slope
assuming $c' = 0$ Equation 12.3 gives

$$F = \frac{(20-10)\times 0.50}{20\times 0.25} = 1.00$$

i.e. the slope is very close to failure in this condition

b) dry slope
Using Equation 12.2

$$F = \frac{0.50}{0.25} = 2.00$$

the slope has more than adequate stability in this condition

c) water table at 2.0 m below ground level
with $c' = 0$ Equation 12.1 simplifies to

$$F = \frac{\tan\phi'}{\tan\beta}\frac{(\gamma z - \gamma_w z + \gamma_w h_w)}{\gamma z}$$

This expression can be used for slip surfaces below the water table:

$z = 2.0$ m $\quad F = \frac{(40-20+20)}{40}\times 2.0 = 2.0$ (dry slope case)

$z = 3.0$ m $\quad F = \frac{(60-30+20)}{60}\times 2.0 = 1.67$

$z = 4.0$ m $\quad F = \frac{(80-40+20)}{80}\times 2.0 = 1.50$

$z = 6.0$ m $\quad F = \frac{(120-60+20)}{120}\times 2.0 = 1.33$

$z = 10.0$ m $\quad F = \frac{(200-100+20)}{200}\times 2.0 = 1.20$

$z = 50$ m $\quad F = \frac{(1000-500+20)}{1000}\times 2.0 = 1.04$

With a constant strength (homogeneous slope) the factor of safety decreases with depth. To assess a likely depth of slip surface any layering of strata should be included or an increase in strength (increasing ϕ') with depth should be considered.

Assuming suction above the water table Equation 12.6 could be used:

$z = 0.5$ m $\qquad F = \left(1 + \dfrac{10 \times 1.5}{20 \times 0.5}\right) \times 2.0 = 5.00$

$z = 1.0$ m $\qquad F = \left(1 + \dfrac{10 \times 1.0}{20 \times 1.0}\right) \times 2.0 = 3.00$

$z = 1.5$ m $\qquad F = \left(1 + \dfrac{10 \times 0.5}{20 \times 1.5}\right) \times 2.0 = 2.33$

Worked Example 12.2 $\phi_u = 0^\circ$ *analysis – method of slices*
A slope is to be cut into a soft clay with undrained shear strength of 30 kN/m² and unit weight of 18 kN/m³. The slope is 8.0 m high and its inclination is 2:1 (horizontal:vertical). Determine the factor of safety for the trial circle shown on Figure 12.20.

Equation 12.9 is used with values of b, α and mid-slice height h determined for each slice. The weight of each slice is obtained from $W = \gamma bh$.

slice no.	b	h	W	α°	$W\sin\alpha$	$b\sec\alpha$
1	0.65	0.15	1.8	–25.7	–0.8	0.72
2	2.0	1.23	43.2	–20.0	–14.8	2.13
3	2.0	2.82	100.8	–1.8	–20.6	2.04
4	2.0	4.06	146.2	–3.9	–9.9	2.01
5	2.0	5.08	182.9	3.9	12.4	2.01
6	2.0	5.82	209.5	11.8	42.8	2.04
7	2.0	6.26	225.4	20.0	77.1	2.13
8	2.0	6.36	229.0	28.6	109.6	2.28
9	2.0	6.02	216.7	38.0	133.0	2.54
10	2.0	0.60	165.6	48.9	124.8	3.04
11	1.7	1.94	58.4	61.7	51.4	3.59
					$\Sigma =$ 505.0	24.53

$$F = \frac{30 \times 24.53}{505.0} = 1.46$$

The included angle θ is 97° and the radius of the circle is 14.6 m so the length of the circular arc is

$$L = R\theta = 14.6 \times 97 \times \frac{\pi}{180} = 24.72 \text{ m}$$

Since this is a more accurate measure of $\sum b\sec\alpha$ a more accurate value of the factor of safety is obtained as

$$F = \frac{30 \times 24.72}{505.0} = 1.47$$

The accuracy of the factor of safety F is affected by the number of slices adopted.
A computer analysis (SLOPE ©) of the same slope using 25 slices gave a value of $F = 1.486$.
This circle does not give the lowest factor of safety so it is not the critical slip circle. A computer run (SLOPE©) obtained the F values for circles with their centres on a grid pattern and all tangential to the depth of 1.6 m below the toe. Contours of factor of safety are plotted on this grid and the circle with the lowest value of F lies at the centre of the contour plot, as shown on Figure 12.20. This gives the minimum factor of safety as 1.430.

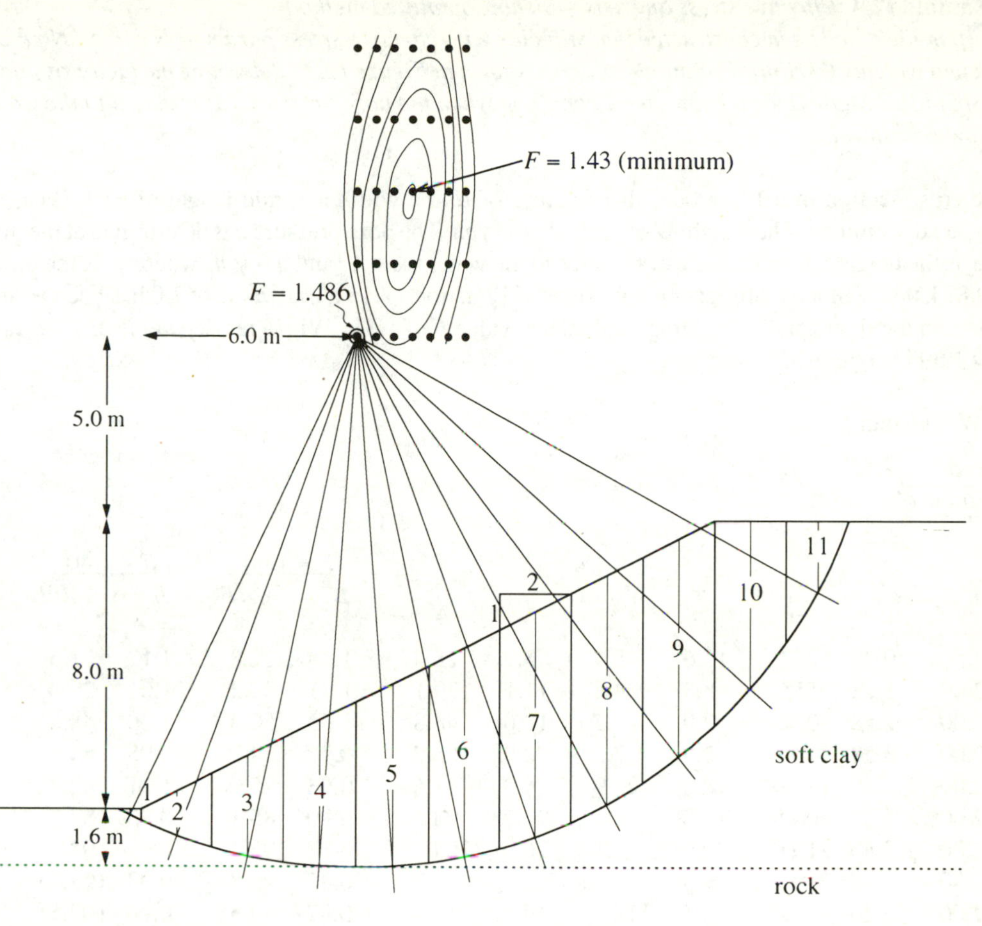

Figure 12.20 *Worked Example 12.2*

Worked Example 12.3 $\phi_u = 0^o$ *analysis – Taylor's method*

For the slope in Example 12.2 determine the minimum factor of safety using Taylor's stability number approach.

Depth factor $D = \dfrac{9.6}{8.0} = 1.2$

From the stability chart, Figure 12.10

$N_s = 0.146$

$$F = \frac{30}{0.146 \times 18 \times 8} = 1.43$$

Worked Example 12.4 *Effective stress analysis – Bishop simplified method*
A 2:1 slope 10 m high, has been constructed in a stiff clay with effective stress parameters $c' = 5$ kN/m² and $\phi' = 30°$ and bulk unit weight 20 kN/m³. For the circular arc shown in Figure 12.21 determine the factor of safety using the Bishop Simplified Method for the long-term condition when the pore pressures are related to the water table at 5.0 m below crest level.

Splitting the cross-section into 12 vertical slices values of α, the width and mid-height of each slice, b and h, respectively, are determined. The weight W of each slice = γbh. The pore pressure u is determined at the mid-point of each slice as the height above the base of the slice to the water table h_w and $u = \gamma_w h_w$ where γ_w is the unit weight of water = 9.81 kN/m³. Since F appears on both sides of Equation 12.15 a trial value of 1.0 has been assumed for the expression on the right hand side giving a calculated value F of 1.31. With a revised value of 1.4 a calculated value of F = 1.394 is obtained.

$$A = c'b + (W - ub)\tan\phi'$$

$$B = \frac{\sec\alpha}{1 + \dfrac{\tan\alpha\tan\phi'}{F}}$$

								F = 1.00		F = 1.40	
slice no.	b	h	W	α	$W\sin\alpha$	u	A	B	$A \times B$	B	$A \times B$
1	2.50	0.40	20.0	–15.6	–5.4	3.9	18.4	1.24	22.8	1.17	21.5
2	2.00	1.38	55.2	–8.8	–8.4	11.1	29.1	1.11	32.3	1.01	29.4
3	2.00	2.58	103.2	–2.9	–5.2	18.0	48.8	1.03	50.3	1.00	48.8
4	2.00	3.58	143.2	2.9	7.2	22.9	66.2	0.97	64.2	0.98	64.9
5	2.00	4.40	176.0	8.8	26.9	26.0	81.6	0.93	75.9	1.01	82.4
6	2.00	5.02	200.8	14.8	51.3	27.2	94.5	0.90	85.1	0.93	87.9
7	2.00	5.40	216.0	20.9	77.1	26.0	104.7	0.88	92.1	0.92	96.3
8	2.00	5.48	219.2	27.3	100.5	21.9	111.3	0.87	96.8	0.93	103.5
9	2.00	5.26	210.4	34.1	118.0	14.8	114.4	0.87	99.5	0.94	107.5
10	2.00	4.68	187.2	41.5	124.0	4.2	113.2	0.88	99.6	0.98	110.9
11	2.00	3.64	145.6	49.9	111.4	0	94.1	0.92	86.6	1.04	97.9
12	1.67	1.57	52.4	59.1	45.0	0	38.6	0.99	38.2	1.15	44.4
					Σ = 642.4				Σ = 843.4		Σ = 895.4

$$F = \frac{843.4}{642.4} = 1.31 \qquad F = \frac{895.4}{642.4} = 1.394$$

Worked Example 12.5 *Effective stress analysis – computer application*
As shown in Example 12.2 a search must be made for the critical slip surface which gives the lowest factor of safety.

For a homogeneous slope this can be achieved by specifying a grid of circle centres and some points for the circles to pass through or to be tangential to a line. Using a commercial software program called SLOPE© marketed by Geosolve the slope in Example 12.4 was analysed using a grid of 25 (5 × 5) circle centres and a number of points beyond the toe for the circles to pass through. The contours of equal factor of safety plotted on the grid of centres gives a series of elliptical loops which lead to the critical circle centre. The lowest factor of safety for each point is plotted on Figure 12.21 and shows that the critical circle cuts at 2.5 m beyond the toe of the slope. The lowest factor of safety was determined as 1.40 with 20 slices and 5 iterations (varying F to find F).

Example 12.4 is a hand calculation for the critical slip circle to check the value of F obtained. It will be appreciated that to search for this critical slip circle can entail much trial and error and the benefits of using a computer program are obvious.

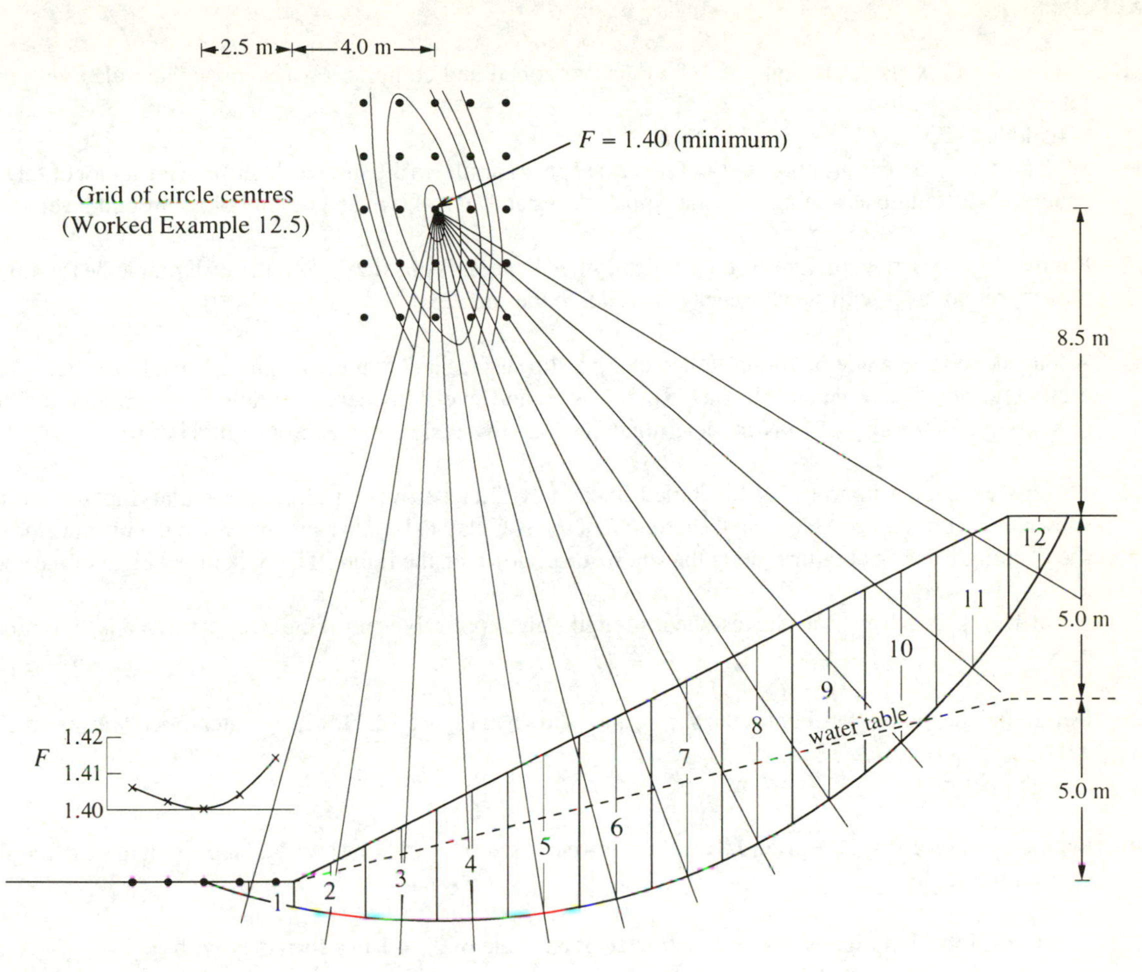

Figure 12.21 *Worked Example 12.4*

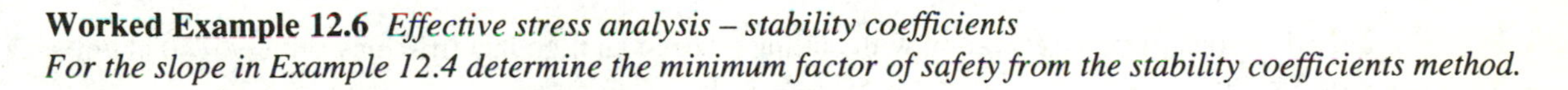

Worked Example 12.6 *Effective stress analysis – stability coefficients*
For the slope in Example 12.4 determine the minimum factor of safety from the stability coefficients method.

Using Equation 12.19 and Table 12.2 for a slope of 2:1

$$\frac{c'}{\gamma H} = \frac{5}{20 \times 10} = 0.025$$

$$\frac{h_w}{H} = \frac{5.0}{10} = 0.50$$

From the Table $a = 0.24$ and $b = 2.00$

$\therefore\ F = 0.24 + 2.00 \times \tan 30^\circ = 1.395 \quad (= 1.40)$

Exercises

12.1 A long slope exists at an angle of 14º to the horizontal and comprises overconsolidated clay with the following properties:
$c' = 5$ kN/m^2 $\phi' = 26º$ $\gamma = 19$ kN/m^3
With the water table at ground level and steady seepage parallel to the surface determine the factor of safety against shear failure assuming potential slip surfaces at 2.0 m, 4.0 m and 6.0 m below ground level.

12.2 For the slope described in Exercise 12.1, determine the factors of safety when the water table lies at 4.0 m below ground level with steady seepage parallel to the surface.

12.3 A long slope at an angle of 10º has failed along a slip surface at 2.5 m below ground level parallel to the ground surface. The water table lies at 0.5 m below ground level with seepage parallel to the ground surface. Assuming $c' = 0$ and $\gamma = 21$ kN/m^3 determine the operative residual value of the angle of friction ϕ_r'.

12.4 A slope excavated in soft clay has failed imediately after excavation along a circular slip surface as shown in Figure 12.22. Assuming the strength of the soft clay to be constant determine the mean value of shear strength acting at failure along the slip surface shown on the figure. The bulk unit weight of the clay is 18.5 kN/m^3.
Note: The approach gives the lowest shear strength value applicable only if the circle analysed is the critical one.

12.5 Determine the factor of safety for the slip surface shown in Figure 12.23, with a water-filled tension crack, 3.0 m deep.
$c_u = 35$ kN/m^2 $\gamma = 20$ kN/m^3

12.6 For the slope profile in Exercise 12.4 use Taylor's curves to determine the lowest shear strength which could support the slope.

12.7 A cutting slope 7 m high has been constructed at an angle of 2.5 : 1 in a firm clay with
$c_u = 55$ kN/m^2 and $\gamma = 19$ kN/m^3.
The bedrock surface lies at 14 m below original ground level. Using Taylor's curves, determine the short-term factor of safety.

12.8 It is required to construct an excavation with cutting slopes 8 m high in a firm clay with $c_u = 40$ kN/m^2 and $\gamma = 20.5$ kN/m^3.
The bedrock surface lies at 12 m below ground level. Using Taylor's curves, determine the slope angle which could be adopted to ensure short-term stability with a factor of safety of 1.5.

12.9 For the slope profile in Figure 12.24 the groundwater regime is represented by steady seepage with pore pressures given by the water table level shown. Determine the factor of safety for long-term conditions on the given slip surface using the Bishop simplified method of slices.
$c' = 3$ kN/m^2 $\phi' = 26º$ $\gamma = 22.2$ kN/m^3

12.10 For the slope and groundwater conditions in Exercise 12.9 determine the minimum factor of safety, using the stability coefficients method, Equation 12.19.

12.11 The slope shown in Figure 12.25 supports a lagoon with water 6 m deep and with a water table inside the slope at 2 m below ground level. Determine the factor of safety on the given slip surface using the Bishop simplified method of slices.
$c' = 5$ kN/m^2 $\phi' = 23°$ $\gamma = 19$ kN/m^3

12.12 The water level in the lagoon in Exercise 12.11 has been drawn-down rapidly to the base level. The groundwater regime is now represented by the water table within the slope as in Exercise 12.11 but with a water table at ground level in the lagoon area. For these slope and groundwater conditions determine the factor of safety on the slip surface shown.

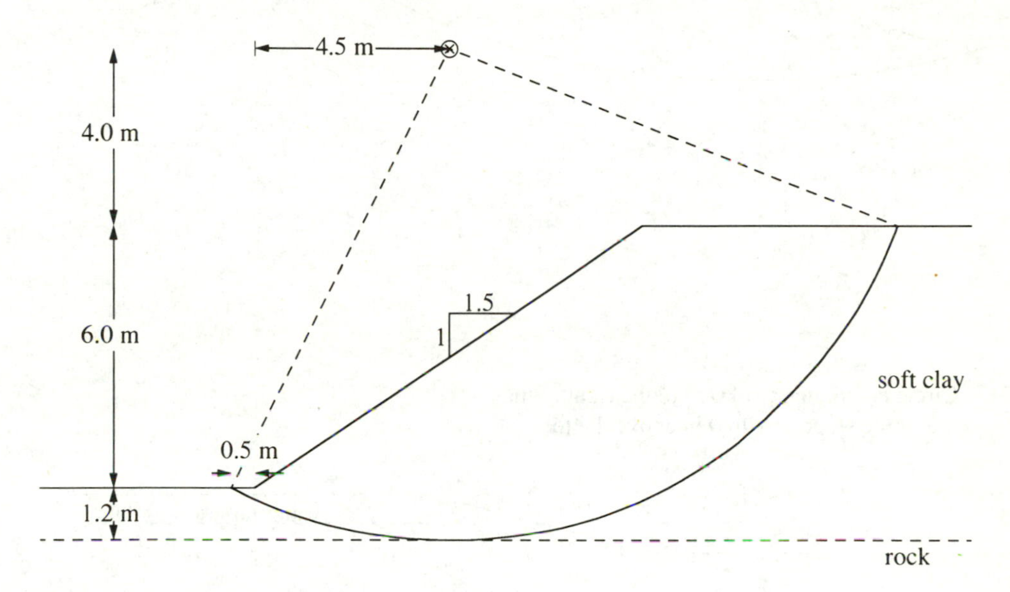

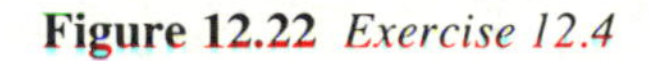
Figure 12.22 *Exercise 12.4*

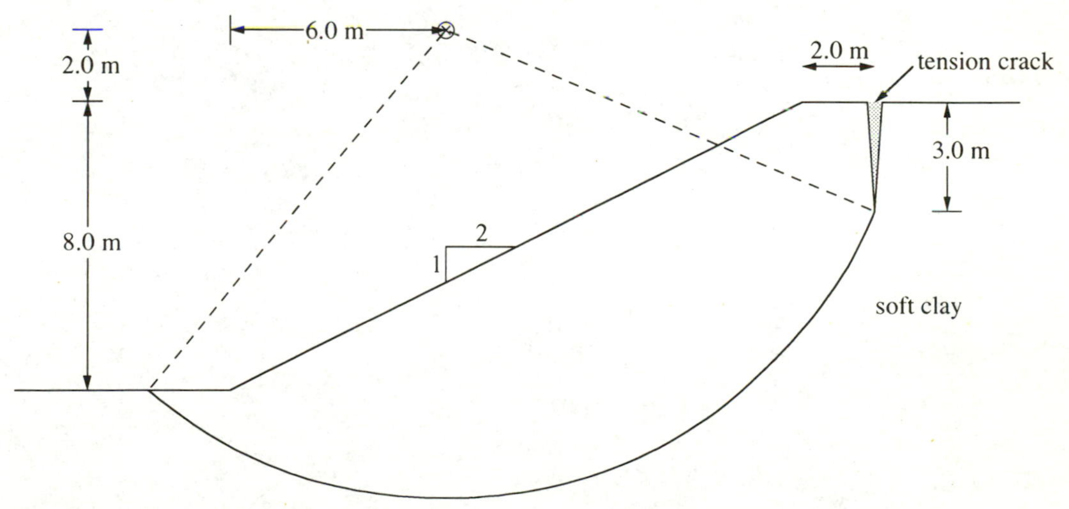

Figure 12.23 *Exercise 12.5*

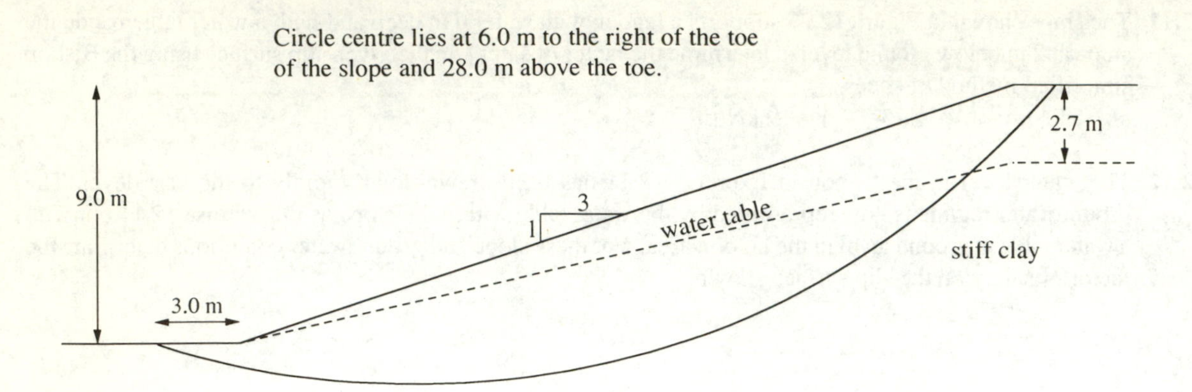

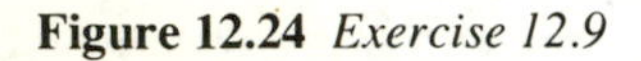
Figure 12.24 *Exercise 12.9*

Circle centre lies at 14.0 m to the right of the toe of the slope and 46.0 m above the toe.

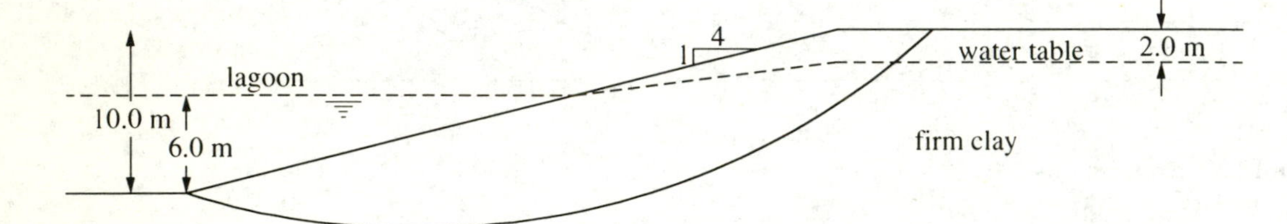

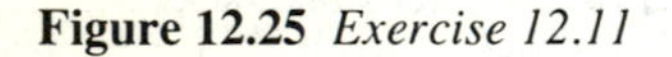
Figure 12.25 *Exercise 12.11*

13 Earthworks and Soil Compaction

Earthworks

Introduction

Earthworks or earth structures are constructed where it is required to alter the existing topography. They comprise excavation and filling and are most commonly formed for highway works such as cuttings and embankments but they may also consist of site levelling such as for industrial or housing estates and excavation and backfilling of quarries, trenches and foundations.

The stability of cutting and embankment slopes is discussed in Chapter 12. In this chapter the construction and compaction of the fill material is considered.

Construction plant *(Figure 13.1)*

The basic functions of construction plant are:

- *excavation*
  To break up and remove soil or rock from the cut areas. Back-acters, rippers, pneumatic breakers and explosives are used for this purpose.
- *loading*
  Some items of plant such as dump trucks require an additional item, a back-acter, to load soil into them whereas scrapers excavate and load themselves.
- *transporting*
  For backfilling pipes, services and foundations the excavated spoil is usually heaped close by so transport is not necessary. On some highway schemes haul distances can be several kilometres when the overall cost and efficiency will be determined by the ability of the plant to run in high gears (faster speeds and more fuel economy) along a haul road. This haul road will simply be the surface of the excavation or fill areas so it must have sufficient strength or trafficability to support the plant. Factors such as gradients, rolling resistance and rutting beneath tyres and tracks should be considered.
  Towed scrapers (tractor-pulled) are most efficient for short hauls, say less than 200–300 m, motorised scrapers are required for intermediate haul distances while dump trucks are most efficient over long haul distances, say greater than 1–2 km.

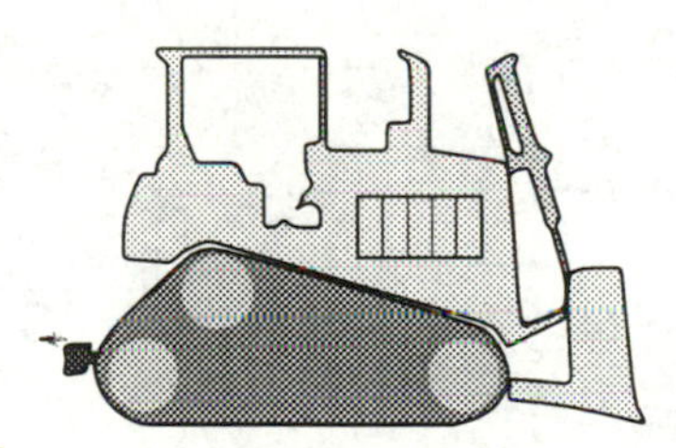
Dozer

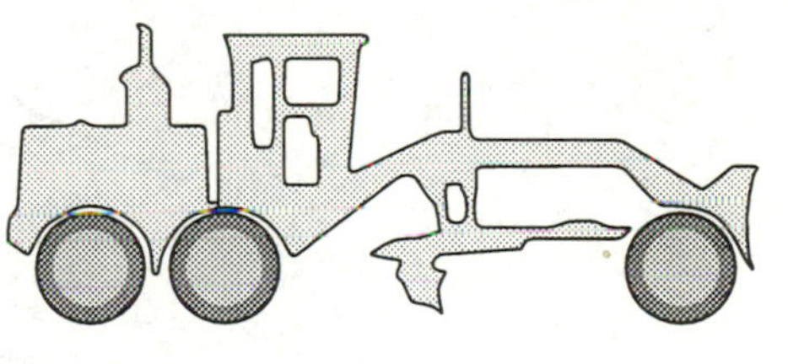
Grader

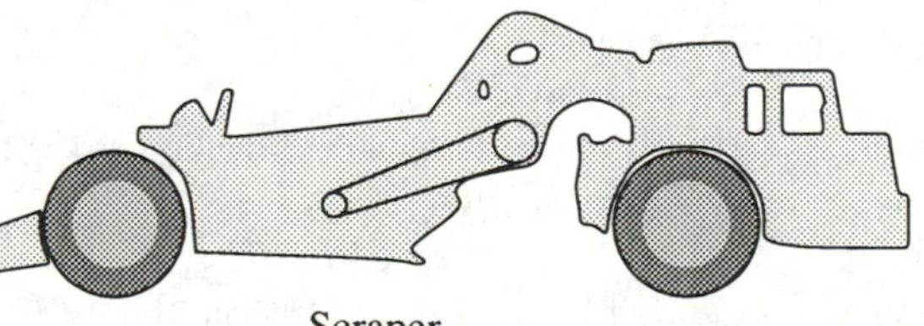
Scraper

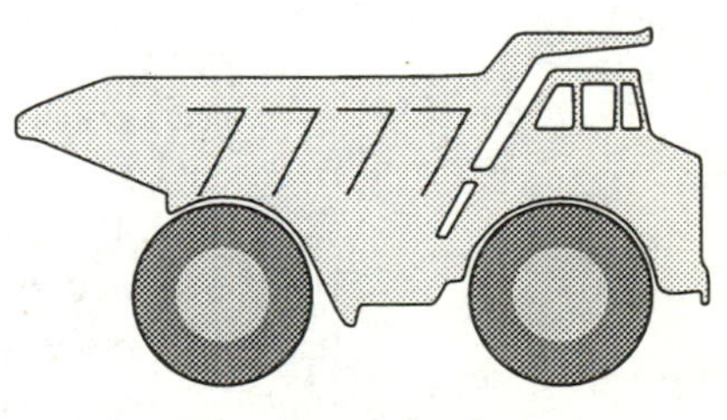
Dump-truck

Figure 13.1 *Typical earthmoving plant*

- *depositing and spreading*
  Some items of plant such as dump trucks can only deposit their loads by tipping in a heap. The soil must then be spread in a thin layer by a dozer or grader in preparation for compaction. Scrapers spread the soil as they unload and provide better mixing of the soils.
- *compaction*
  On large-scale bulk earthworks this will be carried out using rollers running over the surface of the previously spread layers with sufficient compactive effort to provide a stable fill material. These are described later. In more confined excavations such as for backfilling services and foundations, plate compactors, tampers, rammers and dropping-weight compactors may be used.

Fill material used for landscaping purposes may be given little or no compaction since high strengths and low compressibility are not so important. Nominal compaction obtained from the tracks of a dozer is often considered sufficient.

There is a wide variety of earthmoving plant for different purposes and scales of operation. Four basic items are illustrated in Figure 13.1.

Purpose and types of materials *(Figure 13.2)*
Simple earth structures such as flood banks and levées often comprise a homogeneous mass of one soil type. For more important structures different types of materials are required for different purposes, see Figure 13.2, for example. This shows that an earth structure can be more complex than is apparent from the surface.

Table 13.1 lists the types of materials which can form the various parts of a highway structure, summarised from the Specification for Highway Works, 1992.

Material requirements
To ensure stability of an earth fill and to minimise volume changes after construction (swelling and shrinkage) all of the materials used must be of low compressibility, have adequate shear strength after compaction and contain minerals which are not prone to volume and moisture content changes.

In addition they must be inert (unreactive, insoluble), unfrozen (thawing releases excess water), non-degradable (wood tissues, perishable materials), non-hazardous (chemically and physically harmful) and not susceptible to spontaneous combustion such as some unburnt colliery wastes.

For certain specific purposes they may be required to be:

- free-draining – such as for starter layers and drainage backfill so the fines content must be limited to ensure adequate permeability.
- non-crushable – a minimum 10% fines test value is normally specified to ensure that the individual particles (of rock) will not break down, produce more fines and reduce the permeability both under the initial compaction stresses and for long-term durability.

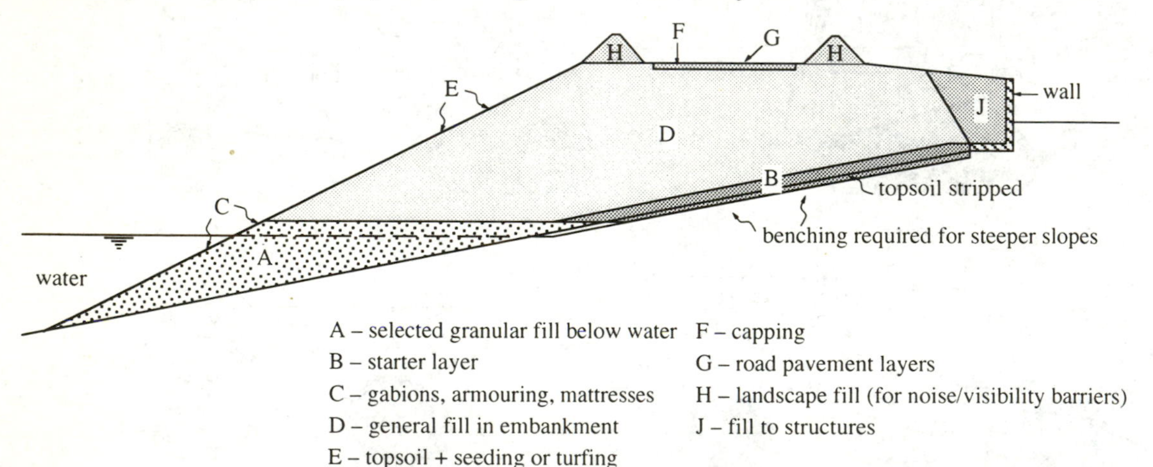

Drainage - required for beneath and alongside the embankment, behind the wall, and alongside the road surface and possibly on the slope surfaces

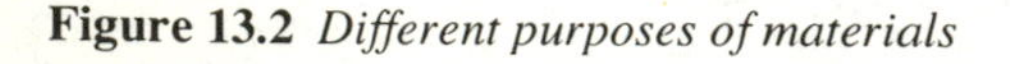
Figure 13.2 *Different purposes of materials*

Table 13.1 *Acceptable earthworks materials*
(Abridged version from Specification for Highway Works, 1992)

CLASS		Material description	Typical use	Typical properties		Compaction requirements*
GENERAL GRANULAR FILL	1A	well graded	General fill	< 125 mm < 15% fines	$U > 10$	Method 2
	1B	Uniformly graded			$U < 10$	Method 3
	1C	Coarse granular		< 500 mm < 15% fines	$U > 5$	Method 5
GENERAL COHESIVE FILL	2A	Wet cohesive	General fill	< 125 mm > 15% fines	$w > PL - 4$	Method 1
	2B	Dry cohesive			$w < PL - 4$	Method 2
	2C	Stony cohesive		< 125 mm 15% > 2 mm		Method 2
	2D	Silty cohesive		< 125 mm > 80% fines		Method 3
	2E	Reclaimed PFA		< 20% FBA		End product 95% ρ_{dmax} (2.5kg)
GENERAL CHALK FILL	3	Chalk	General fill			Method 4
LANDSCAPE FILL	4	Various	Fill to landscape areas			Nominal
TOPSOIL	5A	On site topsoil or turf	Topsoiling			
	5B	Imported topsoil		Complies with B.S. 3882		
	5C	Imported turf	Turfing	Complies with B.S. 3969		
SELECTED GRANULAR FILL	6A	Well graded	Below water	< 500 mm < 5% fines	$U > 10$	No compaction
	6B	Coarse granular	Starter layer	< 500 mm 90% > 125 mm	$U > 5$	Method 5
	6C	Uniformly graded		< 125 mm 90% > 2 mm	$U < 10$	Method 3
	6D		Starter layer below PFA	< 10 mm	$U < 10$	Method 4
	6E		For stabilisation with cement to form capping	< 125 mm LL < 45 PI < 20 org. < 2% fines < 15%	SO4 < 1%	Not applicable
	6F1	Fine grading	Capping	OMC−2 < w < OMC	< 75 mm	Method 6
	6F2	Coarse grading			< 125 mm	
	6G		Gabion filling	< 200 mm min. size > mesh opening		None
	6H		Drainage layer to reinforced earth structure	< 20 mm chemically stable		Method 3
	6I	Well graded	Fill to reinforced earth	< 125 mm < 15% fines chemically stable, frictional	$U > 10$	Method 2
	6J	Uniformly graded			$5 < U < 10$	Method 3
	6K		Lower bedding — Corrugated steel structures	< 20 mm PI < 6 < 10% fines OMC−2 < w < OMC+1 $U > 5$	chemically stable	End product 90% ρ_{dmax} (Vib)
	6L	Uniformly graded	Upper bedding	< 10 mm		None
	6M		Surround	< 75 mm As 6K		As 6K
	6Q	Well graded, uniformly or coarse	Overlying fill	As Class 1A, 1B or 1C		
	6N	Well graded	Fill to structures	< 75 mm minimum strength and permeability	$U > 10$	End product 95% ρ_{dmax} (Vib)
	6P				$U > 5$	
SELECTED COHESIVE FILL	7A	Cohesive	As 6N	15 – 100% fines LL< 45 PI < 25 minimum strength		End product 100% ρ_{dmax} (2.5kg)
	7B	Conditioned PFA	As 6N and to reinforced earth	w controlled minimum strength and permeability		End product 95% ρ_{dmax} (2.5kg)
	7C	Wet cohesive	Fill to reinforced earth	Minimum strength and chemically stable		Method 1
	7D	Stony cohesive				Method 2
	7E		For stabilisation with lime to form capping	> 15% fines PI > 10 org < 2% SO_4 low		Not applicable
	7F	Silty cohesive	For stabilisation with cement to form capping	> 15% fines 80% < 2 mm		
	7G	Conditioned PFA		SO_4 < 1%		
	7H	Wet, dry, stony or silty cohesive and chalk	Overlying fill for corrugated structures	As Class 2A, 2B, 2C, 2D or 3 chemically stable		
MISC. FILL	8	Class 1, 2 or 3	Lower trench fill	Stones, clay lumps must be < 40 mm		
STABILISED MATERIALS	9A	Cement stabilised well graded	Capping	Class 6E + min. 2% cement		Method 6
	9B	Cement stabilised silty cohesive		Class 7F + min. 2% cement MCV < 12		Method 7
	9C	Cement stabilised conditioned PFA		Class 7G + min. 2% cement		As 7B
	9D	Lime stabilised cohesive		Class 7E + min. 2.5% lime		Method 7

*Methods of compaction are given in Table 13.7

- impermeable – for lining canals, ponds or landfill sites a clay layer can be provided which must have low permeability to contain water, be flexible to allow for movements and be plastic to prevent cracking.

Some typical properties required of the various highway construction materials are given in Table 13.1.

Acceptability of fill

A material may be deemed acceptable if it can be incorporated into the permanent works (Specification for Highway Works, 1992).

The general fills comprise the bulk of an earthworks structure. For economic and environmental reasons it is important to use as much on site materials as possible, i.e. use the materials from the cuttings to form the embankments and thereby obtain a balanced earthworks. Otherwise, material may have to be transported to tips off or on site or obtained from borrow pits close to the site, both providing environmental problems. Useful information and discussion on this subject is contained in the Proceedings of the Conference on Clay Fills, 1979.

The criteria for the acceptability of a material depend on three factors:

1 *the nature of the works*
For a highway embankment sufficient strength is required to prevent slope instability and provide an adequate subgrade stiffness (such as the CBR value) for road pavement support. Self-weight settlements within the fill must also be limited so soils with low compressibility must be used. This is usually achieved by using materials which are not excessively moist nor excessively dry. These requirements become less important for landscaping works.

2 *earthmoving efficiency*
In order to maintain adequate trafficability for the earthmoving plant and compactability beneath a roller the soils traversed must have sufficient strength otherwise the job will grind to a halt with excessive rutting and bogging down.
The Transport Research Laboratory have conducted research into this problem and have shown that the depth of rut produced after a single pass of a machine is related to overall efficiency. If the single pass rut depth is less than 50 mm then no difficulties with scraper movement are likely although the rolling resistance will be increased and maximum speed of travel will be reduced. A rut depth of 100 mm or more will represent severe damage to the formation and considerable loss of productivity.
The economics of the use of inferior quality on-site materials must be balanced with the need for feasible operation of plant and its efficient use.
If a material is deemed to be of particularly poor quality it may be used for an inferior type of job such as landscaping or it may be improved such as by drying, otherwise it must be removed from the site.

3 *compactability*
Soils should be compacted at a moisture content near to their optimum moisture content. This is decribed in more detail in the section on soil compaction, below.
If a soil is compacted at moisture contents too dry of its optimum then it is likely that a high air voids content will be left within the soil making it compressible and brittle initially. With a high permeability this soil will increase in moisture content following infiltration from surface water leading to a dramatic loss of strength, particularly if the soil is clayey. All of these are undesirable properties.
A lower limit to moisture content is specified to eliminate these problems. Since the introduction of the moisture condition test a maximum MCV may be specified in the range 12–14 depending on the soil type and degree of compaction required.
If soils are compacted too wet of their optimum then there is a risk of inducing high pore water pressures which could cause instability during construction and excessive consolidation settlements following construction. The upper limit to moisture content (or minimum MCV) is also specified to minimise these problems.

Acceptability of granular soils

Granular soils are generally considered to be in an acceptable condition provided their natural moisture content lies close to the optimum moisture content (OMC) obtained from a compaction test. A range of values is usually quoted as:

$$\text{OMC} \pm x\,\% \tag{13.1}$$

where x may be 0 to 2%. Different optimum moisture contents will be obtained for the three compaction tests available so the test type must be specified.

For soils wetter than this range and, therefore, unacceptable it may be possible, depending on weather conditions, to spread out a layer loosely to allow evaporation until its moisture content reduces to the required value. The effectiveness of this approach will depend on factors such as the climate and the fines content. For soils drier than this range it is feasible to increase their moisture content using sprinklers.

Acceptability of cohesive soils

For cohesive soils a number of methods have been adopted for specifying a lower limit of acceptability, the most common being:

1 ***a maximum moisture content***

The fill must have a moisture content no greater than a given value. On the basis that undrained shear strength increases as moisture content decreases and the test is easy to perform this approach seems attractive. However, on most sites the plasticity (liquid and plastic) limits and the grading characteristics vary so for one value of moisture content a wide range of shear strengths can be obtained, as shown in Figures 13.6 and 13.7. The standard test also requires 24 hours for oven-drying so an assessment cannot be made immediately.

2 ***moisture content related to plastic limit***

A maximum moisture content w determined from:

$$w = PL \times factor \qquad (13.2)$$

has been frequently used as the acceptability criterion in the UK. The factor usually given for scrapers is 1.2, although a factor of 1.3 has been adopted for 'wet' clay fill and a lower factor of 1.0 has been used for Scottish stoney clays. For tracked vehicles this factor could lie between 1.40 and 1.65 (Farrar and Darley, 1975).

These authors have shown that a single factor is not appropriate. Instead, the factor can depend on the size of scraper used, the degree of damage that can be tolerated and the efficiency of plant movement required. Higher factors could also be obtained for soils containing more than 50% fines.

Although a simple test, the poor repeatability of the plastic limit test does not lend itself to use as an acceptability criterion. Sherwood (1971) has shown that if one operator carries out this test a number of times on one soil, up to one-third of all the results can be more than three units above or below the actual value due largely to operator variability.

The plastic limit test is also carried out on material finer than 425 μm (sand size) so comparison with moisture contents from samples containing gravels, especially glacial clays, may lead to false conclusions.

3 ***undrained shear strength***

Rather than using an indirect means (moisture content) to assess strength it is preferable to determine the strength of a clay fill from a re-compacted specimen directly tested either in the unconfined compression or triaxial compression mode.

A result can be obtained quickly, and for highly cohesive, stone-free soil quite simply using either a hand vane of small diameter or the unconfined compression apparatus either on site or in a laboratory. The main disadvantages are when the soil has poor cohesion to form a cylindrical specimen, when stones are present large diameter specimens are necessary and the difficulty of achieving complete removal of air voids (full saturation) after compaction, especially for stiff fissured clays.

Farrar and Darley (1975) and Arrowsmith (1979) have shown the operation of earthmoving plant to be related to the strength of the clay, see Table 13.2.

4 ***moisture condition value*** **(MCV)** *(Figures 13.3 and 13.4)*

Due to the difficulties associated with the above methods for assessing suitability of earthworks material, the Transport Research Laboratory developed the moisture condition test (Parsons, 1976, Parsons and Boden, 1979). The objectives of this test were to provide an immediate result for site use, to be applicable to a wide range of soil types, to eliminate the effect of operator error and to use a large sample for more representative behaviour.

The test (BS 1377:1990, Part 4) consists of determining the compactive effort required to almost fully compact a given mass of soil. 1.5 kg of moist

Table 13.2 *Operation of earthmoving plant on clay soil*

Plant type	Minimum shear strength kN/m² [1]		
	Farrar and Darley (1975) [2]		Arrowsmith (1979) [3]
	feasible operation [4]	efficient operation [5]	
small dozer, wide tracks	20		
small dozer, standard tracks	30		
large dozer, wide tracks	30		
large dozer, standard tracks	35		
Towed and small scrapers(less than $15m^3$)	60	140	35
Medium and large scrapers (over $15m^3$)	100	170	50

1 – Values are vane shear strength which is likely to be higher than triaxial shear strength
2 – Values were obtained in the fill area which may have 'dried' during dry weather
3 – Values were obtained from the cut area and relate to 'feasible' operation. These soils may be stronger in the fill area during dry weather.
4 – Represents deepest rut of 200 mm after a single pass of machine
5 – Represents depth of rut not greater than 50 mm after a single pass

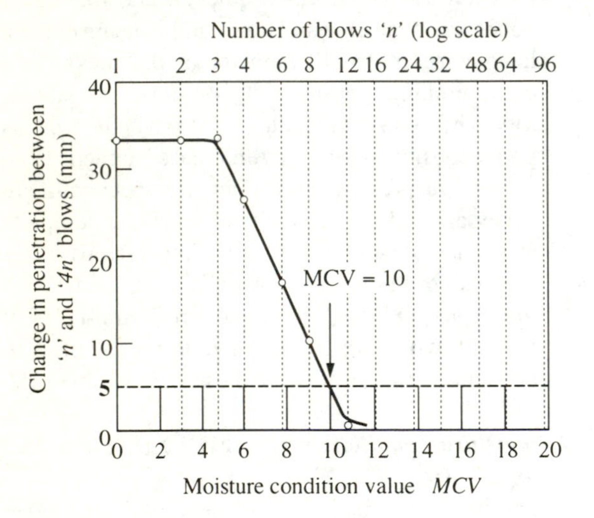

Figure 13.3 *Determination of the moisture condition value of a sample of heavy clay (From Parsons and Boden, 1979)*

soil (with particles or lumps less than 20 mm) is placed loose in a 100 mm internal diameter steel mould and a lightweight disc, 99.2 mm in diameter is placed on top. Compactive effort is provided by the number of blows of a free falling cylindrical rammer of 7 kg mass and 97 mm diameter falling onto the disc with a height of drop of 250 mm. As the number of blows increases the particles and lumps of soil move and remould to expel air between them.

The effect of each blow will be related to the density of soil achieved which is assessed by measuring the penetration of the rammer into the mould. On the basis that the change in density with the penetration of the rammer is directly related to the logarithm of the number of blows the change in penetration (mm) between one number of blows and four times that number is plotted against the logarithm of the number of blows (Figure 13.3).

As a very large number of blows may be required to remove the last remaining air voids a change of penetration of 5 mm was selected to represent the point beyond which no significant change in density occurs. The moisture condition value (MCV) is defined as:

$$MCV = 10 \log_{10} B \tag{13.3}$$

where B is the number of blows corresponding to a change of penetration of 5 mm on the steepest straight line through the points, Figure 13.3.

The test should only be carried out on cohesive soils and granular soils with cohesive fines, otherwise variable and misleading results may be obtained. However, it is less important to assess the 'clean' sands and gravels for suitability as they are often considered as 'all weather' materials.

A moisture condition 'calibration' comprises the determination of the relationship between moisture content and MCV. This is obtained by adjusting the moisture content of the soil (drying and wetting) before carrying out the moisture condition test. Parsons and Boden (1979) showed that the moisture content *w*-MCV calibration usually produces a straight line with an equation of the form:

$$w = a - b \times \text{MCV} \qquad (13.4)$$

where a is the intercept or moisture content w when $MCV = 0$ and b is the slope (Figure 13.4).

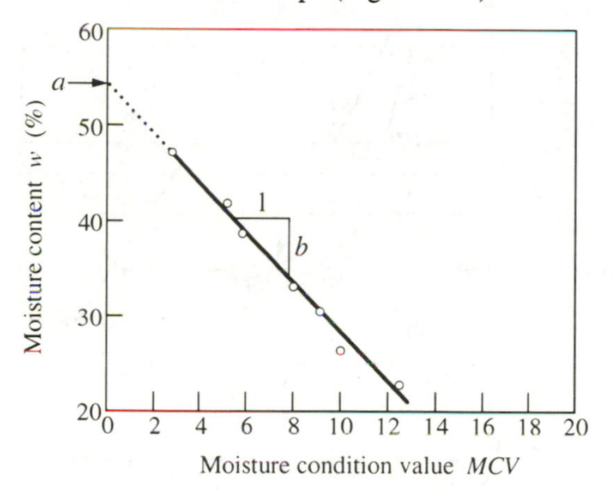

Figure 13.4 *Moisture condition calibration of a heavy clay soil (From Parsons and Boden, 1979)*

Care must be exercised when dealing with soils which are prone to alteration on drying (see Table 2.13 in Chapter 2) and which will crush during compaction. It is then preferable to wet up or dry gradually from the natural condition and to use separate samples for crushable soils.

For cohesive soils a relationship between undrained shear strength (determined by the hand vane) and MCV has been given (Parsons and Boden, 1979) in the form:

$$\log c_u = c + d \times \text{MCV} \qquad (13.5)$$

with values of the coefficients c and d given in Table 13.3.

Table 13.3 *Values of the coefficients c and d (From Parsons and Boden, 1979)*

Soil type	c	d	Number of results	correlation coefficient
Clay - high plasticity	0.74	0.111	40	0.94
Clay - intermediate plasticity	0.77	0.107	44	0.96
Clay - low plasticity	0.91	0.112	14	0.89
Silt - high plasticity	0.70	0.105	15	0.97

As an acceptability criterion it has been found that for UK conditions a minimum MCV of about 8 relates to the limits of strength for trafficability purposes as well as for the stability of an earth structure.

Efficiency of earthmoving *(Figure 13.5)*
The suitability of a soil as it affects the operation of earthmoving plant has been assessed by TRL (Parsons and Darley, 1982). A reasonable relationship was found to exist between the soil condition given by the MCV, the type of plant (including factors such as the number of driven wheels, tyre width, maximum available engine power and total mass) and the efficiency of operation as represented by the speed of travel which could be achieved on the haul road. Some of these factors are illustrated in Figure 13.5.

Single pass rut depth is a measure of the degree of damage likely to be caused in a fill area. Speed of travel on the haul road will affect the cycle time and the overall volume of fill moved each day.

Material problems
The effects of weather are all too obvious on an earthworks project with softening during wet weather, dust (visibility and environmental) problems during hot, dry weather and frost damage during freezing conditions. Two factors are considered here, softening and bulking.

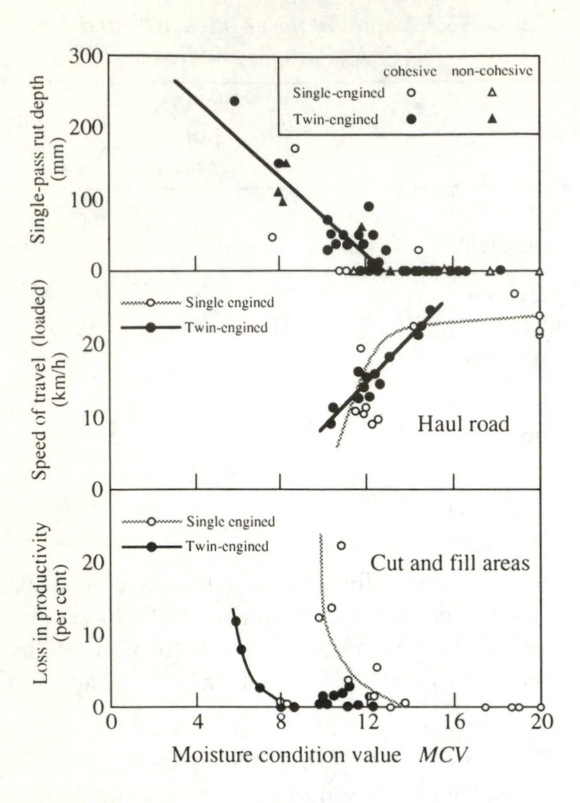

Figure 13.5 *Factors affecting operations with medium scrapers (From Parsons and Boden, 1979)*

Softening *(Figures 13.6 and 13.7)*
For cohesive soils in particular, and granular soils to a lesser extent, their condition will be affected by changes in moisture content. The degree to which they are affected can be termed moisture sensitivity which is a different phenomenon to remoulding sensitivity (see Figure 7.17). A soil exhibiting both types of sensitivity is likely to deteriorate even more rapidly.

Moisture sensitivity is the reduction in strength produced by an increase in moisture content. This is illustrated for clays of different plasticity in Figure 13.6 and for different gravel (or sand) contents in Figure 13.7. For a change in moisture content of 1% the change in shear strength is much greater for low plasticity clays and very gravelly clays so these will be more sensitive to wetting and more prone to softening. Glacial clays have a variable gravel content and are typically of low plasticity so they can be easily weakened by wetting.

Similar to the plots in Figures 13.6 and 13.7 the value of *b* from Equation 13.4 is an indicator of moisture sensitivity. Matheson (1988) has suggested levels of moisture sensitivity related to the *b* value, given in Table 13.4.

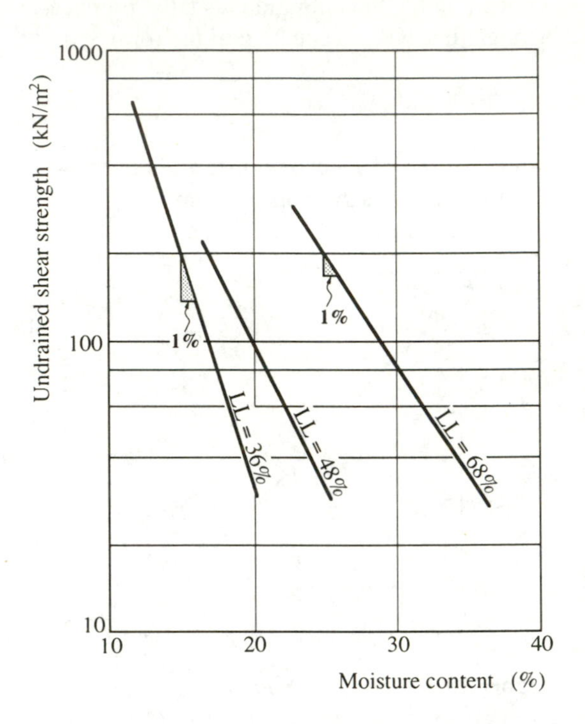

Figure 13.6 *Moisture sensitivity – effect of plasticity (From Clayton, 1979)*

Table 13.4
Moisture sensitivity related to MCV calibration slope b (After Matheson, 1983, 1988)

b	Moisture sensitivity
> 1.0	low
0.5 – 1.0	moderate
0.33 – 0.5	high
< 0.33	very high

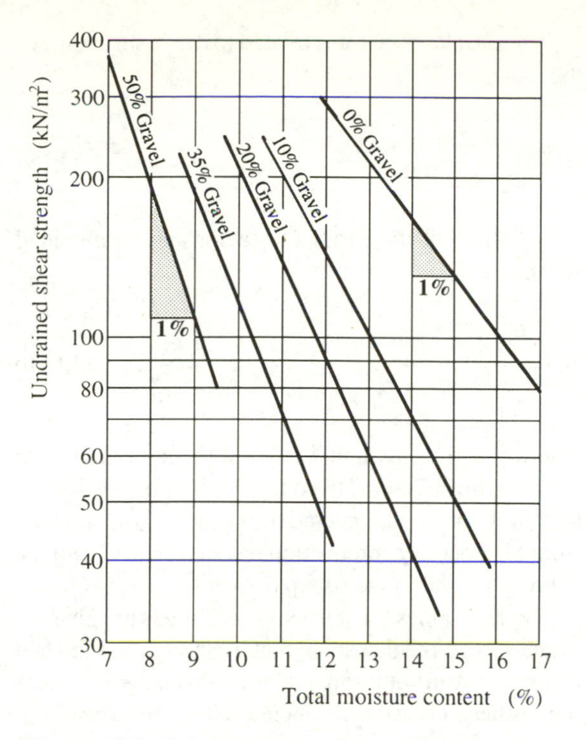

Figure 13.7 *Moisture sensitivity – effect of gravel (From Barnes and Staples, 1988)*

Bulking *(Figure 13.8)*

In estimating the cost of an earthmoving contract it is desirable to achieve a balance between the volumes of soil removed from the cuttings and the volume of soil required to form the embankments. For this purpose, the estimator will use a mass haul diagram, the principles of which are illustrated in many text books on surveying. However, it would be incorrect to compare the volumes from the cuttings and the embankments directly due to the phenomenon of bulking.

Soil in its natural *in situ* location in the cutting is referred to as bank volume, measured in bank m^3. When it is excavated loosening and breaking up into lumps increases the volume occupied when it is referred to as loose volume, loose m^3. Following compaction, at its final location, the soil may occupy a smaller or larger compacted volume, compacted m^3, depending on the soil type and level of compaction. This is illustrated in Figure 13.8.

To allow for these effects factors must be applied to the bank volumes.

For assessing how much bank volume can be removed by one scraper or dump truck each trip the following expression is used:

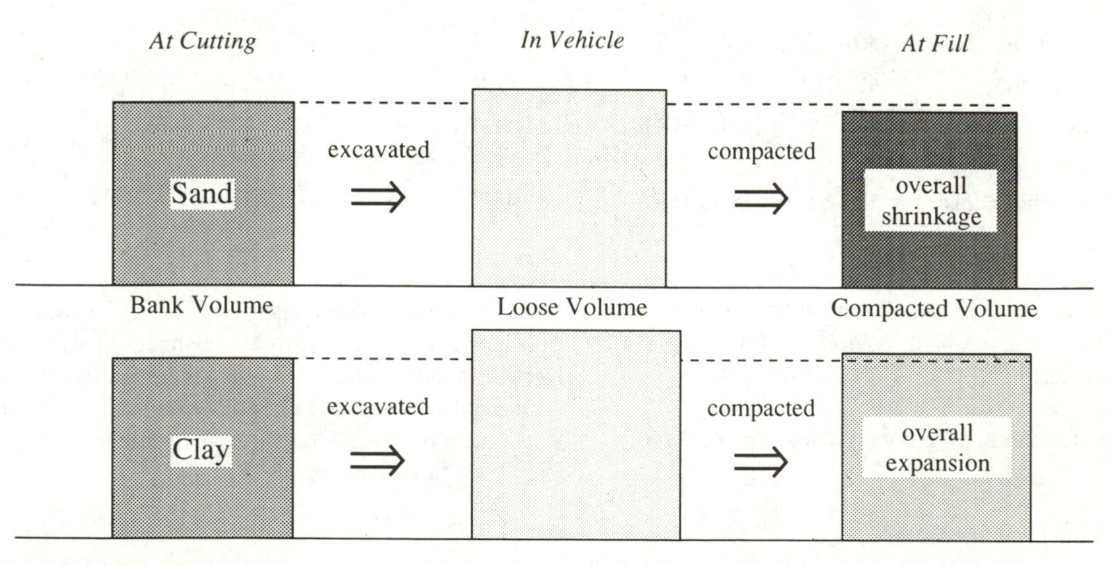

Figure 13.8 *Bulking*

bank m^3 removed by each load =
machine payload (m^3) × LF (13.6)

where LF = load factor.
The load factor is related to the % swell of the material:

$$LF = \frac{100}{100 + \%\ swell} \tag{13.7}$$

Some typical values for soils and rocks are given in Table 13.5.

Table 13.5
Typical values of % swell and load factor

Material type	% swell	Load factor *LF*
Soils		
Sand	10 – 15	0.87 – 0.91
Gravel	12 – 18	0.85 – 0.89
Clay	20 – 30	0.77 – 0.83
Peat	30 – 40	0.71 – 0.77
Topsoil	30 – 40	0.71 – 0.77
Rocks		
Coal	30 – 40	0.71 – 0.77
'soft' rocks e.g. mudstone, shale, chalk	30 – 50	0.67 – 0.77
Sandstone	40 – 70	0.59 – 0.71
Limestone	50 – 70	0.59 – 0.67
igneous rocks e.g. basalt, granite, gneiss	50 – 80	0.56 – 0.67

To assess the compacted volume which could be formed from a given bank volume the following expression is used:

compacted volume, m^3 = bank volume, m^3 × SF (13.8)

where SF = shrinkage factor.

The shrinkage factor is related to the % shrinkage of the material:

$$SF = \frac{100 - \%\ shrinkage}{100} \tag{13.9}$$

An estimate of the shrinkage factor can be obtained from:

$$SF = \frac{\gamma_b}{\gamma_c} \tag{13.10}$$

where γ_b is the bulk unit weight of the material *in situ* in the cutting obtained from field or laboratory density tests and γ_c is the compacted bulk unit weight obtained from a laboratory compaction test or directly from the compacted soil in the fill area.

Typical values for some soils and rocks are given in Table 13.6. Note that some materials such as sand and chalk reduce in volume overall (net shrinkage) whereas most other soils and rocks increase (negative shrinkage or net bulk-up).

Table 13.6
Typical values of % shrinkage and shrinkage factor

Material type	% shrinkage	Shrinkage factor *SF*
Sand	0 – 10	0.90 – 1.00
Chalk	0 – 15	0.85 – 1.00
Most other soils, weak rocks	0 – 10 (–ve)	1.00 – 1.10
Harder rocks	5 – 20 (–ve)	1.05 – 1.20

Further loss of material is to be expected overall on a job due to the creation and maintenance of haul roads, overfilling and trimming of embankments and removal of material following weather deterioration. This can be as much as 10–15% on some sites although 5% is considered more typical.

Soil compaction

Introduction

After a layer of soil has been deposited and spread in the fill area it must be rendered as strong and stiff as possible to ensure stability and minimise settlements. Energy is applied to the soil to remould lumps of clay, move granular particles together and essentially remove as much air as possible. This process is known as compaction.

For nearly all soils the extent to which air can be removed depends on the strength of the clay lumps or the friction between the granular particles which in turn depend on the moisture content of the soil during compaction. The degree of compaction achieved is measured by dry density, ρ_d:

$$\rho_d = \frac{\text{mass of soil particles}}{\text{volume occupied}}$$

This represents the amount of solid soil particles in a given volume. Compaction should aim to achieve as high a dry density as possible.

Factors affecting compaction

The main factors affecting compaction are:

1 ***moisture content*** *(Figures 13.9 and 13.10)*
At low moisture contents the strength of clay lumps and friction between granular particles is high so a given compactive effort will not be able to remove all air voids leaving the soil in an overall compressible state when it is subjected to stresses from further layers of fill or a structure.
At higher moisture contents clay lumps become weaker and friction between granular lumps reduces so the air voids are more easily removed during compaction. The dry density increases until a maximum value ρ_{dmax} is reached at the optimum moisture content (OMC). This is shown on the typical compaction curve, Figure 13.9.
At moisture contents above the optimum value soil particles cannot be as close together because there is more water present in the voids, see Figure 13.10.

2 ***compactive effort*** *(Figure 13.11)*
Applying more energy to a soil will reduce the air voids content and increase the dry density so more compaction energy can be beneficial, especially for soils dry of the optimum value. However, if the soil is already moist and weaker then applying more energy is wasteful since the air can quickly be removed.
Applying high energy to a very moist soil may be damaging since high pore water pressures can be built up which could cause instability during construction and consolidation settlements after construction.

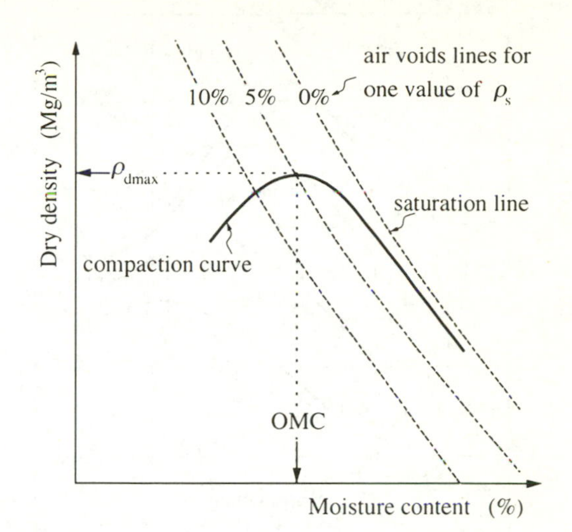

Figure 13.9 *Typical compaction curve*

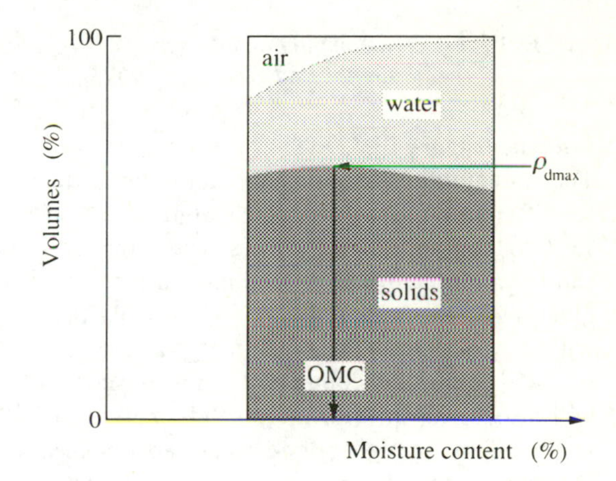

Figure 13.10 *Compaction curve – volumes of solids, water and air*

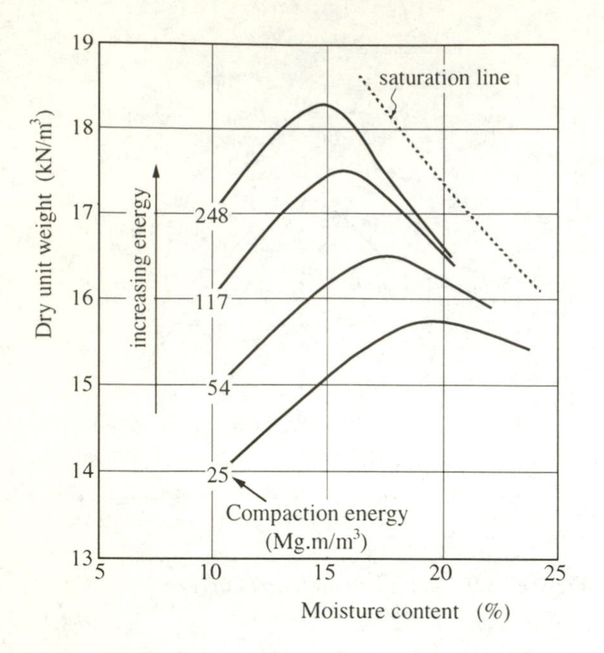

Figure 13.11 *Effect of compaction energy (From Lambe and Whitman, 1969)*

3 ***soil type** (Figure 13.12)*
The strength-moisture content relationship differs for different soils, as illustrated in Figures 13.6 and 13.7, so the compactability or ease with which soils can be compacted will depend on soil type. At a given moisture content a clay with low plasticity will be weaker than a heavy or high plasticity clay so it will be easier to compact. Its moisture content will also be lower with more solids present in a given volume so higher dry densities are obtained. Difficulties have been experienced in adequately compacting stiff fissured high plasticity clays on some projects due to the strength of the clay lumps.

Field compaction

Introduction

Compaction is achieved in the field by traversing a fairly thin layer of soil with an item of compaction plant a sufficient number of times (passes) until a required density is achieved.

The layer thickness and number of passes must be chosen to ensure that the required density is produced throughout the layer with no undesirable density gradients, such as a poorly compacted lower level. Generally, plant of greater weight can transmit compaction energy to lower levels so layer thickness can be increased and the number of passes can be decreased. However, plant which is too heavy can damage a compacted soil surface by applying too much pressure and causing rutting or degradation, and some plant may just be too light to remould stiff clay lumps or move granular particles closer.

Compaction plant

To some extent construction traffic provides compactive effort, particularly crawler tracked vehicles, and for some purposes such as landscaping this may be sufficient. For the main earthworks compaction must be applied more uniformly and of known amount.

Compaction is achieved by specialist items of plant which are designed to apply energy to the soil by means of pressure and where suited this is assisted by a kneading or remoulding action, vibration or impact.

The main types of compaction plant are:

- ***smooth-wheeled rollers***
  These comprise a smooth steel drum roller which is either towed (single or tandem) by a crawler tractor or self-propelled (two or three-roller tandem) at a speed of about 2.5 to 5 km/hour. The mass can be

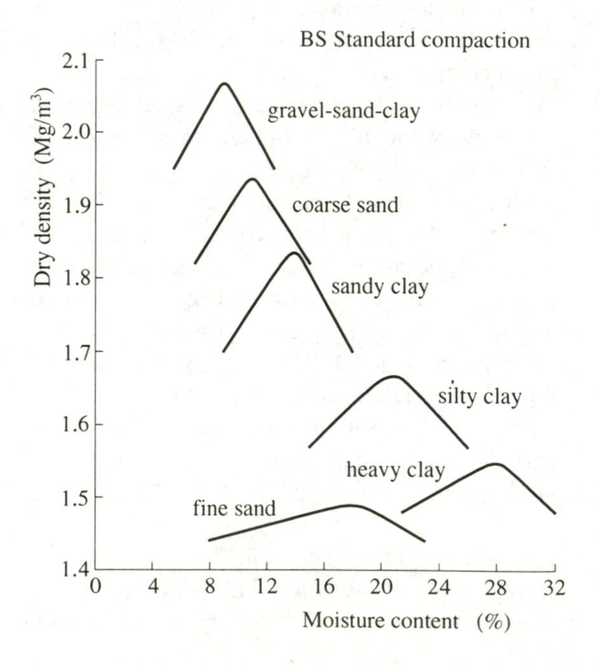

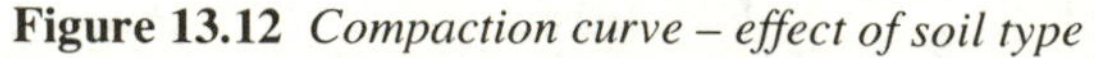
Figure 13.12 *Compaction curve – effect of soil type*

increased by water or sand ballast. They are suited to firm cohesive and well-graded granular fills providing small layer thicknesses (125–150 mm) are adopted. They may become unstable in uniform sands, due to the roller pushing itself into the soil, and are generally unsuited for coarse, granular soils without the assistance of vibrations.
They produce a smooth surface which is useful for encouraging rainfall run-off. However, the smooth surface provides a poor bond between layers leaving the earthworks with a laminated type of structure which would be very undesirable for water-retaining earthworks. With some clay soils and plant types plant-induced shear surfaces can be produced which are polished or slickensided and could promote instability. If these surfaces are detected then it will be necessary to adjust the compaction procedure by applying less effort or scarifying each surface before placing another layer.

- ***vibratory rollers***
  Vibrational energy assists compaction considerably by shaking the soil and reducing inter-particle friction while the pressure applied moves the particles together. A vibratory roller can be any item of plant with a vibratory attachment, even a sheepsfoot roller, although the smooth drum is the most common. They can be towed, self-propelled or manually guided with speeds no greater than about 1.5 to 2.5 km/hour. Better compaction is achieved with lower speeds and with a frequency giving the maximum amplitude. Vibrations are applied either by separate engines mounted on the towing frame or by the rotation of eccentric weights within the drum.
  They are suited to most soil types, though they may be less efficient with moist clays and be unstable in uniform fine sands due to pushing in. Towed rollers rather than self-propelled rollers should be used.

- ***tamping rollers***
  Various shapes of projections or 'feet' have been fitted to a steel drum roller to penetrate the soil layer and produce lateral compaction as well as vertical compaction, high localised pressure and a mixing and kneading action. They provide good interlock between successive layers so they are particularly suited to water-retaining structures such as earth dams.
  They may be towed or self-propelled with one or more drums mounted on one or more axles travelling at speeds of between about 4 and 10 km/hour. The feet are circular, square or rectangular of varying lengths and tapers and must be designed to ensure adequate coverage and penetration of the soil layer without trapping soil and clogging up.
  For sheepsfoot rollers the projections are typically about 180–240 mm long, 70–80 mm square or 75–90 mm diameter with spacings of 200–280 mm. For pad-type tamping rollers the end area of each foot is greater than 100 mm square and the sum of the areas of these feet may occupy up to 25% of the surface of an area swept around the ends of the feet. For sheepsfoot rollers this figure is about 5-10%.
  They are most suited to cohesive soils, particularly soils of low moisture content and fine granular soils although sheepsfoot rollers may be less efficient in the latter due to disturbance of the soil as the projections retract.

- ***grid rollers***
  These consist of an open steel mesh drum ballasted with concrete blocks attached to the frame and towed by a crawler tractor. They provide high localised pressure and are most suited to soft rocks and stiff clays where breaking lumps is beneficial. They are less suited to moist clays and uniform sands where they may become bogged down.

- ***pneumatic-tyred rollers***
  A number of rubber-tyred wheels mounted on one or two axles provide a useful kneading effect. On two axle versions the wheels are off-set to provide complete coverage of the soil layer and with the wobbly-wheel type the wheels are mounted so that they move from side to side increasing the kneading action. The wheels are usually mounted independently or in pairs so that a more uniform pressure can be applied over an uneven surface. The tyres are either small with no tread or large and with tread. The applied pressures and hence the compactive effort is increased by ballast and by adjusting the tyre inflation pressures.
  They are most suited to moist cohesive soils and well-graded granular soils. They produce a relatively smooth compacted surface with poor bonding between layers and may be prone to leaving laminations and plant-induced shear surfaces.

- ***vibrating plate compactors***
  These comprise a vibrating unit mounted on a steel plate which is manually operated. Weights vary up to about 2 tonnes with plate areas up to 1.6 m^2. They operate fairly slowly, less than 1 km/hour and are mostly used in small, confined areas such as backfilling around structures. They are most efficient when compacting granular soils.

- ***vibro-tampers***
  These weigh between 50 and 100 kg and apply compaction by vibrations. They are manually guided, useful in confined spaces and can be used for cohesive and granular soils.

- ***power rammers***
  These are machines actuated by explosions in an internal combustion engine causing the machine to impact on the soil layer. They are manually operated, typically weigh about 100 kg and are only suited to compaction in small, confined areas such as backfilling narrow trenches.

- ***dropping weight compactors***
  These consist of a mass of 200–500 kg lifted by a hoist mechanism and dropped through a height of 1–3 m. They are, therefore, useful for compaction in small, confined areas using cohesive or well-graded granular fill.

Specification of compaction requirements

End-product specification

A logical way of specifying how much compaction effort should be applied to a soil would be to stipulate a required property of the compacted soil such as a minimum density or a maximum air content. Then the state of the finished earthworks and its properties are known. This approach is referred to as an end-product specification and requires *in situ* density tests and laboratory moisture content tests. The two most common approaches are:

1 *relative compaction (Figure 13.13)*
  The required dry density of the earth fill must be greater than a certain proportion (90, 95 or 100%) of

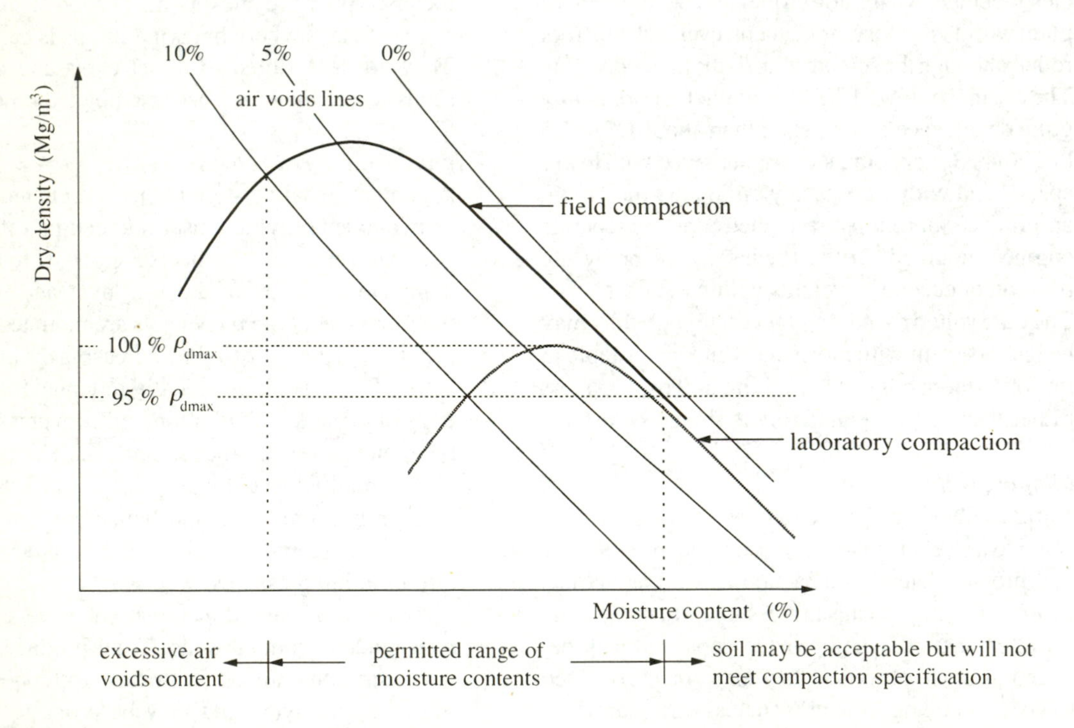

Figure 13.13 *Relative compaction*

the dry density obtained from a laboratory test, usually BS light or vibrating hammer compaction. There are several disadvantages with this method, including:
– the cost and time required for testing. Waiting a day for a moisture content result before deciding to place another layer of fill is unacceptable.
– variability of soil type. A laboratory compaction test should be carried out on the same type of material each time the *in situ* density is determined.
– particle size. The fill material may contain coarse gravel, even cobble sizes but the compaction test can only realistically be carried out on material less than 20 mm size.
– mode of compaction. Soil is compacted in a much different way in the field compared to the laboratory test. The type of compaction applied by each plant item also varies.
– range of moisture content must be specified. Soils compacted in the field on the dry side of laboratory optimum moisture content may achieve the required 95% of the maximum dry density but may still contain an excessive air void content (Figure 13.13). Soils compacted in the field wet of laboratory optimum may be considered acceptable for use but may never achieve 95% of the maximum dry density.

2 *air voids content*
A maximum air voids content A_v of 10% has often been quoted for the bulk of an earth fill with a value of 5% for the top of an embankment, to provide a good foundation support or subgrade for road pavement layers. The *in situ* bulk density must be determined, together with the moisture content and particle density, so that air voids content can be calculated from a formula, see Table 2.15.
The main disadvantages with this approach are:
– statistical variation. It has been estimated that for a permitted probability of 90% (9 out of 10 results give A_v less than 10%) the mean value of A_v obtained from the tests must be 7%. To ensure a representative statistical sample a large number of tests are required and this is prohibitive.
– range of moisture content must be specified for a particular plant item. An upper limit of moisture content is required, otherwise it may be easy to achieve less than 10% air voids with just one pass with the risk of inadequate coverage and poor compaction in the bottom half of a layer. A lower limit is required otherwise a large number of passes will be required to achieve the desired air voids content. These effects are illustrated in Figure 13.11.

Method specification
In view of the major difficulties associated with the end-product specification, and following extensive research by the Transport Research Laboratory, the Department of Transport requires that fill placed for highway construction be compacted using a Method Specification (Anon, 1992) except for a few selected fill materials.

For a given soil type the DTp Specification states a method of compaction to be adopted. For example, for a well-graded granular general fill material (Class 1A) Method 2 compaction must be used (Table 13.1). The methods of compaction are given in the DTp Specification and an abridged version is included in Table 13.7.

For each method of compaction, the types and masses of compaction plant which are unsuitable are stated, so the choice becomes limited to those which are suitable. The method of compaction is given as a maximum depth of compacted layer and a minimum number of passes for each type and mass of suitable plant.

For the top 600 mm of an embankment composed of general granular or cohesive fill the number of passes is doubled to provide a stiffer formation to receive the pavement layers.

The main advantages of this approach are:

- it removes most of the disadvantages of the end-product approach
- the requirements for compaction are more precise
- the contractor's estimate of costs involved should be more accurate
- fill quality control is minimised with much less testing. Some control or compliance testing is advisable.

However, adequate supervision to ensure the work is carried out as specified is essential. This can be time-consuming on large-scale projects. With some soil types there may still be doubt concerning the quality of the compacted fill so some control testing could be needed.

Table 13.7 *Method compaction for earthworks materials: plant and methods (Abridged version from Specification for Highway Works, 1992)*

Type of compaction plant	Ref No	Category	Method 1		Method 2		Method 3		Method 4		Method 5		Method 6			Method 7	
			D	N#	D	N#	D	N#	D	N	D	N	N for D = 110 (mm)	N for D = 150 (mm)	N for D = 250 (mm)	N for D = 150 (mm)	N for D = 250 (mm)
Smooth wheeled roller		Mass per metre width of roll:															
	1	over 2100 kg up to 2700 kg	125	8	125	10	125	10*	175	4	unsuitable		unsuitable	unsuitable	unsuitable	unsuitable	unsuitable
	2	over 2700 kg up to 5400 kg	125	6	125	8	125	8*	200	4	unsuitable		16	unsuitable	unsuitable	unsuitable	unsuitable
	3	over 5400 kg	150	4	150	8	unsuitable		300	4	unsuitable		8	16	unsuitable	12	unsuitable
Grid roller		Mass per metre width of roll:															
	1	over 2700 kg up to 5400 kg	150	10	unsuitable		150	10	250	4	unsuitable		unsuitable	unsuitable	unsuitable	unsuitable	unsuitable
	2	over 5400 kg up to 8000 kg	150	8	125	12	unsuitable		325	4	unsuitable		20	unsuitable	unsuitable	16	unsuitable
	3	over 8000 kg	150	4	150	12	unsuitable		400	4	unsuitable		12	20	unsuitable	8	unsuitable
Tamping roller		Mass per metre width of roll:															
	1	over 4000 kg	225	4	150	12	250	4	350	4	unsuitable		12	20	unsuitable	4	8
Pneumatic-tyred roller		Mass per wheel:															
	1	over 1000 kg up to 1500 kg	125	6	unsuitable		150	10*	240	4	unsuitable		unsuitable	unsuitable	unsuitable	unsuitable	unsuitable
	2	over 1500 kg up to 2000 kg	150	5	unsuitable		unsuitable		300	4	unsuitable		unsuitable	unsuitable	unsuitable	12	unsuitable
	3	over 2000 kg up to 2500 kg	175	4	125	12	unsuitable		350	4	unsuitable		unsuitable	unsuitable	unsuitable	6	unsuitable
	4	over 2500 kg up to 4000 kg	225	4	125	10	unsuitable		400	4	unsuitable		unsuitable	unsuitable	unsuitable	5	unsuitable
	5	over 4000 kg up to 6000 kg	300	4	125	10	unsuitable		unsuitable		unsuitable		12	unsuitable	unsuitable	4	16
	6	over 6000 kg up to 8000 kg	350	4	150	8	unsuitable		unsuitable		unsuitable		12	unsuitable	unsuitable	unsuitable	8
	7	over 8000 kg up to 12000 kg	400	4	150	8	unsuitable		unsuitable		unsuitable		10	16	unsuitable	unsuitable	4
	8	over 12000 kg	450	4	175	6	unsuitable		unsuitable		unsuitable		8	12	unsuitable	unsuitable	4
Vibratory roller		Mass per metre width of a vibratory roll:															
	1	over 270 kg up to 450 kg	unsuitable		75	16	150	16	unsuitable		unsuitable		unsuitable	unsuitable	unsuitable	unsuitable	unsuitable
	2	over 450 kg up to700 kg	unsuitable		75	12	150	12	unsuitable		unsuitable		unsuitable	unsuitable	unsuitable	unsuitable	unsuitable
	3	over 700 kg up to 1300 kg	100	12	125	10	150	6	125	10	unsuitable		16	unsuitable	unsuitable	unsuitable	unsuitable
	4	over 1300 kg up to 1800 kg	125	8	150	8	200	10*	175	4	unsuitable		6	16	unsuitable	unsuitable	unsuitable
	5	over 1800 kg up to 2300 kg	150	4	150	4	225	12*	unsuitable		unsuitable		4	6	12	12	unsuitable
	6	over 2300 kg up to 2900 kg	175	4	175	4	250	10*	unsuitable		400	5	3	5	11	10	unsuitable
	7	over 2900 kg up to 3600 kg	200	4	200	4	275	8*	unsuitable		500	5	3	5	10	10	unsuitable
	8	over 3600 kg up to 4300 kg	225	4	225	4	300	8*	unsuitable		600	5	2	4	8	8	unsuitable
	9	over 4300 kg up to 5000 kg	250	4	250	4	300	6*	unsuitable		700	5	2	4	7	8	unsuitable
	10	over 5000 kg	275	4	275	4	300	4	unsuitable		800	5	2	3	6	6	12

Refer to Table 13.1 for Material Classes related to each method of compaction.

– For Material Classes 1A, 1B, 2A, 2B, 2C or 2D within the top 600 mm of the embankment the number of passes is doubled.

* – The roller must be towed by a track-laying tractor. Self-propelled rollers are not suitable.

D – Maximum depth of compacted layer

N – Minimum number of passes

Control of compaction in the field *(Figure 13.14)*
Where it is necessary to obtain the dry density or air voids content of the compacted fill, the *in situ* bulk density must be determined. Suitable methods include the core-cutter method for cohesive soils, the sand-replacement method for granular soils and the nuclear moisture/density gauge for a wide range of materials. In each case the results will have variable accuracy and the latter methods require a calibration. Where thick layers of fill are placed the density should be determined throughout the layer.

On some projects a field trial may be justified to determine the most efficient type of plant and method of compaction. This could comprise spreading layers of different thicknesses and running the compaction plant over, measuring the dry density of the fill at given numbers of passes. This should determine the most economical method to adopt by providing the maximum layer thickness and minimum number of passes which will give a required dry density, Figure 13.14a.

The soil should be spread initially at its natural moisture content. The exercise could then be repeated with the moisture content of the soil adjusted either by adding or removing moisture, depending on the optimum moisture content and the climate, Figure 13.14b.

Laboratory compaction

Introduction

The compaction characteristics of soils can be assessed using standard laboratory tests. The soil is compacted by dropping a mass or vibrating a weight onto thin layers in a cylindrical mould using an amount of compaction energy per unit volume given by:

$$\text{energy} = \frac{mhbn}{V} \qquad (13.11)$$

where

m = weight of hammer or mass
h = height of drop of hammer
b = number of blows per layer
n = number of layers
V = volume of mould.

The effect of compaction energy is illustrated in Figure 13.11.

Laboratory tests

Three tests are described in BS 1377:1990, Part 4.

- ***Light compaction***
  This test may be referred to as BS light compaction, the Proctor method or the 2.5 kg rammer method. The test is carried out on soil with particles larger than 20 mm removed. The soil is prepared by drying and then adding water so that its moisture content is fairly low and sufficiently mixed to ensure uniform

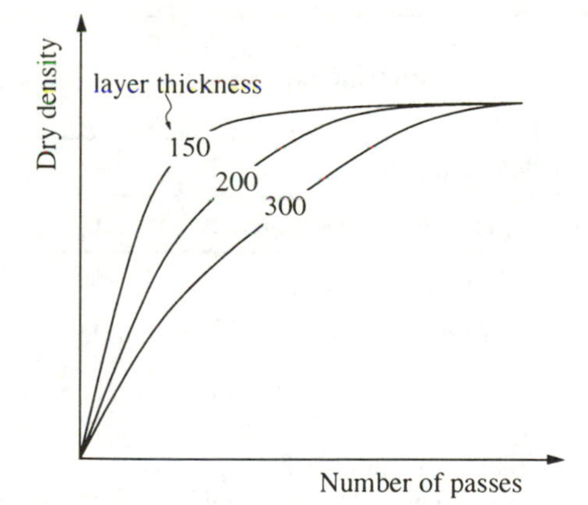

a) At one moisture content

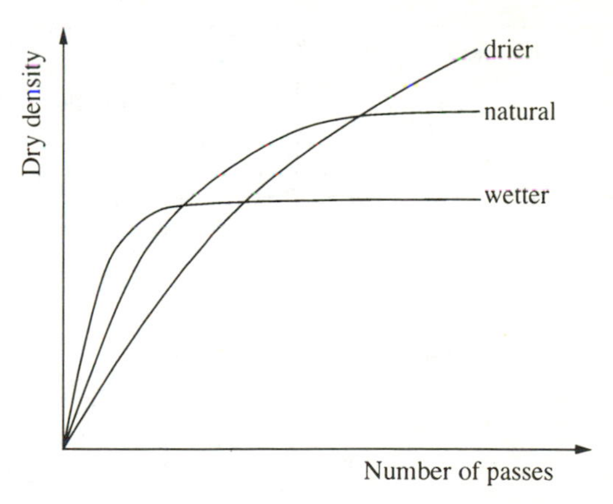

b) For one layer thickness

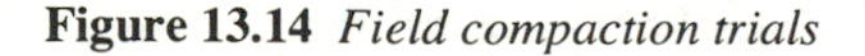

Figure 13.14 *Field compaction trials*

moisture content throughout. Cohesive soils should be chopped into pieces smaller than 20 mm. Preparation procedures for soils which contain particles greater than 20 mm for the 1 litre mould or 37.5 mm for the CBR mould are given in BS 1377:1990 and summarised in Table 13.8.

For soils which may be affected by oven-drying or even air-drying it is preferable to commence the test with the soil at its natural moisture content, and then to adjust the moisture content by air-drying or blow-drying for compaction at lower moisture contents, and mist spraying for compaction at higher moisture contents.

The soil is compacted (in three layers of equal thickness) into a metal mould of 105 mm diameter and of 1 litre or 1000 cm^3 capacity, or into a CBR mould of 152 mm diameter and 2305 cm^3 capacity. Each layer receives blows from a 2.5 kg mass falling freely through a height of 300 mm, 27 blows in the one-litre mould and 62 blows in the CBR mould.

The test should be carried out to ensure that the final compacted surface lies just above the top of the mould, but no more than about 5 mm. Otherwise, the test must be discarded. After trimming the soil surface flush with the top of the mould, so that its volume can be taken as one litre, the mould and soil are weighed and by subtracting the weight of the mould the bulk density or unit weight of the soil can be determined. The soil is removed from the mould and a smaller specimen taken for moisture content determination. The moisture content of the soil is then adjusted, up or down, and the test is repeated to give at least five density values.

The dry density of the soil is calculated (see Table 2.15) and plotted versus moisture content and providing a reasonable 'inverted parabola' shape can be drawn through the points the maximum dry density, ρ_{dmax} and optimum moisture content, OMC are read off, see Figure 13.9.

- ***Heavy compaction***

  This test may be referred to as BS heavy compaction or the 4.5 kg rammer method. The test procedure is the same as for the light compaction test but with five equal layers of soil in the one-litre and CBR moulds and compaction for each layer provided by blows from a 4.5 kg mass falling freely though 450 mm. 27 blows are applied in the one- litre mould and 62 blows are applied in the CBR mould.

Table 13.8 *Sample preparation methods for compaction tests (Adapted from BS 1377:Part 4:1990)*

Grading zone	Minimum % passing test sieves			Type of mould	Preparation method
	20 mm	37.5 mm	63 mm		
(1)	100	100	100	1 litre	Test whole sample
(2)	95	100	100		Remove particles > 20 mm, test remainder
(3)	70	100	100	CBR mould	Test whole sample
(4)	70	95	100		Remove particles > 37.5 mm, test remainder
(5)	70	90	100		1) Remove particles > 37.5 mm. Replace with same quantity of between 20 and 37.5 mm and test this modified sample or 2) Remove particles > 37.5 mm, test remainder of sample. Apply correction for stone content to ρ_{dmax} and OMC
(X)	less than 70	less than 90	less than 100	Test not applicable	Too coarse to be tested

- ***Vibrating hammer***
  Referred to as the vibrating hammer method, this test is more appropriate for granular soils and is not suitable for cohesive soils. The soil for this test can be from grading zones 1 to 5 in Table 13.8. The soil is compacted into a CBR mould, 152 mm diameter and approximately 2305 cm^3 capacity, in three equal layers. Each layer is compacted by placing a circular tamper, 145 mm diameter, on top and vertically vibrating it using a vibrating hammer operating at a frequency of between 25 and 45 Hz for a period of 60 seconds.
  Some particles, especially weak rocks such as chalks and shale, will break down during compaction so this material must be discarded after compaction and fresh samples used each time the moisture content is adjusted.

Air voids lines

On the plot of dry density versus moisture content the air voids lines for 0, 5 and 10% air voids content A_v must be plotted using the expression:

$$\rho_d = \frac{1 - \frac{A_v}{100}}{\frac{1}{\rho_s} + \frac{w}{100\rho_w}} \qquad (13.12)$$

where:
ρ_d = dry density (Mg/m^3)
ρ_s = particle density (Mg/m^3)
ρ_w = water density, assumed to be 1.0 Mg/m^3
A_v = air voids content (%)
w = moisture content (%).

Some soils exhibit more than one 'peak' on the compaction curve so the air voids lines are a means of confirming that the most relevant part of the curves has been obtained, where the air voids content is lowest.

The air voids lines are sensitive to the value of particle density obtained or assumed so it is preferable to determine ρ_s where light or heavy minerals are suspected, rather than assuming a typical value.

Correction for stone content

For *in situ* soils which contain particles larger than 20 mm a 'correction' to the maximum dry density and optimum moisture content values may be required, so that direct comparison of the laboratory test result can be made with the *in situ* soil.

This is because gravel particles (> 20 mm) usually contain little moisture within themselves and exist within a matrix of more moist soil. A good example is a 'boulder' clay or glacial clay which comprises a moist cohesive matrix containing relatively 'dry' coarse gravel or cobble size particles.

The correction is based on the displacement of the soil matrix which has been used in the tests by coarse gravel of particle density ρ_g and assuming the coarse gravel contains no moisture of its own. In BS 1377:1990 the procedure is only suggested for grading zone 5 so if the percentage of the particles passing 37.5 mm is *B*% the 'corrected' dry density can be obtained from:

$$\text{'corrected'}\ \rho_{dmax} = \frac{\rho_{dmax}}{\frac{B}{100} + \frac{\rho_{dmax}}{\rho_g}\left(1 - \frac{B}{100}\right)} \qquad (13.13)$$

where ρ_g is the particle density of the gravel particles and ρ_{dmax} is the maximum dry density obtained from the compaction test with the coarser particles removed.

The 'corrected' optimum moisture content is given by:

$$\text{'corrected' OMC} = \frac{B}{100} \times \text{test OMC} \qquad (13.14)$$

The effect of this correction is to move the compaction curve upwards and to the left.

Worked Example 13.1 *Moisture condition value*

The results of a moisture condition test on a sample of silty clay are given in Figure 13.15. Determine the moisture condition value.

The change in penetration from:

1 blow to 4 blows = 81 – 58 = 23 mm

2 blows to 8 blows = 92.5 – 69 = 23.5 mm and so on

From the plot of change in penetration (natural scale) to number of blows (logarithm scale) the steepest straight line through the points cuts the 5 mm change in penetration line at 14.9 blows

$\therefore$ MCV = 10 $\log_{10}$ 14.9 = 11.7

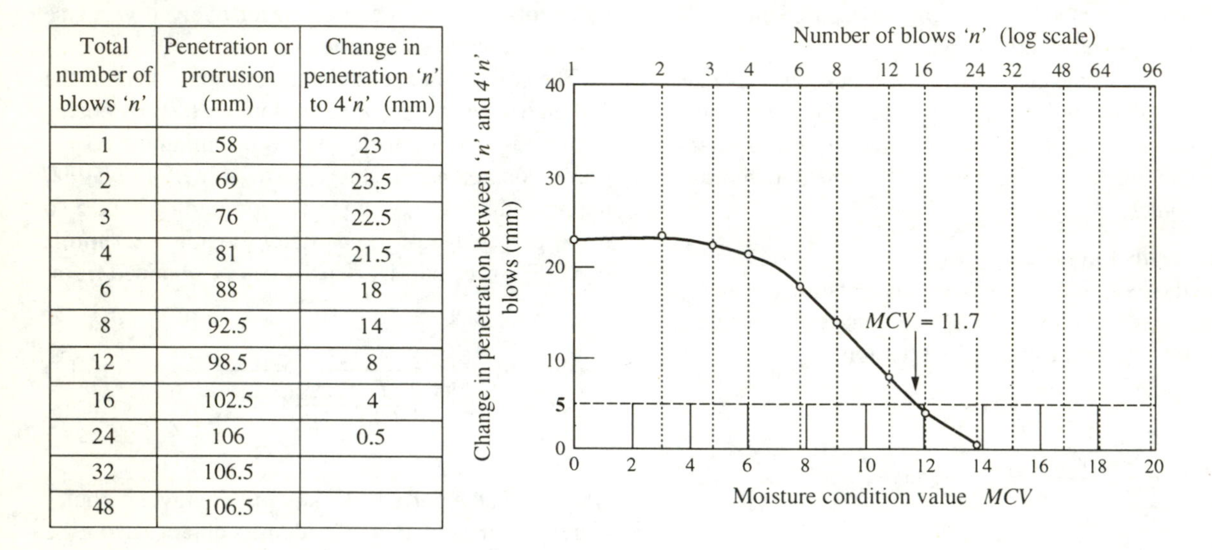

Total number of blows 'n'	Penetration or protrusion (mm)	Change in penetration 'n' to 4'n' (mm)
1	58	23
2	69	23.5
3	76	22.5
4	81	21.5
6	88	18
8	92.5	14
12	98.5	8
16	102.5	4
24	106	0.5
32	106.5	
48	106.5	

Figure 13.15 *Worked Example 13.1*

Worked Example 13.2 *Bulking – sand*

An embankment with a volume of 20 000 m³ is to be constructed of sand taken from a cutting. Assuming a shrinkage factor for the sand of 0.90 and a load factor of 0.85 determine:

a) the volume of cutting required (bank m³)

b) the production for 3 scrapers each with a heaped capacity of 20 m³ and a toatl cycle time of 6 minutes

a) the compacted volume = 20 000 m^3

$$\therefore \text{ bank volume required} = \frac{20\,000}{0.90} = 22\,222 \text{ m}^3$$

b) machine load in bank m^3 = 20 × 0.85 = 17.0 m^3

$$\therefore \text{ total number of loads required} = \frac{22\,222}{17.0} = 1307$$

(**Note:** If bulking was not considered this figure would be 1000)

For average working conditions an efficiency rating given by 50 working minutes per hour is assumed

$$\text{Hourly fleet production} = 17.0 \times \frac{60}{6} \times \frac{50}{6} \times 3 = 425 \text{ bank m}^3\text{/hour}$$

$$\text{Total time required} = \frac{22\,222}{425} = 52.3 \text{ hours}$$

Worked Example 13.3 *Bulking – clay*
For the same project as in Example 13.2 determine the total time required if the embankment is to be constructed of clay with a shrinkage factor of 1.1 and a load factor of 0.8.

Bank volume required $= \dfrac{20000}{1.1} = 18182\ \text{m}^3$

machine load $= 20 \times 0.8 = 16.0\ \text{m}^3$

total number of loads required $= \dfrac{18182}{16.0} = 1136$

Hourly fleet production $= 16.0 \times \dfrac{60}{6} \times \dfrac{50}{6} \times 3 = 400$ bank m^3/hour

Total time required $= \dfrac{18182}{400} = 45.5$ hours

Worked Example 13.4 *Compaction test (Figure 13.16)*
The results of a 2.5 kg rammer compaction test on a sample of silty clay are given below. Plot the compaction curve and determine the maximum dry density and optimum moisture content.
Volume of mould used = 1 litre
Weight of mould + base = 1.368 kg

	1	2	3	4	5	6
Wt. of mould + compacted soil (kg)	3.467	3.482	3.497	3.462	3.436	3.370
Wt. of tin (g)	29.31	29.44	29.50	29.43	29.38	29.59
Wt. of tin + wet soil (g)	156.64	152.65	130.18	159.88	132.71	153.08
Wt. of tin + dry soil (g)	140.70	138.12	119.17	147.11	123.06	143.09

For point 1

Bulk density $= \dfrac{3.467 - 1.368}{1.0} \times \dfrac{1000}{1000} = 2.099\ \text{Mg/m}^3$

Moisture content $= \dfrac{156.64 - 140.70}{140.70 - 29.31} \times 100 = 14.3\%$

Dry density $= \dfrac{2.099}{1 + 0.143} = 1.836\ \text{Mg/m}^3$

The remainder of the results are tabulated below

	1	2	3	4	5	6
Bulk density (Mg/m^3)	2.099	2.114	2.129	2.094	2.068	2.002
Moisture content (%)	14.3	13.4	12.3	10.9	10.3	8.8
Dry density (Mg/m^3)	1.836	1.864	1.896	1.888	1.875	1.840

From the compaction curve, Figure 13.16:
Maximum dry density ≈ 1.902 Mg/m^3
Optimum moisture content ≈ 11.9 %

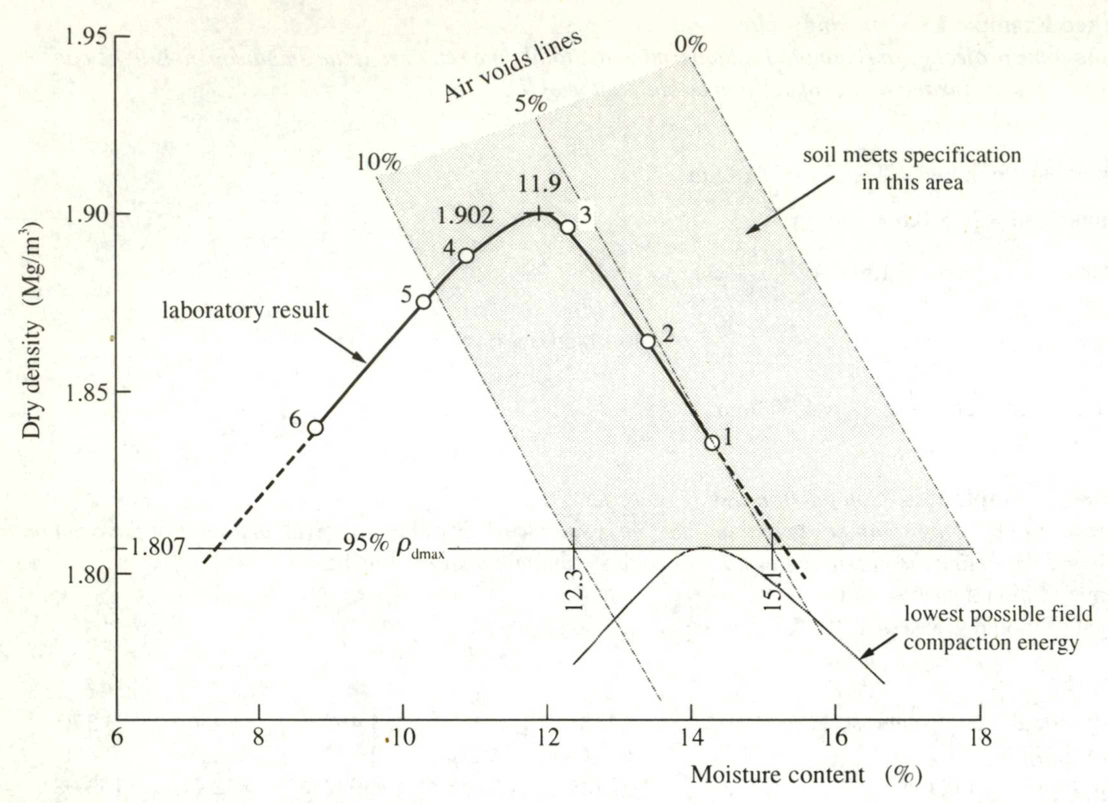

Figure 13.16 *Worked Examples 13.4 to 13.7*

Worked Example 13.5 *Air voids lines.*
Determine the values of dry density at selected moisture content values for air voids contents of 0, 5 and 10% assuming the particle density = 2.67 Mg/m³.

Using Equation 13.12 the values are tabulated below:

Moisture content w (%)	Dry density (Mg/m³)		
	$A_v = 0\%$	$A_v = 5\%$	$A_v = 10\%$
8	2.200	2.090	1.980
10	2.107	2.002	1.896
12	2.022	1.921	1.820
14	1.944	1.847	1.750
16	1.871	1.777	1.684
18	1.803	1.713	1.623

These values are plotted in Figure 13.16.

Worked Example 13.6 *Void ratio, degree of saturation and air voids content at the optimum value*
From the results of Example 13.4 determine the above values.

Expressions to derive these parameters are given in Table 2.15.
Assuming particle density = 2.67 Mg/m³
Bulk density $\rho_b = 1.902 \times (1 + 0.119) = 2.128$ Mg/m³

$$2.128 = \frac{2.67(1+0.119)}{1+e} \text{ giving void ratio } e = 0.404$$

$$\text{Porosity } n = \frac{0.404}{1+0.404} = 0.288$$

$$\text{Degree of saturation } S_r = \frac{wG_s}{e} \times 100 = \frac{0.119 \times 2.67}{0.404} \times 100 = 78.7\%$$

Air voids content $A_v = 0.288\,(1 - 0.787) \times 100 = 6.1\%$

Worked Example 13.7 *Relative compaction*
The minimum dry density to be achieved by field compaction has been specified as 95% of the maximum dry density obtained from the 2.5 kg rammer compaction test. From the results of Examples 13.4 and 13.5 determine the minimum moisture content which must be specified to ensure that no more than 10% air voids may be present. Determine the maximum moisture content w that can be permitted.

$$\text{The minimum dry density} = \frac{95}{100} \times 1.902 = 1.807 \text{ Mg/m}^3$$

Using Equation 13.12:

$$1.807 = \frac{1 - \frac{10}{100}}{\frac{1}{2.67} + \frac{w}{100}} \quad \text{giving } w = 12.3\%$$

Note that this minimum moisture content would only apply if the field compaction energy is less than the laboratory compaction energy. If the field compaction energy is greater then soils of lower moisture contents can be compacted to the specified requirements.
The maximum moisture content can be obtained assuming that wet of optimum compaction achieves says 5% air voids. Then

$$1.807 = \frac{1 - \frac{5}{100}}{\frac{1}{2.67} + \frac{w}{100}} \quad \text{giving } w = 15.1\%$$

Worked Example 13.8 *Correction for stone content*
A gravelly clay contains 25% and 8% gravel retained on 20 mm and 37.5 mm sieves, respectively. A compaction test has been carried out using the CBR mould on material remaining after the gravel greater than 37.5 mm has been removed. The maximum dry density of the material from the compaction curve is 1.965Mg/m³ and the optimum moisture content is 10.8%. Determine the maximum dry density and optimum moisture content of the whole sample including the gravel greater than 37.5 mm. The particle density of the gravel particles is 2.65 Mg/m³.

According to BS 1377:1990 this material would lie within grading zone 5 so a correction for the effect of stone content would be permissible. Assuming the coarse gravel merely displaces the remaining matrix and would have no effect on the compaction test result and that it contains no moisture of its own Equations 13.13 and 13.14 can be used.

$$\text{'corrected'}\rho_{\text{dmax}} = \frac{1.965}{\frac{92}{100} + \frac{1.965}{2.65}\left(1 - \frac{92}{100}\right)} = 2.006\ \text{Mg/m}^3$$

$$\text{'corrected' OMC} = \frac{92}{100} \times 10.8 = 9.9\%$$

The effect of this 'correction' is to move the compaction curve upwards and to the left.

Exercises

13.1 The results of a moisture condition test on a sample of stiff sandy clay are given below. Determine the moisture condition value of this clay.

Total number of blows, *n*	Penetration (mm)
1	43.0
2	56.0
3	63.0
4	68.0
6	75.0
8	80.5
12	87.0
16	91.0
24	95.5
32	98.5
48	100.5
64	101.5
96	102.0
128	102.0

13.2 The results of a MCV calibration are given below. Determine the values of *a* and *b* and assess the moisture sensitivity.

MCV	2.0	3.4	5.2	7.8	10.4	12.9	14.9
water content (%)	30.0	28.9	27.5	25.3	23.6	21.3	19.9

13.3 An embankment of total volume 15 450 m^3 is to be constructed using sand taken from a cutting. The bulk density of the sand *in situ* is 1.86 Mg/m^3 and in the compacted state in the embankment it is 1.98 Mg/m^3. Four dump trucks each with a heaped capacity of 15 m^3 are to be used operating on an average cycle time of 8 minutes with an efficiency rating of 0.75. The bulk density of the sand loaded in the dump trucks is estimated to be 1.67 Mg/m^3. Determine:
a) the volume of cutting required to make the embankment
b) the total number of loads
c) the hourly production rate and total time required.

13.4 The results of a BS light compaction test are given below. Plot the compaction curve and obtain the optimum moisture content and maximum dry density for the soil.

water content (%)	14.3	15.8	17.7	19.2	20.9	22.6
bulk density, (Mg/m^3)	1.967	2.008	2.065	2.088	2.077	2.072

13.5 For the result obtained in Exercise 13.4 determine the void ratio, porosity, degree of saturation and air voids content for the soil at its optimum moisture content. Assume the specific gravity of the particles to be 2.72.

14 Site Investigation

Site investigation

Introduction

Problems associated with the ground conditions can lead to cost over-runs, longer construction periods and possibly expensive litigation. It is essential, therefore, to find out as much as possible about the site and its ground conditions with a thorough site investigation.

Site investigation consists of collecting available information about the site and its environment and carrying out a ground investigation. This information is then used to assess the suitability of the site for designing and constructing the proposed works with regard to stability, serviceability, ease of construction, and acceptable performance balanced by concern for safety, economics and the environment.

Geotechnical design using methods of analysis given in other chapters of this book (such as bearing capacity and settlements of shallow and pile foundations, stability of embankment and cutting slopes, earth pressures on retaining structures) cannot be carried out until the appropriate model of the site is determined and the relevant parameters for each soil type obtained.

As well as being carried out for new works, site investigations may be used for seeking sources of construction materials, quarrying, mineral prospecting, hydrogeology and groundwater abstraction, selecting sites for waste disposal, for assessing the degree of contamination of soils and groundwater, for checking the safety of existing works when constructing new works nearby and forensic investigations to establish the causes of failures such as with slopes, or defects such as with structures affected by excessive ground movements, and to design remedial measures such as for slope stabilisation or underpinning of buildings.

More detailed coverage of this subject is given in the U.K. Code of Practice BS 5930:1981 and books by Clayton, Simons and Mathews (1983) Cottington and Akenhead (1984), Joyce (1982) and Weltman and Head (1983).

Stages of investigation

A degree of discipline must be exercised in carrying out a site investigation, otherwise some information may be overlooked or simply not gathered. To reduce this risk, a staged procedure is usually adopted comprising desk study, site reconnaissance, topographic surveys and hydrographic surveys followed by a detailed ground investigation with sampling, laboratory and in situ testing and groundwater observations.

These stages will overlap and complement each other and may not be considered complete until all of the stages are concluded and inter-related. The planning of the ground investigation will proceed much more efficiently if the previous stages have been carried out thoroughly.

Geotechnical design is often as much an art as a science, relying on experience and empiricism for its effectiveness. Due to the uncertainties concerning methods of analysis, material properties and behaviour a major element of geotechnical works consists of performance monitoring, particularly where savings in costs and construction periods may be gained and where the consequences of the works not achieving their predicted performance may be serious. Re-design at intermediate stages or remedial works are then a viable solution to progressing the works.

Desk study

A desk study is the collection of as much information about the site as possible. A comprehensive list is given in BS 5930:1981 including land surveys, boundaries, site features, topography, natural and artificial drainage, access, flood risk, utilities such as water, drainage, sewerage, electricity, gas, telephone both for adequate supply to the site and as obstructions below ground.

Much useful information can be readily obtained from maps, both recent and old, coastal charts, aerial photographs, memoirs, local authority records and local libraries. The British Geological Survey publishes geological maps and memoirs for most of Great Britain and has available various other geological records including borehole and well records.

Previous uses of the site must be determined even in 'untouched' rural areas where changes in topography, erosion, deposition, diversions of streams, rivers and drainage conditions could affect the project. In more developed areas records of underground mining, mineral extraction, quarrying operations, waste tipping, and demolished properties will give clues to potential hazards and obstructions in the ground. Local residents and former workers may be useful in this respect.

Industrial areas must be examined for underground obstructions and the effects of the industrial processes such as removal of groundwater, changes in temperature, ingress of harmful liquid chemicals and incorporation of harmful solids.

Useful guidance is included in BS 5930:1981 and BRE Digest 318:1987.

Site reconnaissance

A thorough examination of the site should be made by visiting the site and its surroundings as early as possible and preferably in conjunction with the desk study, when the information already obtained can be checked and omissions or uncertainties can be investigated.

A walk-over survey observing the features within and around the proposed works with checks on access, adjacent properties, surface topography, surface water, drainage, present site use, evidence of ground conditions from geomorphology, quarries, cuttings, exposures, type and condition of vegetation, condition of existing structures.

Useful guidance is included in BS 5930:1981 and BRE Digest 348:1987.

Ground investigation

Extent of the ground investigation

With the information obtained from the desk study and site reconnaissance, the variability expected in the ground conditions and the type and scale of the project the amount of ground investigation can be assessed.

Skimping on site investigation is very risky and rarely cost-effective since for nearly all projects most risk lies in the ground conditions.

Typical costs lie around 0.2 to 2 per cent of the total cost of the project (Anon, 1991) whereas contractual claims involving ground conditions which had not been foreseen can cost the client much more than this figure and involve lengthy litigation. BRE Digest 322:1987 recommends that a minimum of 0.2% of the project cost should be spent on ground investigation for low-rise buildings and that the developer should actively participate in the investigation process to ensure that he/she appreciates the risks involved with the ground conditions.

The purpose of the ground investigation is to obtain a three-dimensional cross-section of the site and exploratory holes such as boreholes and trial pits only establish the ground conditions at a point so the cross-sections across the site must be completed by inferring the intermediate conditions. For uniform homogeneous conditions and simple geology this could be achieved fairly confidently with widely spaced exploratory holes but for variable conditions an accurate impression can only be gained from closely spaced boreholes.

For many sites, the degree of variability will not be known so a constant monitoring and review of the investigation is necessary to ensure that the ground conditions are adequately investigated. To ensure this, a useful procedure is to carry out a preliminary and limited investigation to obtain basic, general information which can be used to plan a more intensive investigation.

In some instances, double drilling is carried out by sinking a borehole without undisturbed sampling or testing to determine the soil layering, followed by an adjacent borehole with samples or tests taken at predetermined depths.

The actual spacing and locations of the holes will depend on the nature of the project, its size and its shape. For housing estates, holes may be sunk at 30–100 m apart and not necessarily related to building layout since this may change. Conversely, the properties may be located to suit the ground conditions encountered.

For structures, holes at 10–30 m apart may be appropriate with locations at the corners, some intermediate points and locations of heavy loading or special construction. It should be remembered that more confidence can be placed in interpolation between exploratory holes than extrapolation away from them so a wide coverage is preferable.

For linear structures such as pipelines and sewers holes at special crossings or manholes are usually sunk with less investigation between. If these works are constructed by tunnelling then vertical boreholes will provide limited and inefficient information. If the ground conditions require, horizontal drilling from the face of the tunnel using pilot boreholes, can be cost-effective and reduce the risk of collapse.

Highway construction includes major and minor structures and earthworks of cuttings and embankments, each requiring different consideration, so the investigation must be designed and carried out to cater for these.

Exploratory holes should be sunk at locations away from proposed foundations, tunnels or shafts if possible. Backfilling trial pits will leave a disturbed and softened zone and boreholes may provide connections between different water bearing layers producing groundwater contamination or ingress into the works. It is preferable to backfill boreholes with an impermeable grout.

Depth of exploration

On the cross-section through the site will be superimposed the proposed construction so the depth of exploration must include the depth of the works and the depth of ground which may be affected by the works or the depth of ground which may affect the works. In the case of the investigation of mining areas workings at deep levels may affect the stability of structures which themselves do not stress the workings.

Providing workings are not present a general guide is to take boreholes:

1 in thick compressible strata, to a depth where the changes in stress applied by the works are minimal. For this purpose, a depth of 1.5 times the width of the loaded area is suggested, where the loaded area may be an individual foundation if the foundations are spaced widely or the whole of the structure if the foundations are closely spaced or the structure is supported on a raft foundation. In the case of a pile group the loaded area would tend to be the equivalent raft, see Chapter 10, part way down the pile length. However, since the length of pile would not be known at the investigation stage, if piles are likely to be required then the depth of investigation should be as in 2) below.

2 into relatively incompressible strata, or strata which for the type of construction envisaged will not contribute to settlements or other movements. This is considered as the 'rigid' stratum in methods for stress distribution (see Chapter 5) and settlements (Chapter 9).

3 into sound, unweathered bedrock. A penetration of 2–5 m should be obtained depending on the hardness of the rock and to make certain it is bedrock and not a boulder. Obtaining rock cores by rotary coring is preferable to chiselling in a cable percussion borehole.

Choice of method of investigation

The main factors which affect the choice of method of investigation are:

a) ***Access***

Cable percussion rigs, rotary drilling rigs and mechanical excavators require reasonably smooth, unhindered access to a location with space for erection of the apparatus and sufficient headroom. Otherwise, access must be provided by breaking out or removing obstructions, cutting roadways in hilly or hummocky terrain or forming temporary roads on waterlogged or boggy sites. In some instances, especially on steep slopes the drilling rigs are dismantled and re-erected on scaffold staging or other supports. Some companies specialise in 'mini-site investigation' for very limited access.

b) ***Equipment limitations***

Mechanical excavators are typically limited to depths of 3–4 m and may damage unforeseen services. Side support or trench shoring must be provided if personnel are to enter the pit and this may obscure some of the exposure.

Hand augering is carried out without support to the sides of the hole so it is limited to self-supporting strata without obstructions present such as soft or firm clays. A maximum depth of 5–6 m is then possible.

The cable percussion rig has a winch capacity of 1–2 tonnes so the amount of casing it can pull out of the ground is limited. For deep boreholes it is usual to commence with larger diameter casing and then reduce in diameter to extend the borehole. In this way, maximum depths of about 60 m can be achieved, although deeper boreholes, up to 90 m, can be sunk if the lower ground strata are self-supporting and casing is not needed over this depth.

In bedrock and in some heavily overconsolidated glacial soils such as gravelly clays rotary drilling is the most effective method with core samples obtained.

c) ***Types of ground***

Made ground is best investigated with trial pits since it is at shallow depths and its variability can be assessed. Hand dug pits may be necessary if underground services or buried structures are to be exposed without damage.

Silts, sands and gravels are best investigated with cable percussion boring which can support the soil with casing during drilling and provide the access for carrying out *in situ* SPTs, permeability tests or installing piezometers. The Dutch cone penetration test rig should also be used to provide a continuous, undisturbed plot of the layering and densities of the strata. For soils which contain cobbles and boulders cable percussion boring can be time-consuming and not truly representative because of the chiselling required so if the nature of the scheme warrants large-scale pits with dewatering if necessary, may be the most appropriate for shallow depths.

Cable percussion boreholes are most commonly used in clays to obtain U100 samples and perform *in situ* vane tests although difficulties are often experienced if clays contain cobbles or boulders and care must be taken with soft, sensitive and laminated clays.

Methods of ground investigation

a) *Trial pits*

Trial pits are excavated with a hydraulic back-hoe excavator forming a trench about 3–5 m long with a width equal to the back-acter bucket, 0.6 or 0.9 m and a depth determined by the reach of the hydraulic arms, between 3 and 6 metres, depending on the type of machine.

The sides must never be assumed to be stable. Trench collapse is still unfortunately a major cause of death on construction sites. The sides must be supported at all times if the pits are to be entered. If they are not supported then collapse often occurs due to the surcharge on the side of the pit where the spoil is temporarily stock-piled. Access is often preferable because the bucket smears the sides of the pit and obscures the *in situ* structure.

A trial pit record should describe all four faces if they are different and dip and direction measurements of bedding, joints, fissures, unconformities should be taken if present. 'Undisturbed' samples can be taken using hand-auger rods with 38 mm diameter sampling tubes or preferably 100 mm diameter sampling tubes if the clay is not too stiff and *in situ* vane tests can be carried out. Bulk and smaller disturbed samples can be taken from the pit, the bucket or the spoil heap. Samples of the groundwater must not be forgotten.

b) *Headings or adits*

Headings are excavated using hand-tools and timber supports in a near horizontal direction from the bottom of shafts or from the surface of sloping ground. They are not commonly used for investigation purposes alone due to their cost and slow progress although they are used occasionally for construction of short sections of tunnels.

c) *Hand auger* *(Figure 14.1)*

A hole, usually, 100 mm in diameter, is formed by manually rotating a cross-piece above ground level attached by rods to an auger bucket below ground level. Soil is collected in the auger, removed from the ground and inspected to obtain a record of the strata. A cylindrical or barrel-auger (post-hole or Iwan-type auger) is used to remove clayey soils and a flat circular auger with a flap-valve is used to remove gravel although variable success can be expected with the latter due to the obstructions.

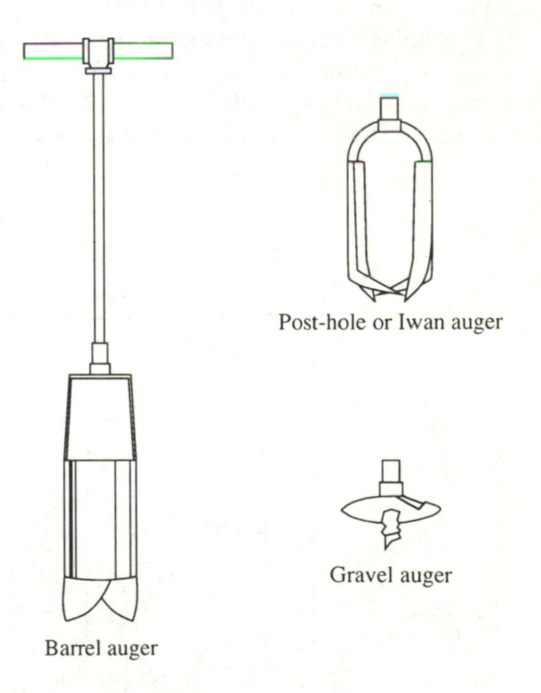

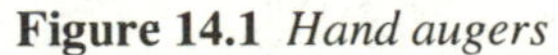

Figure 14.1 *Hand augers*

Disturbed samples are taken from the cuttings and 38 mm diameter sample tubes can be pushed or driven in for 'undisturbed' samples. Vane tests may also be carried out *in situ*. Depths of 5–6 m are achievable if the soil is self-supporting, e.g. soft or firm clay but progress can be halted by one gravel particle or when water-bearing strata are entered.

d) ***Cable percussion boring*** *(Figures 14.2 and 14.3)*
Often referred to as shell and auger boring an auger is rarely used with this method in the UK. The rig consists of a four-leg 'A-frame' or derrick with a diesel-powered winch lifting and releasing a cable running over a pulley wheel so that tools attached to the end of the cable are lifted vertically and dropped free-fall.
Once the rig is erected and stabilised the hole should be commenced by hand excavation to about 1 m depth, to ensure that shallow underground services are avoided. Then a heavy steel tube (clay-cutter or shell) is dropped into this hole a number of times to collect soil inside it and extend the depth of the hole. The tube is removed, the soil is cleaned out and the process repeated.
As the hole progresses the soils may not be self-supporting and soil at the top of the hole may fall in, so steel lining tubes called casing are driven into the hole to support the soil, to provide a guide for the boring tools and to ensure a 'clean' hole. Casing consists of short sections of steel tube connected

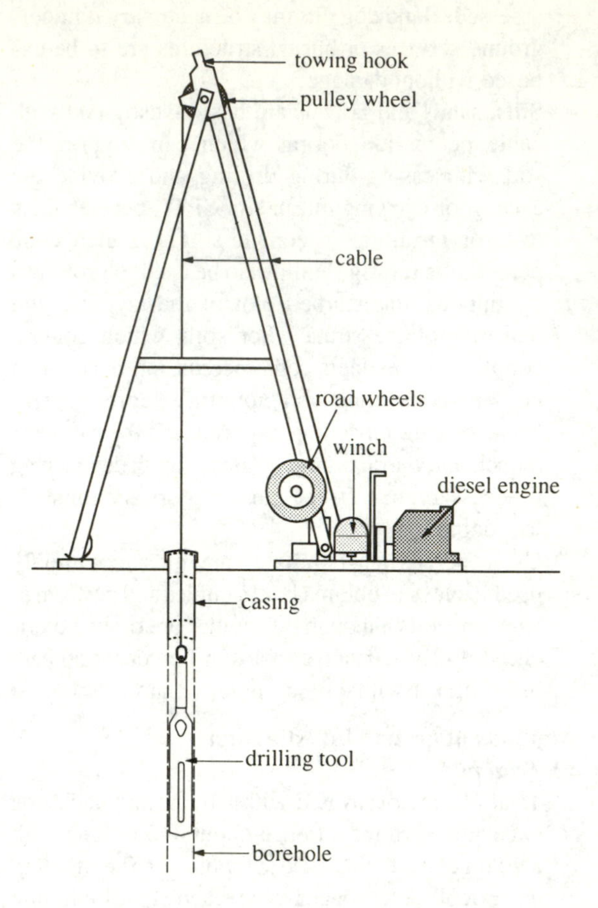

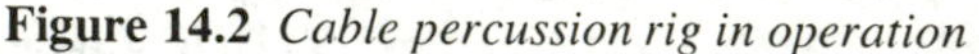

Figure 14.2 *Cable percussion rig in operation*

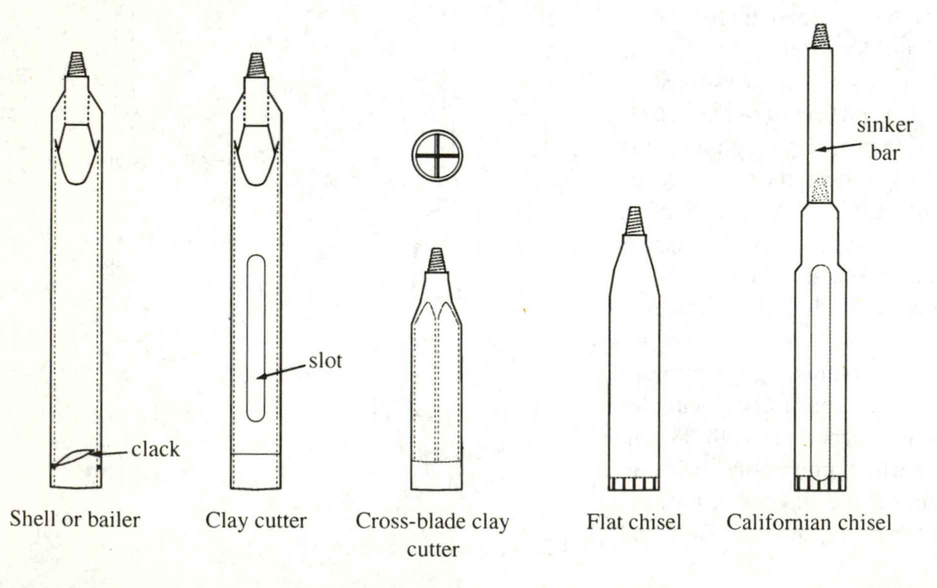

Figure 14.3 *Cable percussion drilling tools*

together by their threaded ends with a protective steel cutting shoe at the base and a driving head at the top. Casing diameters of 150 and 200 mm are most common.

The clay-cutter (Figure 14.3) is an open-ended tube, which allows clay to enter, and with slots in the sides to enable removal of the clay. An alternative is the cross-blade clay cutter which is lighter and requires attached weights called sinker bars, but allows easier removal of the clay.

The shell (Figure 14.3) is used for removing granular soils. It also comprises a steel tube but without slots and with a flap-valve or 'clack' at its lower end. To get sands and gravels into the shell a pumping action is required so water must be added into the borehole to 'fluidise' the sands and gravels. With short up and down movement sand and gravel enters the shell and is retained for removal by the clack. If cobbles or boulders are encountered, they must be broken up by dropping a heavy chisel on them, so they are displaced sideways or removed by the shell.

Where soils are in a loose sensitive state the percussive action of the tools can cause disturbance and excessive piston and suction action may loosen a dense sand so samples or tests must be taken from at least one diameter below the bottom of the borehole to ensure a more representative result. Care is required in taking disturbed samples of granular soils from the shell to prevent loss of finer particles. Addition of water to a borehole precludes the testing of samples for moisture content.

e) ***Mechanical augers** (Figure 14.4)*

Two basic types exist – solid rod bottom augers and hollow stem continuous flight augers.

The bottom augers may comprise either a flight-auger with short helical flights up to about 600 mm in diameter or a bucket auger with cutters on an angled base-plate with diameters between 300 and 1800 mm. Both are rotated at the ends of drill rods, to cut into the soil and retain it on the flights or in the bucket. They are then raised from the borehole and emptied for identification of the strata and taking disturbed samples. Addition of water should not be necessary. To progress the borehole this procedure is repeated.

Undisturbed samples or tests can be taken at intervals from the bottom of the borehole. These boreholes cannot be followed by significant lengths of casing so they are only suited to self-supporting ground such as clayey soils and soils above the water table.

The continuous flight auger method consists of a full length helical flight wrapped around a hollow tube, with a central rod and pilot assembly at the bottom of the auger to prevent soil from entering the auger hollow stem. As the auger rotates, soil rises up the spiral for identification and disturbed sampling. The augers are typically 150 to 250 mm outside diameter, with hollow stem diameters of 75 to 125 mm. These augers can reach to depths of 30 to 50 m in suitable soils.

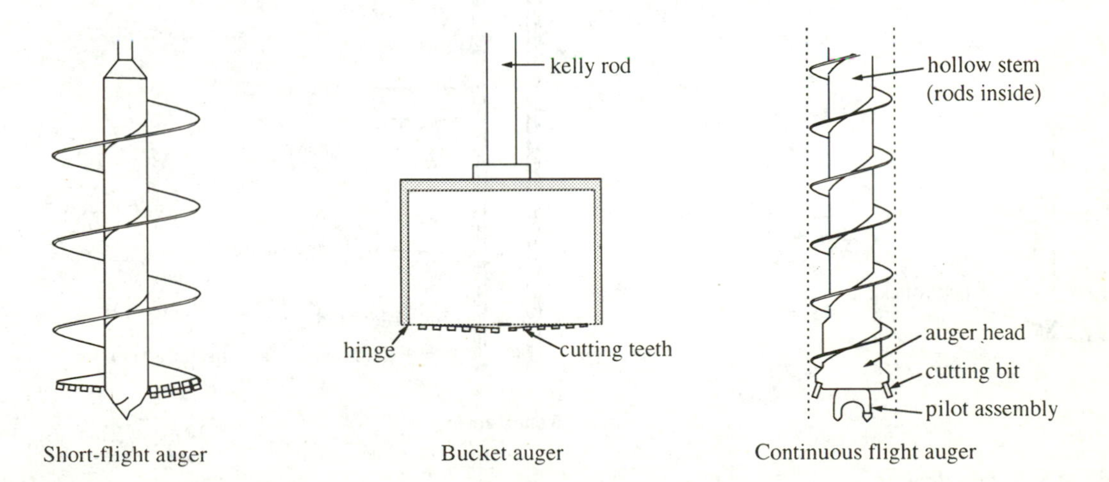

Figure 14.4 *Mechanical augers*

At intervals, the central rod and pilot assembly can be removed to allow sampling tubes to be pushed or driven in at the bottom of the borehole, or standard penetration tests may be carried out. Water balance is not usually maintained so disturbance of the ground due to piping may occur.

The torque required to turn the auger can become considerable, so a heavy duty rotary drilling rig is necessary. These are lorry mounted and will, therefore, be more expensive and require better access conditions than the light cable percussion rig. As the soil is brought to the surface it is difficult to assess its depth in the ground, so changes of strata will not be determined accurately unless frequent undisturbed sampling is carried out.

Penetration of the ground can be severely hampered in coarse cohesionless soils and finer silts and sands can flow up the hollow stem when drilling below the water table, causing excessive disturbance of the ground at the bottom and around the auger.

f) ***Rotary open hole and core drilling*** *(Figure 14.5)*

These techniques are used in harder rock exploration when cable percussion or auger methods can no longer penetrate.

Open hole drilling consists of rotating a rock roller bit at the end of hollow drill rods to cut a cylindrical hole. The drilling fluid, which is usually compressed air or pumped water, passes down the hollow rods to cool the bit and remove cuttings by flushing them back up the outside of the rods. The cuttings and rate of penetration are the only means of identification so the accuracy of descriptions and changes of strata will be limited.

Rotary coring consists of cutting a cylinder of rock using a rotating double-tube core barrel with a coring bit attached to the bottom of the outer barrel. A swivel mechanism at the top of the inner barrel allows it to remain stationary as the outer barrel and coring bit cut an annular hole and provide a cylindrical rock sample.

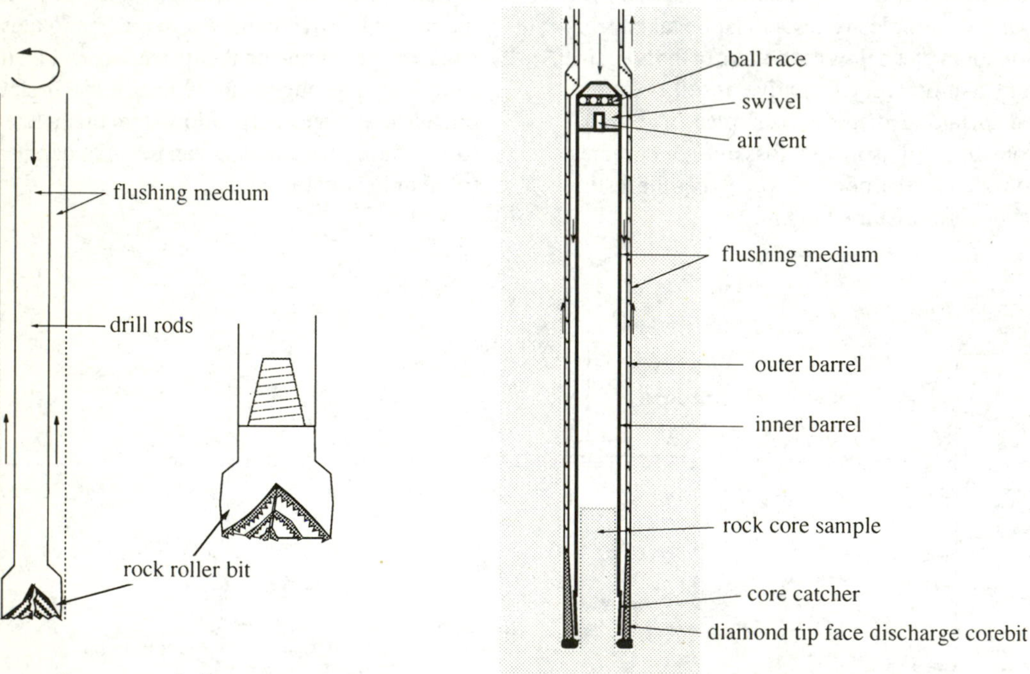

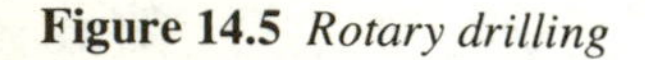

Figure 14.5 *Rotary drilling*

The flushing medium, which may be compressed air, pumped water, mud or foam, passes down the hollow drill rods, between the inner and outer barrels to cool the drill bit, remove the cuttings and flush them up the borehole outside the core barrel. The inner barrel does not rotate and the flushing medium passes outside the inner barrel so that the rock core sample is not disturbed, smeared or eroded. This is especially important with friable weathered rocks.

A plastic liner inside the inner barrel enables the rock core to be removed from the core barrel with minimal disturbance, keeping the rock pieces in their correct positions for inspection.

It is essential to obtain 100% core recovery or as near as possible. Core loss means strata not identified and these are likely to be the weakest materials and hence the most critical. The type of core bit used must be matched to the nature of the rock and larger diameter coring should be adopted in weaker rocks. Core sizes between 18 and 165 mm diameter may be obtained, with H size (76 mm) usually specified as a minimum requirement in the UK.

Undisturbed sampling – sampling quality

Due to the relief of total stresses, possible groundwater changes and cutting action while taking samples it is not possible to obtain truly undisturbed samples from the ground. The aim must be, therefore, to minimise disturbance as much as possible. A useful classification of sample quality is given in Table 14.1, based on the soil properties that can be reliably determined from a sample. Soil properties determined from laboratory tests on lower classes of sample should be treated with some caution.

Difficulties in obtaining good quality undisturbed samples can be experienced with soft clays, sensitive clays, laminated clays, gravelly clays, partially saturated soils, collapsing soils, organic soils, coarse soils and cohesionless soils with even poorer quality likely below a water table.

Types of samples

a) ***Block samples***

These are cut by hand from the soil at the base of an excavation. They can be quite large and carefully selected. However, they can be heavy, difficult to remove, handle and transport so they are time-consuming and liable to be expensive. They are not confined by sampling equipment so the soil must have some self-supporting capability and they must not be allowed to change in moisture content.

b) ***General purpose open-tube sample (U100)*** *(Figure 14.6)*

This is the most commonly adopted method of sampling when using cable percussion boring in all cohesive soils.

The sampler comprises a steel or aluminium alloy tube, 450 mm long, with a cutting shoe threaded on to its lower end. The sampling tube is threaded into a driving assembly with an overdrive space and a non-return valve above the tube.

The borehole is cleaned of loose debris and the sample tube is driven or pushed into the soil at the bottom of the borehole a sufficient distance so that it is full of 'undisturbed' soil, with air and any water passing through the non-return valve. Remaining debris or softened soil enters the overdrive space from where it is discarded. It must not be overdriven as this would compress the soil.

The sampler and tube are then removed from the borehole with the non-return valve closed, providing a suction effect to assist in retaining the soil in the tube. Sometimes soil is not recovered in the tube so a core-catcher device is frequently used. The samples recovered are trimmed at the ends, sealed with wax, labelled top and bottom and retained with screw caps for transport to the laboratory.

The cutting shoe is larger than the inside and outside diameters of the tube to reduce friction or adhesion on the inside for minimum disturbance to the sample and on the outside to facilitate withdrawal of the sampler from the borehole.

The internal diameter of the cutting shoe D_c is about 1% less than the internal diameter of the tube D_s to provide clearance. This must not be exceeded since the sample would be allowed to expand, open up macro-fabric and possibly soften if the sample was taken from a wet borehole. The outside diameter of the cutting shoe D_w should be slightly greater than the outside diameter of the tube, D_T.

The area ratio (Equation 14.1), defined in Figure 14.6, represents the volume of soil displaced during sampling compared to the volume of the sample and is related to disturbance effects so it should be as low as possible.

Table 14.1
Sampling quality (From Rowe, 1972)

Quality class	Properties	Purpose	Typical sampling procedure
1	Remoulded properties Fabric Water content Density and porosity Compressibility Effective stress parameters Total stress parameters Permeability Coefficient of consolidation	Laboratory data on in situ soils	Piston thin walled sampler with water balance
2	Remoulded properties Fabric Water content Density and porosity Compressibility Effective stress parameters Total stress parameters	Laboratory data on in situ insensitive soils	Pressed or driven thin or thick walled sampler with water balance
3	Remoulded properties Fabric A 100% recovery Continuous B 90% recovery Consecutive	Fabric examination and laboratory data on remoulded soils	Pressed or driven thin or thick walled samplers. Water balance in highly permeable soils
4	Remoulded properties	Laboratory data on remoulded soils. Sequence of strata	Bulk and jar samples
5	None	Approximate sequence of strata only	Washings

c) ***Thin-walled tube sampler*** *(Figure 14.6)*
Using a thin-walled tube (Shelby tube) with a sharp integral cutting edge rolled slightly inwards gives an area ratio of about 10% and an inside clearance so these tubes are more suited to softer, sensitive soils. They are usually 38, 75 or 100 mm diameter. They are not suited to stiff soils or soils containing gravel particles when they are liable to damage and cohesionless soils may not be retained in the tube. The length of sample recovered will often be less than the length of sampler pushed into the soil due to friction on the inside of the tube exceeding the bearing capacity of the soil and preventing soil entering the tube.

d) ***Split-barrel SPT sampler*** *(Figure 14.6)*
This sampler is used in conjunction with the standard penetration test. Two half-cylinders are joined together by a threaded open steel cutting shoe at the bottom and threaded onto drilling rods at the top.

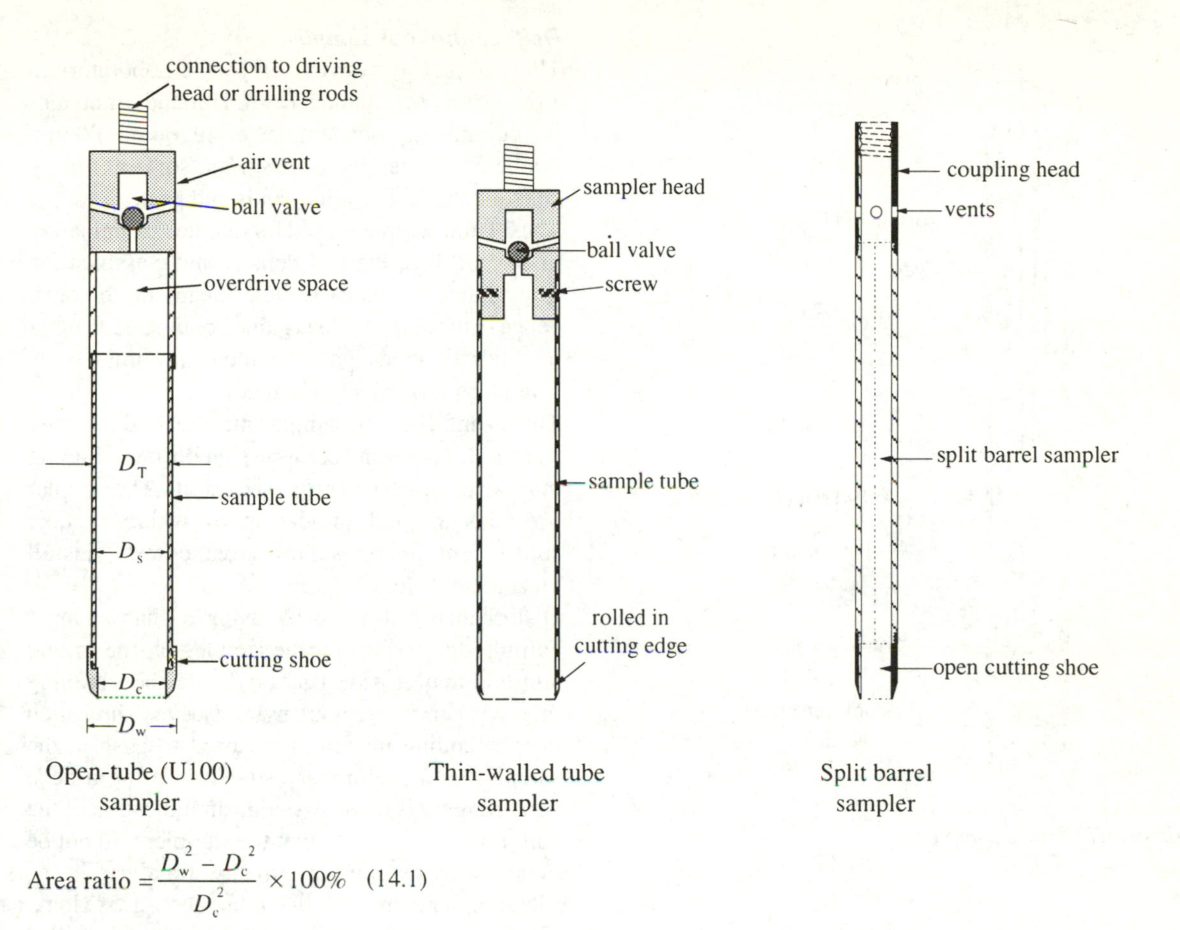

$$\text{Area ratio} = \frac{D_w^2 - D_c^2}{D_c^2} \times 100\% \quad (14.1)$$

Figure 14.6 *Tube samplers*

A sample is obtained following removal from the borehole, by undoing the two semi-cylindrical halves of the sampler. Due to inside friction, a full length sample is seldom obtained and, because the cutting shoe has a thick wall with an outside diameter of 51 mm and an inside diameter of 35 mm, giving an area ratio of more than 100%, the samples are highly disturbed.

e) ***Piston sampler*** *(Figure 14.7)*
For soils which are particularly sensitive or may not be recovered on removal of the sampler, such as very soft clays, silts, muds and sludges, the piston sampler should be used. The purpose of the piston is to retain soil in the sampling tube during removal by means of suction.

The piston is attached to an inner solid rod and the sampling tube and sampler head is attached to hollow sleeve rods. With the piston at the bottom of the tube the piston and tube are locked together. The whole assembly is lowered down the borehole, using extension pieces of solid and hollow rods and then pushed into the bottom of the borehole and beneath the disturbed zone. The rods are unclamped and with the piston held stationary the sampling tube is pushed downwards, not driven.
Short length tubes may be pushed in manually but for longer length tubes and firmer soils a block and tackle arrangement attached to the drilling rig is required. When the sampling tube has been pushed in its full length and no more the piston rods and sampling tube rods are clamped together and twisted

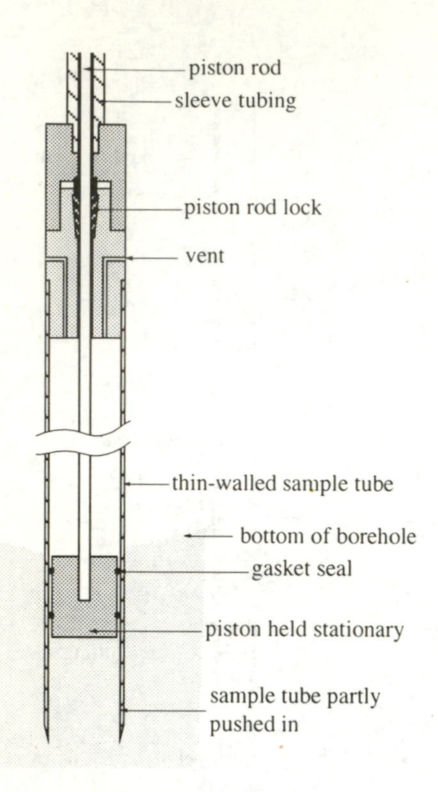

Figure 14.7 *Piston sampler*

to shear the soil and break the suction at the bottom of the tube. The complete assembly is then slowly withdrawn from the borehole.

The sampler is available in diameters between 20 and 125 mm with lengths between 0.15 and 1.5 m but the 100 mm diameter 600 mm long sampler is the most common. When the importance of the project has justified their use, 250 mm diameter samples have been taken for testing in the large diameter hydraulic cell apparatus for consolidation and permeability properties.

The sampler produces good quality, undisturbed specimens because it is pushed beneath any disturbed zone, and it has a low area ratio. It may not withstand pushing into stiff soils and may be damaged if gravel particles are present. Good recovery is achieved due to the piston suction although this may be eliminated when sampling partially saturated soils and cohesionless soils.

f) ***Delft continuous sampler***

This sampler was developed by the Laboratorium Voor Grond Mechanica, Delft, Holland for obtaining continuous core samples of 29, 66 or 100 mm diameter to depths of 18m in soft, normally consolidated clays, silts, sands and peat.

The 29 mm sample is used for detailed examination of the soil layering and density and classification tests. The continuous sample means more confidence in the soil profile is gained compared to other methods of investigation which may miss some layers or fabric of significance.

The 66 and 100 mm samples may be used for other tests, although some compression during sampling may occur which will affect the results. The sampler provides a good indication of where to take subsequent piston samples, vane tests, install piezometers etc.

Disturbance is limited by using a sharp, longer cutting edge, by pushing the sampler into the ground and eliminating side friction inside the sampling tube with an impervious nylon sleeve. The Dutch deep sounding machines are used to push in the sampler so a preliminary sounding using a cone penetrometer is made to obtain an indication of the soil profile and ensure that the sampler will not be damaged by resistant or gravelly strata.

The sampler comprises two tubes about 1.5 m long. The inner tube is a thin-walled plastic tube filled with a bentonite based fluid of similar density to the surrounding ground. This fluid will support the sample and lubricate the space between the sample and the inner tube. Enclosed inside the outer tube is a magazine containing a rolled-up tube of nylon stockinette which is treated with a rubberising fluid to make it impervious and flexible. The end of this nylon sleeve is fixed to the top cap.

As the sampler is pushed into the ground, the cutting shoe cuts a continuous sample which is fed into the stockinette and supported in the plastic inner tube by the bentonite fluid. As the sampler is advanced extension tubes are added until the final depth is reached. The soil is then locked in the sampler by a sample-retaining clamp above the magazine and the whole sampler is withdrawn from the ground. Samples are cut into 1 m lengths and placed in purpose-made core boxes. The stockinette is cut, the samples are split in half, examined, logged and photographed as a continuous record of the soil profile.

Methods of *in situ* testing

a) *Standard penetration test*

This test was developed in the USA in the 1920s to assess the density of sands. It is now the most commonly used *in situ* test in cable percussion boring.

The test measures the penetration resistance of the split-barrel sampler, when it is driven into the soil, at the bottom of a borehole in a standard manner. The *N* value, which is the number of blows required to achieve 300 mm penetration of the soil, indicates the relative density of a sand or gravel, the consistency of other soils such as silts or clays and the strength of weak rocks such as chalk, shale, mudstone.

Over the years, many empirical correlations with other parameters have been obtained and methods of analysis such as for settlements of foundations have been developed empirically using the *N* value. Skempton (1986) has examined various factors which affect the results.

The test is described in BS 1377:1990, Part 9 and ASTM D1586. The split-barrel sampler (Figure 14.6) is attached to stiff drill rods and lowered to the bottom of the borehole. A standard blow consists of dropping a mass of 64 kg free fall through 760 mm onto an anvil at the top of the rods and ensuring that this amount of dynamic energy is transferred to the sampler as much as possible.

The number of blows required to achieve each 75 mm penetration is recorded for a full penetration of 450 mm. The initial 150 mm penetration is referred to as seating drive and the blows required for this penetration are not considered as this zone is in disturbed soil. The next 300 mm of penetration is referred to as the test drive and the number of blows required to achieve this fully is termed the penetration resistance or *N* value.

In dense or hard soils the seating drive may not achieve 150 mm penetration so this is considered as the first 25 blows. If the full 300 mm penetration is not achieved with a further 50 blows the test is terminated and the amount of penetration obtained is recorded.

In very loose soils it is worthwhile continuing penetration beyond 450 mm to confirm that the soil is loose and has not been disturbed by drilling.

Water balance should be maintained for tests carried out below the water table to prevent loosening of the soil due to piping. Often very low results are obtained in sands immediately beneath a clay layer when sub-artesian groundwater flow up the borehole causes extensive disturbance and these results should be discarded. In gravelly soils or weak rocks a solid 60° cone shoe is sometimes used to reduce damage but no sample is recovered and higher *N* values may be obtained.

The test is simple to carry out, quick and inexpensive, so it is frequently used and may be the sole means of assessing the strength of the ground when difficult conditions prevent other samples and tests being carried out. However, the results are sensitive to variations in the test equipment used and the procedure adopted (Thorburn, 1984) and several factors affect the results even when the test is carried out in a standard manner (Skempton, 1986). For these reasons, judgement is required in the assessment of the results and they should always be considered as approximate. If any doubt exists then the results should be supplemented by and compared with other tests, such as the Dutch cone penetration test.

b) *Dutch cone penetrometer* *(Figure 14.8)*

A review of this type of test is given in Meigh (1987). This test was originally devised for determining settlements of sands. Experience with the test has enabled the soil types and layers to be assessed so a continuous strata record with undisturbed testing is available, but no samples are obtained. The results should, nevertheless, be compared with boreholes and visual identification of samples.

With a high rate of penetration and rapid data acquisition and analyses several tests can be carried out in a day to depths of 15–40 m depending on the soil conditions.

The electric friction cone penetrometer test is preferred to the mechanical types of cone penetrometer. A piezometric sensor and filter (piezocone) may be incorporated in the penetrometer tip to record equi-

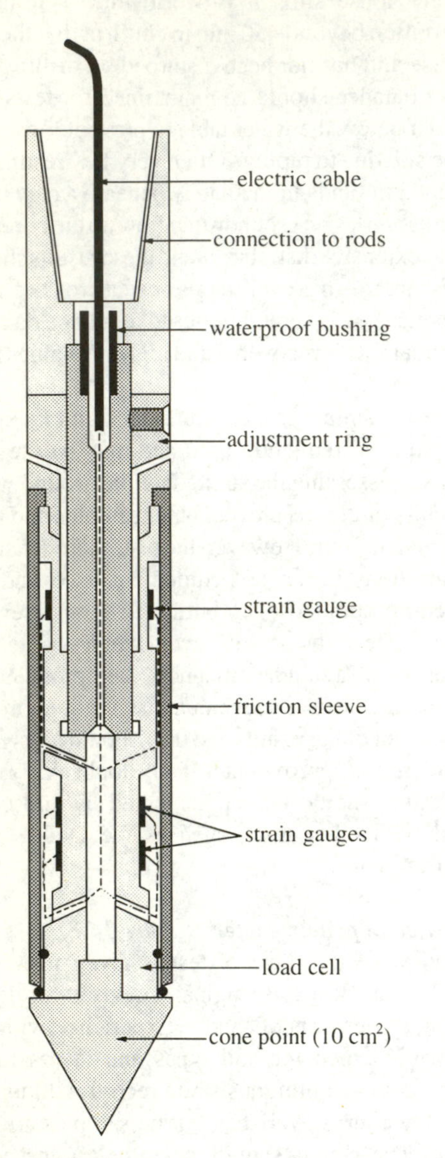

Figure 14.8 *Cone penetrometer*

librium pore water pressures at intervals and excess pore pressures during penetration and some devices include a temperature measuring sensor.

The tip shown in Figure 14.8 comprises a solid 60° cone with a diameter of 35.7 mm and a base area of 1000 mm^2 (10 cm^2), a cylindrical friction sleeve above the cone with a surface area of 150 cm^2, internal sensing devices for measuring the axial thrust on the cone and the frictional force on the sleeve and an inclinometer to check vertical alignment and drift.

The electrical signals are transmitted by an umbilical cable within the hollow push rods to recorders/analysers/plotters at ground level as the tip is pushed into the ground at a constant rate of penetration of 20 mm/second.

The point or cone resistance q_c (axial force/cross-sectional area), the side or sleeve friction resistance f_s (frictional force/sleeve surface area) and the friction ratio R_f

$$R_f = \frac{f_s}{q_c} \qquad (14.2)$$

are plotted continuously with depth.

The maximum thrust which can be applied depends on the type and weight of truck or rig used to mount the penetrometer. Tests usually have to be terminated when dense sands, gravelly soils or rock is encountered or when the vertical mis-alignment is excessive.

c) ***Vane test***

This test is described in BS 1377:1990 Part 9 and ASTM D2573. It consists of rotating cruciform-shape thin steel plate blades in a clay and shearing a cylinder of the soil relative to the surrounding soil. The test is appropriate for very soft to firm saturated clays and is assumed to obtain the undisturbed undrained shear strength.

The height H of the vane blades should be twice the vane width (or diameter D) and the blades must be sharpened and thin enough to give an area ratio of less than 12%. Vane widths of 50 mm and 75 mm are usually adopted for firm and soft clays, respectively.

The vane is attached to rods and pushed about 0.5 m below the bottom of the borehole into undisturbed soil. A torque head is then attached to the top of the rods and the rods rotated at a speed of between 0.1 and 0.2 degrees/second until the soil is sheared. The maximum torque reading T is taken and the vane shear strength is obtained as illustrated in Figure 7.17. The torque required to rotate the rods on their own should be measured and this value deducted.

The vane could then be rotated rapidly at least ten times to remould the soil on the blade edges.

Repeating the test will provide the remoulded shear strength and a measure of sensitivity of the soil.
The test is quick and simple and plots of vane shear strength with depth can be obtained to determine the rate of increase of strength. This is particularly important when laboratory tests may indicate no increase of strength, due to the greater stress relief and disturbance which can occur with conventional sampling from boreholes at depth.
Bjerrum (1972) showed that the vane shear strength is usually larger than the *in situ* shear strength. He proposed the relationship:

$$\textit{in situ}\ c_u = \mu \text{ vane } c_u \qquad (14.3)$$

and gave μ values as shown on Figure 7.17. However, the μ value is dependent on more than the mineralogy of the clay or plasticity index and varies considerably as shown by the results of Ladd *et al* (1977).

d) ***Pressuremeter test*** *(Figures 14.9 and 14.10)*
This test was developed in the 1950s by Louis Menard. The test is not included in the British Standard but is described in ASTM D4719 and pressuremeter testing is described in general in Mair and Wood (1987).
The pressuremeter consists of three parts, top and bottom cylindrical guard cells and an intermediate water-filled measuring cell. The apparatus is lowered to a test level in a borehole and the top and bottom guard cells are inflated to minimise end-effects on the measuring cell which is, in turn, inflated to apply radial stresses to the sides of the borehole. The inflation is provided by water pressurised by nitrogen or carbon dioxide.
The apparatus must be calibrated so that corrections can be applied for expansion resistance of the measuring cell and volume increase of the connecting tubes. The volume change of the central test cell is measured as the pressure is increased in stages and a pressure-volume change plot obtained. At an intermediate stage the pressure can be reduced and re-applied to obtain a reloading modulus.
Each pressure increment is maintained for 1 or 2 minutes, recording the volume change at 15, 30, 60 and 120 seconds. The difference in volume change readings between 30 and 60 seconds is referred to as 'creep' and a creep curve can be useful in distinguishing the pseudo-elastic and plastic phases.

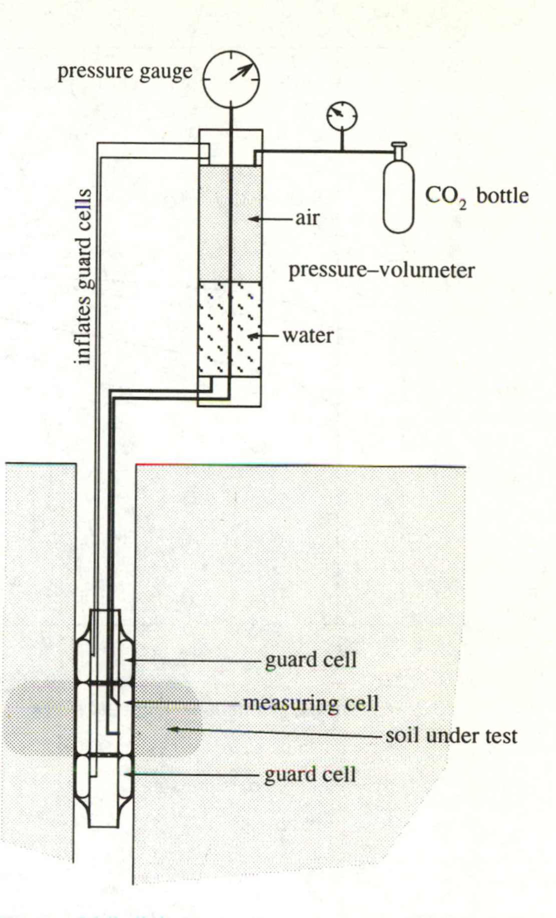

Figure 14.9 *Schematic representation of the Menard type pressuremeter*

A typical result is shown in Figure 14.10. The value p_o gives a measure of the *in situ* horizontal total stress, the shear modulus G and elastic modulus E_p can be estimated from the pressure-volume gradient in the elastic region and the limit pressure p_L can be used to obtain an estimate of the undrained shear strength c_u.
The pressuremeter tests a fairly large volume of soil in its *in situ* condition and may be the most suitable method for ground which is difficult to sample, such as glacial clays and weak rocks. However, the test is carried out in a pre-formed borehole where some

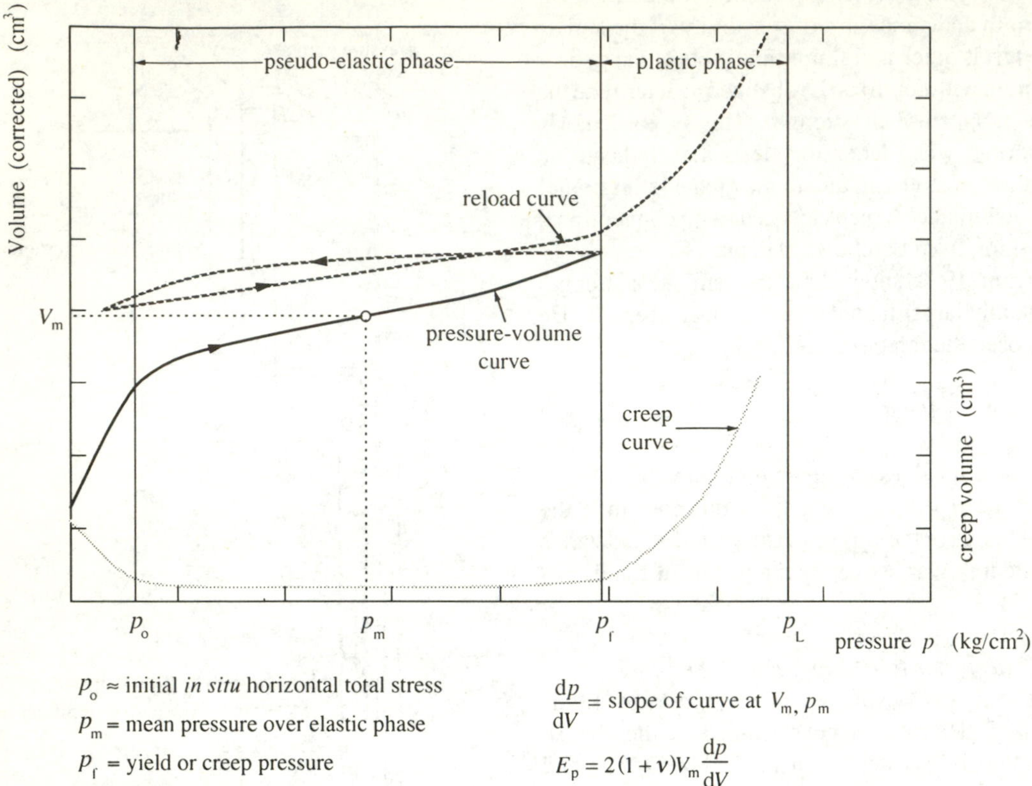

Figure 14.10 *Pressuremeter test result*

self-supporting capacity is required. The apparatus tests an annulus of soil around the borehole so the method of forming the hole and preparation, with varying degrees of disturbance, will have an effect on the result obtained. The test will provide parameters related to the horizontal direction which may not be relevant for vertical loading by structures, especially if the ground is anisotropic.

To reduce the effects of soil disturbance self-boring pressuremeters have been developed (Wroth *et al*, 1973, Baguelin *et al*, 1973). As this pressuremeter is pushed into the soil a rotating cutter and a flushing medium remove the spoil. At required depths the pressuremeter is expanded using gas pressure and its expansion is measured by three spring feelers which sense the deformation of the membrane and enable measurement of radial deformation. The device requires skilled operators to insert without causing excessive disturbance, and will not penetrate soils containing obstructions such as gravel particles.

Groundwater observations

The mechanics of soils is fundamentally dependent on the effective stresses present. Since these cannot be measured directly they are obtained from an estimate of total stress and measurement of pore water pressure. Thus it is imperative that the latter be carried out. Groundwater movements may also affect civil engineering works and the quality or chemistry of the groundwater is important if it is to be used for abstraction or where it may affect material durability.

The importance of careful observation and recording of groundwater encountered during drilling cannot be over-emphasised. Depths of entries, rates of seepage, any change in level after 15 to 20 minutes, when water is sealed off by casing, when water is added, depths of casing, morning and evening water levels should all be recorded. From the impression of groundwater conditions observed during boring, the depths for piezometers, response zones and seals can be more appropriately specified. Very slow seepages should not be ignored because in low permeability soils high pore pressure conditions may be missed.

Groundwater levels should never be considered as static occurrences. They may vary with seasonal, tidal or weather conditions and they could be higher or lower than the level obtained during the period of on-site work. Monitoring water levels over appropriate periods of time should, therefore, be carried out to establish equilibrium water levels and their variations. Some instruments for monitoring water levels are described below.

a) *Standpipe* *(Figure 14.11)*

This is the simplest method of measuring water levels. It comprises plastic tubing, perforated along its whole length or a lower section, and backfilled with a gravel or sand filter material. Considerable volumes of groundwater will be required to saturate the backfill and to fill the tube up to equilibrium level so depending on the permeability of the soil this could take a long time, known as the response time.

Water level recordings are made using an electrical device called a dipmeter probe. This responds to the completion of an electrical circuit causing a light to illuminate or a sound to be heard when the probe lowered down the standpipe enters the water.

The main disadvantages of the standpipe are its slow response time and no distinction is made between groundwater from different layers.

b) *Standpipe piezometer* *(Figure 14.11)*

A porous plastic or ceramic cylindrical element or tip is attached to the bottom of unperforated plastic tubing and placed at a pre-determined depth to monitor water levels at that depth only. This is achieved by providing a short response zone of sand backfill around the piezometer tip with a seal above and below.

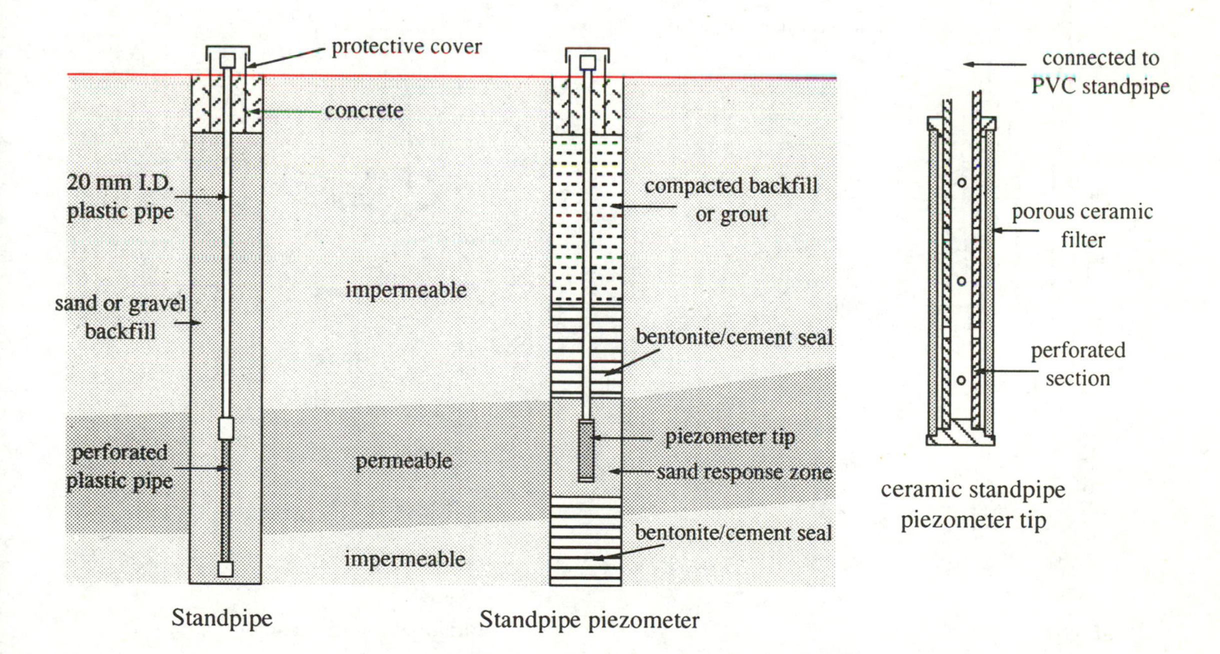

Figure 14.11 *Standpipe and standpipe piezometer installations*

The seal may comprise bentonite or a bentonite cement mixture in the form of granules, pellets, balls or pumpable grout. It is preferable to backfill the whole of the borehole above and below the sand response zone with grout.

If more than one piezometer is installed in a borehole, there may be doubt concerning the separation of the two response zones by adequate seals, so it is preferable to install piezometers at different levels in separate boreholes. To speed up the response time, the sand backfill should be placed in a near saturated condition and water can be added to the standpipe tube up to the anticipated water level.

c) ***Hydraulic piezometer*** *(Figure 14.12)*

A porous ceramic tip is installed in a compacted fill or earth structure and is connected to a mercury-water manometer, a Bourdon gauge or a pressure transducer at a location remote from the piezometer tip for direct readings of pore water pressure.

The tubes connecting the tip to the readout must be completely full of water so twin tubing is used to allow periodic flushing of the system with de-aired water. Pressure readings from each tube can be taken as a check. The tubing comprises nylon covered with polythene so that it is impervious to both air and water. Pressures in the range of – 5 to + 200 m head of water can be measured with this device for long-term monitoring.

If the ground is fully saturated then a ceramic with a large pore size (mean diameter of 60 microns) and low air entry value can be used. However, if air or gas is present this ceramic will not be suitable as the gases will enter the tip and then only pore *air* pressure will be measured.

For partially saturated ground such as compacted fill or when negative pore pressures are anticipated a finer ceramic (mean pore size of 1 micron) with a high air entry value should be used. This type will permit water to pass through but a high air pressure

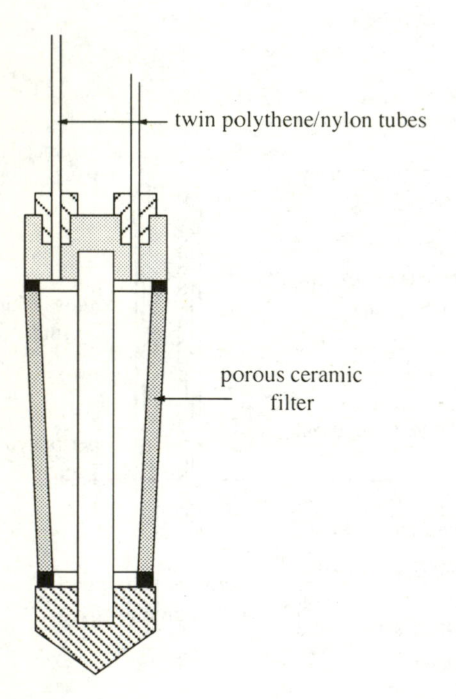

Hydraulic piezometer

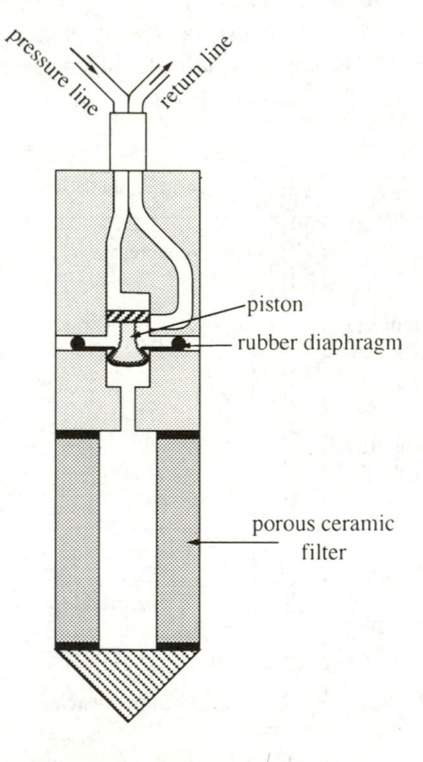

Pneumatic piezometer

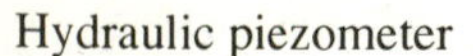

Figure 14.12 *Piezometers*

(air entry value) will be required to cause air to penetrate the ceramic so the air should be excluded. The response time of this piezometer is short, so it is useful for measuring pore pressure changes due to fill placement, superimposed loads or tidal variation.

d) ***Pneumatic piezometer*** *(Figure 14.12)*
This piezometer is used in similar applications as for the hydraulic piezometer and comprises a porous ceramic tip attached to two tubes. One is attached to a gas supply and pressure measurement device and the other is attached to a flow indicator at a remote location at ground level.
Compressed air or nitrogen is passed down one tube (the pressure line) until the pressure is equal to the pore water pressure in the tip. At this stage a flexible rubber diaphragm-type valve is activated and the compressed air can pass up the return line to the flow indicator. These piezometers have a short response time but cannot normally measure negative pore pressures, and cannot be de-aired following installation.

c) ***Vibrating wire piezometer***
This piezometer contains a tensioned stainless steel wire attached to a diaphragm. One side of the diaphragm is in contact with the groundwater pressure inside a porous ceramic tip with the other side of the diaphragm connected to atmospheric pressure by an air line. As the pore water pressure deflects the diaphragm the tension in the wire and, hence, its frequency of vibration changes. This frequency is recorded and calibrated to pressure readings.
The piezometer gives a rapid response time but requires calibration which cannot be checked after installation for the standard type and may give misleading results because it cannot be de-aired after installation.

Investigation of contaminated land

In many countries the demise of certain industrial activities and poor control of pollution of current processes has left the ground in a contaminated state. Hazards to public health and construction materials must be assessed so that appropriate remedial measures can be carried out.

A thorough desk study, site reconnaissance and walk-over survey must be undertaken to identify likely problems and to improve the planning of the ground investigation. Types, numbers and locations of sampling points must be considered carefully to ensure adequate coverage. Solids, liquids and gases should be sampled and analysed independently and their concentrations compared with trigger concentrations such as those given in ICRCL (1983).

More detailed advice can be obtained from DD175:1988, Leach and Goodger (1991) and Cairney (1987).

Site investigation reports

Introduction

Information must be collected and collated and assessments of the ground conditions must be made so that the works can be designed, methods of construction can be determined and the overall costs and risks can be estimated. Thus various parties, the designer, the contractor and the client will require information and its interpretation but their needs, responsibilities and concerns will differ.

It is unlikely that these parties will see the samples taken or observe the tests carried out, especially the main contractor, and in many cases several months, if not years, elapse between the site investigation period and construction period when the samples have either been thrown away or are unrepresentative. It is essential, therefore, that the site investigation report provides an accurate, clear, unambiguous and complete picture of the site conditions at and below ground level.

Site investigation should be treated as an on-going process, with information collected affecting decisions about how to proceed. Thus a fixed lump sum contract approach can be counter-productive although persuading many clients that this is the case can prove difficult. To assist in this process for larger schemes a preliminary site investigation can prove invaluable in assessing viability for the scheme, types and methods of construction and identifying problem areas before more detailed investigation is carried out.

Many reports are separated into two parts or separate volumes:

- factual report
- interpretative report.

Factual report

This should comprise the factual information collected during the ground investigation and is the document provided by the site investigation contractor. A brief written section should be included at the beginning of the report consisting of:

- ***Introduction***
  The site name, its location and a brief description. The name of the client, architect, consulting engineer or other interested parties. The purpose of the investigation and the party responsible for planning the investigation and issuing instructions.

- ***Scope of the work***
  The types of investigation carried out, the depths of investigation, numbers and locations of exploratory holes, frequency and type of in situ testing, sampling, groundwater observations and monitoring devices. The period of the investigation, delays incurred and problems encountered such as access, obstructions, equipment limitations, sample recovery, inappropriate testing, weather conditions.

- ***Anticipated geology***
  From available geological records a summary of the geology and other relevant conditions of the site are described since these could affect the scope of the work.

- ***Ground conditions encountered***
  A brief summary of the sequence of deposits, their nature, thicknesses and variability. The groundwater conditions should also be described although this may be more open to interpretation.

- ***Signatures***
  of persons within the site investigation company responsible for the organisation and preparation of the report.

Appendices

These will form the bulk of the report and will include the following:

- site location plan
  usually an extract from an OS map.

- site plan
  Plans to identify existing and proposed features and give locations of the exploratory holes. On some jobs, it can be useful to give dimensions in relation to existing features or preferably surveyed co-ordinates since some features may be removed after the investigation.

- records of boreholes *(Figure 14.13)*, trial pits, hand augers, probing, hand excavations
  Descriptions of the strata on these records must be complete, clear and unambiguous. Guidance is given on standardised descriptive terms in BS 5930:1981 and in chapter 2 of this book.
  The borehole record or log is often the least complete piece of information yet it is the the most important. It is compiled from the strata boundaries and simple descriptions given by the driller on his daily log, descriptions of the jar and bulk samples by a geologist or engineer and the results of in situ and laboratory tests. A typical borehole record is given in Figure 14.13.
  The samples will have been taken at specified intervals, with gaps in knowledge (apart from the driller's observations) and often the U100s which have not been extruded for testing are discarded without inspection. The client pays for obtaining these samples so a little more expense incurred to obtain descriptions of these better quality samples will be cost-effective.

- records of *in situ* test results
  These comprise both the information collected on site and the calulated test result or parameter. They could include the SPT, cone penetrometer plots, vane tests, pressuremeter tests, permeability (open borehole, packer or piezometer) tests. This information is best presented in a standard form and to the same scale so that information can be readily compared.

- groundwater observations
  The observations during the drilling period in the exploratory hole are reported on the hole records. The readings obtained subsequently in standpipes or piezometers should be tabulated or plotted on a time base.

Name of Site Investigation Company	**BOREHOLE RECORD**	BOREHOLE 3
Client: Department of Transport	Site: Proposed By-pass	Sheet 1 of 1; Job Ref. 94/56

Casing depth / Daily progress	Water level	Samples in situ tests and coring runs – Depth (m) From	To	Type	N value or core recov.	Depth (m)	Description of strata	Reduced level	Legend	287.043 N 654.529 E
							Ground level (metres OD)	9.50		
12.7.94						0.30	Topsoil	9.20		
	17.7	1.00	1.45	U			Soft becoming firm with depth grey organic silty CLAY with thin bands or partings of silt			bentonite/cement
		1.50		D						
		2.00	2.45	U		(4.70)	Vertical root holes in upper horizon			
		2.50		D			Layer of peat at 3.50-4.00 m			
		4.00	4.45	U						
		4.50		D			(Estuarine Clay)			
		5.00	5.45	SD	N=5	5.00		4.50		
(6.00)							Medium dense brown fine and medium SAND and fine to coarse GRAVEL			sand
13.7.94	17.7	6.00	6.45	SD	N=15	(3.00)	Gravel mostly subrounded to rounded			
		7.50	7.95	SD	N=25		(Flood Plain Gravel)			
		8.00	8.45	U		8.00	Very stiff fissured brown silty CLAY with much fine to coarse subangular to subrounded gravel and occasional cobbles	1.50		bentonite/cement
		8.50		D						
		10.00	10.45	U						
		10.50		D		(6.00)	Lenses of sand at 10.2 and 12.5 m			
		12.00	12.45	U						
		12.50		D			(Glacial Clay)			
		14.00	14.45	SD	N=76	14.00		−4.50		
(10.00)							Greyish brown highly becoming moderately weathered weak becoming moderately strong thinly bedded fine to medium grained SANDSTONE			sand
15.7.94		15.00	16.50	H	T=78 S=47 R=32	(>4.0)	Bedding subhorizontal			
		16.50	18.00	H	T=95 S=75 R=45	18.00	Frequent joints at 45º (Coal Measures) (Westphalian B)	−8.50		
						END	Piezometer sealed at 6.00 m, water level at 2.00 m 17.7.94 Piezometer sealed at 15.00 m, water level at 7.00 m 17.7.94			

T – Total core recovery
S – Solid core recovery
R – Rock quality designation

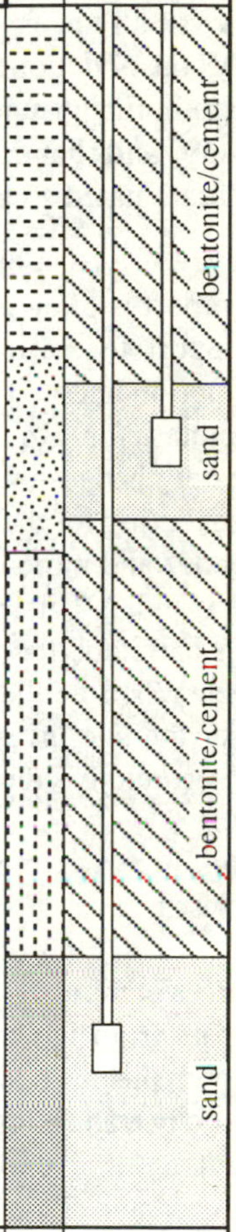

Boring equipment and methods
Cable percussion, 150 mm, GL – 15.00 m
Rotary coring, water flush 15.00 – 18.00 m

Logged by: GEB Scale 1:100

Remarks
Chiselling sandstone 1 hour
Very fast seepage at 5.00 m, level rose to 2.20 m after 20 mins.
Sealed off by casing at 10.00 m
Moderate seepage at 14.00 m, level rose to 12.00 m after 20 mins
Water level at 2.00 m on morning of 13.7.94 and at 8.00 m on 15.7.94

Figure 14.13 *Typical Borehole record*

- laboratory test results
  The schedule of laboratory testing is usually specified by the client or his engineer or proposed by the contractor for approval.
  The results are also presented in a standard form and could include classification tests (moisture content, bulk density, particle density, liquid and plastic limits, particle size distribution), shear strength tests (undrained c_u, $\phi_u = 0$ and effective stress c' and ϕ'), consolidation tests (m_v and c_v), compaction tests (optimum moisture content, maximum dry density, MCV) and chemical tests (sulphates, chlorides, pH, contaminated soils).

Interpretative report
This report is often prepared separately from the factual report for ease of reading and assimilation of information and may be written by the site investigation contractor. However, it is more often prepared separately because it is prepared by the geotechnical adviser acting for the client, architect or engineer who has better and more up-to-date information about the project and is better placed to incorporate changes which inevitably occur during the development of a project. When the client does not appoint a geotechnical adviser the site investigation contractor will prepare the interpretative report either on a lump sum or fee basis.

This report is mostly a written document and it comprises:

- ***Introduction***
  Apart from the information given in the factual report the introduction should state the terms of reference and engagement, the brief from the client with a fuller explanation of the purpose of the project and a clear understanding of the client's requirements and any limitations or gaps in the information gathered.

- ***Walk-over survey, desk study, anticipated geology***
  These could be combined or separate sections and would describe the information obtained, its consequences and how it has affected and been addressed by the site investigation.

- ***Site model** (Figure 14.14)*
  This is essential for design and construction purposes. The geometry of the ground conditions is best displayed on cross-sections through the site with the ground investigation information superimposed such as borehole records or cone penetrometer plots.
  An interpolation of the ground conditions should be carried out between these locations based on an understanding of the geology of the site. The engineer/geologist should demonstrate enough confidence in his/her knowledge and the adequacy of the investigation to interpolate with lines showing boundaries between deposits and strata. Alternative interpolations may be included to illustrate possible complexities. Lines showing water table levels may also be interpreted.
  A typical soil section using the information obtained from Borehole 3 (in Figure 14.13) is plotted in Figure 14.14 showing how borehole information can be interpreted to produce a site model.
  Once the layers have been determined parameters for design purposes can be allocated to each layer. Test results plotted on the sections can be useful in showing variations with depth or across the site. Plots of soil properties with depth, on their own, should be treated with caution if there is no consideration for the different soil types, the deposits are dipping or the ground surface is sloping.

- ***Advice and recommendations***
  This section condenses all of the information gathered, calculations and analyses using methods described in other chapters in this book, and the purpose of the scheme into advice and recommendations concerning the design, construction, monitoring and potential risks involved.
  This may include recommendations for earthworks, foundations, walls, basements, pavements, tunnels, dams and other types of structures and comments and advice on construction problems and expedients such as groundwater control, excavation stability and support, ground improvement, sources of materials, chemical or contamination hazards. This section must identify the problems of the site in relation to the project needs.

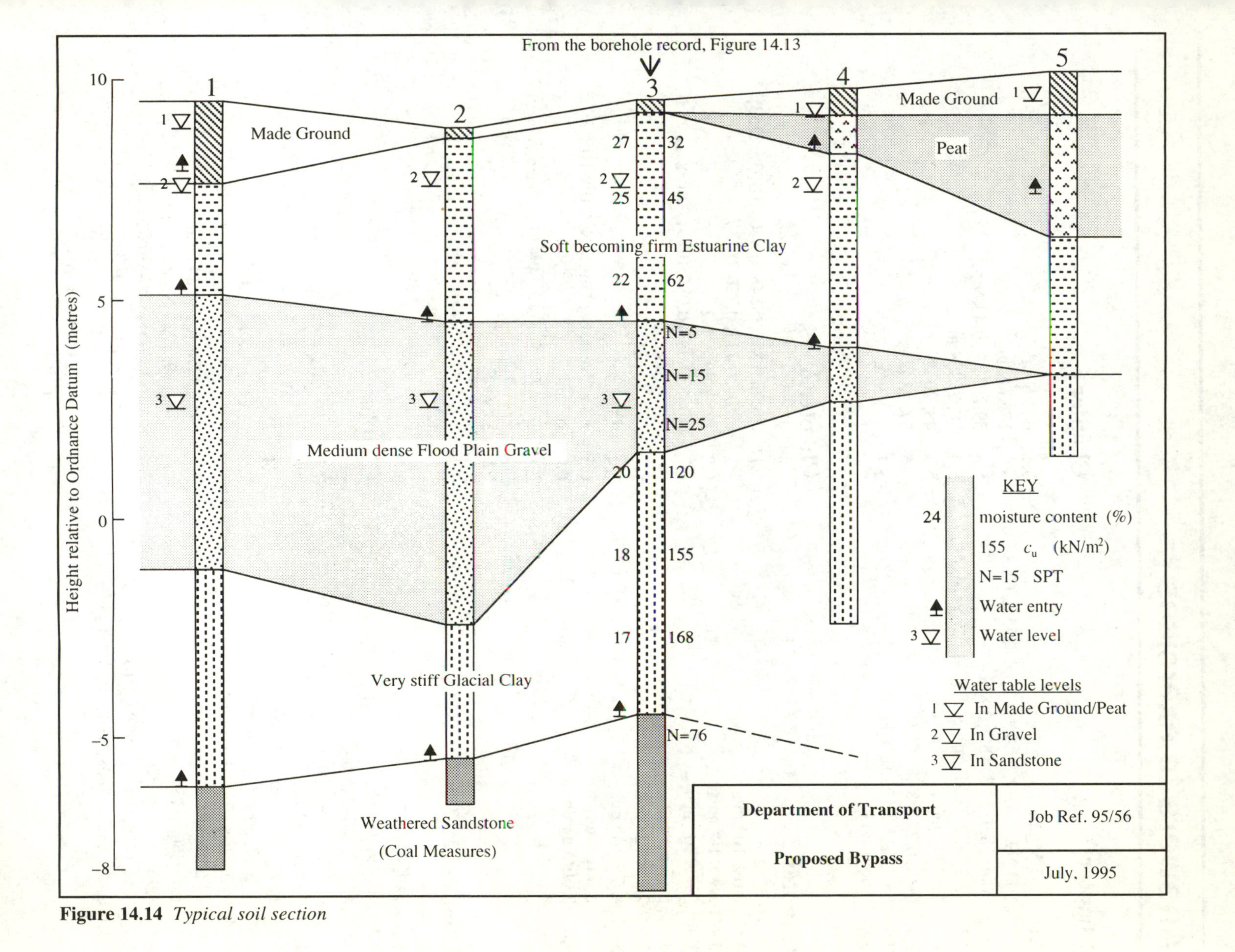

Figure 14.14 *Typical soil section*

Answers to exercises

Chapter 2

2.1 2.65 Mg/m^3
2.2 100, 99.1, 95.9, 93.8, 90.2, 84.4, 75.2, 63.8, 50.4, 38.9, 27.8, 15.2, 3.8 %
2.3 12.2 %, 2.22 Mg/m^3, 21.8 kN/m^3, 1.98Mgm^3, 19.4 kN/m^3
2.4 0.38, 0.27, 88 %, 3.3 %
2.5 55 %, 27 %, 28 %, CH
2.6 0.29, 0.71

Chapter 3

3.1 8.6×10^{-5} m/s
3.2 2.7×10^{-8} m/s
3.3 2.4×10^{-7} m/s
3.4 60.5 m^3/day
3.5 2.9×10^{-5} m/s
3.6 8.6×10^{-4} m/s
3.7 5 litres/min/m
3.8 23 litres/min/m
3.9 4.0, 2.3
3.10 22.5 litres/hour

Chapter 4

4.1 139, 98, 41 kN/m^2
4.2 90, 49, 41 kN/m^2
4.3 a) 105, 29.4, 75.6 kN/m^2
b) 195, 78.4, 116.6 kN/m^2
4.4 42.8, 67.4 kN/m^2
4.5 1) 41.1, 69.1, 97.1 kN/m^2
2) 57.7, 69.1, 97.1 kN/m^2
3) 57.7, 85.7, 113.7 kN/m^2
4.6 19.6 kN/m^2
4.7 1.0, 1.50, 1.18
4.8 0.96
4.9 0.34, 0.58, 0.74, 0.88, 0.95, 0.98
4.10 0, 0.22, 0.16, 0.08, 0.01, –0.07, –0.13

Chapter 5

5.1 62 kN/m^2
5.2 172 kN/m^2
5.3 59.2, 37.2, 9.4 kN/m^2
5.4 50, 21 kN/m^2
5.5 48, 20 kN/m^2
5.6 57, 23 kN/m^2
5.7 45 kN/m^2

Chapter 6

6.1 0.856, 0.852, 0.848, 0.840, 0.803, 0.753, 0.704, 0.712, 0.726, 0.737
6.2 125 kN/m^2, 2.2
6.3 0.086, 0.087, 0.087, 0.201, 0.138, 0.070 m^2/MN
0.09, 0.15 m^2/MN
6.4 0.16
6.5 3.3 m^2/year, 0.20 m^2/MN, 2.0×10^{-10} m/s
6.6 70 mm
6.7 a) 12 kN/m^2 b) 42 kN/m^2
6.8 46 mm
6.9 53 weeks
6.10 83 %
6.11 1.35 m

Chapter 7

7.1 57.7, 20.2 kN/m^2
7.3 24 kN/m^2
7.4 21.5º
7.5 15 kN/m^2, 39º
7.6 1.5
7.7 248.1, 238.1, 244.3 kN/m^2
124, 119, 122 kN/m^2
7.8 $a = 16$ kN/m^2 $\alpha = 30.5º$
$c' = 20$ kN/m^2 $\phi' = 36º$
7.9 $a = 10$ kN/m^2 $\alpha = 26.6º$
$c' = 11.5$ kN/m^2 $\phi' = 30º$

7.10 $\lambda = 0.14$ $N = 2.47$
7.11 0.87
7.12 184 kN/m^2
7.13 0.60, 0.50
7.14 a) 65, 97 kN/m^2 b) 129, 195 kN/m^2

Chapter 8

8.1 a) 586 b) 686 kN/m^2
8.2 2393 kN
8.3 3.1 m
8.4 9.3
8.5 a) 2380 kN/m^2 b) 9.2
8.6 a) 4065 kN/m^2 b) 15.7
8.7 a) 4.2 b) 8.0
8.8 a) 3.1 b) 4.9

Chapter 9

9.1 28 mm, 11 mm
9.2 1) 14 mm 2) 9 mm
9.3 1) 21 mm 2) 23 mm
9.4 42 mm, 12 mm
9.5 1) 19 mm 2) 29 mm
9.6 164 mm
9.7 165 mm
9.8 41 mm, 14 mm
9.9 84 mm, 177 mm
9.10 19 mm, 28 mm
9.11 22 mm, 33 mm

Chapter 10

10.1 1086 kN
10.2 16.4 m
10.3 1477 kN
10.4 9498 kN
10.5 a) 4570 kN b) 5890 kN
10.6 1206 kN
10.7 729 kN

Chapter 11

11.1 96 kN, 1.43 m
11.2 806 kN, 1.76 m
11.3 82 kN, 1.49 m
11.4 1) 86 kN, 1.39 m 2) 18.5 kN
11.5 1) 6.4 2) 2.4 3) 172, 54 kN/m^2
11.6 1) 6.5 m 2) 312 kN
11.7 1) 2.5 mm 2) 1.8

Chapter 12

12.1 1.51, 1.23, 1.13
12.2 3.53, 2.24, 1.81
12.3 15.7º
12.4 17 kN/m^2
12.5 1.38
12.6 17 kN/m^2
12.7 2.5
12.8 30º
12.9 1.34
12.10 1.30
12.11 1.90
12.12 1.30

Chapter 13

13.1 14.4
13.2 32, 0.8
13.3 1) 16447 m^3
2) 1221
3) 303 bankm^3/hour, 54.3 hours
13.4 18.5%, 1.76 Mg/m^3
13.5 0.55, 0.35, 92%, 2.7%

References

Alpan, I. (1967). The empirical evaluation of the coefficient K_o and K_{or}. *Soils and Foundations* **7**, No. 1, p. 31.

Anon. (1977). The description of rock masses for engineering purposes. *Report by the Geological Society Engineering Group Working Party*, vol. 10, p. 355–388.

Anon. (1990). Tropical residual soils. *Geological Society Engineering Group Working Party Report*, vol. 23, No. 1, p. 1–111.

Arrowsmith, E.J. (1979). Roadwork fills - a materials engineer's viewpoint. *Proc Conf. Clay Fills, ICE*, London, p. 25–36.

ASTM, (1992). *Annual book of ASTM standards*, vol. 04.08. American Society for Testing and Materials, Philadelphia, Pa.

Atkinson, J.H., Bransby, P.L. (1978). *The mechanics of soils - an introduction to critical state soil mechanics*. McGraw-Hill Book Company (UK) Limited.

Baguelin, F., Jezequel, J.F., Le Mehaute, A. (1974). Self-boring placement of soil characteristics measurement. *Proc. Conf. on subsurface exploration for underground excavation and heavy construction*, ASCE, Henniker, New Hampshire, p. 312–332.

Barnes, G.E. (1992). Stability coefficients for highway cutting slope design. *Ground Engineering*, **2**, No. 4, p. 26–31.

Barnes, G.E. and Staples, S.G. (1988). The acceptability of clay fill as affected by stone content. *Ground Engineering*, 21, No. 1, p. 22–30.

Barron, R.A. (1948). Consolidation of fine-grained soils by drain wells. *Trans. ASCE*, **113**, Paper 2346, p. 718–54.

Berezantzev, V.G., Khristoforov, V. and Golubkov, V. (1961). Load bearing capacity and deformation of piled foundations. *Proc. 5th Int. Conf. Soil Mechanics and Foundation Engineering, Paris*, **2**, p. 11–12.

Bishop, A.W. (1955). The use of the slip circle in the stability analysis of slopes. *Geotechnique*, **5**, No. 1, p. 7–17.

Bishop, A.W. and Al-Dhahir, Z.A. (1970). Some comparisons between laboratory tests, in situ tests and full scale performance, with special reference to permeability and coefficient of consolidation. *Proc. Conf. In Situ Investigations in Soil and Rocks, ICE, London*, p. 251–264.

Bishop, A.W. and Bjerrum, L. (1960). The relevance of the triaxial test to the solution of stability problems. *Proc. Research Conf. Shear Strength of cohesive soils, ASCE, Boulder, Colorado*, p. 437–501.

Bishop, A.W., Green, G.E., Garga, V.K., Andresen, A. and Brown, J.D. (1971). A new ring shear apparatus and its application to the measurement of residual strength. *Geotechnique*, **21**, No. 4, pp. 273–328.

Bishop, A.W. and Henkel, D.J. (1957). *The measurement of soil properties in the triaxial test*. Edward Arnold (Publishers) Ltd. London.

Bishop, A.W. and Henkel, D.J. (1962). *The measurement of soil properties in the triaxial test*. Edward Arnold (Publishers) Ltd. London. 2nd Edition.

Bishop, A.W. and Morgenstern, N. (1960). Stability coefficients for earth slopes. *Geotechnique*, **10**, No. 4, p. 129–150.

Bjerrum (1972). Embankments on soft ground. *Proc. ASCE Conf. Performance of Earth and earth-supported structures*, **2**, p. 1–54.

Boussinesq, J. (1885). *Application des potentials à l'étude de l'équilibre et du mouvement des solids elastiques*, Gauthier-Villars, Paris.

BRE Digest 240 (1980). *Low-rise buildings on shrinkable clay soils : Part 1*. Building Research Establishment Digest, HMSO, London.

BRE Digest 241 (1990). *Low-rise buildings on shrinkable clay soils : Part 2*. Building Research Establishment Digest, HMSO, London.

BRE Digest 242 (1980). *Low-rise buildings on shrinkable clay soils : Part 3*. Building Research Establishment Digest, HMSO, London.

BRE Digest 251 (1990). *Assessment of damage in low-rise buildings*. Building Research Establishment Digest, HMSO, London.

BRE Digest 274 (1991). *Fill Part 1 : Classification and load carrying characteristics.* Building Research Establishment, HMSO, London.

BRE Digest 298 (1987). *The influence of trees on house foundations in clay soils.* Building Research Establishment Digest, HMSO, London.

BRE Digest 318 (1987). *Site investigation for low-rise buildings: desk studies.* Building Research Establishment Digest, HMSO, London.

BRE Digest 322 (1987). *Site investigation for low-rise buildings : procurement.* Building Research Establishment Digest, HMSO, London.

BRE Digest 343 (1989). *Simple measuring and monitoring of movement in low-rise buildings. Part 1: cracks.* Building Research Establishment Digest, HMSO, London.

BRE Digest 344 (1989). *Simple measuring and monitoring of movement in low-rise buildings. Part 2: settlement, heave, out-of-plumb.* Building Research Establishment Digest, HMSO, London.

BRE Digest 348 (1989). *Site investigation for low-rise buildings : the walk-over survey.* Building Research Establishment Digest, HMSO, London.

BRE Digest 361 (1991). *Why do buildings crack?* Building Research Establishment Digest, HMSO, London.

Brinch Hansen, J. (1970). A revised and extended formula for bearing capacity. *Bulletin No. 28, Danish Geotechnical Institute, Copenhagen*, p. 5–11.

British Standards Institution (1981). BS 5930 : *Code of practice for site investigations*, BSI, London.

British Standards Institution (1981). BS. 6031: *Code of practice for earthworks*, BSI, London.

British Standards Institution (1986). BS 8004 : *Code of practice for foundations*, BSI, London.

British Standards Institution (1987). BS 8002: *Draft code of practice for earth retaining structures*, BSI, London.

British Standards Institution (1988). DD 175 : 1988. *Draft for development : Code of practice for the identification of potentially contaminated land and its investigation.* BSI, London.

British Standards Institution (1989). BS 8081 : *Ground anchorages.* BSI London.

British Standards Institution (1990). BS 1377 : *British Standard Methods of test for soils for civil engineering purposes*, BSI, London.

Part 1 : General requirements and sample preparation.
Part 2 : Classification tests.
Part 3 : Chemical and electro-chemical tests.
Part 4 : Compaction - related tests.
Part 5 : Compressibility, permeability and durability tests.
Part 6 : Consolidation and permeability tests in hydraulic cells and with pore pressure measurement.
Part 7 : Shear strength tests (total stress).
Part 8 : Shear strength tests (effective stress).
Part 9 : In situ tests.

Bromhead, E.N., (1978). A simple ring-shear apparatus. *Ground Engineering* **12,** 5, p. 40–44.

Brooker, E.H. and Ireland, H.O. (1965). Earth pressures at rest related to stress history. *Canadian Geotechnical Journal,* **2**, no. 1, pp. 1–15.

Burland, J.B. (1970). Discussion on session A. *Proc. Conf. In Situ Investigations in Soils and Rocks. British Geotechnical Society, London*, p. 61–62.

Burland, J.B. (1973). Shaft friction of piles in clay - a simple fundamental approach. *Ground Engineering,* **6**, no. 3, May, p. 30–32.

Burland, J.B., Broms, B.B. and de Mello, V.F.B. (1978). Behaviour of foundations and structures. *Proc. 9th Int. Conf. Soil Mechanics and Foundation Engineering, Tokyo,* Session 2, p. 495–546. Also published as BRE Current Paper CP51/78.

Burland, J.B. and Burbridge, M.C., (1985). Settlement of foundations on sand and gravel, *Proc. ICE,* **78** (Dec.), p. 1325–81.

Burland, J.B., Butler, F.A. and Dunican, P., (1966). The behaviour and design of large diameter bored piles in stiff clay. *Proc. Symp. Large Bored Piles, ICE, London*, p. 51–71.

Burland, J.B. and Wroth, C.P., (1974). Settlement of buildings and associated damage. State of the art review, *Proc. Conf. on Settlement of Structures, Cambridge,* p. 611–654, Pentech Press, London.

Butler, F.G. (1974). Heavily overconsolidated clays, *Proc. Conf. on Settlement of Structures, Cambridge,* p. 531–578, Pentech Press, London.

Butterfield, R. and Banerjee, P.K. (1971). A rigid disc embedded in elastic half space. *Journal S.E. Asian Soc. Soil Engineering,* **2**, No. 1, p. 35–52.

Cairney, T. (1993). *Contaminated land : problems and solutions.* Blackie Academic and Professional, Glasgow.

Caquot, A. and Kerisel, J. (1948). *Tables for the calculation of passive pressure, active pressure and bearing capacity of foundations.* Gauthier-Villars, Paris.

Casagrande, A. (1936). The determination of the preconsolidation load and its practical significance. *Proc. 1st Int. Conf. Soil Mechanics and Foundation Engineering, Cambridge, Mass. USA,* **3**, p. 60–64.

Cedergren, H. (1989). *Seepage, drainage and flow nets.* 3rd ed., Wiley, New York.

Chandler, R.J. (1972). Lias clay : weathering processes and their effect on shear strength. *Geotechnique,* **22**, No. 3, p. 403–431.

Chandler, R.J. and Apted, J. (1988). The effect of weathering on the strength of London Clay. *QJEG,* **21**, p. 59–68.

Chandler, R.J. and Davis, A.G. (1973). Further work on the engineering properties of Keuper Marl. CIRIA Report 47, October 1973.

Chandler, R.J. and Peiris, R.A. (1989). Further extensions to the Bishop and Morgenstern slope stability charts. *Ground Engineering,* **22**, No. 4, 33–38.

Christian, J.T. and Carrier, W.D. (1978). Janbu, Bjerrum and Kjaernsli's chart reinterpreted. *Canadian Geotechnical Journal,* **15**, p. 123–128.

Civil Engineering Code of Practice No. 2 (1951). *Earth retaining structures.* Institution of Structural Engineers.

Clayton, C.R.I. (1979). Two aspects of the use of the moisture condition apparatus. *Ground Engineering,* May, p. 44–48.

Clayton, C.R.I., Simons, N.E. and Matthews, M.C. (1982). *Site Investigation.* Granada Publishing Ltd, St Albans, Herts.

Cole, K.W. (1972). Uplift of piles due to driving displacement. *Civil Engineering and Public Works Review,* March, p. 263–269.

Collins, K. and McGown, A. (1974). The form and function of microfabric features in a variety of natural soils. *Geotechnique,* **24**, No. 2, p. 223–254.

Cooke, R.W. (1974). Settlement of friction pile foundations. *Proc. Conf. on Tall Buildings, Kuala Lumpur,* No. 3, p. 7–19. Also BRE Current Paper CP 12/75.

Cooke, R.W. and Price, G. (1973). Strains and displacements around friction piles. *Proc. 8th Int. Conf Soil Mechanics and Foundation Engineering, Moscow,* **2-1**, p. 53–60. Also BRE Current Paper CP 28/73.

Cooke, R.W., Price, G. and Tarr, K. (1980). Jacked piles in London Clay : interactional group behaviour under working conditions. *Geotechnique,* **30**, No. 2, p. 97–136.

Coulomb, C.A. (1776). *Essai sur une application des règles des maximis et minimis à quelque problèmes de statique relatif à l'architecure.* Memoirs Divers Savants, Academie Science, vol. 7. Paris.

D'Appolonia, D.J., Poulos, H.G. and Ladd, C.C. (1971). Initial settlement of structures on clay. *Proc. ASCE,* **97**, No. SM10, p. 1359–1377.

Department of Transport (1978). *Reinforced earth retaining walls and bridge abutments for embankments.* Technical Memorandum (Bridges) BE 3/78, London.

Department of Transport, Scottish Office, Welsh Office, Department of the Environment for Northern Ireland (1992). *Specification for Highway Works.* Four volumes, HMSO, London.

Eyles, N. and Sladen, J.A. (1981). Stratigraphy and geotechnical properties of weathered lodgement till in Northumberland, England. *QJEG,* **14**, p. 129–141.

Fadum, R.E. (1948). Influence values for estimating stresses in elastic foundations. *Proc. 2nd. Int. Conf. Soil Mechanics and Foundation Engineering, Rotterdam,* **3**, p. 77–84.

Farrar, D.M. and Darley, P. (1975). The operation of earthmoving plant on wet fill. Department of the Environment, *TRRL Report LR 688*, Crowthorne, Berks.

Fox, E.N. (1948). The mean elastic settlement of a uniformly loaded area at a depth below the ground surface. *Proc. 2nd Int. Conf. Soil Mechanics and Foundation Engineering, Rotterdam,* **1**, p. 129.

Fox, L. (1948). Computations of traffic stresses in a simple road structure. *Proc. 2nd Int. Conf. Soil Mechanics and Foundation Engineering, Rotterdam,* **2**, p. 236–246.

Fleming, W.G.K. and Sliwinski, Z.J. (1977). *The use and influence of bentonite in bored pile construction.* CIRIA/DOE Report PG3.

Fleming, W.G.K., Weltman, A.J., Randolph, M.F. and Elson, W.R. (1985). *Piling Engineering.* Surrey University Press/Halstead Press.

Fraser, R.A. and Wardle, I.J. (1976). Numerical analysis of rectangular rafts on layered foundations. *Geotechnique,* **26**, No. 4, p. 613–630.

Gibson, R.E. (1963). An analysis of system flexibility and its effect on time-lag in pore water pressure measurements. *Geotechnique,* **13**, p. 1–11.

Gibson, R.E. (1966). A note on the constant head test to measure soil permeability in situ. *Geotechnique,* **16**, No. 3, p. 256–259.

Giroud, J.P. (1970). Stresses under linearly loaded rectangular area. *Journal Soil Mech. Found. Div., ASCE,* **96**, No. SM1, p. 263–268.

Hambly, E.C. (1979). *Bridge foundations and substructures.* Building Research Establishment.

Head, K.H. (1982). *Manual of soil laboratory testing. Volume 2 : permeability, shear strength and compressibility.* Pentech Press.

Head, K.H. (1986). *Manual of soil laboratory testing. Volume 3 : effective stress testing.* Pentech Press.

Head, K.H. (1989). *Soil technician's handbook.* Pentech Press, London.

Head, K.H. (1992). *Manual of soil laboratory testing. Volume 1: soil classification and compaction tests,* Second Edition. Pentech Press, London.

Holtz, R.D., Jamiolkowski, M.B., Lancelotta, R. and Pedroni, R. (1991). *Prefabricated vertical drains: Design and Performance.* CIRIA Ground Engineering Report, Butterworth-Heinemann, Oxford.

Hooper, J.A. (1979). *Review of behaviour of piled raft foundations.* CIRIA Report R83.

Huntington, W.C. (1957). *Earth pressures and retaining walls.* John Wiley, New York.

Hvorslev, M.J. (1951). *Time lag and soil permeability in ground water observations.* U.S. Waterways Experiment Station Bulletin 36, Vicksburg.

Institution of Civil Engineers (1979). *Proceedings of the Conference on Clay Fills.*

Institution of Civil Engineers (1991). *Inadequate site investigation,* Ground Board, ICE, Thomas Telford Ltd., London.

Institution of Structural Engineers (1975). *Design and constrution of deep basements.* IStructE, London.

Institution of Structural Engineers (1986). *Soil-structure interaction : The real behaviour of structures.*

ICRCL (1987). *Guidance on the assessment and redevelopment of contaminated land.* Interdepartmental Committee on the Redevelopment of Contaminated Land. Guidance Note 59/83, 2nd edition, 1987.

Irvine, D.J. and Smith, R.J.H. (1983). *Trenching practice.* CIRIA Report 97, London.

Jaky, J. (1944). The coefficient of earth pressure at rest. *Journal for Society of Hungarian Architects and Engineers, Budapest, Hungary,* October 1944, pp. 355–358.

Jaky, J. (1948). Pressure in silos. *Proc. 2nd Int. Conf. Soil Mechanics and Foundation Engineering,* **1**, p. 103–107.

Jamiolkowski, M., Lancellotta, R., Pasqualini, E., Marchetti, S. and Nova, R. (1979). Design parameters for soft clays. General Report, *Proc. 7th European Conf. on Soil Mechanics and Foundation Engineering,* **5**, p. 27–57.

Janbu, N. (1973). Slope stability computations. In *Hirschfield and Poulos (eds.). Embankment Dam Engineering, Casagrande Memorial Volume,* John Wiley and Sons, New York, 47–86.

Janbu, N., Bjerrum, L. and Kjaernsli, B. (1956). *Veiledning ved lfsning av fundamenteringsoppgaver.* Norwegian Geotechnical Institute, Oslo, Publication 16, p. 30–32.

Jewell, R.A. and Wroth, C.P. (1987). Direct shear tests on reinforced sand. *Geotechnique,* **37**, No. 1, p. 53–68.

Ladd, C.C., Foott, R., Ishihara, K., Schlosser, F. and Poulos, H.G. (1977). Stress-deformation and strength characteristics : state-of-the-art report. *Proc. 9th Int. Conf. Soil Mechanics and Foundation Engineering, Tokyo,* **2**, p. 421–494.

Landva, A.O., Korpijaakko, E.O. and Pheeney, P.E. (1983). Geotechnical classification of Peats and Organic Soils. *ASTM STP 820 Testing of Peats and Organic Soils.* American Society for Testing and Materials, p. 37–51.

Landva, A.O. and Pheeney, P.E. (1980). Peat fabric and Structure. *Canadian Geotechnical Journal,* **17**, p. 416–435.

Leach, B.H. and Goodger, H.K. (1991). *Building on derelict land.* CIRIA Report SP 78.

Lupini, J.F., Skinner, A.E. and Vaughan, P.R. (1981). The drained residual strength of soils. *Geotechnique,* **31**, No. 2, pp. 181–213.

Mair, R.J. and Wood, D.M. (1987). *Pressuremeter testing : methods and interpretation.* CIRIA Ground Engineering Report : in-situ testing, Butterworths, London.

Marsland, A. (1971). The shear strength of stiff fissured clays. *Proc. Roscoe Memorial Symposium, Cambridge,* March, 1971. Also published as BRE current paper CP 21/71.

Matheson, G.D. (1988). *The use and application of the moisture condition apparatus in testing soil suitability for earthworking.* S.D.D. Applications Guide No. 1, TRRL Scothill Branch.

Mayne, P.W. and Kulwahy, F.H. (1982). K_o - OCR relationships in soil. *ASCE, JGED,* **108**, pp. 851–872.

Meigh, A.C. (1976). The Triassic rocks, with particular reference to predicted and observed performance of some major foundations. 16th Rankine Lecture, *Geotechnique,* **26**, No.3, p. 391–452.

Meigh, A.C. (1987). *Cone penetration testing : methods and interpretation.* CIRIA Ground Engineering Report : in-situ testing, Butterworths, London.

Mesri, G. (1973). Coefficient of secondary compression. *Proc. ASCE,* **99**, SM1, p. 123–137.

Meyerhof, G.G. (1953). The bearing capacity of foundations under eccentric and inclined loads. *Proc. 3rd Int. Conf. Soil Mechanics and Foundation Engineering, Zurich,* **1**, p. 440–445.

Meyerhof, G.G. (1976). Bearing capacity and settlement of pile foundations. *Proc. ASCE, Journal Geot. Eng. Div.,* **102**, G.T.3., p. 197–228.

Milovic, D.M. and Tournier, J.P. (1971). Stresses and displacements due to rectangular load on a layer of finite thickness *Soils and Foundations,* **11**, No. 1, March, p. 1–27.

Morgenstern, N.R. and Price, V.E. (1967). The analysis of the stability of general slip surfaces. *Geotechnique,* **15**, No. 1, p. 79–93.

NAVFAC DM7. (1971). *Design Manual: soil mechanics, foundations and earth structures,* US Department of the Navy, Washington, DC

NAVFAC DM7. (1982). *Design Manual: Soil mechanics, foundations and earth structures, soil dynamics, deep stabilisation and special geotechnical construction.* Design Manual 7, US Department of the Navy, Alexandria, Va.

Newmark, N.M. (1942). *Influence charts for computation of stresses in elastic foundations.* Engineering Experiment Station Bulletin No. 338, University of Illinois, Urbana, Ill.

NHBC (1992). *Building near trees.* National House-Building Council, Practice Note 3.

Norbury, D.R., Child, G.H. and Spinks, T.W. (1986). A critical review of Section 8 (BS 5930) - soil and rock description. *Geological Society, Engineering Geology Special Publication,* No. 2, p. 331–342.

O'Connor, M.J. and Mitchell, R.J. (1977). An extension of the Bishop and Morgenstern slope stability charts. *Can. Geot. Journal,* **14**, p. 144–151.

Padfield, C.J. and Mair, R.J. (1984). *Design of retaining walls embedded in stiff clays.* CIRIA Report 104, London.

Parsons, A.W. (1976). The rapid measurement of the moisture condition of earthwork material. *TRRL Report LR 750,* Department of the Environment, Crowthorne, Berks.

Parsons, A.W. and Boden, J.B. (1979). The moisture condition test and its potential applications in earthworks. *TRRL Supplementary Report 522,* Department of Transport, Crowthorne, Berks.

Parsons, A.W. and Darley, P. (1982). The effect of soil conditions on the operation of earthmoving plant. *TRRL Report LR 1034,* Department of Transport, Crowthorne, Berks.

Peck, R.B., Hanson, W.E. and Thornburn, T.H. (1974). *Foundation engineering.* 2nd edition, John Wiley, New York

Poulos, H.G. (1968). Analysis of the settlement of pile groups. *Geotechnique,* **18**, No. 4, p. 449–471.

Poulos, H.G. and Davis, E.H. (1974). *Elastic solutions for soil and rock mechanics.* John Wiley and Sons, New York.

Poulos, H.G. and Davis, E.H. (1980). *Pile foundation analysis and design.* John Wiley and Sons, New York.

Poulos, H.G. and Mattes (1971). Settlement and load distribution of pile groups. *Aust. Geomechanics Journal,* **G1**, No. 1, p. 18–28.

Randolph, M.F. and Wroth, C.P. (1982). Recent developments in understanding the axial capacity of piles in clay. *Ground Engineering,* **15**, No. 7, p. 17–25.

Roscoe, K.H., Schofield, A.N. and Wroth, C.P. (1958). On the yielding of soils. *Geotechnique,* **8**, No. 1, p. 22–53.

Rowe, P.W. (1962). The stress-dilatancy relation for static equilibrium of an assembly of particles in contact. *Proc. Royal Society of London, Series A,* p. 500–527.

Rowe, P.W. (1968). The influence of geological features of clay deposits on the design and performance of sand drains. *Proc. Inst. Civil Engineers, Suppl. 1,* p. 1–72.

Rowe, P.W. and Barden, L. (1966). A new consolidation cell, *Geotechnique,* **16**, No. 2, 162–170.

Schmertmann, J.H. (1955). The undisturbed consolidation behaviour of clay. *Trans. ASCE,* **120**, p. 1201–1227.

Schmertmann, J.H. (1970). Static cone to compute static settlement over sand. *Proc. ASCE,* **96**, No. SM 3, Paper 7302, p. 1011–1043.

Schmertmann, J.H., Hartmann, J.P. and Brown, P.R. (1978). Improved strain influence factor diagrams. *Proc. ASCE,* **104**, No. GT 8, p. 1131–1135.

Schofield, A.N. and Wroth, C.P. (1968). *Critical state soil mechanics.* McGraw-Hill, London.

Semple, R.M. and Rigden, J. (1986). Shaft capacity of driven pipe piles in clay. *Ground Engineering,* January, p. 11–19.

Serota, S. and Lowther, G. (1973). SPT practice meets critical review. *Ground Engineering,* **6**, No. 1, p. 20–25.

Sherwood, P.T. (1971). The reproducibility of the results of soil classification and compaction tests. *TRRL, Report LR 339,* Department of Transport, Crowthorne, Berkshire.

Site Investigation Steering Group (1993). *Site investigation in construction.* Thomas Telford Limited, London.

Part 1 – Without site investigation ground is a hazard.

Part 2 – Planning, procurement and quality management.

Part 3 - Specification for Ground Investigation.

Part 4 - Guidelines for the safe investigation by drilling of landfills and contaminated land.

Skempton, A.W. (1953). The post glacial clays of the Thames Estuary. *Proc. 3rd Int. Conf. Soil Mechanics and Foundation Engineering, Zurich,* **1**, p. 302.

Skempton, A.W. (1953). The colloidal "activity" of clays. *Proc. 3rd Int. Conf. Soil Mechanics and Foundation Engineering, Zurich,* **1**, p. 57–61.

Skempton, A.W. (1954). The pore pressure coefficients A and B. *Geotechnique,* **4**, No. 4.

Skempton, A.W. (1959). Cast in-situ bored piles in London Clay. *Geotechnique,* **9**, p. 153–178.

Skempton, A.W. (1961). Horizontal stresses in an overconsolidated eocene clay. *Proc. 5th ICSMFE, Paris,* **1**, p. 352–357.

Skempton, A.W. (1964). Long-term stability of slopes. *Geotechnique,* **14**, No. 2, p. 77–101.

Skempton, A.W. (1966). Summing up. *Proc. Symp. on Large Bored Piles, London.*

Skempton, A.W. (1986). Standard penetration test procedures and the effects in sands of overburden pressure, relative density, particle size, ageing and overconsolidation. *Geotechnique,* **36**, No. 3, p. 425–447.

Skempton, A.W. and Bjerrum, L. (1957). A contribution to the settlement of foundations on clay. *Geotechnique,* **7**, No. 4, p. 168–178.

Skempton, A.W. and Hutchinson, J.N. (1969). Stability of natural slopes and embankment foundations. *Proc. 7th Int. Conf. Soil Mechanics and Foundation Engineering, Mexico City,* State-of-the-Art Volume, p. 291–340.

Skempton, A.W. and MacDonald, D.H. (1956). Allowable settlements of buildings. *Proc. ICE, part 3,* **5**, p. 727–768.

Stroud, M.A. and Butler, F.G. (1975). The standard penetration test and the engineering properties of glacial materials. *Proc. Symp. Engineering Properties of Glacial Materials.* Midlands Soil Mechanics and Foundations Society.

Taylor, D.W. (1937). Stability of earth slopes. *Journal Boston Soc. Civ. Engrs.,* **24**, p. 197–246.

Taylor, D.W. (1948). *Fundamentals of soil mechanics.* John Wiley and Sons, New York.

Terzaghi, K. (1943). *Theoretical soil mechanics,* John Wiley and Sons, New York.

Terzaghi, K. (1954). Anchored bulkheads. *Trans. ASCE,* **119**, paper 2720, p. 1243-1281.

Terzaghi, K. and Peck, R.B. (1967). *Soil mechanics in engineering practice.* 2nd edition, John Wiley and Sons, New York, first edition published 1948.

Tomlinson, M.J. (1970). *The adhesion of piles driven in stiff clay.* CIRIA Research Report No. 26.

Tomlinson, M.J. (1971). Some effects of pile driving on skin friction. *Proc. Conf. on Behaviour of Piles, ICE, London,* p. 107–114.

Tomlinson, M.J. (1986). *Foundation design and construction.* Fifth edition. Longman Scientific and Technical.

Tomlinson, M.J. (1987). *Pile design and construction practice.* Third edition, Viewpoint Publications, Palladian Publications Limited.

Ueshita, K. and Meyerhof, G.G. (1968). Surface displacement of an elastic layer under uniformly distributed loads. *Highway Research Board Record,* No. 228, p. 1–10.

Vesic, A.S. (1967). *A study of bearing capacity of deep foundations.* Final Report, Project B-189, Georgia Inst. Tech., Atlanta, Georgia, p. 231–6.

Vesic, A.S. (1969). Experiments with instrumented pile groups in sand. *ASTM, STP 444,* p. 177–222.

Vesic, A.S. (1975). Bearing capacity of shallow foundations. In *Foundation Engineering Handbook,* Edited by Winterkorn, H.F. and Fang, H-Y., Van Nostrand Reinhold Company.

Vesic, A.S. (1977). *Design of piled foundations.* N.C.H.R.P. Synthesis of Highway Practice, No. 42, Transportation Research Board, Washington, D.C.

Vijayvergiya, V.N. and Focht, J.A.J. (1972). A new way to predict the capacity of piles in clay. *4th Annual Offshore Tech. Conf., Houston,* **2**, p. 865–874.

Weltman, A.J. (1980). *Pile load testing procedures.* CIRIA/DOE Report P.67.

Weltman, A.J. and Healy, P.R. (1978). *Piling in 'boulder clay' and other glacial tills.* CIRIA/DOE Piling Development Group Report PG 5, Nov. 1978.

Whitaker, T. (1957). Experiments with model piles in groups. *Geotechnique*, **7**, No. 1, p. 147–167.

Whitaker, T. (1970). *The design of piled foundations.* Oxford : Pergamon.

Whitaker, T. and Cooke, R.W. (1966). An investigation of the shaft and base resistances of large bored piles in London Clay. *Proc. Symp. on Large Bored Piles, London,* p. 7–49.

Wilkinson, W.B. (1968). Constant head in situ permeability tests in clay strata. *Geotechnique,* **18**, p. 172–194.

Wroth, C.P. (1979). Correlations of some engineering properties of soils. *Proc. 2nd. Int. Conf. on Behaviour of Offshore Structures, London,* **1**, p. 121–132.

Wroth, C.P. and Hughes, J.M.O. (1973). An instrument for the in-situ measurement of the properties of soft clays. *Proc. 8th Int. Conf. Soil Mechanics and Foundation Engineering, Moscow,* **102**, p. 487–494.

Wynne, C.P. (1988). *A review of bearing pile types.* CIRIA/PSA Report PG1, second edition.

Index